# Continuation and Bifurcations:
# Numerical Techniques and Applications

# NATO ASI Series

## Advanced Science Institutes Series

*A Series presenting the results of activities sponsored by the NATO Science Committee, which aims at the dissemination of advanced scientific and technological knowledge, with a view to strengthening links between scientific communities.*

The Series is published by an international board of publishers in conjunction with the NATO Scientific Affairs Division

| | | |
|---|---|---|
| **A** | **Life Sciences** | Plenum Publishing Corporation |
| **B** | **Physics** | London and New York |
| | | |
| **C** | **Mathematical** | Kluwer Academic Publishers |
| | **and Physical Sciences** | Dordrecht, Boston and London |
| **D** | **Behavioural and Social Sciences** | |
| **E** | **Applied Sciences** | |
| | | |
| **F** | **Computer and Systems Sciences** | Springer-Verlag |
| **G** | **Ecological Sciences** | Berlin, Heidelberg, New York, London, |
| **H** | **Cell Biology** | Paris and Tokyo |

**Series C: Mathematical and Physical Sciences - Vol. 313**

# Continuation and Bifurcations: Numerical Techniques and Applications

edited by

## Dirk Roose

Department of Computer Science,
Katholieke Universiteit Leuven,
Leuven, Belgium

## Bart De Dier

Avery STDE,
Turnhout, Belgium

and

## Alastair Spence

School of Mathematical Sciences,
University of Bath,
Bath, U.K.

**Kluwer Academic Publishers**

Dordrecht / Boston / London

Published in cooperation with NATO Scientific Affairs Division

Proceedings of the NATO Advanced Research Workshop on
Continuation and Bifurcations: Numerical Techniques and Applications
Leuven, Belgium
September 18–22, 1989

**Library of Congress Cataloging in Publication Data**

NATO Advanced Research Workshop on Continuation and Bifurcations:
  Numerical Techniques and Applications (1989 : Louvain, Belgium)
    Continuation and bifurcations : numerical techniques and
  applications / edited by Dirk Roose, Bart de Dier, and Alastair
  Spence.
        p.    cm. -- (NATO ASI series. Series C, Mathematical and
  physical sciences ; vol. 313)
    "Proceedings of the NATO Advanced Research Workshop on
  Continuation and Bifurcations: Numerical Techniques and
  Applications, Leuven, Belgium, September 18-22, 1989."
    "Published in cooperation with NATO Scientific Affairs Division."
    Includes bibliographical references and index.
    ISBN-13:978-94-010-6781-2        e-ISBN-13:978-94-009-0659-4
    DOI:10.1007/978-94-009-0659-4

    1. Continuation methods--Congresses.  2. Bifurcation theory-
  -Congresses.    I. Roose, Dirk.   II. Dier, Bart de, 1961-    .
  III. Spence, Alastair, 1948-    .  IV. North Atlantic Treaty
  Organization. Scientific Affairs Division. V. Title.  VI. Series:
  NATO ASI series.  Series C, Mathematical and physical sciences ; no.
  313.
  QA377.N36   1989
  515'.353--dc20                                                90-40289

ISBN-13:978-94-010-6781-2

Published by Kluwer Academic Publishers,
P.O. Box 17, 3300 AA Dordrecht, The Netherlands.

Kluwer Academic Publishers incorporates the publishing programmes of
D. Reidel, Martinus Nijhoff, Dr W. Junk and MTP Press.

Sold and distributed in the U.S.A. and Canada
by Kluwer Academic Publishers,
101 Philip Drive, Norwell, MA 02061, U.S.A.

In all other countries, sold and distributed
by Kluwer Academic Publishers Group,
P.O. Box 322, 3300 AH Dordrecht, The Netherlands.

# Contents

# Preface

In September 1989, a NATO Advanced Research Workshop on "Continuation and Bifurcations : Numerical Techniques and Applications" was held at the Katholieke Universiteit Leuven, Belgium. Participants came from 10 countries in Europe and North America and were mainly from universities and research institutes. This proceedings volume contains 26 of the 38 papers which were presented at the meeting. Abstracts of most other contributions are also included.

The central theme of the workshop was the solution of parameter dependent nonlinear problems using numerical continuation. More specifically the aims can be stated as : to describe typical bifurcation problems in scientific, engineering and industrial problems ; to discuss current mathematical ideas and new developments in numerical analysis and numerical techniques and to describe and evaluate program packages and to discuss future needs with respect to software.

The interests of the participants extended over the complete spectrum of theory, numerical analysis, software and applications, and this spread is reflected both in the composition of this volume and in several of the papers. For example, there are contributions on the application of Centre Manifold and Liapunov-Schmidt theory to derive low dimensional systems which can be analysed by normal form theory for dynamical systems or singularity theory. On the numerical analysis front there are contributions, for example, on the computation of homoclinic and heteroclinic orbits, on the detection of Hopf bifurcations, on the computation of bifurcations in the presence of symmetry, on the calculation of rotating waves, and on the use of inertial manifolds in the bifurcation analysis of the Kuramoto-Sivashinski equation. Also included are descriptions of software packages for use on personal computers, and contributions on the use of symbolic manipulation codes. Several interesting applications are described, including separation in 3-D Navier Stokes flows, chaos in electrical circuits, the dynamics of passive optical systems and Marangoni convection in crystal growth.

The editors wish to thank the participants who made the workshop so successful, and the NATO Scientific Affairs Division, the National Science Foundation of Belgium (N.F.W.O.), the Ministry of Education of the Flemish Government and the K. U. Leuven for their generous sponsorship of the workshop. Finally, the editors acknowledge the opportunity given by Kluwer to publish the proceedings in the NATO ASI Series.

Dirk Roose                   Bart De Dier                   Alastair Spence

# List of Participants

P.J. Aston, School of Mathematics, University of Bath, Claverton Down, Bath BA2 7AY, United Kingdom

G. Bader, Institut für Angewandte Mathematik, Im Neuenheimer Feld 294, D–6900 Heidelberg, Germany

P.G. Bakker, Luchtvaart en Ruimtevaarttechniek, T. U. Delft, Kluyverweg 1, NL–2629 HS Delft, The Netherlands

M. Beckers, Dept. of Computer Science, K.U.Leuven, Celestijnenlaan 200A, B–3030 Heverlee-Leuven, Belgium

W.J. Beyn, Fakultät für Mathematik, Universität Konstanz, Postfach 5560, D–7750 Konstanz, Germany

H. Byrne, Numerical analysis Group, Oxford University, 8–11 Keble Road, Oxford OX1 3QD, United Kingdom

L. Chua, Dept. of Electrical Engineering and Computer Science, University of California at Berkeley, Berkeley, CA 94720, U.S.A.

R. Cools, Dept. of Computer Science, K.U.Leuven, Celestijnenlaan 200A, B–3030 Heverlee-Leuven, Belgium

G. Dangelmayr, Institut für Informationsverarbeitung, Köstlinstraße 6, D–7400 Tübingen, Germany

B. De Dier, Avery STDE, Tieblokkenlaan 1, B–2300 Turnhout, Belgium

G. Degrande, Dept. Bouwkunde, K.U.Leuven, De Croylaan 2, B–3030 Heverlee-Leuven, Belgium

J. Degreve, Dept. of Chemical Engineering, K.U.Leuven, De Croylaan 2, B–3030 Heverlee-Leuven, Belgium

M. Dellnitz, Institut für Angewandte Mathematik, Universität Hamburg, Bundesstraße 55, D–2000 Hamburg, Germany

J.C. Eilbeck, Dept. of Mathematics, Heriot-Watt University, Riccarton, Edinburgh EH14 4AS, United Kingdom

N. Furter, Dept. of Mathematics, University of Warwick, CV4 7AL Coventry, United Kingdom

T.J. Garratt, School of Mathematics, University of Bath, Claverton Down, Bath BA2 7AY, United Kingdom

W. Govaerts, Seminarie voor Hogere Analyse, R.U.Gent, Krijgslaan 281, B-9000 Gent, Belgium

M. Holodniok, Dept. of Chemical Engineering, Prague Institute of Chemical Technology, Suchbatarova 5, 166 28 Praha 6, Czechoslovakia

P.L. Houtekamer, K.N.M.I., Afdeling DM, Postbus 201, NL-3730 AE de Bilt, The Netherlands

I. Hoveijn, Mathematisch Instituut, Rijksuniversiteit Utrecht, Budapestlaan 6, NL-3508 TA Utrecht, The Netherlands

M. Impey, University of Bristol, School of Mathematics, Bristol BS8 1TW, United Kingdom

J.P. Kernévez, Département de Génie Informatique, U.T.C., B.P. 649, F-60206 Compiègne Cedex, France

I.G. Kevrekidis, Dept. of Chemical Engineering, Princeton University, Princeton, NJ 08544, U.S.A.

A. Khibnik, Research Computing Centre, USSR Academy of Sciences, Pushchino, Moscow Region, 142292 USSR

M. Kleczka, Institut B für Mechanik, Universität Stuttgart, Pfaffenwaldring 9, D-7000 Stuttgart 80, Germany

P. Kunkel, Fachbereich Mathematik, Universität Oldenburg, Postfach 2503, D-2900 Oldenburg, Germany

T. Küpper, Institut für Angewandte Mathematik, Universität Hannover, Welfengarten 1, D-3000 Hannover 1, Germany

Y. Kuznetsov, Research Computing Centre, USSR Academy of Sciences, Pushchino, Moscow Region, USSR 142292

Mei Zhen, Fachbereich Mathematik, Universität Marburg, Lahnberge, D-3550 Marburg/Lahn, Germany

H.D. Mittelmann, Dept. of Mathematics, Arizona State University, Tempe, AZ 85287, U.S.A.

G. Moore, Imperial College of Science, Technology and Medicine, Huxley Building, Queens Gate, London SW7 2BZ, United Kingdom

G. Nicolis, Faculté des Sciences, Université Libre de Bruxelles, Campus Plaine, C.P. 231, 1050 Brussels, Belgium

J. Orság, Dept. of Chemical Engineering, Prague Institute of Chemical Technology, Suchbatarova 5, 166 28 Praha 6, Czechoslovakia

E. Ponce-Nunez, Departamento de Matematica Aplicada, Universidad de Sevilla, Escuela Superior de Ingenieros Industriales, Avda Reina Mercedes S/N, SP-41012 Sevilla, Spain

C. Price, ESAT / MI2, K.U.Leuven, Kard. Mercierlaan 94, B-3030 Heverlee-Leuven, Belgium

E. Riks, N.L.R., Vestiging Noordoosterpolder, Voorsterweg 31 - P.B. 153, NL-8316 PR Marknesse, The Netherlands

A. Rodriguez-Luis, Departamento de Matematica Aplicada, Universidad de Sevilla, Escuela Superior de Ingenieros Industriales, Avda Reina Mercedes S/N, SP-41012 Sevilla, Spain

D. Roose, Dept. of Computer Science, K.U.Leuven, Celestijnenlaan 200A, B-3030 Heverlee-Leuven, Belgium

P. Rosendorf, Dept. of Chemical Engineering, Prague Institute of Chemical Technology, Suchbatarova 5, 166 28 Praha 6, Czechoslovakia

H. Schippers, N.L.R., Informatics Division, Voorsterweg 31 - P.B. 153, NL-8316 PR Marknesse, The Netherlands

R. Seydel, Dept. of Mathematics, University of Pittsburgh, Pittsburgh, PA 15260, U.S.A. (on leave from : Institut für Angewandte Mathematik, Universität Würzburg, Germany)

A. Spence, School of Mathematics, University of Bath, Claverton Down, Bath BA2 7AY, United Kingdom

A. Steindl, Institute of Mechanics, Technical University Vienna, Wiedner Hauptstraße 8-10, A-1040 Wien, Austria

A.M. Stuart, School of Mathematics, University of Bath, Claverton Down, Bath BA2 7AY, United Kingdom

H. Troger, Institut für Mechanik, Technische Universität Wien, Wiedner Hauptstraße 8 - 10 / 325, A-1040 Wien, Austria

K. Ulrich, Institut für Angewandte Mathematik, Universität Hannover, Welfengarten 1, D-3000 Hannover 1, Germany

L. Vandenberghe, ESAT 01, K.U.Leuven, Kard. Mercierlaan 94, B-3030 Heverlee-Leuven, Belgium

A. Vanderbauwhede, Instituut voor Theoretische Mechanika, R.U.Gent, Krijgslaan 281 (S9), B-9000 Gent, Belgium

B. Werner, Institut für Angewandte Mathematik, Universität Hamburg, Bundesstraße 55, D-2000 Hamburg, Germany

C. Wilmers, Technische Universität Hamburg-Harburg, Arbeitsbereich Meerestechnik II, Gißendorferstraße 42, D-2100 Hamburg 90, Germany

W. Wu, School of Mathematics, University of Bath, Claverton Down, Bath BA2 7AY, United Kingdom

N. Yamamoto, Kansai University, Suita, Osaka, Japan

# BIFURCATION TO ROTATING WAVES FROM NON-TRIVIAL STEADY-STATES

P. J. ASTON
*Department of Mathematics*
*University of Surrey*
*Guildford GU2 5XH*
*United Kingdom*

A. SPENCE and W. WU
*School of Mathematical Sciences*
*University of Bath*
*Bath BA2 7AY*
*United Kingdom*

ABSTRACT. This paper considers bifurcation problems which are equivariant with respect to the rotations (or translations) $r_\alpha$, $\alpha \in [0, 2\pi)$ and the two reflections $s_1$ and $s_2$ which generate a group that we call $O^2(2)$. The rotational equivariance forces the linearisation of the problem to have a zero eigenvalue at every non-trivial steady-state solution. We show that when this zero eigenvalue has algebraic multiplicity two but geometric multiplicity one then bifurcation to rotating (or travelling) waves occurs, subject to a non-degeneracy condition. This result is obtained by reformulating the problem as a standard steady-state bifurcation in the presence of a reflectional symmetry. Finally, the generic form of bifurcation from rotating waves is considered.

## 1. Introduction

We consider bifurcation from a *non-trivial* branch of steady-state solutions to rotating (or travelling) wave solutions of the time-dependent nonlinear problem

$$\frac{dX}{dt} + g(X, \lambda) = 0 , \qquad t \geq 0 \tag{1.1}$$

where $g$ is a $C^2$-mapping from $H \times \mathbb{R}$ into $H$, a Hilbert space with inner product $< , >$. In our previous paper (Aston, Spence and Wu (1989)) we assume that $g$ is $O(2)$-equivariant, that is,

$$\gamma g(X, \lambda) = g(\gamma X, \lambda) \quad \forall \gamma \in O(2), \ X \in H \tag{1.2}$$

where $O(2)$ is the group generated by the rotations $r_\alpha$, $\alpha \in [0, 2\pi)$ and a reflection $s$. One example of this arises when the steady-state problem $g(X, \lambda) = 0$ represents a boundary value problem in one space dimension, say $\sigma$, with periodic boundary conditions. In this case, $g$ is often equivariant with respect to $r_\alpha$, which acts by rotation or translation as

$$r_\alpha X(\sigma, t) = X(\sigma + \alpha, t), \quad \alpha \in [0, 2\pi),$$

and *one* of the two reflections $s_1$ and $s_2$ which act by

$$s_1 X(\sigma, t) = X(-\sigma, t)$$

$$s_2 X(\sigma, t) = -X(-\sigma, t).$$

In this paper, we extend our previous results to the case when $g$ is equivariant with respect to $r_\alpha$ and *both* reflections $s_1$ and $s_2$. This situation arises in reaction-diffusion equations when the

1

*D. Roose et al. (eds.), Continuation and Bifurcations: Numerical Techniques and Applications,* 1–8.
© 1990 *Kluwer Academic Publishers.*

nonlinear reaction function is odd in $X$.

Other authors have considered bifurcation from the *trivial* solution to rotating waves which arises as a Hopf bifurcation (see for example Iooss (1984)). However, there is an additional complication when considering bifurcation from a non-trivial branch of steady-state solutions, namely that $g_X(X, \lambda)$ is singular at every solution point (see Lemma 2.1). We shall show how this difficulty may be overcome by using a phase condition which enables us to reformulate the problem as a standard, steady-state bifurcation in the presence of a reflectional symmetry. We then show (see Theorem 3.3) that bifurcation to rotating waves occurs when $g_X(X, \lambda)$ has a zero eigenvalue with geometric multiplicity one and algebraic multiplicity two, subject to a non-degeneracy condition. This is in direct contrast to Hopf bifurcation where a complex conjugate pair of eigenvalues must cross the imaginary axis to produce a branch of time-periodic solutions.

No numerical results are presented here but we refer the reader to Aston *et al* (1989) where results for the Kuramoto-Sivashinsky equation are presented. Other numerical results for bifurcation to rotating waves from non-trivial steady-state solutions are given in Scovel, Kevrekidis and Nicolaenko (1988) and Kevrekidis, Nicolaenko and Scovel (1989).

Finally, we assume throughout that $H$ is *finite-dimensional* although the extension to infinite dimensions described in Aston *et al* (1989) applies here also.

## 2. Preliminary Theory

We now set up the problem in a framework suitable for the bifurcation analysis of Section 3. We are interested in solutions of (1.1), where $g$ is equivariant with respect to the rotations $r_\alpha$, $\alpha \in [0, 2\pi)$ and the reflections $s_1$ and $s_2$ which satisfy the relations

$$s_i r_\alpha = r_{2\pi-\alpha} s_i , \qquad i=1,2 \tag{2.1a}$$

$$s_1 s_2 = s_2 s_1 . \tag{2.1b}$$

We define $s_{12} := s_1 s_2$ and we refer to the group generated by $r_\alpha$, $s_1$ and $s_2$ as $O^2(2)$. Let us now introduce our terminology. We call $(x, \lambda)$ a *steady-state* solution of (1.1) if

$$g(x, \lambda) = 0 \tag{2.2}$$

and $(X(t), \lambda)$ a *rotating wave* solution of (1.1) if

$$X(t) = r_{ct} x \tag{2.3}$$

where $c \in \mathbb{R}$ is the velocity of the wave and $x \in H$ is independent of time. We define the following subspaces of $H$:

$$H_s = \{x \in H : s_1 x = x\}$$

$$H_a = \{x \in H : s_1 x = -x\}$$

$$H^\Sigma = \{x \in H : \sigma x = x, \ \forall \sigma \in \Sigma\}$$

where $\Sigma$ is any subgroup of $O^2(2)$. We will refer to $x \in H^\Sigma$ as $\Sigma$-*symmetric*. We also define

$$r_\alpha' x := \frac{d}{d\alpha}(r_\alpha x) ,$$

$$A := r_0' .$$

Then $r_\alpha'$ is a bounded linear operator on $H$ since the group action is smooth on finite-dimensional Hilbert spaces (Knapp (1986)).

We assume, without loss of generality (Golubitsky, Stewart and Schaeffer (1988, p31)), that the inner product $< \, , \, >$ is $O^2(2)$-invariant, that is

$$<\gamma x, \, \gamma y> = <x, \, y> \quad \forall x, y \in H, \ \gamma \in O^2(2).$$

For the sake of simplicity, we assume that $H^{O^2(2)} = \{0\}$ so that (1.1) has the trivial, steady-state solution $x=0$ for all $\lambda \in \mathbf{R}$. Applying the theory outlined in Aston (1990) to bifurcation from the $O^2(2)$-symmetric trivial solution shows that steady-state, symmetry-breaking bifurcation can occur where the resulting primary branches of solutions are symmetric with respect to $r_{2\pi/n}$, $s_1$ and $s_2 r_{\pi/n}$ for some $n \in \mathbf{Z}^+$ (by the Equivariant Branching Lemma). Use of the relations (2.1) gives $(s_{12} r_{\pi/n})^2 = r_{2\pi/n}$ and so this group is precisely the dihedral group $\tilde{D}_{2n}$ generated by the "rotation" $s_{12} r_{\pi/n}$ and the reflection $s_1$. The primary branches of solutions thus lie in $H^{\tilde{D}_{2n}} \times \mathbf{R}$. The main result of this paper (Theorem 3.3) is to show that bifurcation from such a primary branch of steady-state solutions to a branch of rotating wave solutions of the form (2.3) occurs when the $s_1$-symmetry is broken. This result can be proved in the context of steady-state bifurcation theory using the equation which we now derive.

It is easily shown, by differentiating (2.1a) with respect to $\alpha$, that the linear operator $A$ anti-commutes with both $s_1$ and $s_2$ and hence commutes with $s_{12}$. Similarly, $A$ commutes with $r_\alpha$. It follows from this and the equivariance of $g$ that a rotating wave solution $(r_{ct} x, \lambda)$ of (1.1) satisfies the "steady-state" equation

$$\tilde{g}(x, c, \lambda) := g(x, \lambda) + cAx = 0 , \tag{2.4}$$

and that $\tilde{g}$ is equivariant with respect to $r_\alpha$, $\alpha \in [0, 2\pi)$ and $s_{12}$ only. Thus, if $(x, c, \lambda)$ is a solution of (2.4), then $(r_\alpha x, c, \lambda)$ is also a solution for all $\alpha \in [0, 2\pi)$ due to the $r_\alpha$-equivariance of $\tilde{g}$, giving rise to an *orbit* of *conjugate solutions*. In order to eliminate this nonuniqueness, we introduce a phase condition of the form

$$<l, \, x> = 0, \quad l \in H. \tag{2.5}$$

We then rewrite (2.4) and (2.5) as

$$G(y, \lambda) = 0, \quad G: Y \times \mathbf{R} \to Y,$$

$$G(y, \lambda) := \begin{bmatrix} g(x, \lambda) + cAx \\ <l, \ x> \end{bmatrix} , \tag{2.6}$$

$$y = (x, c) \in Y := H \times \mathbf{R}.$$

and define an inner product on $Y$ by

$$y_1 \cdot y_2 = <x_1, \ x_2> + c_1 c_2$$

where $y_i = (x_i, c_i)$, $i = 1, 2$. (See Aston *et al* (1989) for details of the analysis using a more general phase condition.)

Differentiating the $r_\alpha$-equivariance condition for $\tilde{g}$ with respect to $\alpha$ leads to the following result.

**Lemma 2.1**

If $(x, c, \lambda)$ is a solution of (2.4), then

$$\tilde{g}_x(x, c, \lambda) Ax = 0 . \tag{2.7}$$

When considering bifurcation from the trivial solution $x=0$, this result provides no information since $Ax=0$. However, since we are considering bifurcation from a non-trivial branch of (steady-state) solutions, $Ax\neq0$ and so $\tilde{g}_x(x, 0, \lambda)=g_x(x, \lambda)$ has a non-trivial null space at every non-trivial steady-state solution of (1.1). Also, since we are considering a branch of solutions with $x\in H^{\tilde{D}_{2n}}$, then $\text{Null}(g_x(x, \lambda))$ must be $\tilde{D}_{2n}$-invariant (see Aston (1990)) and so $\gamma Ax\in\text{Null}(g_x(x, \lambda))$ for all $\gamma\in\tilde{D}_{2n}$. However, since $s_1$ anti-commutes with $A$, we have $s_1Ax=-As_1x=-Ax$. Similarly, $s_{12}r_\pi/_nAx=Ax$. Thus, the one-dimensional space spanned by $Ax$ is $\tilde{D}_{2n}$-invariant and so we obtain no new null-vectors. The phase condition (2.5) can be viewed as a means of eliminating this singularity in the system (2.6). In particular, if $(x, c, \lambda)$ is a solution of (2.4), then $(Ax, 0)$ will not be in the null space of $G_y((x, c), \lambda)$ provided that the non-degeneracy condition

$$<l, Ax> \neq 0 \tag{2.8}$$

is satisfied.

Finally, the system (2.6) inherits certain equivariance properties from $\tilde{g}$. If we define group actions on $Y$ in terms of those on $H$ by

$$S_1y := (s_1x, -c),$$

$$S_2y := (s_2x, -c),$$

$$R_\pi/_ny := (r_\pi/_nx, c),$$

where $y=(x, c)\in Y$, then we have the following result which is proved using the invariance of the inner product.

**Lemma 2.2**

If $l\in H^{Z_{2n}}\cap H_a$, where $Z_{2n}$ is the cyclic group generated by $r_\pi/_n$, then $G$ is equivariant with respect to $S_1$ and $S_{12}R_\pi/_n$ where $S_{12}:=S_1S_2$.

Note that solutions of (2.6) which satisfy $S_1y=y$ must have $c=0$ and automatically satisfy the phase condition and so they consist of steady-state solutions of (1.1) contained in $H_s$. Thus, the primary branch solutions referred to earlier consist precisely of $\tilde{D}_{2n}$-symmetric solutions of (2.6) where $\tilde{D}_{2n}$ is the group generated by $S_1$ and $S_{12}R_\pi/_n$. Note that $S_{12}R_\pi/_ny = (s_{12}r_\pi/_nx, c)$ and so $S_{12}R_\pi/_n$-symmetric solutions do not necessarily have $c=0$.


## 3. Analysis of bifurcation to rotating waves

We now apply standard symmetry-breaking bifurcation techniques to $G(y, \lambda)=0$ defined by (2.6) to prove the existence of a branch of rotating waves bifurcating from a non-trivial branch of steady-state solutions contained in $H^{\tilde{D}_{2n}}\times R$ for some $n\in Z^+$, subject to a non-degeneracy condition.

The first step is to restrict the problem to an appropriate fixed point subspace of $Y$. Since the linear operator $A$ commutes with $r_\alpha$, $\alpha\in[0, 2\pi)$ and anti-commutes with $s_1$ and $s_2$, it follows that if $x\in H^{\tilde{D}_{2n}}$, then the one-dimensional subspace of $H$ spanned by $Ax$ is $\tilde{D}_{2n}$-invariant and so it must also be $\tilde{D}_{2n}$-irreducible since it has no proper subspaces. Thus, $Ax$ is contained in one of the isotypic components of $H$ and so only the corresponding "block" of $g_x(x, \lambda)$ will be singular at every steady-state solution $(x, \lambda) \in H^{\tilde{D}_{2n}} \times \mathbf{R}$ of (2.1) (see Aston (1990)). If $x \in H^{\tilde{D}_{2n}}$, then the irreducible representation of $\tilde{D}_{2n}$ on the subspace spanned by $Ax$ is

$$s_1 = [-1] , \qquad s_{12} r_{\pi/n} = [1] ,$$

and so the corresponding isotypic component is $H^{\tilde{Z}_{2n}} \cap H_a$ where $\tilde{Z}_{2n}$ is the cyclic group of order $2n$ generated by $s_{12} r_{\pi/n}$. As we are only interested in bifurcation associated with this isotypic component, we take our setting to be

$$\tilde{Y} := Y^{\tilde{Z}_{2n}} = H^{\tilde{Z}_{2n}} \times \mathbf{R} \tag{3.1a}$$

and restrict $G$ accordingly. Thus, we consider the system

$$G(y, \lambda) = 0, \qquad G : \tilde{Y} \times \mathbf{R} \to \tilde{Y},$$

$$G(y, \lambda) := \begin{bmatrix} g(x, \lambda) + cAx \\ <l, x> \end{bmatrix}, \tag{3.1b}$$

$$y = (x, c) \in \tilde{Y}, \qquad l \in H^{\tilde{Z}_{2n}} \cap H_{aa}.$$

Since $s_{12} r_{\pi/n}$ acts as the identity on $H^{\tilde{Z}_{2n}}$, the only non-trivial action of $\tilde{D}_{2n}$ on $\tilde{Y}$ is the reflection $S_1$. Using this reflection, we decompose $\tilde{Y}$ as $\tilde{Y} = \tilde{Y}_s \oplus \tilde{Y}_a$ where

$$\tilde{Y}_s := \{ y \in \tilde{Y} : S_1 y = y \} = H^{\tilde{D}_{2n}} \times \{0\}$$

$$\tilde{Y}_a := \{ y \in \tilde{Y} : S_1 y = -y \} = (H^{\tilde{Z}_{2n}} \cap H_a) \times \mathbf{R} .$$

Clearly, solutions of (3.1) with $y \in \tilde{Y}_s$ are steady-state, $\tilde{D}_{2n}$-symmetric solutions of (1.1). Since $\tilde{Y}_s$ and $\tilde{Y}_a$ are invariant under $G_y(y_s, \lambda)$ for all $y_s \in \tilde{Y}_s$, we denote the restrictions of $G_y(y_s, \lambda)$ to $\tilde{Y}_s$ and $\tilde{Y}_a$ by $G_y^s(y_s, \lambda)$ and $G_y^a(y_s, \lambda)$ respectively.

The analysis now follows similar lines to that of Aston et al (1989) and so we only summarise the main results briefly without proofs. Thus, we now assume that there exists a steady-state solution $(x_0, \lambda_0) \in H^{\tilde{D}_{2n}} \times \mathbf{R}$ of (2.1). We have already seen that at such a point $Ax_0 \in \text{Null}(g_x^0)$, where we use the notation $g_x^0 := g_x(x_0, \lambda_0)$, and so we now make the further assumption that the zero eigenvalue of $g_x^0$ has geometric multiplicity one and algebraic multiplicity two. This assumption can be summarised by the following conditions, where $^*$ denotes the adjoint operator :

$$\text{Null}(g_x^0) = \text{span}\{Ax_0\} \tag{3.2a}$$

$$\text{Null}((g_x^0)^*) = \text{span}\{\psi_1\} \tag{3.2b}$$

$$<\psi_1, Ax_0> = 0 \tag{3.2c}$$

$$<\psi_1, \zeta_1> \neq 0 \tag{3.2d}$$

where

$$g_x^0 \zeta_1 + Ax_0 = 0, \qquad <l, \zeta_1> = 0 . \tag{3.2e}$$

Finally, we also assume that $l \in H^{\tilde{Z}_{2n}} \cap H_a$ is chosen so that

$$<l, Ax_0> \neq 0 . \tag{3.2f}$$

Now symmetry-breaking bifurcation can occur if $G_y^a(y_0, \lambda_0)$ has a non-trivial null-space, where $y_0 = (x_0, 0)$.

**Theorem 3.1**

Assume that (3.2a), (3.2b) and (3.2f) hold at $(x_0, \lambda_0) \in H^{\tilde{D}_{2n}} \times \mathbf{R}$.

a) If $<\psi_1, Ax_0> \neq 0$, then $G_y^a(y_0, \lambda_0)$ is non-singular.

b) If $<\psi_1, Ax_0> = 0$, then

$$G_y^a(y_0, \lambda_0)\Phi = 0, \qquad G_y^a(y_0, \lambda_0)^*\Psi = 0$$

where

$$\Phi = \begin{bmatrix} \zeta_1 \\ 1 \end{bmatrix}, \qquad \Psi = \begin{bmatrix} \psi_1 \\ 0 \end{bmatrix}.$$

This result says that if (3.2c) is satisfied at $(x_0, \lambda_0)$, then bifurcation is possible. It is well known (Werner and Spence (1984)) that symmetry-breaking bifurcation will occur if

$$B_0 := \Psi.G_{y\lambda}^0\Phi + \Psi.G_{yy}^0\Phi V \neq 0 \tag{3.3}$$

where $V \in \tilde{Y}_s$ is the solution of

$$G_y^0 V + G_\lambda^0 = 0 .$$

Assuming in addition that (3.2d) holds, this non-degeneracy condition can be interpreted as an eigenvalue crossing condition as shown by the following result.

**Theorem 3.2**

If (3.2) holds at $(x_0, \lambda_0) \in H^{\tilde{D}_{2n}} \times \mathbf{R}$, then there exists $\varphi \in H^{\tilde{Z}_{2n}} \cap H_a$ and $\sigma \in \mathbf{R}$ which are smooth functions of $\lambda$ satisfying

$$g_x(x(\lambda), \lambda)\varphi(\lambda) = \sigma(\lambda)\varphi(\lambda)$$

with $\sigma(\lambda_0)=0$, $\varphi(\lambda_0)=Ax_0$. Moreover,

$$B_0 \neq 0 \iff \frac{d}{d\lambda}\sigma(\lambda)|_{\lambda=\lambda_0} \neq 0 .$$

Clearly, when such a bifurcation occurs, the branch of rotating wave solutions of (2.6) will be contained in $\tilde{Y} \times \mathbf{R} = H^{\tilde{Z}_{2n}} \times \mathbf{R}$. These results lead to our main theorem which is as follows.

**Theorem 3.3**

Bifurcation from a branch of steady-state solutions of (2.6) contained in $H^{\tilde{D}_{2n}} \times \{0\} \times \mathbf{R}$ to a branch of rotating wave solutions of (2.6) contained in $H^{\tilde{Z}_{2n}} \times \mathbf{R}^2$ will occur if an eigenvalue of $g_x(x, \lambda)|_{H^{\tilde{Z}_{2n}} \cap H_a}$ passes through zero with non-zero velocity. Moreover, the bifurcation is a symmetric pitchfork where, if $(x, c, \lambda)$ is a rotating wave solution then $(s_1 x, -c, \lambda)$ is also a solution. This corresponds to a reflected wave rotating in the opposite direction.

It follows immediately from this result that a stable branch of steady-state solutions (in the sense of orbital stability, see Golubitsky *et al* (1988, p87)) loses stability at such a bifurcation to rotating waves.

The condition (3.2d) ensures that $\Psi.\Phi \neq 0$ which implies that $G_y^a(y_0, \lambda_0)$ has an algebraically simple eigenvalue. This is important computationally since we may infer that $\det G_y^a(y_s(\lambda), \lambda)$, $y_s(\lambda) \in \tilde{Y}_s$ changes sign as $\lambda$ passes through $\lambda_0$, which provides an easily computable indicator for the detection of a bifucation point. The bifurcation point can then be solved for directly using the extended system

$$F(z)=0, \quad F:Z \rightarrow Z,$$

$$F(z):= \begin{bmatrix} g(x,\lambda) \\ g_x(x,\lambda)\varphi+Ax \\ <l,\varphi> \end{bmatrix},$$

$$z=(x,\varphi,\lambda) \in Z:=H^{\tilde{D}_{2n}} \times (H^{\tilde{Z}_{2n}} \cap H_a) \times \mathbf{R}.$$

This system has $(x_0, \zeta_1, \lambda_0)$ as an isolated solution if $B_0 \neq 0$.

Finally, we note that the condition (3.2d) which ensures that the algebraic multiplicity of the zero eigenvalue of $g_x(x, \lambda)$ is equal to two is a generic condition. This follows from the fact that if $x \in H^{\tilde{D}_{2n}}$, then all the generalised eigenspaces of $g_x(x, \lambda)$ are $\tilde{D}_{2n}$-invariant. Further, the generalised eigenspaces of real eigenvalues are generically irreducible (cf. Golubitsky et al (1988, p84)). Now the isotypic component $H^{\tilde{Z}_{2n}} \cap H_a$ is associated with a one-dimensional irreducible representation and so every irreducible subspace is one-dimensional. Thus, generically, only one additional eigenvalue will become zero at a particular value of $\lambda$.

## 4. Bifurcation from rotating waves

In this Section, we consider possible modes of bifurcation from the branch of rotating wave solutions of (2.6) contained in $Y^{\tilde{Z}_{2n}} \times \mathbf{R}$. Since $\tilde{g}_x(x, c, \lambda)$ has a non-trivial null space containing $Ax$ at every rotating wave solution (by Lemma 2.1), it is natural to consider what happens when the zero eigenvalue has algebraic multiplicity 2 at an isolated point in analogy with the theory of Section 3. It is easily proved, in an identical way to Theorem 3.1, that in this case $G_y^0$ has a one-dimensional null space which is contained in $Y^{\tilde{Z}_{2n}}$. Since no symmetry is broken, this case corresponds generically to a turning point with respect to $\lambda$.

Clearly, any steady-state bifurcation of (2.6) will result in another branch of rotating wave solutions. Generically, steady-state bifurcation is associated only with non-trivial, absolutely irreducible representations (see Golubitsky et al (1988, p82)). The only (non-trivial) absolutely irreducible representation of $\tilde{Z}_{2n}$ is the one-dimensional representation

$$S_{12}R_{\pi/n} = [-1] .$$

If bifurcation of this type occurs, the $S_{12}R_{\pi/n}$ symmetry is broken resulting in a branch of solutions which is symmetric with respect to $(S_{12}R_{\pi/n})^2 = R_{2\pi/n}$. Thus, the bifurcating branch will be contained in $Y^{Z_n} \times \mathbf{R}$ where $Z_n$ is the cyclic group generated by $R_{2\pi/n}$.

Similarly, the only (non-trivial) absolutely irreducible representation of $Z_n$ is

$$R_{2\pi/n} = [-1]$$

if $n$ is even, which corresponds to period-doubling bifurcation. Thus, any further bifurcations to rotating waves will be period-doubling bifurcations.

Hopf type bifurcation can also occur resulting in motion on an invariant torus (see Iooss (1984) and Aston et al (1989)).

8

**References**

Aston, P.J. (1990). Introduction to symmetry-breaking bifurcation theory. These proceedings.

Aston, P.J., Spence, A. and Wu, W. (1989). Bifurcation to rotating waves in equations with O(2) symmetry. Submitted to *SIAM J. Appl. Math.*

Golubitsky, M., Stewart, I.N. and Schaeffer, D.G. (1988). *Singularities and Groups in Bifurcation Theory, Vol 2* Appl. Math. Sci. **69**, Springer, New York.

Iooss, G. (1984). Bifurcation and transition to turbulence in hydrodynamics. In *Bifurcation Theory and Applications*, ed. *Salvadori, L., Lecture Notes in Math.* **1057**, 152-201, Springer, New York.

Kevrekidis, I.G., Nicolaenko, B. and Scovel, J.C. (1989). Back in the saddle again : A computer assisted study of the Kuramoto-Sivashinsky equation. Submitted to *SIAM J. Appl. Math.*

Knapp, A.W. (1986). Representation Theory of Semisimple Groups : An Overview Based on Examples. *Princeton University Press, Princeton, N.J.*

Scovel, J.C., Kevrekidis, I.G. and Nicolaenko, B. (1988). Scaling laws and the prediction of bifurcations in systems modelling pattern formation. *Phys. Letts. A,* **130**, 73-80.

Werner, B. and Spence, A. (1984). The computation of symmetry-breaking bifurcation points. *SIAM J. Num. Anal.* **21**, 388-399.

# USE OF APPROXIMATE INERTIAL MANIFOLDS
# IN BIFURCATION CALCULATIONS

H.S. BROWN[1], M.S. JOLLY[2], I.G. KEVREKIDIS[1] AND E.S. TITI[3]

[1]*Department of Chemical Engineering, Princeton University*
*Princeton, NJ 08544, USA*

[2]*Department of Mathematics, Indiana University*
*Bloomington, IN 47405 USA*

[3]*Department of Mathematics, University of California - Irvine*
*Irvine, CA 92717, USA*

ABSTRACT. The theory of inertial manifolds provides a rigorous connection between certain evolution PDEs and low-dimensional dynamical systems. The restriction of the flow of the PDE to the inertial manifold is given by a finite set of ODEs called an inertial form, which captures all the long time dynamic behavior of the PDE. We give a brief survey of several numerical schemes to approximate inertial manifolds and hence inertial forms. We then implement these approximate inertial forms and use them to construct bifurcation diagrams of two model PDEs (the Kuramoto-Sivashinsky equation and a coupled reaction diffusion system). Some numerical issues arising in these schemes are discussed.

## 1. Introduction

Continuing advances in computer technology make possible the numerical bifurcation analysis of problems whose size was once prohibitive. Nevertheless, as larger, more realistic models are studied, with dynamic behavior of increasing complexity, hardware limitations are always encountered. Thus, there will always be a need for more effective discretizations of Partial Differential Equations (PDEs), yielding lower-dimensional, yet accurate systems.

This is often possible for *dissipative* nonlinear evolutionary equations. Such equations generate infinite dimensional dynamical systems with trajectories that enter and eventually remain in an *absorbing ball* contained in the phase space [see e.g. Hale (1988)]. The evolution of this absorbing ball as a set leads to the notion of the *global attractor A*, strictly defined as the $\omega$-limit set of the ball. The global attractor, under additional assumptions, is nonempty, compact, and invariant [Billotti and Lasalle (1971)]. Such a set would for example contain all (stable or unstable) steady states, limit cycles, invariant tori, strange attractors, objects we usually compute in a numerical bifurcation analysis.

9

*D. Roose et al. (eds.), Continuation and Bifurcations: Numerical Techniques and Applications, 9–23.*

While it is known in certain cases that the set $\mathcal{A}$ has finite Hausdorff dimension (see Constantin and Foias (1985), Foias and Témam (1979), Mallet-Paret (1976), Mañé (1981)), it may be quite complicated topologically and attract solutions very slowly. For this reason it is desirable to embed $\mathcal{A}$ into an *inertial manifold*; a finite dimensional positively invariant Lipschitz manifold which attracts all trajectories at an exponential rate [see e.g. Foias *et al.* (1988d)]. The flow restricted to such a manifold is equivalent to that of a finite system of ordinary differential equations called an *inertial form*.

An inertial form is then an optimal finite-dimensional representation of the PDE as far as long-term dynamics -and hence bifurcation analysis- are concerned. Since, however, these forms are not known explicitly, they must be approximated before the numerical bifurcation calculations can be performed.

There is a growing list of dissipative PDEs modeling physical systems for which the existence of inertial manifolds has been established, and which are therefore candidates for this type of bifurcation analysis. These systems include the Cahn-Hilliard (phase transitions) [Constantin *et al.* (1988), Nicolaenko *et al.* (1989)], Ginzburg-Landau (hydrodynamic instabilities) [Constantin (1987), Doering *et al.* (1988), Ghidaglia and Héron(1987)], Swift-Hohenberg (convection) [Taboada (1989)], several reaction diffusion equations [Constantin *et al.* (1988), Jolly (1989), Mallet-Paret and Sell (1988)], damped Hamiltonian systems [Nicolaenko (1987)], as well as the Kuramoto-Sivashinsky equation (interfacial instabilities, wrinkled flame fronts) [Constantin *et al.*(1988,1989), Foias *et al.*(1988c,d), Témam (1988b)]. It is interesting that it is still an open question whether the very equations which motivated the theory, the Navier-Stokes, have an inertial manifold.

## 2. The Examples

In this paper we use approximate inertial forms to perform bifurcation analysis of two dissipative PDEs in one space dimension. The first is the Kuramoto-Sivashinsky equation

$$\frac{\partial u}{\partial t} + 4\frac{\partial^4 u}{\partial x^4} + \alpha\left[\frac{\partial^2 u}{\partial x^2} + u\frac{\partial u}{\partial x}\right] = 0, \qquad (KSE)$$

subject to boundary conditions

$$u(x,t) = u(x + 2\pi, t), \quad u(x,t) = -u(-x,t).$$

Our second example is a set of two coupled reaction-diffusion equations arising in the context of catalytic chemical reactions,

$$\frac{\partial y_1}{\partial t} = \frac{\partial^2 y_1}{\partial x^2} + Bi[(1 - y_1) - r(y_1, y_2)]$$
$$\frac{\partial y_2}{\partial t} = \gamma\frac{\partial^2 y_2}{\partial x^2} + Bi[\alpha(\mu - y_2) - r(y_1, y_2)], \qquad (RDE)$$

where

$$r(y_1, y_2) = \frac{Day_1 y_2}{(1 + K_1 y_1 + K_2 y_2)^2}$$

where $Bi$, $Da$, $K_1$, $K_2$, $\alpha$, $\gamma$ and $\mu$ are positive parameters, and the corresponding boundary conditions are

$$\frac{\partial y_i}{\partial x} = 0, \quad (i = 1 \text{ or } 2 \text{ at } x = 0 \text{ and } 1).$$

These equations, discussed in detail by Dabholkar, Balakotaiah and Luss (1989), model concentration patterns arising on an isothermal catalytic wire for a bimolecular surface reaction with Langmuir-Hinshelwood adsorption-desorption kinetics.

We now outline the application of the existence theory of inertial manifolds for this particular reaction diffusion system. By Smoller (1983) (Corollary 14.9 page 202) one can show that the domain:

$$D = \{(y_1, y_2) : y_1 \geq 0 \text{ and } y_2 \geq 0\}$$

is an invariant region for the solution of $(RDE)$. Using this result, one can show that the set:

$$B = \{(y_1, y_2) : 0 \leq y_1 \leq 1 \text{ and } 0 \leq y_2 \leq \mu\}$$

is absorbing in $L^\infty$. As a result one can easily deduce from Témam (1988b) (Chapter III, section 1.2 pages 91-102) that the corresponding dynamical system has a compact, global attractor. Applying the Constantin-Foias-Témam technique (see for instance Témam (1988b), Constantin et al. (1989) or Constantin and Foias (1985)) one can estimate the Hausdorff and fractal dimensions of this global attractor in terms of the physical constants. Moreover, one can show that the eigenvalues of the linear diffusion operator satisfy a certain gap condition [Constantin (1989)], which is used to establish the existence of an inertial manifold, following for example Constantin et al. (1989).

### 3. Approximate Inertial Forms

We describe approximate inertial manifolds (AIMs) in terms of the general evolutionary equation

$$\frac{du}{dt} + Au + F(u) = 0, \ u \in H. \tag{1}$$

The leading linear term $A$, is assumed to have a spectrum consisting of positive eigenvalues $\lambda_1 \leq \lambda_2 \leq \ldots$, and a complete orthonormal set of eigenfunctions. For the KSE, for example, we take $A = \frac{\partial^4}{\partial x^4}$ and define $F$ pointwise by the remaining terms. Let $P = P_m$ denote the orthogonal projector onto the span of the first $m$ eigenfunctions of $A$, and let $Q = Q_m = I - P_m$ be that onto the orthogonal complement. In most approaches to date, the inertial manifold is represented as the graph of a function $\Phi : P_m H \to Q_m H$. The inertial form then reads

$$\frac{dp}{dt} + Ap + PF(p + \Phi(p)) = 0, \quad p \in PH.$$

A number of AIMs in the literature [Fabes et al. (1988), Foias et al.(1988b 1989a,b,c), Foias and Témam (1988a,b,c), Jauberteau et al. (1989a,b) Marion (1989a,b), Marion and Témam (1989), Sell (1989), Témam (1988a,1989) and Titi (1988,1989)] also characterize the AIM as the graph of a function $\Phi_a : PH \to QH$. We then speak of the *(explicit) approximate inertial form* ((E)AIF), given by

$$\frac{dp}{dt} + Ap + PF(p + \Phi_a(p)) = 0. \qquad (2)$$

A common measurement of error is given in terms of the $L^2$ distance to the global attractor:

$$E_\mathcal{A}(\Phi_a) = \max_{u=p+q\in\mathcal{A}} |\Phi_a(p) - q| \le k\lambda_{m+1}^{-\gamma}, \qquad (3)$$

for some positive constants $k$ and $\gamma$, both independent of $m$, the dimension of the AIM.

A common approach is to find the AIM as the fixed point of a contraction mapping. Consider the *implicit approximate inertial form* (IAIF)

$$\frac{dp}{dt} + Ap + PF(p + q) = 0, \ p \in PH, \quad Aq + QF(p+q) = 0, \ q \in QH, \qquad (4)$$

which effectively restricts (1) to a smooth AIM passing through all the steady states. This AIM is actually the graph of an analytic function $\Phi^s$, which is implicitly given by the fixed points for the family of mappings

$$T_p(q) = -A^{-1}QF(p+q),$$

provided $m$ is large enough [Foias and Témam (1978), Foias and Saut (1983)]. Setting $\Phi_0 \equiv 0$, which is the linear AIM used in a traditional Galerkin method, and taking the first and second iterates, yields explicit functions

$$\Phi_1(p) = -A^{-1}QF(p), \text{ and } \Phi_2(p) = -A^{-1}QF(p + \Phi_1(p)), \qquad (5)$$

respectively. The AIM given by $\Phi_1$ corresponds to the AIM introduced in Foias *et al.* (1988b) under a different motivation, while that given by $\Phi_2$ was developed in Titi (1988,1989) for the Navier-Stokes equations and later applied to the KSE in Jolly *et al.* (1989a). The error between $\Phi^s$ and $\Phi_2$ is of the same order as $E_\mathcal{A}(\Phi^s)$, suggesting that two iterates are sufficient.

A similar procedure, but one motivated by an implicit Euler integration step, is used in Foias *et al.* (1989a) to produce the IAIF

$$\frac{dp}{dt} + Ap + PF(p+q) = 0, \ p \in PH,$$
$$(I + \tau A)q + \tau QF(p+q) = 0, \ q \in QH, \qquad (6)$$

Of special interest is the first iterate, with trivial initial guess, of the associated contraction mapping

$$T_p^{(\tau)}(q) = -\tau(I + \tau A)^{-1}QF(p+q).$$

For appropriately chosen $\tau$ this yields an explicit function

$$\Psi_{1,\tau}(p) = -\tau(I + \tau A)^{-1}QF(p), \qquad (7)$$

giving an AIM with an error comparable to that for the fixed point of $T^{(\tau)}$. The AIM given by (7) was tested in computations in Foias *et al.* (1988a) and Jolly *et al.* (1989b).

**Table 1.** Estimates of order $\lambda_{m+1}^{-7}$ for
approximate inertial manifolds applied to the KSE

| approximation | $\Phi_a(p)$ | $\gamma$ |
|---|---|---|
| linear (Galerkin) | 0 | 1 |
| $\Phi_1$ | $-A^{-1}QF(p)$ | 3/2 |
| Euler ($\Psi_{1,\tau}$) | $-\tau(I+\tau AQ)^{-1}QF(p)$ | 7/4 |
| pseudo-steady ($\Phi_2$) | $-A^{-1}QF(p+\Phi_1(p))$ | 2 |

The errors for the three explicit AIMs presented here (four counting the flat traditional Galerkin, which also is an AIM) are given in Table 1 for the KSE. A more complete table, including error estimates for some additional approximate inertial forms can be found in Jolly *et al.* (1989a). We are currently working on obtaining such estimates for the reaction-diffusion problem discussed here and similar coupled reaction diffusion systems Brown *et al.* (1990).

An alternative measurement due to Sell (1989) offers a direct correspondence to qualitative behavior. In this approach, called *approximation dynamics*, one constructs a perturbed evolution equation

$$\frac{dv}{dt} + Av + F(v) + E(v) = 0, \ v \in H \tag{8}$$

for which $\Phi_a$ is a true inertial manifold and measures the error in terms of a $C^1$ norm on $E$. This error, referred to as the *order* of the AIM, decreases with the dimension $m$, for the AIMs $\Phi_1$, $\Phi_2$, and $\Psi_{1,\tau}$ defined in (5) and (7). One also refers to the *order* of a hyperbolic set $\mathcal{K}$, denoted ord($\mathcal{K}$), which is a measurement of the strength of hyperbolicity defined in terms of the linearized flow (the smaller the order, the stronger the hyperbolicity). It is shown in Sell (1989) that for a given hyperbolic invariant set $\mathcal{K}$ for (1) there exists an $\eta_0 = \eta_0(\text{ord}(\mathcal{K})) > 0$, such that the flow on any AIM of order $\eta \le \eta_0$ has a hyperbolic invariant set $\mathcal{K}_\eta$ which is "homomorphic" to $\mathcal{K}$. Moreover, there is a converse statement [also in Sell (1989)], which involves similar inequalities and allows one to determine if the behavior seen with the AIF

$$\frac{dp}{dt} + Ap + PF(p + \Phi_a(p)) + PE(p + \Phi_a(p)) = 0,$$

is actually present for the PDE. Equation (8) is not necessarily unique, and often one may take $PE \equiv 0$ so that the AIF matches (2). However, we mention a situation below where it is desirable to have a nontrivial perturbation in the $P$-component. The theory of approximation dynamics is in the same spirit as the work in Constantin *et al.* (1984), Titi (1987), where a criterion to determine the existence of stable steady states of the Navier-Stokes equations from the computed solution of a Galerkin approximation was presented.

## 4. Applications

We now illustrate the qualitative performance of a nonlinear AIM in obtaining bifurcation information for steady and time-dependent solutions of the KSE. The steady state results are summarized in figure 1. Figure 1a is a steady state bifurcation diagram obtained using a 12 mode traditional Galerkin approximation which is accurate for values of the bifurcation parameter $\alpha$ through 70. The branches of steady states which are born at the trivial solution at $\alpha = 4, 16, 36, 64$ are referred to as the *unimodal*, *bimodal*, *trimodal* and *quadrimodal* branches respectively, for the number of spatial humps close to their onset. Each branch actually represents a superimposed pair of symmetric branches. We do not describe here the complicated network of secondary bifurcations also present [ see Armbruster *et al.* (1988a,b), Chang (1986), Chen and Chang (1986), Green and Kim (1988), Hyman and Nicolaenko (1986), Hyman *et al.* (1986), Kevrekidis *et al.* (1989), LaQuey *et al.* (1975), Michelson (1986), Scovel *et al.* (1988)]. Figure 1a as well as the other bifurcation diagrams in this paper were obtained using AUTO [Doedel (1981)], while phase portraits were obtained using SCIGMA [Kevrekidis and Jolly (1987)]. A number of steady state bifurcations (open squares) and Hopf bifurcations (darkened squares) are marked in Figure 1a. The labeling scheme for the steady state bifurcations, based on scaling laws for the KSE, is discussed in Scovel *et al.* (1988).

As $\alpha$ increases, so does the complexity of the behavior, i.e. the dimension of the global attractor, and of any inertial manifold containing it. Therefore, the number of modes of the AIFs used should increase as $\alpha$ increases. The lowest current estimate of the dimension of inertial manifolds for the KSE is obtained in Constantin *et al.* (1989) and is of the form $\dim(\mathcal{M}) \leq c\alpha^{3/2}$, where $c$ is a constant which is independent of the parameter $\alpha$. Rigorous, *a priori* estimates such as these are asymptotic for large $\alpha$, and not practical for small $\alpha$. Instead we are guided in our choice of the number of modes used in the AIFs, by prior computational results. A lower bound is provided by the dimension of the number of positive eigenvalues of the linearization at the trivial solution, since its unstable manifold is contained in the global attractor, and hence in any inertial manifold. Three-mode AIFs, like the ones we use here, could therefore not be accurate above $\alpha = 64$. It should be mentioned, however, that whether or not three modes are sufficient for all $\alpha \leq 64$, is far from proved rigorously. Figures 1c,d show the bifurcation diagrams obtained using three and four traditional Galerkin modes respectively. It is obvious that the three mode traditional Galerkin fails to capture the bimodal branch (bifurcating at $\alpha = 16$) accurately. The four mode traditional Galerkin truncation does capture the behavior of the bimodal branch, including the Hopf bifurcation on it at approximately $\alpha = 30$. It also captures the so called *bi-tri* branch, connecting the bimodal branch to the trimodal branch. It is, however, inaccurate in capturing the dependence of the trimodal branch on the parameter.

Figure 1b shows the bifurcation diagram obtained using a three mode Euler AIF ($\Psi_{1,\tau}$ with $\tau = .1$). Note that as $\tau \to \infty$, one has that $\Psi_{1,\tau} \to \Phi_1$. We have also produced the bifurcation diagram for $\Phi_1$, and found it to be nearly indistinguishable from that in figure 1b. We see that for these two simple AIMs, even for such low values of $m$, there is a qualitative and quantitative advantage over the traditional Galerkin approximation with the same number of modes.

One discrepancy in the otherwise accurate diagram in figure 1b is the location of the Hopf bifurcation point on the bi-tri branch: it occurs "after" the turning point on the bi-tri branch rather than "before" it. In Jolly *et al.* (1989a) we show that this artifact is corrected using slightly more elaborate, but still three-mode, AIFs (like the pseudo-steady ($\Phi_2$) AIF).

To demonstrate that the AIFs are capable of capturing not only the steady state,

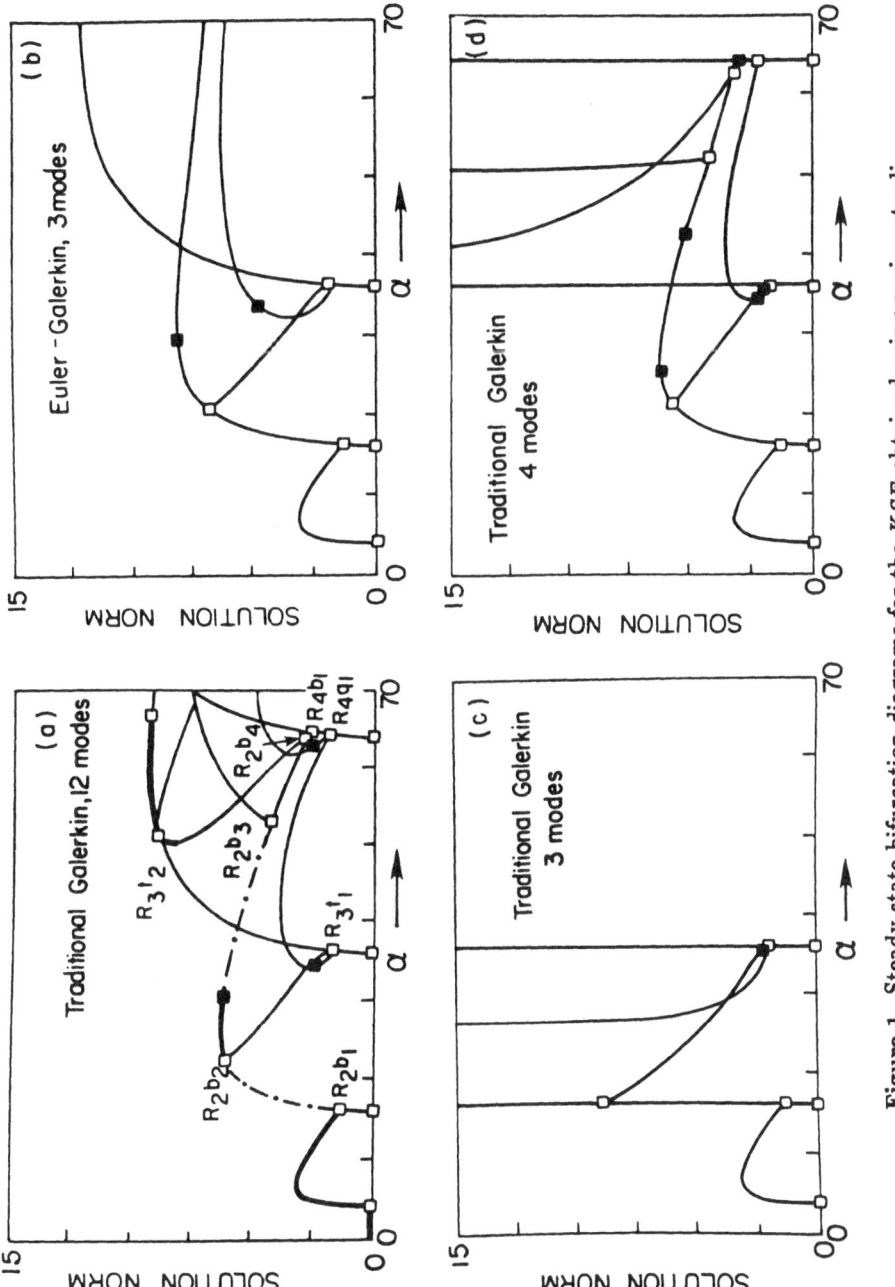

Figure 1. Steady state bifurcation diagrams for the *KSE* obtained using various traditional Galerkin discretizations (1a, 1c, 1d) and a three-mode Euler-Galerkin AIF (1b).

but also the dynamic behavior of the PDE accurately, we have included continuation calculations of a few of the limit cycle branches occurring in this parameter interval. We chose the limit cycle branch emanating from the Hopf bifurcation off of the bi-tri steady state branch. Figure 2a shows the behavior of this oscillatory branch obtained through a traditional nine-mode Galerkin truncation. The branch undergoes a period-doubling bifurcation, and subsequently shows the typical approach to a Sil'nikov saddle-focus homoclinic connection (a global bifurcation). The branch undergoes several successive turning point bifurcations, while its period approaches infinity. There also exist secondary period-doublings along this typical "corkscrew" picture, in agreement with what is generally expected in the neighborhood of such a connection (see for example Glendinning and Sparrow (1984) or Wiggins (1988)). These have not been included here. We see two different saddle-focus connections. The primary limit cycle branch eventually connects with the "upper" part of the bi-tri branch of steady states. At the same time, the limit cycle branch resulting from the first period-doubling has its own saddle-focus connection (and the associated corkscrew bifurcation diagram) with the "lower" part of the bi-tri steady state branch.

Figure 2b shows the same elements of a bifurcation diagram obtained through the AIM given by $\Phi_1$ in Table 1. We see that, even though the original Hopf bifurcation was not quantitatively accurate, the limit cycle branch emanating subcritically from it has a turning point, and continues towards higher values of $\alpha$, in agreement with we believe to be the true bifurcation diagram (figure 2a). We do indeed observe the first period doubling bifurcation, as well as the two distinct Sil'nikov corkscrews for the two limit cycle branches. In figures 2c and 2e we have included phase portraits of limit cycles close to the upper (2c) and lower (2e) homoclinic connections obtained using the nine-mode traditional Galerkin method. We have included for comparison phase portraits of limit cycles obtained via the $\Phi_1$ AIM close to the same saddle connections (figures 2d and 2f).

The above results indicate that it is indeed possible to produce accurate bifurcation diagrams using less equations in an AIF than in the traditional (Galerkin) method. An additional benefit, which may not be obvious, is that when using fewer equations the exponential blowup of the solutions when integrating backwards in time is less severe. This allows us, for example, to obtain good pictures of the stable manifold of a saddle-focus participating in a Sil'nikov loop.

It should be mentioned, at this point, that while these methods yield smaller dynamical systems, there is extra effort involved in evaluating the corresponding vectorfields. This could be outweighed, however, by savings in repeatedly inverting smaller Jacobians in bifurcation calculations or implicit and semi-implicit integration schemes.

One of the most basic dynamic properties of the original evolution equation one would want an AIF to preserve is that of dissipation. This issue is addressed for several AIFs applied to the KSE in Jolly $et$ $al.$ (1989b). There it was proved that the implicit AIFs (3) and (5) are indeed dissipative. Also included in Jolly $et$ $al.$ (1989b) is numerical evidence to indicate that such is not the case for the explicit AIFs, given by (2) with $\Phi_a = \Phi_2$, and $\Phi_a = \Psi_{1,\tau}$. By adjusting small terms in these explicit approximate inertial forms, the dissipation can be recovered, while maintaining the same order of approximation, and all the correct qualitative numerical results of the original, unadjusted AIF. Note that an adjusted AIF

$$\frac{dp}{dt} + Ap + PF(p + \Phi_a(p)) + PE(p) = 0,$$

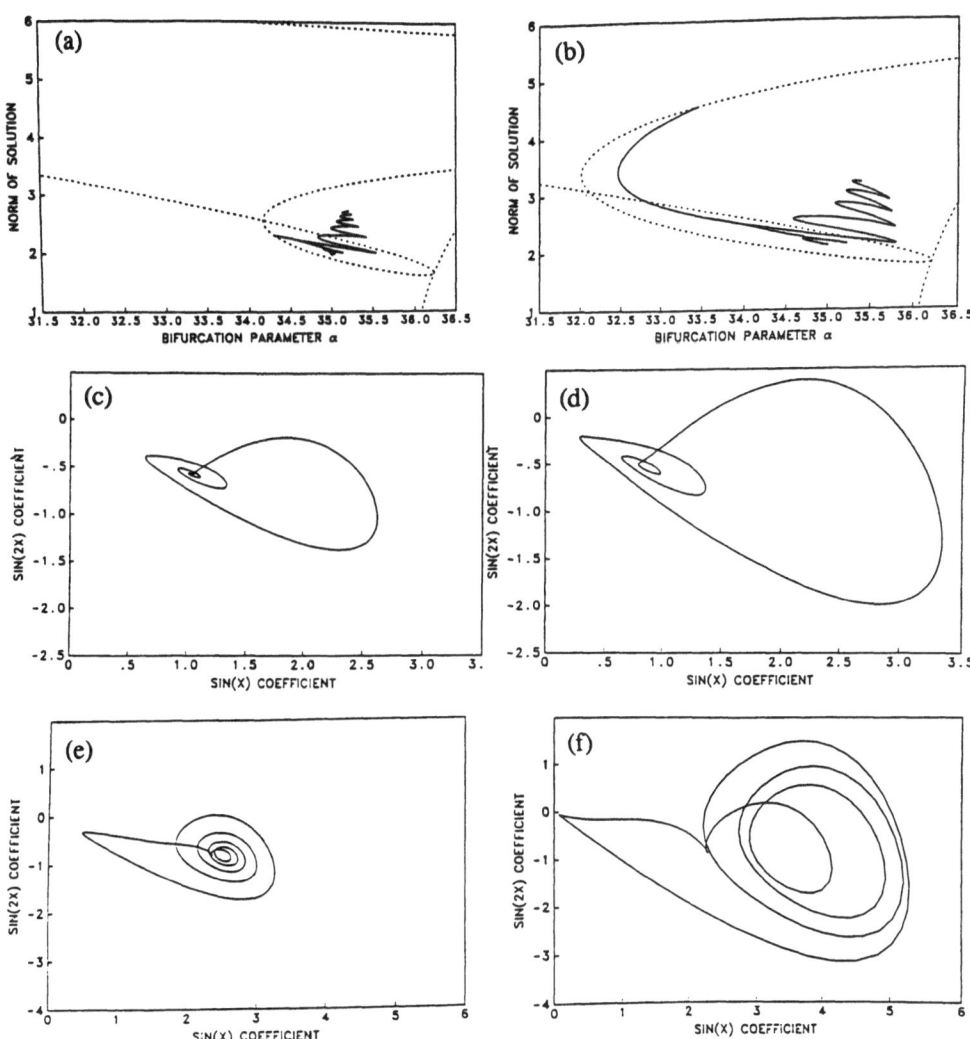

**Figure 2.** Sil'nikov behavior for the *KSE* obtained using a traditional nine-mode Galerkin disretization (2a, 2c, 2e) and a three-mode $\Phi_1$ AIF (2b, 2d, 2f). Figures 2a and 2b show the bifurcation diagrams (solid lines: limit cycle branches, broken lines: steady states). Representative projections of the phase portraits of limit cycles close to the "upper" (2c and 2d) and the "lower" (2e and 2f) Sil'nikov loop are also shown.

also falls under the theory of approximation dynamics described in Sell (1989).

We have also applied the AIM given by $\Phi_1$ in the steady state bifurcation analysis of $(RDE)$. Figure 3a shows a detailed bifurcation diagram obtained with 9 modes in each component, with inflation up to 36 modes in a (partial) dealiasing. The three-mode AIM and the three-mode traditional Galerkin were compared. They were both incapable of capturing the behavior above $Bi \approx 18$. Their accuracy in capturing the connection of the "unimodal" with the "bimodal" branch (figure 3b) was comparable.

## 5. Discussion and Conclusions

Having illustrated the construction and performance of certain AIFs, we now discuss two important issues. The first concerns the appropriate choice of dimension of the AIM for the problem under investigation. As we discussed above for the KSE, this number may vary as the parameters of the system change. Sharp rigorous estimates of this dimension are difficult to obtain. Numerical evidence (for example, calculations of fractal dimensions or Lyapunov exponents of chaotic attractors, the indices of saddle steady states or limit cycles) obtained through sufficiently large traditional discretizations of the PDE, seems to be the best practical guide in this choice. Even for systems which might not possess a true inertial manifold, AIMs are well motivated, and are currently being developed and applied in the study of the Navier-Stokes equations [Jauberteau *et al.* (1989a,b), Foias *et al.* (1989b), Titi (1988,1989)].

Another important issue, which we believe will generate an interesting interplay between the theory of AIMs and numerical analysis, is the relation between aliasing errors and the use of AIFs for spectral methods. For a polynomial nonlinearity, the term $F(p)$, where $p = P_m u$, can be calculated to within machine accuracy using for example a Discrete Fourier Transform (DFT), with $p$ inflated or padded to include a sufficient number of zero components in the higher modes. For example, in a quadratic nonlinearity, like that in the KSE, one inflates through the $2m^{th}$ mode. For general nonlinearities, however, and in particular for the rational polynomial appearing in $(RDE)$, $F(p)$ would have an infinite number of nontrivial components. In a *pseudo-spectral* method one applies the DFT to $p$ (possibly inflated to some extent). This results in an approximation $\hat{P}_m F(p)$ of $P_m F(p)$ where $\hat{P}_m$ is given by

$$\hat{P}_m = ( I_m \quad I_m \quad \cdots )$$

where $I_m$ is the $m$ by $m$ identity matrix. The terms in $\hat{P}_m F(p) - P_m F(p)$ are called *aliasing errors*, and detailed investigations of their effects have been discussed in Orszag (1971, 1972). All AIFs involve the parameterization of the higher modes in terms of the lower modes, implying that some inflation towards dealiasing in the nested evaluations of the nonlinearity $F$ is necessary. We are currently examining this effect in the context of $(RDE)$.

We have seen, in the case of $(RDE)$, that a nonlinear AIM $(\Phi_1)$ will not necessarily outperform the traditional Galerkin approximation which uses a linear AIM with the same number of modes. This may be associated with the slower growth of the eigenvalues of $(RDE)$ as compared to the KSE. In general, estimates of the type contained in (3) could explain the difference in performance of various AIFs. On the other hand, at such

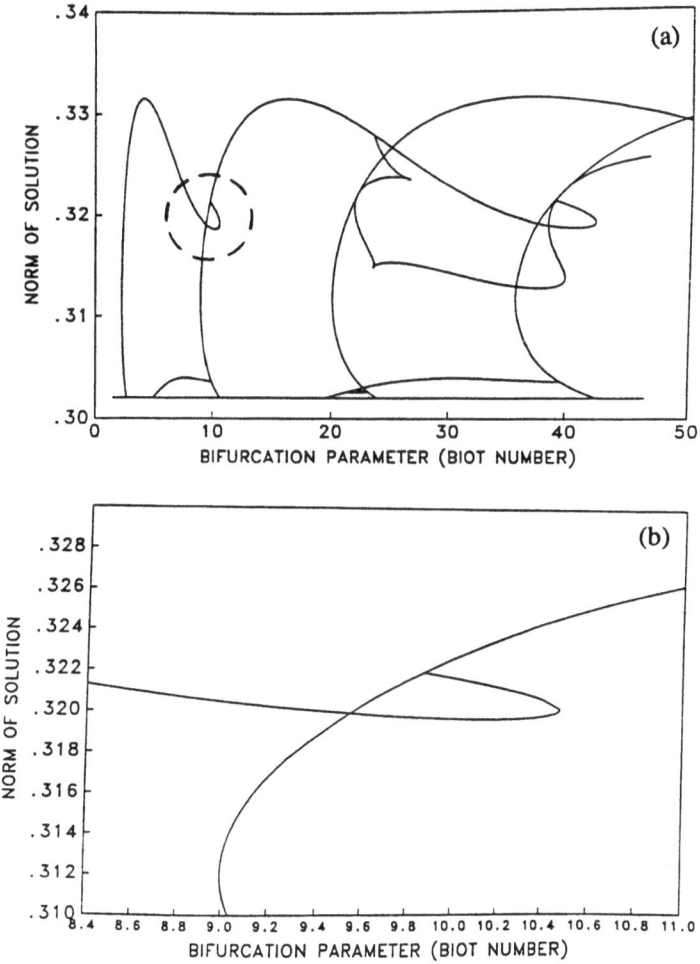

**Figure 3.** (3a). Steady state bifurcation diagram for the *RDE* using a traditional Galerkin pseudospectral discretization (9 modes in each component, inflated with zeroes up to 36 modes for (partial) dealiasing). (3b). A blowup of the connection between the unimodal and bimodal branches, obtained using a three-mode $\Phi_1$ AIF (again inflated with zeroes up to 24 modes for partial dealiasing).

low values of $m$ as we consider here, these estimates can only be considered as rough guidelines.

A promising research direction is the application of AIMs in the context of finite-difference approximations. Foias, Témam and Titi are currently developing schemes along these lines [Foias et al. (1989c)]. This may prove particularly appropriate for problems like ($RDE$), where aliasing errors are significant.

**Acknowledgements.** This work was partially supported by NSF grant ECS-8717787, a Packard Foundation Fellowship, and the U.S. Army Research Office through the Mathematical Sciences Institute of Cornell University. Part of this work was performed when one of the authors (M.S.J) was visiting the Institute for Mathematics and its Applications at the University of Minnesota.

### References

Armbruster, D., J. Guckenheimer, and P. J. Holmes (1988a), *Heteroclinic cycles and modulated travelling waves in systems with O(2) symmetry*, Physica, **29D**, pp. 257-282.

Armbruster, D., J. Guckenheimer, and P. J. Holmes (1988b), *Kuramoto-Sivashinsky dynamics on the center-unstable manifold*, SIAM J. Appl. Math., (to appear).

Billotti, J. E. and J. P. LaSalle (1971), *Dissipative periodic processes*, Bull. Amer. Math. Soc., 77, pp. 1082-1088.

Brown, H.S., M.S. Jolly, I.G. Kevrekidis and E.S. Titi (1990), *Implementation of approximate inertial manifolds for systems of coupled reaction-diffusion equations*, in preparation, .

Chang, H.-C. (1986), *Travelling waves on fluid interfaces: normal form analysis of the Kuramoto-Sivashinsky equation*, Phys. Fluids, 29, pp. 3142-3147.

Chen, L.-H. and H.-C.Chang (1986), *Nonlinear waves on thin film surfaces- II. Bifurcation analyses of the long-wave equation*, Chem. Eng. Sci., 41, pp. 2477-2486.

Constantin, P. (1989), *Private communication.*

Constantin, P. (1987), in: *Proceedings of the AMS/SIAM Summer Conference on the Connection between Finite and Infinite Dimensional Systems*, Boulder, Col., 1987, C. Foias, B. Nicolaenko, and R. Temam eds.Contemporary Mathematics (to appear).

Constantin, P. and C. Foias (1985), *Global Lyapunov exponents, Kaplan-Yorke formulas and the dimension of the attractor for 2D Navier-Stokes equations*, Comm. Pure Appl. Math.,38,pp. 1-27.

Constantin, P., C. Foias, B. Nicolaenko, R. Témam (1988), *Integral Manifolds and Inertial Manifolds for Dissipative Partial Differential Equations*, Appl. Math. Sciences, No. 70 Springer Verlag, New York .

Constantin, P., C. Foias, B. Nicolaenko, R. Témam (1989), *Spectral barriers and inertial manifolds for dissipative partial differential equations*, J. Dynamics and Differential Equations, 1, pp. 45-73.

Constantin, P., C. Foias, R. Témam (1984), *On the large time Galerkin Approximation of the Navier-Stokes equations*, SIAM J. Numerical Analysis, 21, pp. 615-634.

Dhabolkar, V. R., V. Balakotaiah and D. Luss, *Stationary concentration profiles on an isothermal catalytic wire*, Chem. Eng. Sci., 44, pp. 1915-1928.

Doedel, E. J. (1981), *AUTO: a program for the bifurcation analysis of autonomous systems*, Cong. Num., **30**, pp. 265–285.

Doering, C. R., J. D. Gibbon, D. D. Holm, and B. Nicolaenko (1988), *Low dimensional behavior in the complex Ginzburg-Landau equation*, Nonlinearity, 1, pp. 279-309.

Fabes, E., M. Luskin and G. R. Sell (1988), *Construction of inertial manifolds by elliptic regularization*, IMA Preprint No. 459 .

Foias, C., M.S. Jolly, I.G. Kevrekidis, G.R. Sell, and E. S. Titi (1988a), *On the computation of inertial manifolds*, Physics Letters A, **131**, pp. 433-436.

Foias, C., O. Manley, and R. Témam (1988b), *Modelling of the interaction of small and large eddies in two dimensional turbulent flows*, Math. Modelling Numerical Anal., **22**, pp. 93-118.

Foias, C., O. Manley, and R. Témam (1989b), *Approximate inertial manifolds and effective viscosity in turbulent flows*, (in preparation).

Foias, C., B. Nicolaenko, G.R. Sell, R. Témam (1988c), *Inertial manifolds for the Kuramoto-Sivashinsky equation and an estimate of their lowest dimensions*, J. Math. Pures Appl., **67**, pp. 197-226.

Foias, C., G.R. Sell and R. Témam (1988d), *Inertial manifolds for nonlinear evolutionary equations*, J. Differential Equations, **73**, pp. 309-353.

Foias, C. and J.C. Saut (1983), *Remarques sur les équations de Navier-Stokes stationnaires*, Annali Scuola Norm. Sup.-Pisa, Ser. IV, **10**, pp. 169-177.

Foias, C., G.R. Sell and E. S. Titi (1989a), *Exponential tracking and approximation of inertial manifolds for dissipative equations*, J. Dynamics and Differential Equations, 1, pp. 199-224.

Foias, C. and R. Témam (1978), *Remarques sur les équations de Navier-Stokes stationnaires et les phénomènes successifs de bifurcation*, Annali Scuola Norm. Sup.-Pisa, Ser. IV, 5, pp. 29-63.

Foias, C. and R. Témam (1979), *Some analytic and geometric properties of the solutions of the Navier-Stokes equations*, J. Math. Pures Appl., **58**, pp. 339-368.

Foias, C. and R. Témam (1988a), *The algebraic approximation of attractors: The finite dimensional case*, Physica D, **32**,, pp. 163-182.

Foias, C. and R. Témam (1988b), *Approximation algébrique des attracteurs. I. Le cas de la dimension finie*, C. R. Acad. Sci., Paris, **307**, pp. Série I. 5-8 .

Foias, C. and R. Témam (1988c), *Approximation algébrique des attracteurs. II. Le cas de la dimension infinie*, C. R. Acad. Sci., Paris, **307**, pp. Série I. 67-70 .

Foias, C., R. Témam and E.S. Titi (1989c), *Inertial manifold interpretation of the finite difference method.*, (in preparation).

Ghidaglia , J. M. and B. Héron (1987), *Dimension of the attractor associated to the Ginzburg-Landau equation*, Physica, **28D**, pp. 282-304.

Glendinning, P. and C. Sparrow (1984), *Local and global behavior near homoclinic orbits*, J. Stat. Phys., **35**, pp. 645-696.

Green, J. M. and J.-S. Kim (1988), *The steady states of the Kuramoto-Sivashinsky equation*, Physica, **33D**, pp. 99-120.

Hale , J. K. (1988), *Asymptotic behavior of dissipative systems*, Math. Surveys and Monographs, **25**, AMS, Providence, R.I. .

Hyman, J.M. and B. Nicolaenko (1986), *The Kuramoto-Sivashinsky equation: A bridge between PDEs and dynamical systems*, Physica, **18D**, pp. 113-126.

Hyman, J.M., B. Nicolaenko, and S. Zaleski (1986), *Order and complexity in the Kuramoto-*

*Sivashinsky model of weakly turbulent interfaces*, Physica, **23D**, pp. 265-292.

Jauberteau, F., C. Rosier, and R. Témam (1989a), *A nonlinear Galerkin method for the Navier-Stokes equations*, Proc. Conf. on "Spectral and High Order Methods for Partial Differential Equations" ICOSAHOM '89, Como, Italie, (to appear).

Jauberteau, F., C. Rosier, and R. Témam (1989b), *The nonlinear Galerkin method in computational fluid dynamics*, Applied Numerical Mathematics, (to appear).

Jolly, M.S. (1989), *Explicit construction of an inertial manifold for a reaction diffusion equation*, J. Differential Equations, **78**, pp. 220-261.

Jolly, M.S., I.G. Kevrekidis, and E.S. Titi (1989a), *Approximate inertial manifolds for the Kuramoto-Sivashinsky equation: analysis and computations*, Physica D, (in press).

Jolly, M.S., I.G. Kevrekidis, and E.S. Titi (1989b), *Preserving dissipation in approximate inertial forms*, J. Dynamics and Differential Equations, (to appear).

Kevrekidis, I.G. and M. Jolly (1987), *On the use of interactive graphics in the numerical study of chemical dynamics* , paper no. 22c, presented to the 1987 Annual AIChE Meeting, New York, Nov. 1987.

Kevrekidis, I.G., B. Nicolaenko, and C. Scovel (1989), *Back in the saddle again: a computer assisted study of the Kuramoto-Sivashinsky equation*, SIAM J. Appl. Math., (in press).

LaQuey, R. E., S.M. Mahajan, P.H. Rutherford, and W.M. Tang (1975), *Nonlinear saturation of the trapped-ion mode*, Phys. Rev. Lett., **34**, pp. 391-394.

Mallet-Paret, J.(1976), *Negatively invariant sets of compact maps and an extension of a theorem of Cartwright*, J. Differential Equations, **22**, pp. 331-348.

Mallet-Paret, J. and G. R. Sell (1988), *Inertial manifolds for reaction diffusion equations in higher space dimensions*, J. Amer. Math. Soc., **1**, pp. 805-866.

Mañé , R. (1981), *On the dimension of the compact invariant sets of certain nonlinear maps*, pp. 230-242 *Lecture Notes in Math., vol. 898*, Springer Verlag, New York .

Marion, M. (1989a), *Approximate inertial manifolds for the pattern formation Cahn Hilliard equations*, Mathematical Modelling and Numer. Anal., **23**, pp. 463-488 .

Marion, M. (1989b), *Approximate inertial manifolds for reaction diffusion equations in high space dimension*, J. Dynamics and Differential Equations, **1**, pp. 245-267 .

Marion, M. and R. Témam (1989), *Nonlinear Galerkin methods*, SIAM J. Numer. Anal., **26**, pp. 1139-1157.

Michelson, D. (1986), *Steady solutions of the Kuramoto-Sivashinsky equation*, Physica, **19D**, pp. 89-111.

Nicolaenko, B. (1987), in: *Proceedings of the AMS/SIAM Summer Conference on the Connection between Finite and Infinite Dimensional Systems*, Boulder, Col., 1987, C. Foias, B. Nicolaenko, and R. Temam eds. Contemporary Mathematics (to appear).

Nicolaenko, B., B. Scheurer, R. Témam (1989), *Some global dynamical properties of a class of pattern formation equations*, Commun. in Partial Differential Equations, **14**, pp. 245-297.

Orszag, S. A. (1971), *Numerical simulation of incompressible flows within simple boundaries: Accuracy*, J. Fluid Mech., **49**, pp. 75-112.

Orszag, S. A. (1972), *Comparison of pseudospectral and spectral approximation*, Studies in Appl. Math., **51**, pp. 253-259.

Scovel, C., I.G. Kevrekidis, and B. Nicolaenko (1988), *Scaling laws and the prediction of bifurcations in systems modeling pattern formation*, Phys. Lett. A, **130**, pp. 73-80.

Sell, G. R., (1989), *Approximation dynamics: hyperbolic sets and inertial manifolds*, University of Minnesota Supercomputer Institute, PreprintNo. 89/39 .

Smoller, J. (1983), *Shock Waves and Reaction-Diffusion Equations*, Springer-Verlag, New York

Taboada, M. (1989), *Finite dimensional asymptotic behavior for the Swift-Hohenberg model of convection*, Nonlinear Analysis, TMA, (to appear).

Témam, R. (1988a), *Variétés inertiélles approximatives pour les équations de Navier-Stokes bidimensionnelles*, C. R. Acad. Sci. Paris, Serie II, **306**, pp. 399-402.

Témam, R. (1988b), *Infinite Dimensional Dynamical Systems in Mechanics and Physics*, Springer Verlag, New York .

Témam, R. (1989), *Induced trajectories and approximate inertial manifolds*, Mathematical Modelling and Num. Anal. **23**, pp. 541-561.

Titi, E. S. (1987), *On a criterion for locating stable stationary solutions to the Navier-Stokes equations*, Nonlinear Anal., Theory, Methods & Appl., 11, pp. 1085-1102.

Titi, E. S. (1988), *Une variété approximante de l'attracteur universel des équations de Navier-Stokes, non linéaire, de dimension finie*, C. R. Acad. Sci., Paris, **307**, pp. Série I. 383-385.

Titi, E. S. (1989), *On approximate inertial manifolds to the Navier Stokes equations*, MSI Preprint 88-119, also in J. Math. Anal. & Appl., (to appear).

Wiggins, S. (1988), *Global Bifurcations and Chaos*, Appl. Math. Sciences, No. 73, Springer Verlag, New York .

# UNDERSTANDING STEADY-STATE BIFURCATION DIAGRAMS FOR A MODEL REACTION-DIFFUSION SYSTEM

J. C. EILBECK
*Department of Mathematics*
*Heriot-Watt University*
*Edinburgh EH14 4AS, UK*

J. E. FURTER
*Mathematics Institute*
*University of Warwick*
*Coventry CV4 7AL, UK*

ABSTRACT. We consider the various tools that can be used to construct and understand bifurcation diagrams for steady-state solutions of reaction-diffusion equations, with the Sel'kov model as a specific example. Apart from the usual linear analysis, other useful analytical tools are scaling laws and singularity theory. On the numerical side we discuss the use of spectral collocation methods.

## 1. Introduction

The study of reaction-diffusion (RD) systems has been the subject of extensive research for many years. Due to the richness of their solution set, detailed investigations are often done through numerical computations rather than to go through the difficulties of analysis. In particular collocation methods, with path following, are very efficient to compute these complicated bifurcation diagrams (c.f. Fig. 2, 3). Nevertheless it is useful to organize the description in some way, in particular this has been one of the purpose of several papers of Fujii, Mimura, Hosono, Nishiura , et al. (c.f. [20, 13, 14]).

In this paper we would like to present a partial study of the stationary solutions of a model RD equation: the Sel'kov scheme for glycolysis. This scheme has been used by Hunding [18, 19] in papers on morphogenesis, and contains a version of the Gray-Scott model [21]. We shall use a mixture of analytical and numerical methods. The idea is to use singularity theory applied to bifurcation problems (c.f. Golubitsky and Schaeffer, [15, 16]) to understand some distinguished local behaviour and then to use numerical means to relate these local organizing centers. We concentrate on the behaviour of the first and second mode for the 1-D case, our main bifurcation parameter being the size of the region. This choice is motivated by some scaling properties of the problem and by the fact that for this interaction the local singularities have a large domain of "influence".

After some description of the model and the basic equations, we shall discuss the different ingredients necessary for our analysis. These ingredients are linear analysis, symmetries, numerical methods, singularity theory and the "singular" and "shadow" walls. Most of this material has already appeared or will appear elsewhere, so we shall only give brief descriptions here. It is worth noticing that for most of that material there are more general trends and theories applying to a wide range of problems.

*D. Roose et al. (eds.), Continuation and Bifurcations: Numerical Techniques and Applications, 25–41.*
© 1990 *Kluwer Academic Publishers.*

## 2. Statement of the problem

The Sel'kov model is obtained via the classical approach for an enzyme reaction with substrate inhibition and product activation. In some physically relevant limit, which we do not wish to explain here, we get the following system

$$u_t = d_1 \Delta u + f(u, v)$$

$$(1)$$

$$v_t = d_2 \Delta v + g(u, v),$$

in some bounded $\Omega \subset \mathbf{R}^n$. We equip the equations with no-flux boundary conditions $\partial u / \partial n = \partial v / \partial n = 0$. $f$ and $g$ are given by $f(u, v) = 1 - uv^p$, $g(u, v) = c(uv^p - v)$, where $c > 0$ and $p > 1$ are real parameters.

Our discussion of the steady states for (1) is based on the changes occurring as a function of the variation of the size of the domain. In particular we focus on the 1-D case. The boundary conditions are therefore $u'(0) = u'(L) = v'(0) = v'(L) = 0$. It is then clear that we can do a second change of coordinates bringing (1) in a simpler form. Let $\lambda = L^2/d_1$ and $d = d_2/cd_1$, the equations for the steady states of (1) are equivalent to the following system (with $g(u, v)$ redefined to take out the factor $c$)

$$u_{xx} + \lambda f(u, v) = 0$$
$$, u'(0) = u'(1) = v'(0) = v'(1) = 0 \qquad (2)$$
$$dv_{xx} + \lambda g(u, v) = 0$$

## 3. Symmetries

Before starting any discussion of the bifurcation diagrams we must take into account the following properties of problems of the type (2).

Those problems have a simple $\mathbf{Z}(2)$-symmetry, $x \mapsto 1 - x$, but actually there is more. It is known that there is a 1-1 correspondence between the solutions of (2) and those of the same equations with periodic boundary conditions (c.f. [13,4] for recent surveys of the question). This latest problem has now an additional $\mathbf{S}^1$ symmetry corresponding to the phase shift $x \mapsto x + \theta$, giving it an $\mathbf{O}(2)$-equivariance. The solutions of our problem (2) lie in a slice of this $\mathbf{O}(2)$-equivariant problem, bringing in "hidden" symmetries. For instance, this accounts for the $\mathbf{Z}(2)$-equivariance of all bifurcations, not just for the odd modes.

Moreover the particular dependence on the bifurcation parameter brings in an additional structure: the scaling law, in particular useful for a reduction of the number of branches we need to investigate. We refer to Fujii et al. [13], Scovel et al. [22]) and Aston [2].

The idea is that if $(u, v, \lambda)$ is a solution of (2) then $(u_n, v_n, n^2\lambda)$ is a solution of (2) where $\hat{u}_n = \hat{u}(nx)$, where $\hat{}$ designates the periodic extension to the problem with periodic boundary conditions. In particular this means that if we plot the bifurcation diagrams on a log scale for the parameter $\lambda$, every curve is exactly replicated with shifts of $2 \log n, n = 2, 3, \ldots$ [8].

## 4. Linear analysis

The easiest way to study problems like (1) or (2) is first to identify trivial, i.e. constant, states and then to look for bifurcations from these trivial branches. In our case there is a unique branch of trivial states $(1, 1, \lambda)$.

The next step is to study the linearisation at $(1, 1, \lambda)$. In our case it is given by

$$
\begin{aligned}
u_{xx} + \lambda(-u - pv) &= 0 \\
dv_{xx} + \lambda(u + (p-1)v) &= 0
\end{aligned}
, u'(0) = u'(1) = v'(0) = v'(1) = 0. \tag{3}
$$

It is a simple problem to find the values of $\lambda$ where there are nontrivial solutions of (3) and to find the associated eigenvectors. The bifurcation points are at $\lambda_n^{\pm} = (n\pi)^2 k^{\pm}$ with the eigenvectors $\Phi_n^{\pm} = (\cos n\pi x)\phi^{\pm}$, where $k^{\pm} = \frac{1}{2}(p - 1 - d \pm \sqrt{[(p - 1 - d)^2 - 4d]})$ and $\phi^{\pm} = (-pk^{\pm}, 1 + k^{\pm})$. We assume $c < 1/(p - 1)$ to insure that there is no Hopf bifurcations from the trivial branch. It is easy to see that for $d > (\sqrt{p} - 1)^2$ there is no bifurcation points, then at $d = (\sqrt{p} - 1)^2$ two appear for every mode (scaling law), separating as $d$ decreases towards 0. The kernels are 1-D, apart from the values $d = d_{n,m}$ (complicated formulas) where two bifurcation points $\lambda_n^+$ and $\lambda_m^-$ collide creating a 2-D kernel of mode interaction.

Before continuing we would like to present Fig. 1 which helps us to organize the investigation and the different methods we can use. This diagram is a corrected version of an old

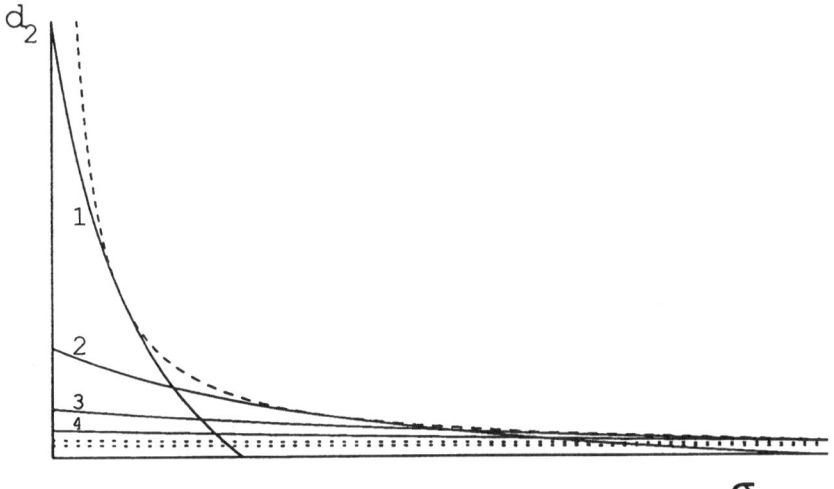

Figure 1: Stability diagram, —- stability curves, - - - - envelope

one appearing in [13]. Instead of the usual $(d_1, d_2)$ - stability diagram for the linearisation of (1), it is a $(\sigma = 1/d_1, d_2)$ stability diagram. The stability curves are the hyperbolas

$$d_2 = \gamma_n(\sigma) = \frac{c}{(n\pi)^2} \left[ \frac{(p-1)(n\pi)^2 - \sigma}{(n\pi)^2 + \sigma} \right].$$

For $n = 1, 2, \ldots$ as shown on the figure. They satisfy the scaling property $\gamma_n(\sigma) = \frac{1}{n^2}\gamma_1(\frac{\sigma}{n^2})$ and they have an envelope in the hyperbola $\sigma d_2 = m = c(\sqrt{p} - 1)^2$. We can think of the solution set of (1) as being a set of manifolds bifurcating from the plane $((1,1), \sigma, d_2)$ along the stability curves. Hence the $\lambda$-bifurcation diagrams of (2) are 1-D slices of that solution set along hyperbolas $\sigma d_2 = d$, in some sense parallel to $\sigma d_2 = m$, with the speed of $\sigma$ proportional to $\lambda$. The $d_2$ axis $\sigma = 0$ is called the "shadow wall" and the $\sigma$ axis $d_2 = 0$ is called the "singular wall".

Fig. 1 helps us to review how the different methods are fitting in.

1. Local analytical techniques.

    (a) along the stability curves, the usual 1-D bifurcation [15, 16].

    (b) at the intersection points we have 2-D mode interaction [1].

2. "Global" analytical techniques.

    (a) in the limit $d_1 \to \infty$, the "shadow" system in the "shadow" wall $\sigma = 0$ [13].

    (b) in the limit $d_2 \to 0$, the "singular" problem SLEP method, [13]. also in our case the problem is regular.

    (c) scaling laws.

    (d) classical analysis (comparison theorems to prove non-existence, etc.)

3. Numerical techniques.

    (a) path following.

    (b) solvers of elliptic systems (finite elements, spectral-collocation methods, etc.)

## 5. Spectral collocation methods and path-following algorithms

Spectral methods for partial differential equations were first extensively discussed in the monograph by Gottlieb and Orszag [17]. Since then there has been many theoretical and practical developments, well documented by the excellent recent textbook by Canuto et al. [3], hereafter referred to as CHQZ.

Here we follow the approach of earlier papers [9, 10, 8]. We approximate the exact solution by a truncated eigenfunction expansion

$$\begin{pmatrix} u(x) \\ v(x) \end{pmatrix} \approx \begin{pmatrix} u^N(x) \\ v^N(x) \end{pmatrix} = \sum_{j=0}^{N} \begin{pmatrix} c_j^{(1)} \\ c_j^{(2)} \end{pmatrix} \cdot \phi_j(x)$$

where the $\phi_j(x)$ satisfy $\Delta\phi_n(x) = \lambda_n\phi_n(x)$. One other common choice for $\phi_j(x)$, Chebyshev polynomials, is discussed below. The collocation approach leads to a set of $(2N+2)$ equations for the coefficients

$$\begin{pmatrix} d_1 B & 0 \\ 0 & d_2 B \end{pmatrix} \begin{pmatrix} c^{(1)} \\ c^{(2)} \end{pmatrix} + \lambda \begin{pmatrix} F \\ G \end{pmatrix} = \begin{pmatrix} 0 \\ 0 \end{pmatrix}$$

where $B = \{b_{ij}\} = \{\lambda_j\phi_j(x_i)\}$, the $x_i = i/N$, $i = 0,\ldots,N$ are a set of collocation points, $F_i = F(u_i^N, v_i^N)$, $u_i^N = u^N(x_i)$, etc. The column vector $u^N = \{u_i^N\}$ can be formed from the vector $c^{(1)} = \{c_i^{(1)}\}$ by the matrix multiplication $u^N = Ac^{(1)}$, where $A = \{a_{ij}\} = \{\phi_j(x_i)\}$, with a similar calculation for $v^N$. Note that our choice of basis functions and collocation points are different from the set discussed in CHQZ, since we are considering an even periodic extension, as discussed below. When discussing trigonometric polynomials, CHQZ generally use the set $\exp^{\sqrt{-1}kx}$ for $-N/2 \le k < N/2$ for even $N$, and $x_k = 2\pi k/N$, $k = 0,\ldots,N-1$. These minor differences lead to some small changes in some of the estimates. Starting from a solution for () for a particular value of $\lambda$ (for example the constant solution), we can trace out solution curves for other values of $\lambda$ using path-following techniques (cf. [10]).

Since these techniques require solving () at a large number of points, it is important to consider questions of accuracy and efficiency for the spectral collocation methods considered here. General considerations of accuracy and efficiency for spectral methods applied to many linear and some nonlinear problems are discussed in CHQZ. A crucial factor in the accuracy of such methods is the smoothness of the solutions of the equations under consideration, and in the case of trigonometric expansion functions, the smoothness of the periodic extension of such solutions. Fortunately, for the problem here, a classical bootstrap argument shows that:

**Theorem 1** *If the functions $f$ and $g$ in equation (2) are infinitely differentiable with respect to both their arguments, and $u(x)$, $v(x)$ are classical ($u, v \in C^2([0,1])$) solutions to (2), then $u(x), v(x) \in C^\infty([0,1])$. Also the even periodic extensions to $u, v$ lie in the space of periodic functions $C_p^\infty([-1,1])$.*

These continuity results suggest that the method may achieve super-algebraic accuracy, i.e. the method may converge faster than any power of $N$. This property is also known as spectral or infinite-order accuracy or exponential convergence, c.f. CHQZ. Proving this result is more difficult, since the equations are nonlinear and have none of the nice properties such as coercivity conditions which can help in such cases. Indeed, we do not expect a general convergence result since the solutions of (2) are in general non-unique, and at turning and bifurcation points the equations are singular. However our problem comes into the range of Prop. 11.3 in CHQZ, which ensures superalgebraic convergence at regular points of the bifurcation diagram.

For most general boundary-value problems, the periodic extension to the exact solution are not smooth, so trigonometric polynomials will not give exponential convergence (cf. CHQZ). One example is given by the corresponding Dirichlet problem for (1). For this problem, although trigonometric polynomials give good results for small $N$, asymptotically they give only algebraic convergence (cf. numerical experiments by [12, 11]). In this case, the use of Chebyshev functions for the expansion for large values of $N$ is recommended and seems to give super-algebraic convergence.

The fact that trigonometric polynomials rather that Chebyshev polynomials can be used for the Neumann problems discussed here is most useful. At a basic level, calculations on the time-dependent problem using fast transforms are 15% more efficient using trigonometric rather than Chebyshev polynomials (CHQZ). For the steady-state solutions discussed here, it is a trivial matter to *recognise* symmetries in a trigonometric expansion, but except for the simplest 2-fold symmetry this is not easy with a Chebyshev expansion or other techniques such as finite difference or finite element methods. Once symmetries have been recognised, of course, any method can be programmed to take advantage of a known symmetry, but again it is simpler to switch between calculations with different symmetries in the trigonometric representation. Also, since we know the exact positions of the bifurcation points along the constant solution from linear theory, it is comforting to know that the first $N$ bifurcation points from this branch will be given *exactly* by a trigonometric expansion. This is not the case for a Chebyshev expansion or any other techniques.

We have already mentioned some advantages of spectral methods over the finite difference approach of De Dier et al. [6, 7, 5]. CHQZ point out the large gains in efficiency of spectral methods as compared to the finite difference or finite element calculations. Another important advantage when large and complicated bifurcation diagrams are under study is that spectral methods give simple and natural data compression techniques for the solutions along the branches–only a few coefficients need to be stored to reconstruct the entire solution at any point. This saving is particulary useful in the Fourier representation for branches with a high degree of symmetry.

Fig. 2 shows a number of bifurcation diagrams in the 1-D case, with various values of $d$, from top to bottom $d = 0.128, d = 0.11, d = 0.0766$.

Numerically the extension from 1-D to 2-D is not trivial, even allowing for the extra computing time required. In 1-D there is an obvious choice of spectral functions and collocation points – the first $n + 1$ trigonometric cosine polynomials and $n + 1$ collocation points equally spaced on the unit interval.

When we go to the rectangle, it is natural that the class of functions we should use is the tensor product of the 1-D expansion functions (CHQZ). The problem comes in deciding which members of this class we should pick, as we are naturally restrained to a finite number. In the rectangular case our set of basis functions are $\phi_{nm}(x, y) = \cos n\pi x \cos m\pi y \sqrt{2}$ for some finite lattice of $n, m$ values. If we are working with the time-dependent problem, it is clear that our set of $n, m$ values should be a rectangle in $n, m$ space, $(0 \leq n \leq N_x) \times (0 \leq m \leq N_y)$, with a corresponding rectangular grid of collocation points in $x, y$ space. With this choice the fast transforms involved are especially efficient. However if we are dealing only with the steady-state problem, we are concerned with maximising the accuracy of our approximation for a fixed number $N + 1$ of eigenfunctions. In this case a rectangular grid of $n, m$ points may not be the best choice. For example, if we wish to ensure that the first $N$ bifurcations from the constant solution are reproduced exactly, we should choose the first $N$ spectral functions in order of increasing eigenvalue. Since this eigenvalue is proportional to $n^2 + 2m^2$, we will need a triangular shaped region in $n, m$ space. If we stick to a rectangular grid of collocation points in $x, y$ space we obtain a singular collocation matrix $A$, i.e. the different spectral functions in each space dimension are not resolved due to aliasing effects.

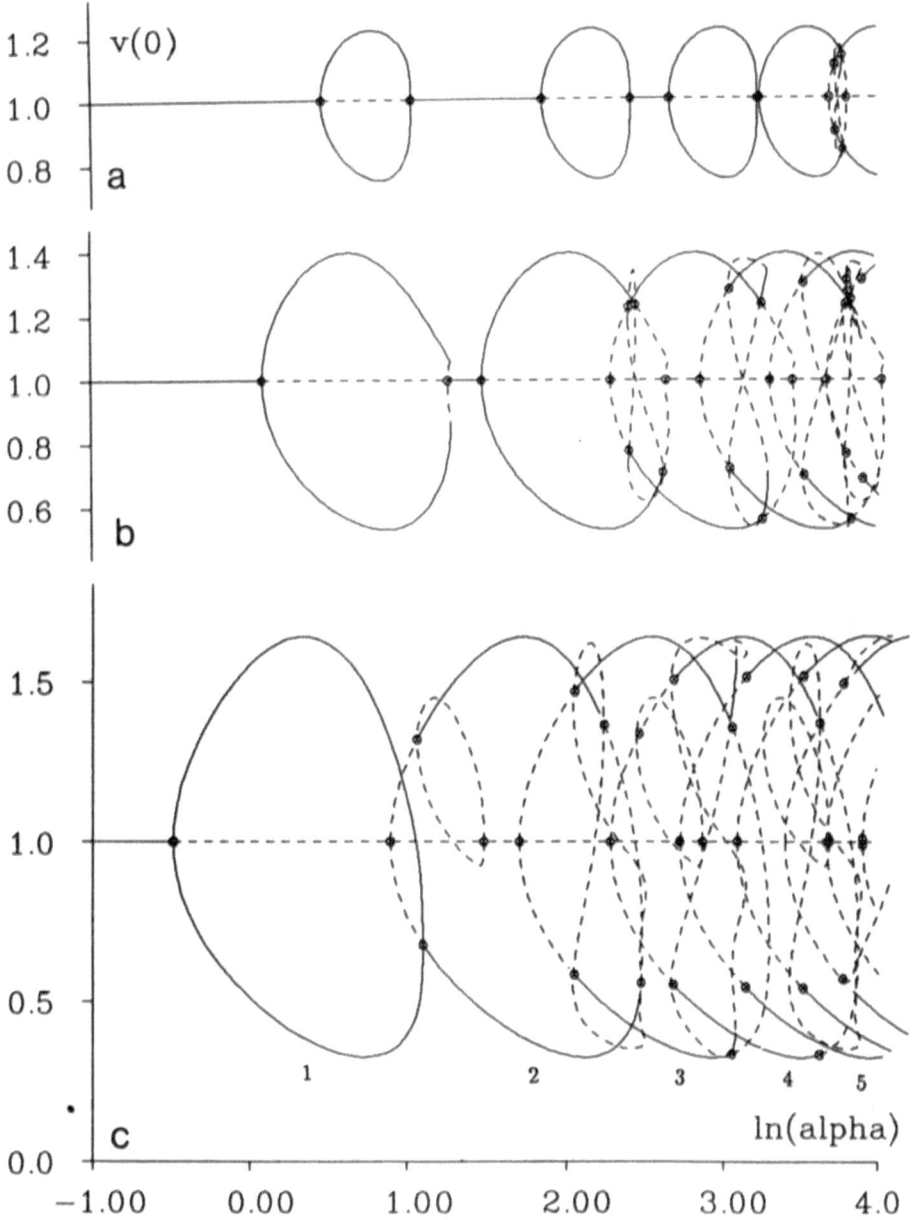

Figure 2: Some bifurcation diagrams for the 1-D Sel'kov model

Even considered as a simple problem in function approximation rather than in the solution of differential equations, little seems to be known about this problem in two or more space dimensions. There may be some natural nonrectangular choice of collocation points, but a preliminary search has failed to discover any. One possible approach would be to fix the collocation points numerically to maximise some norm or determinant of the collocation matrix. This problem requires further investigation.

We have avoided such problems by sticking to the simple choice of the tensor product of the constant solution plus the first $N_x$ eigenfunctions in the $x$ direction and the first $N_y$ eigenfunctions in the $y$ direction, to give a total of $(N_x + 1) \times (N_y + 1)$ eigenfunctions. For our calculations we took $N_x = 10$ and $N_y = 6$, giving 77 spectral functions and $154 \times 154$ matrices in the 2-component calculation. The resulting bifurcation curves are shown in Fig. 3, together with some plots of the resulting solutions $v(x, y)$ at various points.

The full computation, including the stability calculations, took about 40 hours on a Sun 3/60.

## 6. Singularity theory results

Concerning our analysis of the one space dimensional problem, there are two important local situations to look at: The 1-D bifurcation from the trivial branch and the (1,2) mode interaction. We shall give the background and results for these two cases, more detailed computations will appear elsewhere.

The problem (2) has two unfolding parameters, $p$ and $d$, it is therefore clear we should describe the different $\lambda$-diagrams in relation with their position in a $(p, d)$-plane. For computational reasons, and in order to enlarge the narrow region of interest, we choose to use the coordinate system $t = \sqrt{d}$ and $p/(1 + t)^2$. The results are shown in Fig. 4, and the diagrams for the different regions are given in Fig. 5 and 6. Note that in Fig. 4, only the lines separating regions 1/2, 7/8, 2/3, 6/19 and the line through $EA$ have been calculated – other lines are only schematic at the present time.

They will be briefly discussed in the last part of the paper.

By a Lyapounov-Schmidt reduction the 1-D bifurcation equation near $(1, 1, \lambda_n^{\pm})$ is given by the following formula:

$$h(x, \lambda) = PF\left((1, 1) + x\Phi_n^{\pm} + w(x, \lambda), \lambda + \lambda_n^{\pm}\right),$$

where $F$ correspond to the elliptic system in (2), $P$ is the projector onto the kernel of the linearisation identified to $\mathbf{R}$, i.e. let $\phi_{\pm}^{*} = (-cpk^{\pm}, 1 + k^{\pm})$, $\Phi_n^{\pm *} = (\cos n\pi x)\phi_{\pm}^{*}$ and $<, >$ being the usual scalar product in $\mathbf{R}^2$, then if $z \in (L^2(0, 1))^2$,

$$Pz = 2(< \phi_{\pm}, \phi_{\pm}^{*} >)^{-1} \int_0^1 < z, \Phi_n^{\pm *} > ds,$$

$w$ is given implicitly, near $w(0, 0) = 0$, by the equation

$$(I - P)F\left((1, 1) + x\Phi_n^{\pm} + w(x, \lambda), \lambda_n^{\pm} + \lambda\right).$$

From the hidden $\mathbf{O}(2)$-symmetry we know that $h$ is $\mathbf{Z}(2)$-equivariant. Moreover it is easy to prove that $h_{x\lambda}(0, 0) \neq 0$ for any $d$ away from the envelope, i.e. if $d \neq (\sqrt{p} - 1)^2$, $(p \neq (1 + t)^2)$.

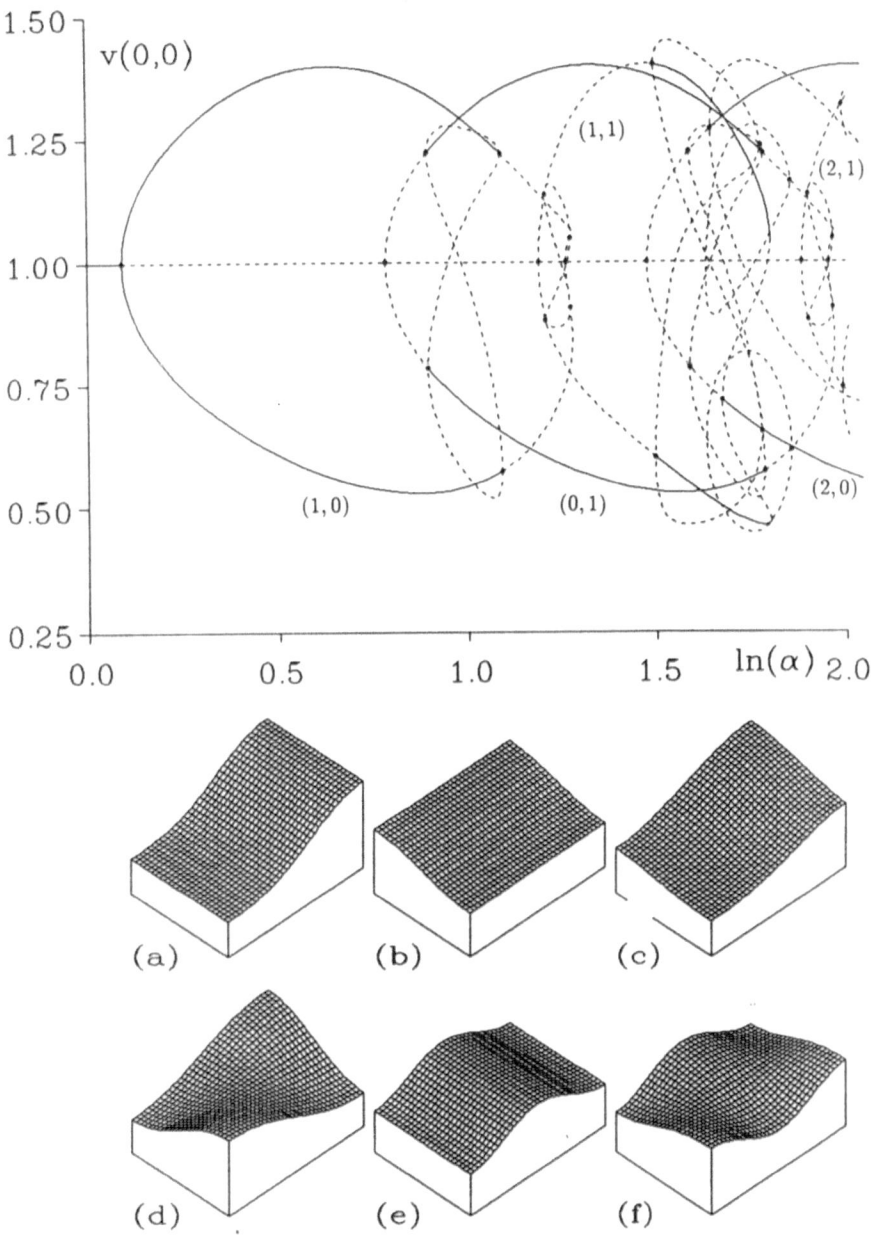

Figure 3: 2-D bifurcation and solution plots, $d = 0.11$

34

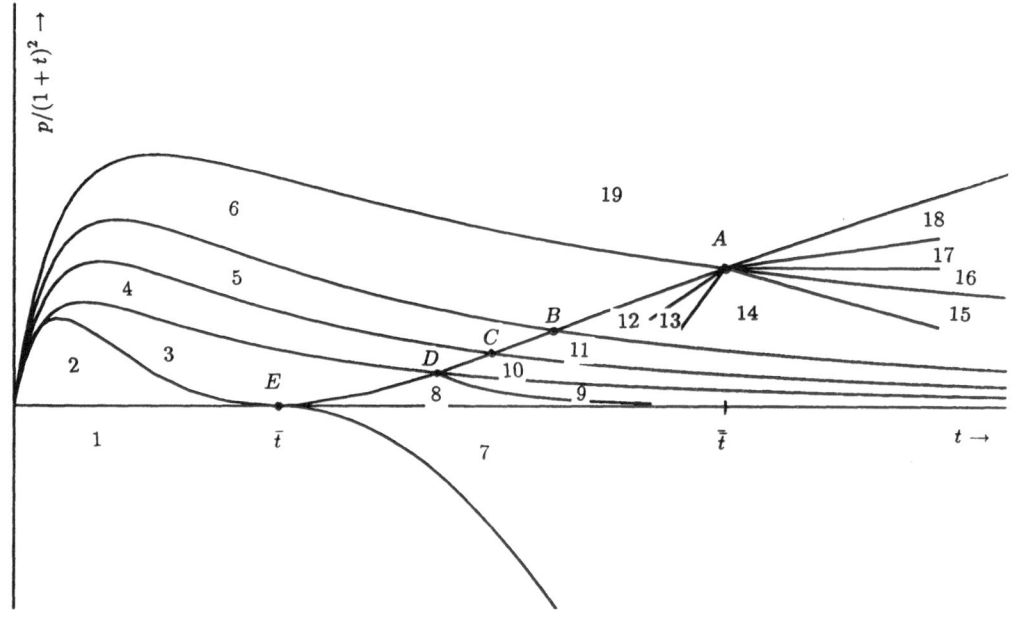

Figure 4: Regions with different types of bifurcation diagrams

The classification for $\mathbf{Z}(2)$-equivariant problems tells us that in this case we have to look for the first integer $k$ for which $h_{x^{2k+1}}(0,0) \neq 0$. If such a finite $k$ exists, the problem $h$ is contact equivalent to $\mathrm{sign}(h_{x^{2k+1}}(0,0))x^{2k+1} \mp \lambda x$. Such problems and their universal unfolding are well-known.

A more interesting question is what happens when $p = (1+t)^2$. Then $h_{x\lambda}(0,0) = 0$ but $h_{x\lambda^2}(0,0) < 0$. In that case the classification theorem tells us that the simplest case satisfying that condition is $\pm x^3 - \lambda^2 x$, of $\mathbf{Z}(2)$-codimension 1. Being able to play with a second parameter we can set $h_{x^3}(0,0) = 0$. Using Mathematica we can compute that $h_{x^3}(0,0)$ is proportional to $(4t^4 + 5t^3 - 11t^2 - 12t - 18)$ which has a unique root at $t = \bar{t} \approx 1.8433$ corresponding to the point $E$ on Fig. 4. Using Mathematica again we find the normal form $-x^5 + 2m\lambda x^3 - x\lambda^2$ with $m \approx 0.1322$. This singularity is of topological codimension 2 and is unfolded by $p$ and $t$, its local behaviour around $E$ is given by Fig. 7.4 in Golubitsky-Schaeffer ([15], pg 277).

A similar analysis for the (1,2)-mode interaction, along the lines of [1], brings the following result. Along the curve $p = (1+t)^2 + t/2$, where the bifurcation diagram present a (1,2)-mode interaction there is a unique codimension-2 singularity, point $A$, at $t = \bar{\bar{t}} \approx 5.0139$. The singularity is completely unfolded by $p$ and $t$ and the normal form unfolding is given by

$$-\lambda x - xy - \alpha x = 0 \qquad (4)$$
$$\lambda y - x^2 - y^2 + \beta y^3 = 0 \qquad (5)$$

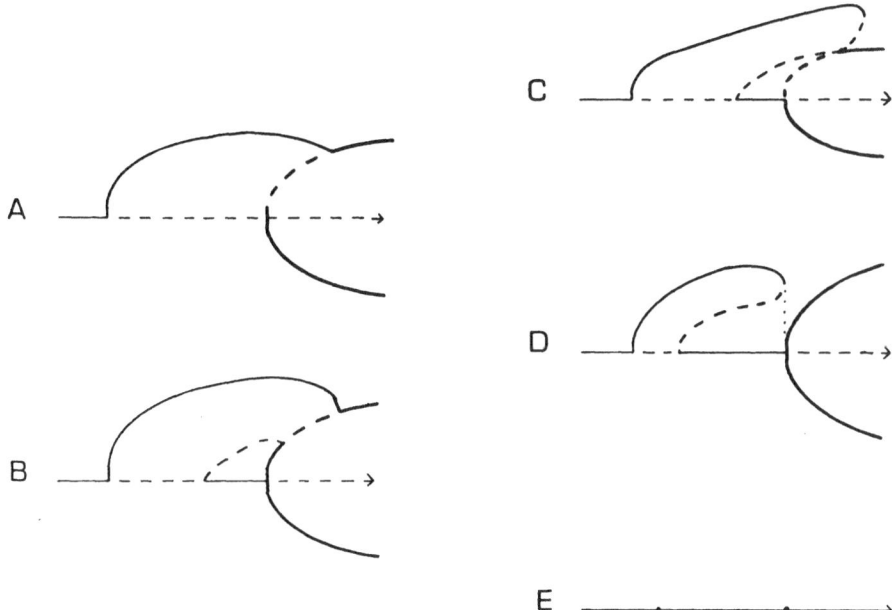

Figure 5: Codimension 2 diagrams

where $x, y$ are the 1, resp. 2, mode components and $\alpha, \beta$ the unfolding parameters. The description of the unfolding can be found in Armburster-Dangelmayr [1], although some minor mistakes have to be corrected.

## 7. "Singular" and "shadow" walls

This analysis has two aims. First it gives the asymptotic limit as $d \to 0$ of the bifurcation diagrams. In our analysis the two walls combine, the shadow system directing the first bifurcation points, $\lambda_n^-$, and the singular system directing the second, $\lambda_n^+$. The second aim is, following Nishiura, to get the restabilization of every mode near these walls through the SLEP methods. Here we shall only describe the first part: the asymptotic limits. As $\sigma \to 0$, $u$ tends to a constant, hence (2) becomes

$$
\begin{aligned}
v_{xx} + \lambda g(u_0, v) &= 0 \\
&\qquad , v'(0) = v'(1) = 0, \lambda = cL^2/d_2, \\
f(u_0, v) &= 0
\end{aligned}
\tag{6}
$$

which in turn is equivalent to the following single-non-local equation

$$
v_{xx} + \lambda \left[ -v + v^p (\textstyle\int v^p)^{-1} \right] = 0, \quad v'(0) = v'(1) = 0.
$$

36

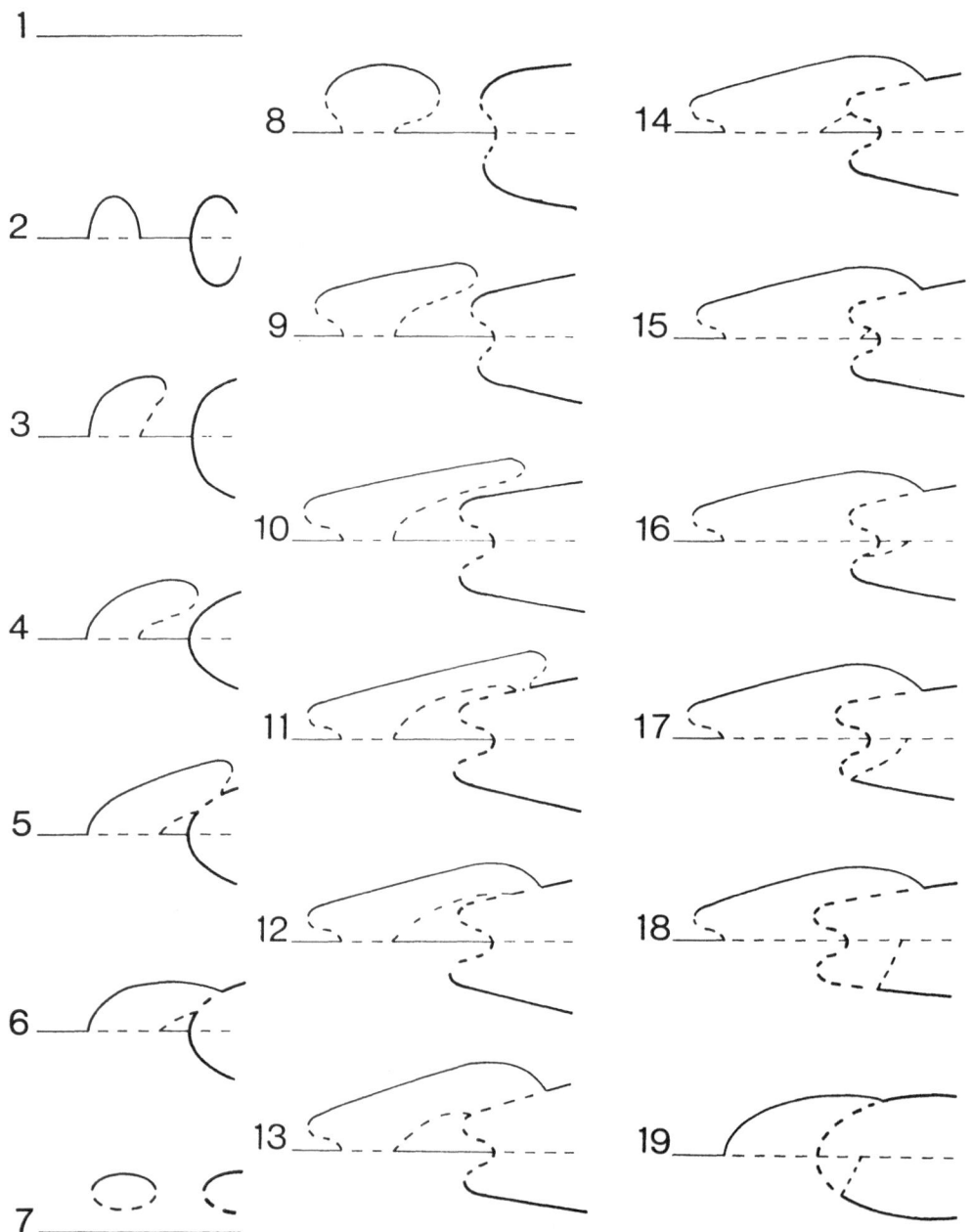

Figure 6: Bifurcation diagrams for the (1,2) modes

The bifurcation diagrams of () then looks like a supercritical pitchfork in $\lambda$ (subcritical in $d_2$).

In the singular wall (2) becomes

$$d_1 u + L^2 f(u, v_0) = 0$$
$$, u'(0) = u'(1) = 0. \tag{7}$$
$$g(u, v_0) = 0$$

In general (7) is a singular perturbation problem, however in this case it is regular and equivalent to the single equation:

$$u_{xx} + \lambda(1 - u^{-1/(p-1)}), \quad u'(0) = u'(1) = 0, \quad \lambda = L^2/d_1.$$

Using different analytical tools we get the well-known behaviour of the subcritical quintic and its unfolding, i.e. $x^5 + \text{sign}(p-3).x^3 + \lambda x = 0$. In particular $p$ affects the asymptotic behaviour through the sign of $(p-3)$.

## 8. Comments and final remarks

We would first like to describe our results shown in Fig. 4 and later figures. In Fig. 4, we have represented 19 regions with different (1,2)-mode interactions. There are 5 codimension 2 organizing centres, the two most important, $A$ and $E$, have already been described. The remaining three, $B$, $C$ and $D$ are multi-local singularities (i.e. the bifurcation diagram possess simultaneous singularities at different places, in this case two).

In Fig. 5, these degenerate diagrams are drawn. By analytical means, in addition to the two parabolas sustaining $A$ and $E$, it is possible to get curves of change of criticality of the bifurcation from the trivial branch. Those curves separate the regions 2 and 3, bind (1,0) and (3,0), and go through the codimension two points. Multilocal phenomena have to be approached numerically. One of the main feature is the number of changes in criticality for primary and secondary bifurcations. For the Sel'kov model we believe that those changes occur for very small relative variations of the parameters. Nevertheless they are important to understand how the interactions occur to fit different generic patterns.

In Fig. 6, the 19 diagrams are drawn qualitatively. The second mode branch is in the plane of the paper. The first mode branches, with thinner lines, are to be considered as "projections" into that plane of pairs, because of the $\mathbf{Z}(2)$-symmetry, of curves lying in a surface starting orthogonally to the two branch then bending towards it away from the trivial branch. Figs. 7 and 8 give numerical plots of of the bifurcation curves as we cross some regions. In studying Fig. 7 note that Fig. 2b corresponds to a Region 3 bifurcation picture.

As a last comment we should discuss some dynamical aspects of the problem. The coefficient $c$ does not intervene in the bifurcation of stationary states but is essential in the dynamics. We have chosen $c$ in such a way as to avoid Hopf bifurcations from the trivial branch. Moreover, for the Sel'kov model, no Takens-Bogdanov phenomena exist in that range. So the interesting problems will occur near the emergence of those Hopf points for $c = 1/(p-1)$. An other possible source of secondary Hopf points is near the mode

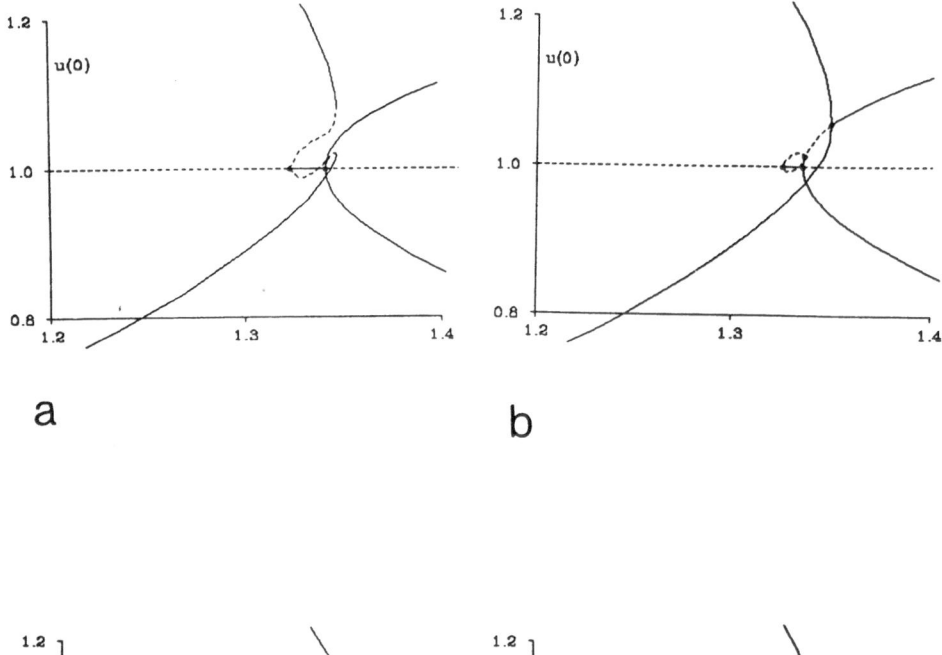

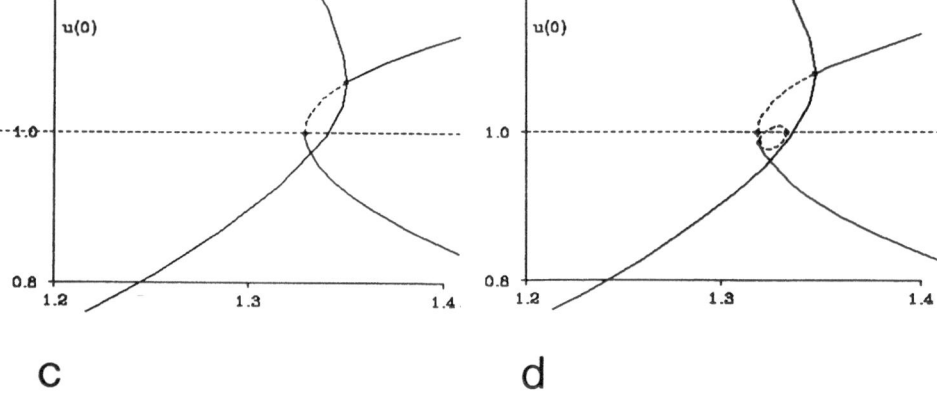

Figure 7: (a) region 4, (b) 5/6 boundary, (c) 6/19 boundary, (d) region 19

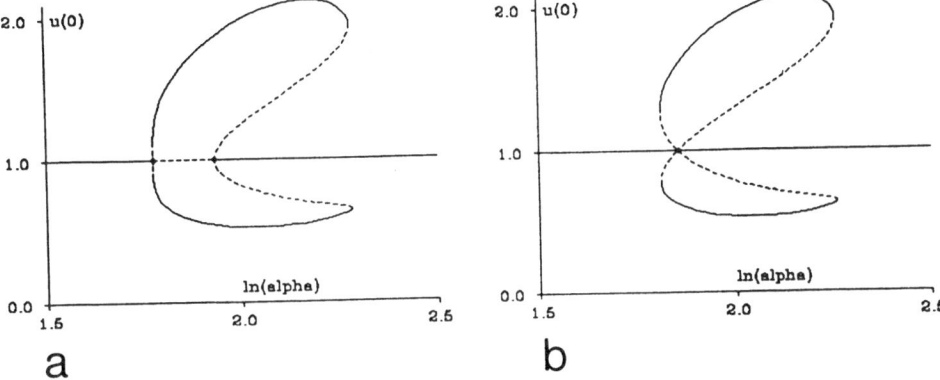

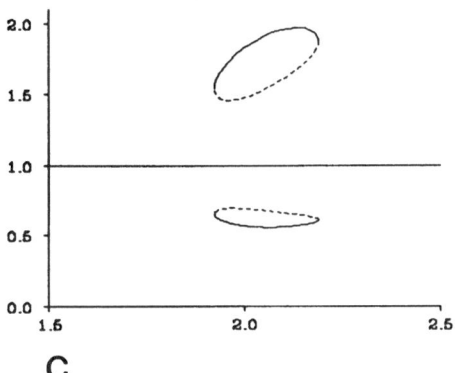

Figure 8: (a) region 8, (b) 8/9 boundary, (c) region 9

interaction. again it is not the case here, but for some other scheme, like the Brusselator, these possibilities exist.

### Acknowledgements

We are grateful to the SERC for research funding under the Nonlinear System Initiative.

# References

[1] D Armburster and G Dangelmayr. Corank-two bifurcations for the brusselator with non-flux boundary conditions. *Dynam. Stab. Syst.*, 1:187–200, 1986.

[2] P Aston. Scaling laws and bifurcation. (preprint, University of Bath), 1989.

[3] C Canuto, M Y Hussaini, A Quarteroni, and T A Zang. *Spectral Methods in Fluid Mechanics*. Springer-Verlag, Berlin, 1987.

[4] J Crawford, M Golubitsky, G Gomes, E Knobloch, and I Stewart. Boundary conditions as symmetry constraints. (preprint, University of Warwick), 1989.

[5] B De Dier, V Hlavacèk, and P Van Rompay. Analysis of dissipative structures in a two-dimensional autocatalytic system: the Brusselator model. (preprint, Katholieke Universitat Leuven), 1987.

[6] B De Dier and D Roose. Determination of bifurcation points and catastrophies for the brusselator model with two parameters. In T Küpper, R Seydel, and H Troger, editors, *Bifurcation: Analysis, Algorithms, Applications*, pages 38–46, Basel, 1987. Birkhaüser.

[7] B De Dier, F Walraven, R Janssen, P Van Rompay, and V Hlavacèk. Bifurcation and stability analysis of a one-dimensional diffusion-autocatalytic reaction system. *Z. Naturforsch*, 42a:994–1004, 1987.

[8] K Duncan and J C Eilbeck. Numerical studies of symmetry-breaking bifurcations in reaction-diffusion systems. In L M Ricciardi, editor, *Biomathematics and Related Computational Problems*, pages 439–448, Dordrecht, 1988. Kluwer.

[9] J C Eilbeck. A collocation approach to the numerical calculation of simple gradients in reaction-diffusion systems. *J. Math. Biol.*, 16:233–249, 1983.

[10] J C Eilbeck. The pseudo-spectral method and path following in reaction-diffusion bifurcation studies. *SIAM J. Sci. Statist. Comput.*, 7:599–610, 1986.

[11] J C Eilbeck. Numerical studies of bifurcation in reaction-diffusion models using pseudo-spectral and path-following methods. In T Küpper, R Seydel, and H Troger, editors, *Bifurcation: Analysis, Algorithms, Applications*, pages 47–60, Basel, 1987. Birkhaüser.

[12] J C Eilbeck and V S Manoranjan. A comparison of basis functions for the pseudo-spectral method for a model reaction-diffusion problem. *J. Comp. Appl. Math.*, 15:371–378, 1986.

[13] H Fujii, M Mimura, and Y Nishiura. A picture of the global bifurcation diagram in ecological interacting and diffusing system. *Physica D.*, 5:1–42, 1982.

[14] H Fujii, Y Nishiura, and Y Hosono. On the structure of multiple existence of stable stationary solutions in systems of RD equations. *Studies in Maths and its Appl.*, 18:157–219, 1986.

[15] M Golubitsky and D Schaeffer. *Singularities and groups in bifurcation theory*, volume 51 of *Appl. Math. Science.* Springer-Verlag, Berlin, 1985.

[16] M Golubitsky, D Schaeffer, and I Stewart. *Singularities and groups in bifurcation theory II*, volume 69 of *Appl. Math. Science.* Springer-Verlag, Berlin, 1988.

[17] D Gottlieb and S A Orszag. *Numerical Analysis of Spectral Methods: Theory and Applications.* SIAM-CMBS, Philadelphia, 1977.

[18] A Hunding. Dissipative structures in reaction-diffusion systems: numerical determination of bifurcations in the sphere. *J. Chem. Phys.*, 72:5241–5248, 1980.

[19] A Hunding and P Sorensen. Size adaptation in turing prepatterns. *J. Math. Biol.*, 26:27–39, 1988.

[20] M Mimura, M Tabata, and Y Hosono. Multiple solutions of two-point boundary value problems of Neumann type with a small parameter. *SIAM J. Math. Anal.*, 11:613, 1980.

[21] S Scott and P Gray. Chemical reactions in isothermal systems:oscillations and instabilities. In S.Sarkar, editor, *Non Linear Phenomena and Chaos*, pages 70–96. Adam Hilger, Bristol, 1986.

[22] K C Scovel, I G Kevrekidis, and B Nicolaenko. Scaling laws and the prediction of bifurcations in systems modelling pattern formation. *Phys. Lett A.*, 130:73–80, 1988.

# BIFURCATIONS, CHAOS AND SELF-ORGANIZATION IN REACTION-DIFFUSION SYSTEMS

G. NICOLIS and P. GASPARD
Faculté des Sciences
Université Libre de Bruxelles
Campus Plaine, C.P.231
Boulevard du Triomphe
B-1050 Bruxelles, Belgium

ABSTRACT. We review recent work about temporal and spatio-temporal self-organized structures in macroscopic physico-chemical systems and their analysis in the light of recent advances in nonlinear mathematics. Chaos and mixed-mode oscillations in far-from-equilibrium chemical reactions are first described. We show how homoclinic tangencies of Sil'nikov type and associated to a cycle can help us in elucidating the complex bifurcation sequences and the related chaotic time evolution observed in these systems. A chemical kinetic example of diffusion-induced spatio-temporal chaos is presented. Finally, we discuss the conditions for pattern formation via spatial symmetry breaking in uniform and non-uniform environments.

## 1. Introduction

Physical sciences provide us with abundant evidence of how the use of the concepts and techniques of nonlinear dynamics and bifurcation theory has suddenly led to the solution of long standing problems. Conversely, over the last twenty years they have frequently been a source of inspiration from which new insights contributing to advances in mathematics have emerged. In this chapter we would like to present a number of selected topics from the broad area of reaction-diffusion systems illustrating this close interaction.

A reaction-diffusion system is a system described by the following set of equations :

$$\frac{\partial X_i}{\partial t} = v_i \left( X_i, ..., X_n ; \lambda, \mu, ... \right) + D_i \, \nabla^2 X_i \, ,$$

$$(1.1)$$

where $X_i$ is a set of variables describing the composition of a mixture; $v_i$ is the rate of a change of $X_i$ due to the chemical reactions; $\lambda$, $\mu$,...are parameters (rate constants, etc.) built in the system; and $D_i$ is the diffusion coefficient of species $i$ in the medium (for simplicity cross-diffusion effects are neglected, a restriction that is valid as long as the reacting mixture behaves as an ideal system) [1].

There are two important features of Eqs.(1.1) that make chemical kinetics special among other fields of physical sciences.

*D. Roose et al. (eds.), Continuation and Bifurcations: Numerical Techniques and Applications*, 43–70.
© 1990 *Kluwer Academic Publishers.*

(i) $v_i$ are, typically, nonlinear functions of the composition variables owing to the presence of cooperative effects like autocatalysis or inhibition, which are ubiquitous in chemistry. These effects are acting spontaneously everywhere in the system. They therefore subsist in the limit of a uniform medium, in which diffusion can be discarded. This latter limit can be achieved in the laboratory through an effective stirring mechanism. Eqs.(1.1) reduce then to a system of nonlinear ODE's

$$\frac{dX_i}{dt} = v_i \left( X_i , ..., X_n ; \lambda , \mu , ... \right) ,$$

$$(1.2)$$

which contain practically all the tremendous variety of complex behaviours suggested by the mathematical analysis of nonlinear dynamical systems with a small number of degrees of freedom. In other words complexity in chemistry is not necessarily induced by the spatial degrees of freedom like in hydrodynamics : it may arise solely from the intrinsic nonlinearities of the kinetics. As a result chemical kinetics provides one of the few authentic physical illustrations of low-dimensional dynamical systems.

(ii) Suppose next that the system is not stirred : diffusion is automatically switched on, and states corresponding to an inhomogenous distribution of $X_i$ in space are created. Before attempting any quantitative evaluation of the properties of such states, let us follow a simple dimensional argument. Among the parameters present in Eqs. (1.1), let us focus our attention on a typical diffusion coefficient, $D$, and a typical rate constant, $k$. From these two quantities we can construct a combination $l_c = (D/k)^{1/2}$ having the dimensions of length and being of completely intrinsic origin. This suggests that reaction-diffusion systems should be capable of undergoing spontaneous symmetry-breaking transitions leading to states endowed with intrinsic characteristic length (see Fig.1). This is to be contrasted with hydrodynamics, where length scales involve, among other factors, the system size.

Having identified the main features that distinguish reaction-diffusion systems from other systems giving rise to bifurcations and nonlinear dynamics, we shall now proceed to a brief account of some selected topics, by placing special emphasis on open questions and recent developments. We shall deal, successively, with mixed-mode oscillations and homoclinic chaos in well-stirred systems; symmetry-breaking and diffusion-induced chaos in a uniform environment; chemical instabilities and bifurcations in a non-uniform environment.

## 2. Mixed-Mode Oscillations and Homoclinic Chaos in Well-stirred Systems

When a chemical reaction takes place in an open, well-stirred reaction cell, the time evolution of the concentration is described by a system of

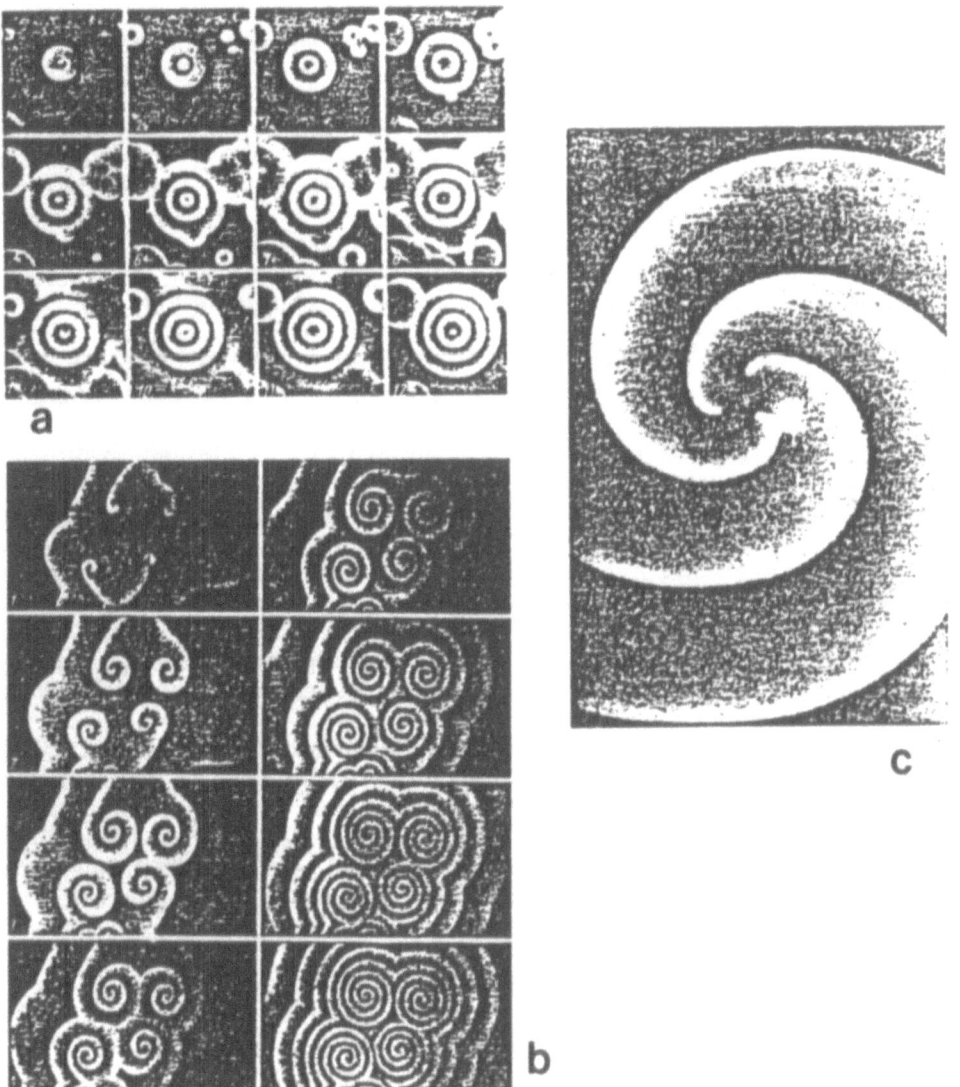

Fig. 1. Spatio-temporal structures in the Belousov-Zhabotinskii reaction in a thin liquid layer: (a) circular waves; (b) emerging pairs of uni-arm spiral waves; (c) a tri-arm spiral wave.

nonlinear first order ordinary differential equations (Eqs.(1.2)). In this case the evolution can be embedded in a finite-dimensional phase space of dimension equal to the number of chemical species involved in the reaction. With the advances of computer technology and, in particular, to graphic terminal technology phase space methods have led to important progress in the analysis of far-from-equilibrium chemical reactions in time-dependent dynamical regimes which have been classified into periodic, quasiperiodic, and chaotic behaviours [2]. Moreover several bifurcation mechanisms for transition from periodic to chaotic behaviours have been observed and studied in detail such as the Feigenbaum period doubling cascade, the intermittency transition, the quasiperiodic route to chaos and the homoclinic tangencies [2]. The latter, discovered in the mid-sixties and early seventies by Sil'nikov and coworkers, have in fact been among the first known examples of bifurcations leading to chaos [3,4].

Homoclinic tangencies are very common in three dimensional ordinary differential equations, in particular in chemical kinetics. A homoclinic tangency will occur in the phase space if two conditions are satisfied:

1) These exists an invariant set, i.e. a fixed point, a periodic orbit or a Cantor set of orbits, which is of saddle type [5]. Accordingly trajectories escape from the invariant set along the unstable manifold while other trajectories are convergent to the invariant set along the stable manifold.

2) The unstable manifold is tangent to the stable manifold at a critical parameter value $\mu = \mu_H$.

This latter condition implies that bifurcations occur in the vicinity of a homoclinic tangency. Not all homoclinic tangencies lead to chaotic behaviours. For instance when a tangency occurs in two dimensions it generates a single periodic orbit [6]. However, general conditions for emergence of homoclinic chaos have been proved [3,4]. We shall now turn to a description of the several possible cases of homoclinic tangencies.


## 2.1 HOMOCLINIC ORBIT ASSOCIATED TO A FIXED POINT

2.1.1. *General results.* The geometry of this homoclinic bifurcation is depicted in Fig.2. We assume that there exists a fixed point which is a saddle-focus with linear stability eigenvalues $(\rho \pm i\omega, \lambda)$. The stable manifold is two-dimensional while the unstable one is one-dimensional. At the critical parameter value $\mu = \mu_H$, the unstable manifold is included into the stable one and forms the homoclinic orbit. However, this situation breaks down away from criticality.

We then have the following general results :

(i) Sil'nikov theorem [3] : provided $|\rho/\lambda| < 1$, there exist uncountably many non periodic orbits.

(ii) As shown in Fig.3, the orbits perform successively (..., $p^{i-1}$, $p^i$, $p^{i+1}$, ...) half-turns around the saddle-focus, interrupted by bursts.

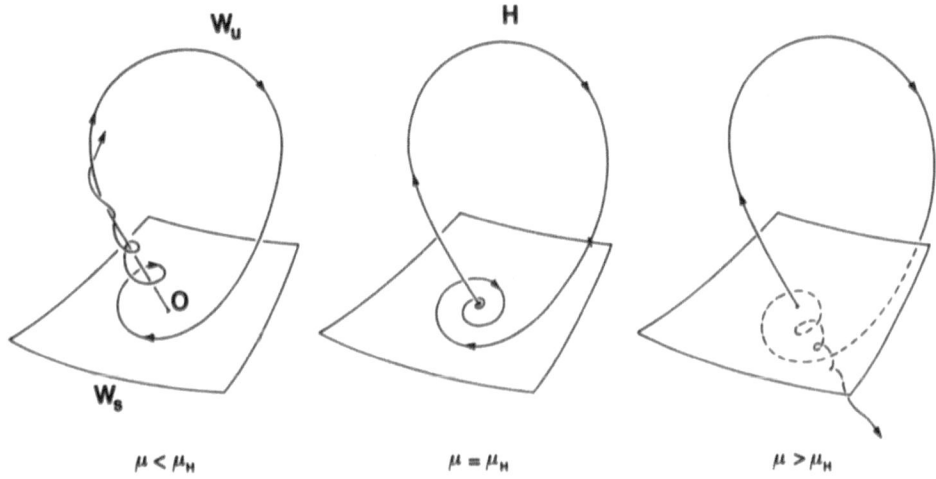

Fig.2. Phase portrait of a homoclinic tangency to a saddle-focus (P) in a three-dimensional phase space, before $(\mu < \mu_H)$, at $(\mu = \mu_H)$, and after $(\mu > \mu_H)$ the bifurcation. $W_s$ and $W_u$ are the stable and unstable manifolds of P. The homoclinic orbit (H) exists at the tangency $(\mu = \mu_H)$.

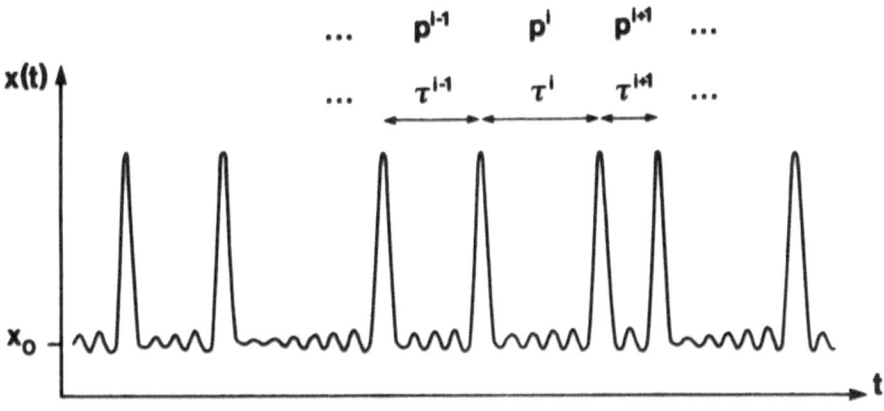

Fig.3. Typical time evolution of a kinetic variable $x(t)$ in Sil'nikov chaos. The small growing oscillations ($p^i$) are interrupted by bursts, forming a random sequence of time intervals ($\tau_i$). The bursts correspond to a circuit in the vicinity of the underlying homoclinic orbit.

Accordingly, they were named mixed-mode oscillations, because they are composed of a mixture of small peaks followed by a large peak. The time spent between two bursts is

$$\tau^i = \frac{\pi \, p^i}{\omega} + \tau_H.$$

where $\tau_H$ is the time for a burst. Most of these orbits are chaotic, but there also exist the single-circuit periodic orbits of the type

$$(..., p, p, p, ...)$$

and of period $T = \pi p / \omega + \tau_H$. These orbits are of relaxation-type and very different from the quasi-sinusoidal oscillations born in the Hopf bifurcation.

(iii) The above single-circuit periodic orbits appear in pairs via tangent bifurcations. The bifurcation points $\{\mu_n^{\pm}\}$ accumulate geometrically to $\mu = \mu_H$ according to

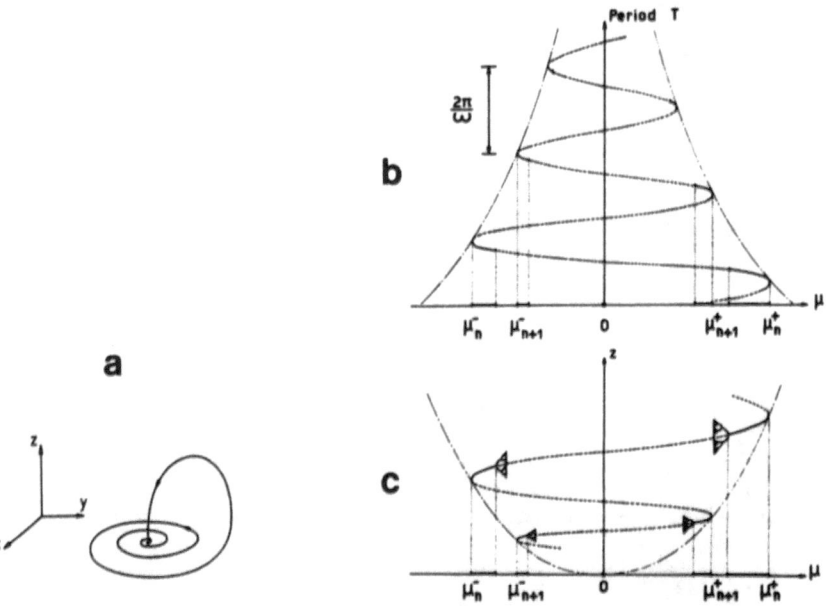

Fig.4. (a) The homoclinic orbit at $\mu = \mu_H$ in the phase space. (b) Bifurcation diagram of the period $T$ of the single-circuit periodic orbit versus the parameter $\mu$. (c) Bifurcation diagram of the $z$-coordinate of the single-circuit periodic orbit in a section plane versus the parameter $\mu$, showing the tangent bifurcations $\{\mu_n^{\pm}\}$ where a pair of saddle-node periodic orbits are born and which are followed by Feigenbaum cascades of period-doublings.

$$\lim_{n \to \infty} \frac{\mu_n^{\pm} - \mu_{n-1}^{\pm}}{\mu_{n+1}^{\pm} - \mu_n^{\pm}} = e^{2\pi |\rho/\omega|} ,$$

$$(2.1)$$

and their period increases to infinity as $\mu \to \mu_H$ (see Fig.4) [7].

*Remarks.* 1°- Sil'nikov's homoclinicity to a saddle-focus is generic in two-parameter space near a doubly degenerate bifurcation of a fixed point with eigenvalues $(\pm i\omega, 0)$ in the case where the codimension two vector field is described by the normal form

$$\dot{q} = i\omega q + (\alpha + i\beta) zq + O(3),$$

$$\dot{z} = -z^2 - |q|^2 + O(3),$$

where $q = x + iy$ and $0 < \alpha < 2$. Topological chaos is thus allowed arbitrarily close to the critical vector field, together with the aforementioned bifurcations [8,9].

    2°- The bifurcation accumulation rate (2.1) at the homoclinic tangency is to be compared with the accumulation rate of period doublings in the Feigenbaum cascade where

$$\lim_{n \to \infty} \frac{\mu_n - \mu_{n-1}}{\mu_{n+1} - \mu_n} = 4.6692016... ,$$

$$(2.2)$$

which is a universal number [2]. In contrast, the quantitative features of the homoclinic tangency depend on the eigenvalues of the saddle-focus.

2.1.2. *Method of construction.* A homoclinic orbit can be numerically constructed in a given nonlinear system by numerical continuation of the local stable and unstable manifolds and by tuning the control parameter in order to induce a crossing of the so constructed global manifolds. Such a numerical method has been developed in Refs. 9 and 10 and is very efficient to localize homoclinic tangencies. A different method to achieve this purpose is the numerical continuation of the single-circuit periodic orbits of Fig.4 in the vicinity of the homoclinic orbit. As its period increases the periodic orbit converges towards the homoclinic orbit.

    It appears that the AUTO method of Doedel [11] presents some shortcomings. This method requires to keep in memory the complete periodic orbit. As the homoclinic orbit is approached, the period goes to infinity and the memory limit is exceeded. To overcome this problem, Sparrow [12] proposed to keep of the periodic orbit only the coordinates of its intersection with a Poincaré surface of section

$$\mathbf{a} \cdot \mathbf{X}^* = b .$$

$$(2.3)$$

The intersection $\mathbf{X}^*$ obeys to

$$\mathbf{X}^* = \Phi\ (\ \tau^*, \mathbf{X}^*, \mu\ ), \tag{2.4}$$

where $\Phi$ is the Poincaré map which must be constructed numerically and $\tau^*$ is the period. The fixed point solution of (2.4) is numerically constructed by a Newton-Raphson method. The fixed point is numerically continued as the control parameter $\mu$ is varied. This method has been implemented in a FORTRAN code named PERIOD in Ref.9 and applied to different nonlinear dynamical system models.

2.1.3. *Examples of homoclinic tangency.* Sil'nikov homoclinic tangency (and its heteroclinic variant) has been observed in the Rössler model [10], in the Lorenz model and the associated Haken model of laser [12], in spin-wave turbulence model [13], in double-diffusive convection models [14], in three variable Lotka-Volterra models [15], in models of nonlinear wave modulation [15], in nerve impulse propagation model [16] in models of heterogeneous catalysis [17], as well as in isothermal chemical kinetic models [10].
   In particular, the following kinetic model [10], satisfying the mass-action law, has been shown to give rise to homoclinic chaos :

$$\dot{X} = X\ (\ dX - fY - Z + g\ ),$$

$$\dot{Y} = Y\ (\ X + sZ - l\ ), \tag{2.5}$$

$$\dot{Z} = \frac{1}{\varepsilon}\ (\ X - aZ^3 + bZ^2 - cZ\ ).$$

This system presents a homoclinic tangency to a saddle-focus at the parameter values $a=0.5$, $b=3$, $c=5$, $\varepsilon=0.01$, $f=0.5$, $g=0.6$, $s=0.3$. Fig.5 shows the homoclinic orbit at $d=0.51$ and $l=1.339$, embedded in a chaotic attractor. The single-circuit periodic orbit has been numerically continued with the program PERIOD for $d=0.43$ and the result is shown in Fig.6 in a plot of its period versus the parameter $l$ [9]. The homoclinic tangency occurs at $l=1.339$. At homoclinicity, the stability eigenvalues of the saddle-focus are ($\rho=0.081$, $\omega=1.064$, $\lambda=-332.4$). Note that the stiffness of the differential equations (2.5) due to the smallness of parameter $\varepsilon=0.01$ causes a distortion of the damped oscillations with respect to the theoretical damped sinusoidal curve of Fig.4. This kinetic model provided the first evidence for the possibility of Sil'nikov homoclinic chaos and the associated mixed-mode oscillations in far-from-equilibrium chemical reactions.
   Complex bifurcation sequence of mixed-mode oscillation and a Sil'nikov homoclinic orbit was also observed in a model of the Belousov-Zhabotinskii chemical reaction by Showalter, Noyes, and Bar-Eli who

51

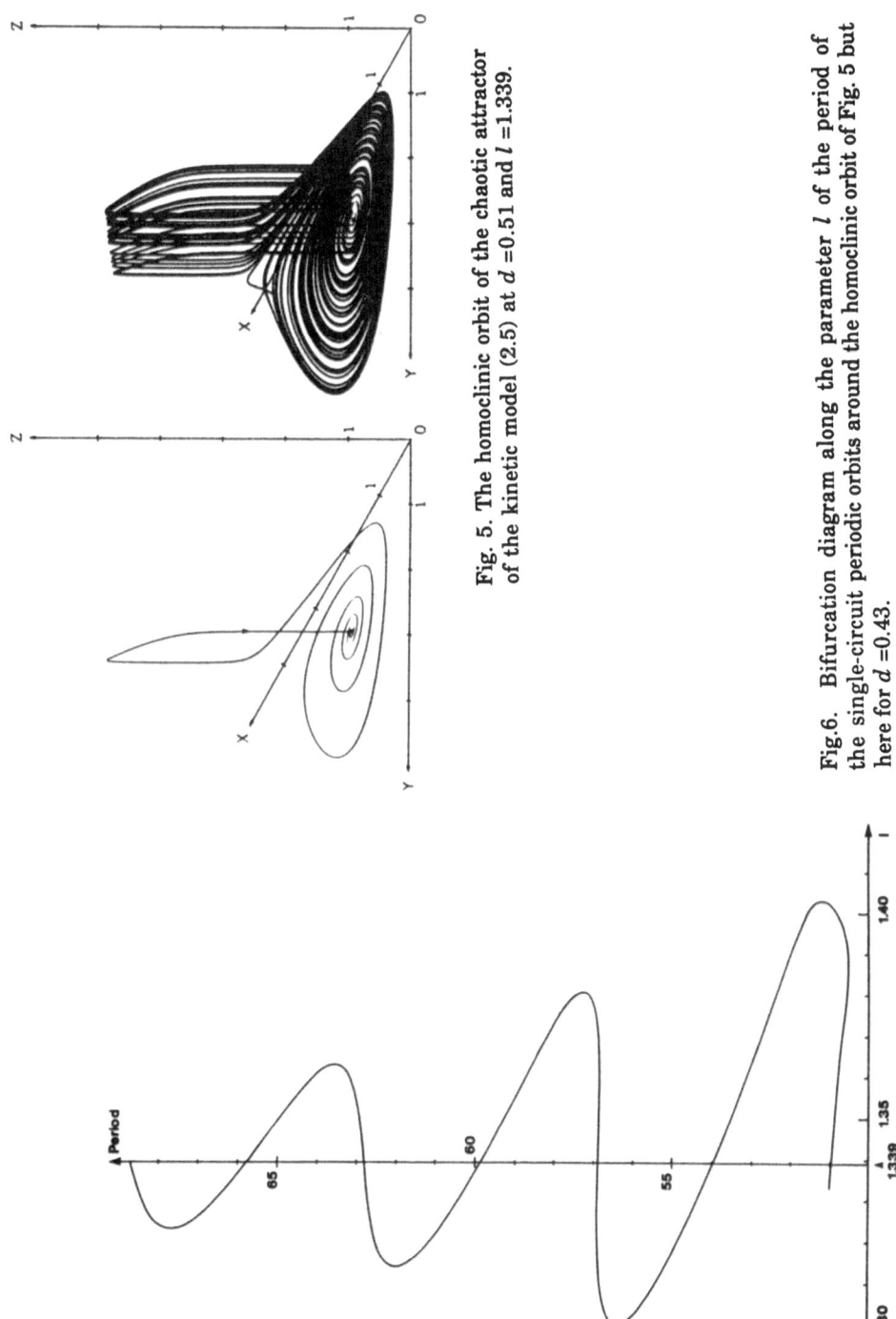

Fig. 5. The homoclinic orbit of the chaotic attractor of the kinetic model (2.5) at $d = 0.51$ and $l = 1.339$.

Fig. 6. Bifurcation diagram along the parameter $l$ of the period of the single-circuit periodic orbits around the homoclinic orbit of Fig. 5 but here for $d = 0.43$.

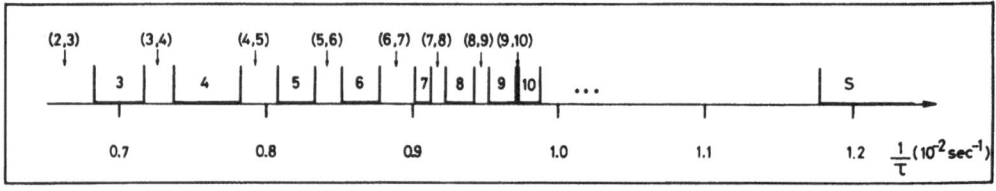

Fig.7. Complex bifurcation sequence of mixed-mode oscillations for the model (2.6) of the Belousov-Zhabotinskii reaction by Showalter, Noyes, and Bar-Eli (from Ref. 19).

proposed the following kinetic scheme

$$A + Y \;\leftrightarrow\; X + P \,, \qquad\qquad X + Y \;\leftrightarrow\; 2\,P \,,$$

$$A + X \;\leftrightarrow\; 2\,W \,, \qquad\qquad C + W \;\leftrightarrow\; X + Z \,,$$

$$2\,X \;\leftrightarrow\; A + P \,, \qquad\qquad Z \;\rightarrow\; g\,Y + C \,, \qquad (2.6)$$

with $A = BrO_3^-$, $C = M_{red}$, $P = HOBr$, $W = BrO_2^\bullet$, $X = HBrO_2$, $Y = Br^-$, $Z = M_{ox}$. The bifurcation sequence of mixed-mode oscillation for varying inverse resident time ($\tau$) is depicted in Fig.7 [19]. The periodic windows are labelled by single integers $n$ when the oscillation is composed of $n$ small peaks followed by a large one and by pairs of integers $(m,n)$ when the oscillation contains $m$ small peaks followed by a large peak, followed then by $n$ small peaks, etc... The number of small peaks in the pattern increases with $1/\tau$ till a transition to a stationary behaviour ($S$) in a complex bifurcation sequence very similar to those occurring near the homoclinic tangency as shown in Fig.4.

In this seven-variable model, it was striking to observe that a homoclinic orbit associated to a saddle-focus persists throughout the whole bifurcation sequence either because of the large stiffness of the system or because of a decoupling into subsystems. This behaviour is reminiscent of Sil'nikov homoclinic chaos although the high dimensionality of the system allows for other possible related mechanisms [20].

Experimental evidence of homoclinic chaos was provided by Argoul,

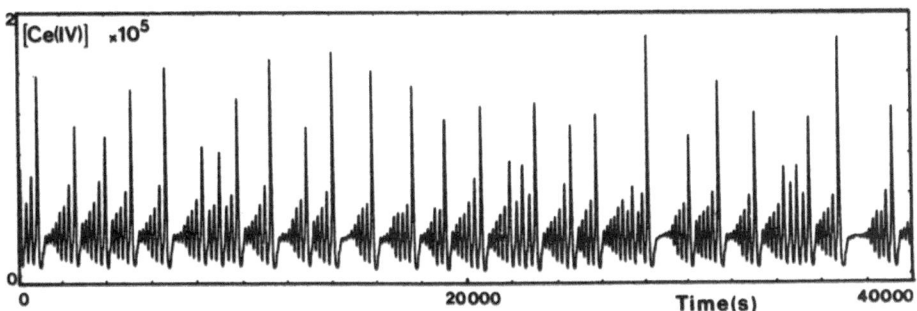

Fig.8. Experimental time evolution of the concentration of $Ce^{IV}$ in the Belousov-Zhabotinskii reaction, showing a typical homoclinic chaos: an oscillatory instability causes the growth of the small oscillations followed by the homoclinic reinjection (adapted from Ref. 21).

Arnéodo, and Richetti in the Belousov-Zhabotinskii reaction [21]. They observed a homoclinic reinjection mechanism to a fixed point of saddle-focus type (see Fig.8). Homoclinic chaos was also experimentally observed in heterogeneous catalysis as well as in electrochemical deposition [17].

We conclude this section with the remark that homoclinic chaos appears to be a mechanism which is shared by a large variety of dynamical systems. They have in common the fact that a stationary instability competes with an oscillatory one leading to chaos.

## 2.2. HOMOCLINIC TANGENCY RELATED TO A PERIODIC SOLUTION

When the homoclinic reinjection is associated to a periodic orbit of saddle type rather than to a saddle-focus, a similar homoclinic tangency occurs with related chaotic behaviours. However, new features appear due to the different topology in the phase space (see Fig.9). Two cases occur depending on the sign of the Lyapunov numbers of the cycle $\{\lambda_u, \lambda_s\}$ [4,5] :

(i) Case I : when $\lambda_u$ and $\lambda_s$ are positive the bifurcation diagram is similar to the Sil'nikov saddle-focus case.

(ii) Case II: when $\lambda_u$ and $\lambda_s$ are negative then the unstable and stable manifolds of the periodic orbit form Möbius strips and the periodic windows of even and odd period bifurcate on two *different* sides of the critical parameter value $\mu=\mu_H$.

Such a mechanism was numerically observed in a model of acetal-

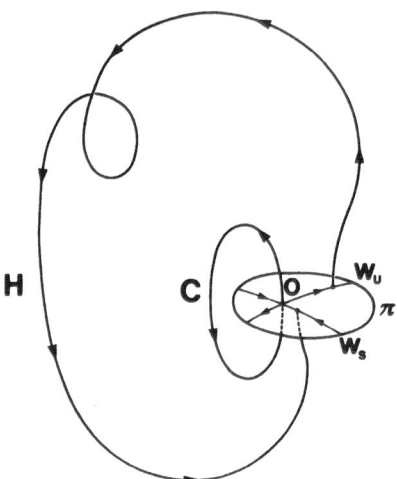

Fig. 9. Phase portrait of a homoclinic orbit (H) associated with a cycle (C) of saddle type. The intersections with the section plane $\pi$ of the stable ($W_s$) and the unstable ($W_u$) manifolds of the cycle are represented.

dehyde combustion

initiation:     $Y \rightarrow X$,

branching:      $X + Y \rightarrow 2X + \text{heat}$,
                $A + X \rightarrow 2X + \text{heat}$,     (2.7)

termination:    $X \rightarrow S_1 + \text{heat}$,
                $X \rightarrow S_2$,

where Y denotes a fuel molecule and X a free radical molecule. The third variable is the temperature which appears in the rate constants $\{k_i(T)\}$ according to Arrhenius law [22].

Fig.10 shows the bifurcation diagram of the period of periodic oscillations versus the temperature as control parameter [5]. The linear behaviour in the logarithmic scale of Fig.10 is an evidence for an accumulation of the periodic windows at the homoclinic tangency according to the geometric law

$$\mu_n \approx \frac{1}{(\lambda_u)^n} ,$$

(2.8)

55

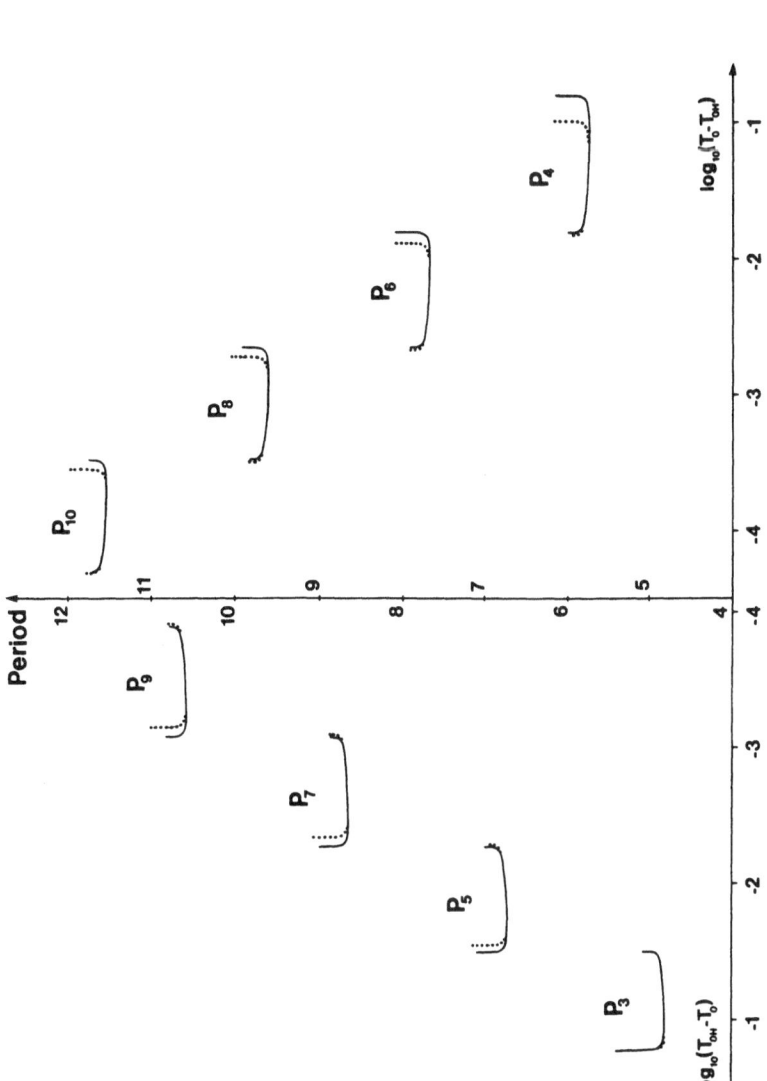

Fig. 10. Bifurcation diagram of the period of the periodic attractors ($P_n$) of the Wang and Mou thermocombustion model (2.7) versus the ambient temperature ($T_0$) in logarithmic scale, showing evidence for the scaling law (2.8) (see Ref. 5).

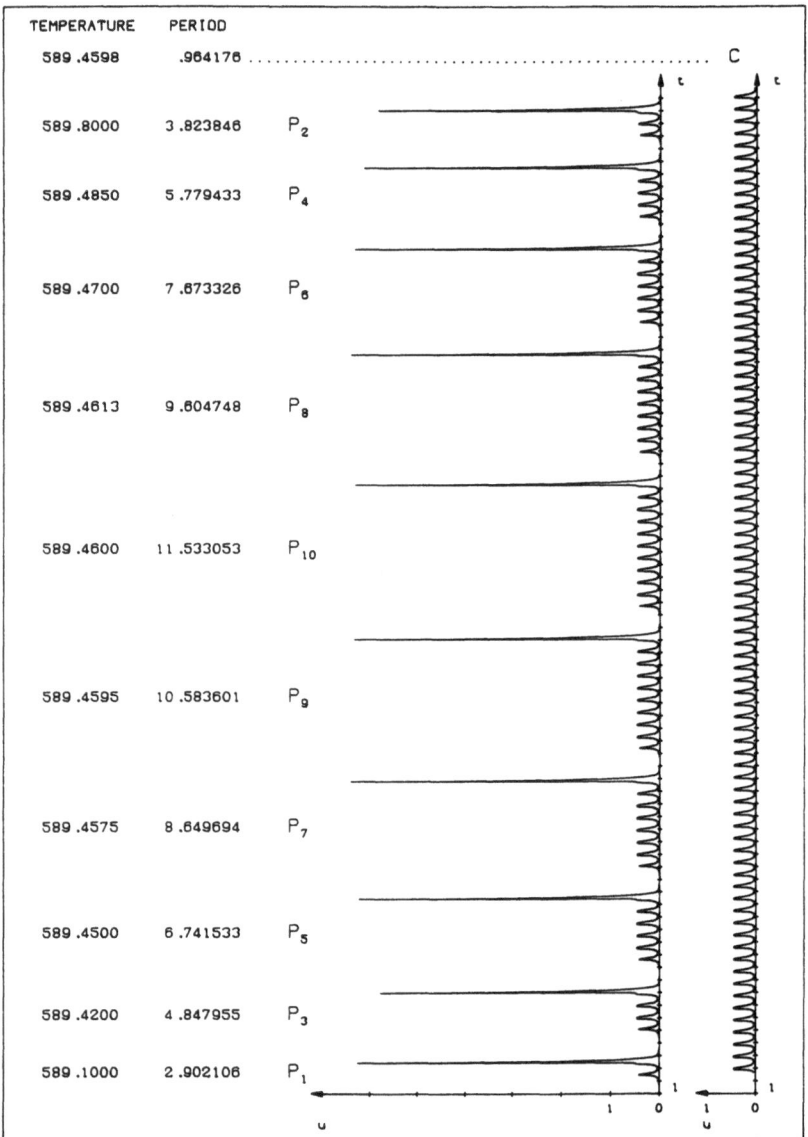

Fig. 11. Shape of the periodic orbits corresponding to Fig. 10 compared with the shape of the unstable cycle (C). The kinetic variable $u$ is the time dependent relative temperature, $(T-T_0)/T_0$, with respect to the ambient temperature $(T_0)$ reported in the left column. The even periods are located beyond the homoclinic tangency and the odd periods below.

with the unstable eigenvalue of the cycle $\lambda_u$ = -2.566. The periodic oscillations $P_n$ are presented in Fig.11 where their mixed-mode character is
apparent. As their period increases they become tangent to the saddle-type
cycle represented at right-hand and the mixed-mode oscillations transform then into a homoclinic orbit [5].

A similar complex bifurcation sequence was observed in experiments on acetaldehyde thermocombustion [23]. Homoclinic tangencies to
a cycle can be common in nonlinear models. In particular, numerical
evidence exists for their appearance in three-variable biochemical model
involving the coupling of two autocatalytic enzymatic reactions [24].

## 3. Symmetry Breaking and Diffusion-induced Chaos in a Uniform Environment

### 3.1. FAR-FROM-EQUILIBRIUM SELF-ORGANIZATION

3.1.1. *Generalities.* In absence of stirring, far-from-equilibrium patterns,
called dissipative structures can appear in a chemically reacting system
as a result of diffusion-induced instabilities [1]. Their existence requires
that : (a) an oscillatory dissipative structure does not emerge before the
stationary one; and (b) the diffusion coefficients of the active species be
sufficiently different. This latter condition is restrictive and is at the
origin of the fact that most experimentally observed dissipative structures,
in particular, in the Belousov-Zhabotinskii reaction, are spatio-temporal.
It is only very recently that experimental evidence in gel reactors has been
provided for genuine stationary dissipative structures with space
symmetry breaking [25]. Historically, this mechanism was first proposed
by Turing in 1952 as a model of morphogenesis.

For a reaction-diffusion system like

$$\frac{\partial \mathbf{X}}{\partial t} = \mathbf{v}(\mathbf{X}, \mu) + D \cdot \nabla^2 \mathbf{X},$$

$$(3.1)$$

the dissipative structure emerges from the uniform state solution,

$$\mathbf{v}(\mathbf{X}_0, \mu) = 0,$$

$$(3.2)$$

given the boundary conditions, $\mathbf{X} = \mathbf{X}_0$ or $\mathbf{n} \cdot \nabla \mathbf{X} = 0$. The linear stability
analysis provides the critical parameter values where small amplitude
perturbations around $\mathbf{X}_0$ become unstable,

$$\mathbf{X} = \mathbf{X}_0 + \mathbf{x}(\mathbf{r}, t),$$

$$(3.3)$$

58

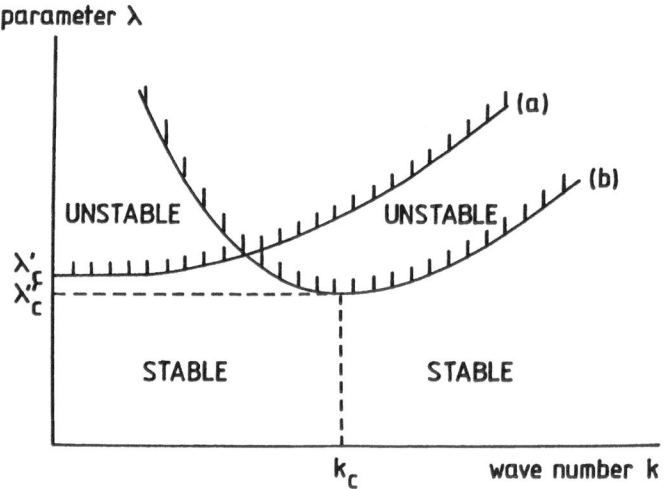

Fig.12. Lines of marginal stability in the plane of the parameter $\lambda$ versus the wavenumber $k_m$, for the temporal instability starting at ($\lambda_c'$, $k_m = 0$) and for the space symmetry breaking instability starting at ($\lambda_c''$, $k_m = k_{m_c}$).

with $\mathbf{x}$ solution of

$$\frac{\partial \mathbf{x}}{\partial t} = \left( \frac{\partial \mathbf{v}}{\partial \mathbf{X}} (\mathbf{X}_0) + D \cdot \nabla^2 \right) \mathbf{x} .$$

(3.4)

This equation admits solutions of the form

$$\mathbf{x} = \mathbf{C}_m \, e^{\omega_m t} \, \varphi_m (\mathbf{r}) ,$$

(3.5)

where

$$\nabla^2 \varphi_m = -k_m^2 \varphi_m .$$

(3.6)

The characteristic equation is then

$$\left( \frac{\partial \mathbf{v}}{\partial \mathbf{X}} (\mathbf{X}_0) - k_m^2 D - \omega_m \, \mathbf{I} \right) \cdot \mathbf{C}_m = 0 .$$

(3.7)

Requiring that Re $\omega_m = 0$, we find the parameter values of marginal stability. Two typical cases are depicted in Fig.12.

3.1.2. *An Example.* A much studied model, known as the Brusselator, illustrates this phenomenon :

| | | | |
|---|---|---|---|
| A | $\rightarrow$ | X, | (R1) |
| B + X | $\rightarrow$ | Y + C, | (R2) |
| 2 X + Y | $\rightarrow$ | 3 X, | (R3) |
| X | $\rightarrow$ | D. | (R4) |

$$(3.8)$$

In this reaction, species X produces Y in R2 which in turn activates the production of X in R3. Species B acts as an excitator. Reactions R2 and R3 will be accelerated if concentration of B increases. In rescaled variables, the equation reaction-diffusion are

$$\frac{\partial X}{\partial t} = A - (B+1) X + X^2 Y + D_1 \nabla^2 X \ ,$$

$$\frac{\partial Y}{\partial t} = B X - X^2 Y + D_2 \nabla^2 Y \ .$$

$$(3.9)$$

The stationary state is

$$X_0 = A \ , \qquad\qquad Y_0 = B/A \ . \qquad\qquad (3.10)$$

Assuming a time-independent structure, fluctuations around this uniform state are in general of the form

$$X(r) = X_0 + x(r) \ , \qquad\qquad Y(r) = Y_0 + y(r) \ , \qquad\qquad (3.11)$$

with

$$x(r), \quad y(r) \approx e^{-r/l_{corr}} \quad \cos 2\pi r /\lambda \ , \qquad\qquad (3.12)$$

in a one-dimensional system. $l_{corr}$ is the correlation length of the fluctuation, i.e. the distance over which a point-like perturbation will spread. $\lambda$ is the spatial period of the oscillations which modulate the decaying envelope of the perturbation. Solving the linearized Eqs.(3.9) allows one to determine these lengths as

$$l_{corr} \cong \left( \frac{4 D_1}{B_c - B} \right)^{1/2} \ , \qquad\qquad (3.13)$$

60

$$\lambda \cong 2\pi \frac{(D_1 D_2)^{1/4}}{A^{1/2}} ,$$
(3.14)

for the concentration $B$ near and below the critical concentration

$$B_c = [1 + A (D_1/D_2)^{1/2}]^2 .$$
(3.15)

When this critical value is reached, the correlation length (3.13) diverges and the uniform state becomes unstable with respect to fluctuations of the form (3.12), which grow and invade the system. Saturation is reached due to the nonlinearity of Eqs.(3.9). This fixes the amplitude of the fluctuations in the nonuniform dissipative structure which are then given by

$$\begin{pmatrix} x(r) \\ y(r) \end{pmatrix} \cong \left( \frac{B - B_c}{\Phi} \right)^{1/2} \begin{pmatrix} c_x \\ c_y \end{pmatrix} \cos 2\pi r /\lambda ,$$
(3.16)

just above the threshold ($B \geq B_c$) with constant $\Phi$, $c_x$, and $c_y$ [1]. According to the sign of $\Phi$, the bifurcation is subcritical or supercritical. This stationary instability may be in competition with an oscillatory instability occurring at the critical parameter value

$$B_c' = 1 + A^2 .$$
(3.17)

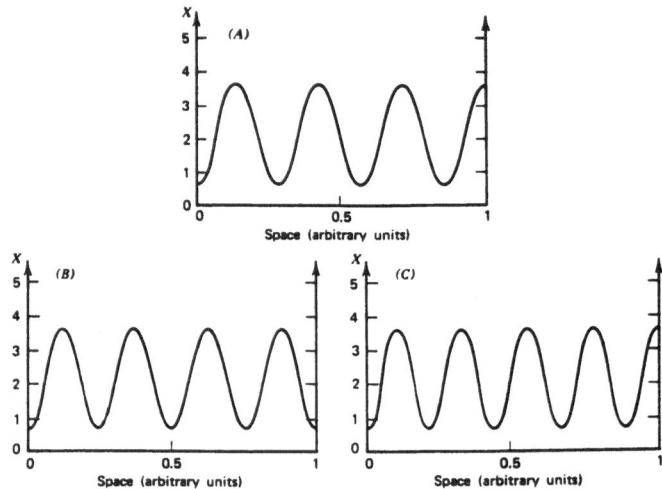

Fig. 13. Different steady state profiles coexisting for the same value of parameters but obtained from different initial conditions in the Brusselator model (3.9).

For $D_1 \geq D_2$ or for $D_2 > D_1$ and $2\sqrt{D_1 D_2}/(D_2 - D_1) > A$, the oscillatory instability occurs before the stationary one and a temporal structure emerges rather than the space symmetry-breaking Turing structure. Above these thresholds, several spatial modes are excited and the structure becomes spatio-temporal (see Fig.13). Note that several nonuniform steady states may coexist for the same boundary conditions for given parameter values as shown in Fig.13. When the length of the system is changed, these different solutions may interchange their stability resulting in bifurcation cascades. The length $L$ of the system (also called aspect ratio) is thus an important bifurcation parameter. In particular, experiments in the Rayleigh-Bénard convective cell with a varying aspect ratio have produced a large variety of patterns diferring both in wavelength and in shape.

3.1.3. *Normal forms.* It is remarkable that close to the critical thresholds, the nonlinear interaction of the various modes is described by universal normal forms. For a stationary instability, where the critical wave number $k_c \neq 0$, and Im $\omega_l = 0$, the normal form is (in Fourier space notation),

$$\frac{dz_k}{dt} = \left[ (\mu - \mu_c) - \left( k^2 - k_c^2 \right)^2 \right] z_k$$
$$- \sum_{k_1} u(k, k_1) z_{k_1} z_{k-k_1}$$
$$- \sum_{k_1 k_2} v(k, k_1, k_2) z_{k_1} z_{k_2} z_{k-k_1-k_2} \ . \tag{3.18}$$

For an oscillatory instability with $k_c = 0$ but Im $\omega_c \neq 0$, we have

$$\frac{dz_k}{dt} = \left[ (\mu - \mu_c) - ck^2 + i\Omega_k \right] z_k$$
$$- \sum_{k_1 k_2} u(k, k_1, k_2) z_{k_1} z_{k_2}^* z_{k-k_1-k_2} \ . \tag{3.19}$$

Because of their high dimensional character these nonlinear differential equation systems may admit chaotic solutions, resulting in the new phenomenon of diffusion-induced chaos.

3.2. DIFFUSION-INDUCED CHAOS

When the local oscillators of a spatially distributed system are weakly perturbed by diffusion, Kuramoto showed that the dynamics can be reduced to a partial differential equation ruling the variation $\psi(r,t) = \Phi(r,t) - t$ of the phase $\Phi(r,t)$ with respect to their uniform motion. All the parameters can be scaled out of the equation which takes the universal form

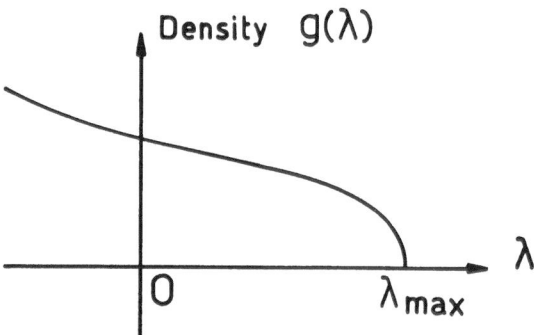

Fig. 14. Schematic density of the distribution of the Lyapunov exponents in diffusion-induced chaos like for (3.20).

$$\partial_t \psi = - \nabla^2 \psi + \left(\nabla\psi\right)^2 - \nabla^2 \nabla^2 \psi \,. \tag{3.20}$$

The same equation arises in combustion theory of turbulent flame front as shown by Sivashinsky [26].

Spatio-temporal chaos has been observed in the numerical integration of this system in one spatial dimension. These chaotic behaviours are characterized by a density function $g(\lambda)$ for the Lyapunov exponents giving the number of Lyapunov exponents taking a value in $[\lambda, \lambda+d\lambda]$ as $Lg(\lambda)d\lambda$ where $L$ is the length of the system (see Fig.14) [27]. If the Pesin formula is of application [28] as we may conjecture, the Kolmogorov-Sinai entropy per unit time and unit length will be given by

$$h_{KS} = \int_0^{\lambda_{max}} g(\lambda)\,d\lambda \,,$$

$$\tag{3.21}$$

as a fundamental quantity for the accumulation rate of data necessary to follow the time evolution of the system in one unit length without ambiguity. For the Kuramoto-Sivashinsky equation, the KS entropy has a value of the order of $10^{-2}$ bits per unit time and spatial length so that the chaos is mild in this system [27]. Still the very existence of a positive $h_{KS}$ highlights the need to resort to a statistical mechanical description of the system, involving such quantities as the correlation functions,

$$\langle \psi(r_1, t_1) \, \psi(r_2, t_2) \, \cdots \, \psi(r_n, t_n) \rangle \,,$$

among the local oscillators [26].

## 3.3. NUMERICAL RESULTS

The direct numerical integration of the Brusselator is necessary to demonstrate the diffusion-induced chaos in the original system rather than on the reduced Kuramoto-Sivashinsky equation. For this purpose, Orzag's pseudo-spectral integration method [29] has been applied to the Brusselator system (3.9) for parameter values $A=2$, $B=5.5$, $D_1=1$, $D_2=0$ and in a box of size $L=50$. Good convergence was obtained by retaining up to about 20 modes. Figs.15a, b, c, d depict successively the instantaneous profile of $X$ and its time dependence in the middle and in the two ends of the box. Time series analyses lead to the conclusion that the chaotic behaviour is generated by a low-dimensional attractor of dimension close to 4-5 [30].

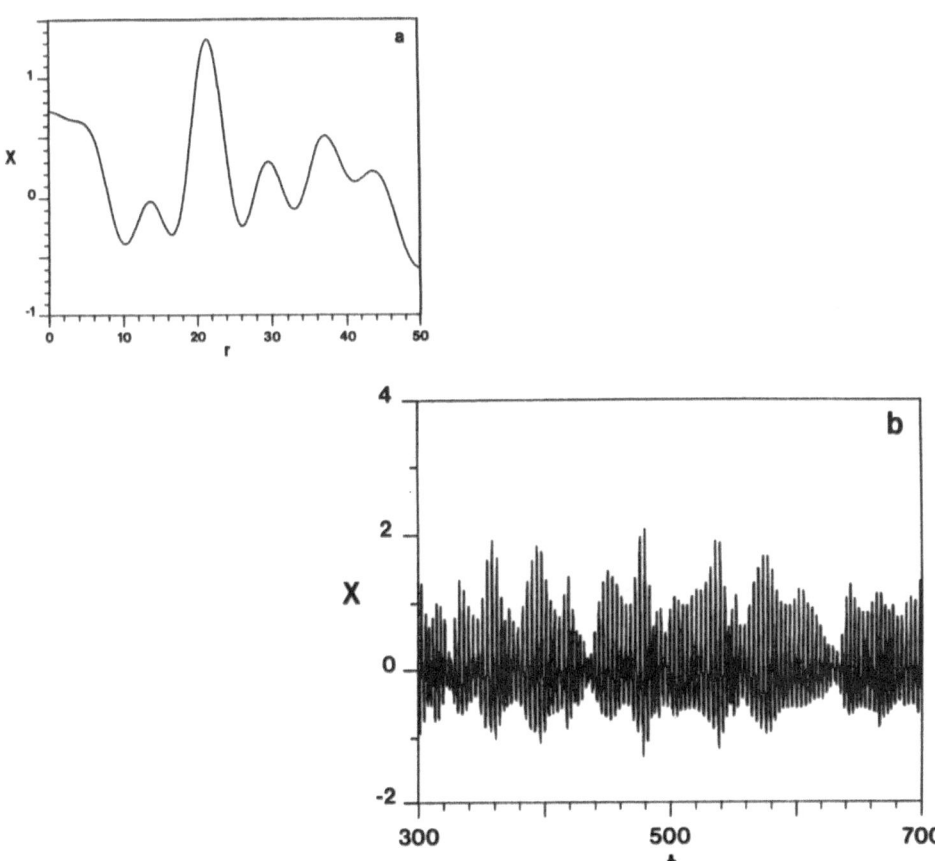

Fig. 15. Numerical results on diffusion-induced chaos in the Brusselator (3.9) for the parameter values described in text. (a) Spatial profile of concentration $X$ at fixed time. (b) Concentration $X$ at position $r=5$ versus time.

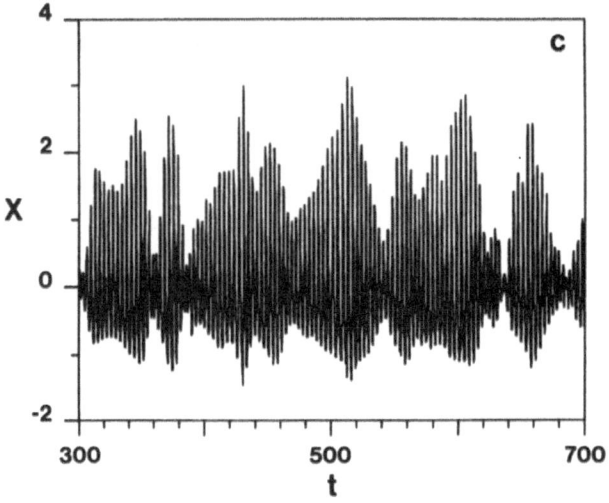

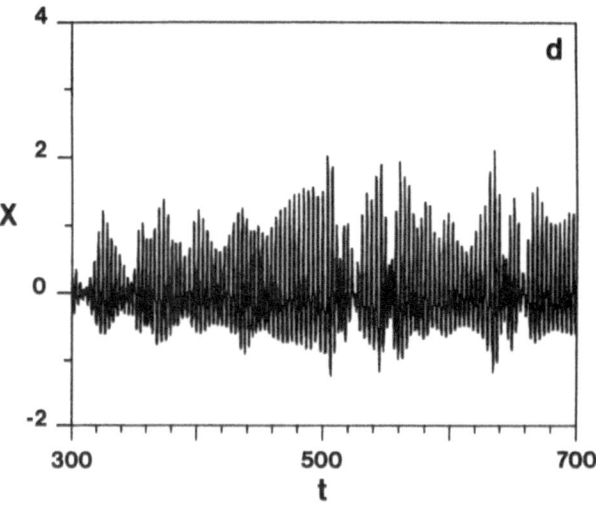

Fig. 15 (cont'd)  Concentration $X$ versus time at positions: (c) $r$=25; (d) $r$=45.

## 4. Bifurcations and Pattern Formation in a Non-uniform Environment

Self-organization and the emergence of dissipative structures often take place in a non-uniform medium. New features then arise such as localized non-uniform states or phase gradient waves. The first phenomenon is best illustrated in the Brusselator model (3.9) when the concentration of the reactive species A has some spatial variation for instance due to its diffusion from the boundaries [1]. The reaction-diffusion equations for this case are now

$$\frac{\partial A}{\partial t} = -A + D_0 \frac{\partial^2 A}{\partial r^2} ,$$

$$\frac{\partial X}{\partial t} = A - (B+1)X + X^2 Y + D_1 \frac{\partial^2 X}{\partial r^2} ,$$

$$\frac{\partial Y}{\partial t} = BX - X^2 Y + D_2 \frac{\partial^2 Y}{\partial r^2} ,$$

$$\tag{4.1}$$

with the boundary conditions that at both sides of the box [0,L], the chemical species A,X,Y have the fixed concentrations $A°$, $A°$, and $B/A°$ respectively.

Analytical results are obtained for this nonlinear problem using a two-scaling method based on the WKB approximation with a matching at the turning points [1]. Fig.16 shows typical non-uniform steady dissipative structures for the described situation. A structured pattern is forming on a subinterval of the box where the concentration $A$ reaches a low enough value so that the spatial instability may occur. Fig.17 shows a similar pattern formation but in a box where the conditions at the left-hand side are different from the conditions at right-hand. Still the pattern is time-independent. Oscillatory structures are also possible [1].

Experimental results were obtained for similar mechanisms occurring in the Couette flow reactor where a coupling between turbulent convection and the Belousov-Zhabotinskii chemical reaction gives birth to a fascinating variety of spatio-temporal fronts [31, 32] (see Fig. 18). The Couette flow is in a regime where a weak diffusion is generated between the convective rolls along the cylinder. A chemical gradient is maintained between the reactors at each side of the convective cell. A front is formed in the Couette cell which undergoes different kinds of instabilities. The front can be stationary or simply time periodic. More complex time oscillations are also possible, as shown in Fig. 18. These oscillations are similar to the mixed-mode oscillations described in Sect.2, which suggests the possibility to reduce the dynamics of the front to a 3-dimensional ordinary differential equation system with a homoclinic orbit to a saddle-focus or a limit cycle. The possibility of chaotic fronts is still an open question.

66

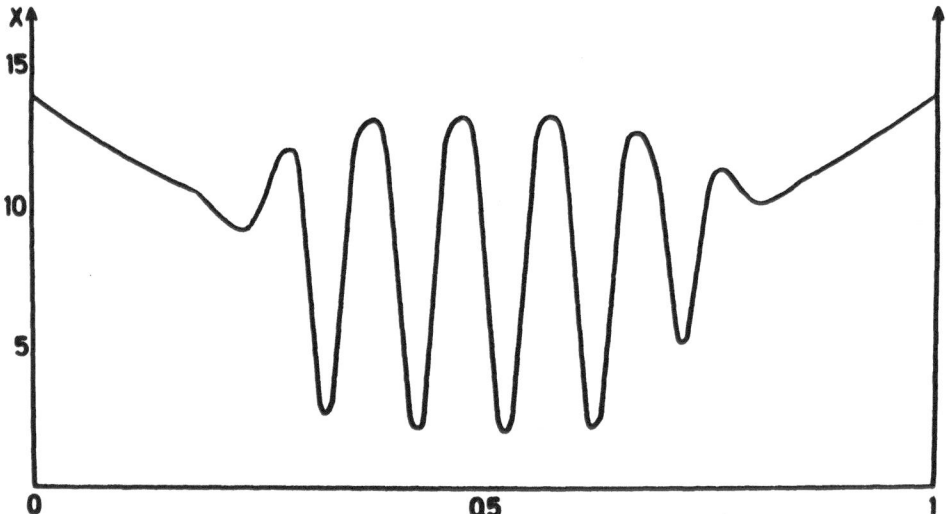

Fig. 16. Steady pattern in a non-uniform environment for model (4.1) at parameter values, $D$=0.1972, $D_1$=1.052 $10^{-3}$, $D_2$=5.26 $10^{-3}$, $Y°=B/A°$=2.14, $A°$=14, $B$=30. The pattern appears in the region where the concentration $A$ reaches the critical threshold for self-organization.

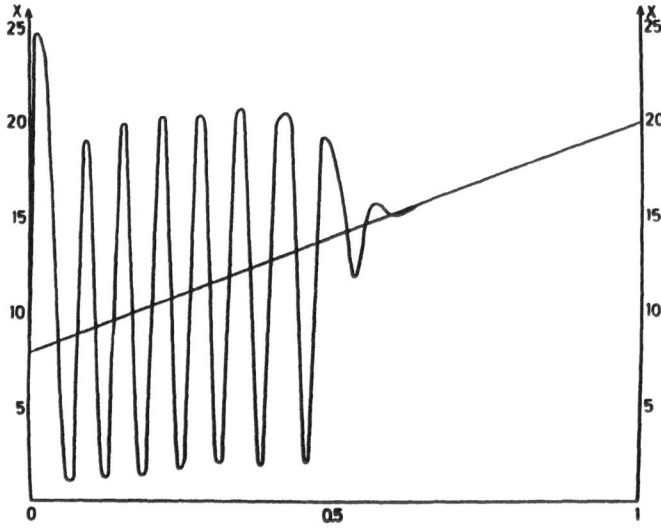

Fig. 17. Same as Fig. 16 but for asymmetric boundary conditions.

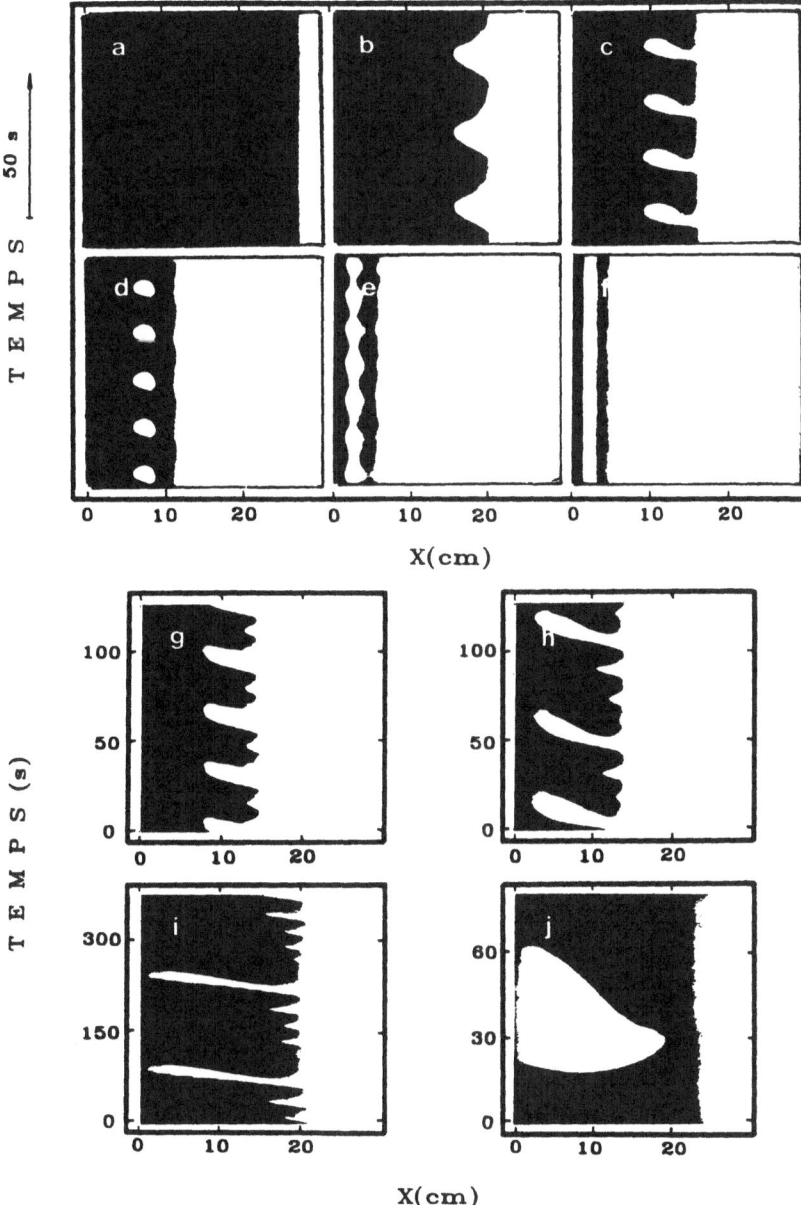

Fig. 18. Experimental observations by Qi Ouyang of complex oscillations of the reactive front for the $ClO_2^-$-$I^-$-$CH_2(COOH)_2$-$H^+$-thiodene chemical reaction in a Couette flow reactor. The figures show space-time diagrams of the front motion for various boundary concentrations of $ClO_2^-$ (adapted from Ref. 32).

68

ACKNOWLEDGEMENT

P. Gaspard is "Chercheur Qualifié" at the National Fund for Scientific Research (Belgium).

## References

[1]   Nicolis, G. and Prigogine I. (1977) "Self-Organization in Nonequilibrium Systems : From Dissipative Structures to Order through Fluctuations", Wiley, New York.

[2]   Bergé, P., Pomeau, Y. and Vidal, C. (1984) "L'ordre dans le chaos", Hermann, Paris.

[3]   Sil'nikov, L.P. (1965) Sov. Math. Dokl **6**, 163; (1970) Math. USSR Sbornik **10**, 91.

[4]   Gavrilov, N.K. and Sil'nikov, L.P. (1972) Math. USSR Sbornik **17**, 467; (1973) **19**, 139.

[5]   Gaspard, P. and Wang, X.-J. (1987) J. Stat. Phys. **48**, 151.

[6]   Gaspard, P. (1990) J. Phys. Chem. **94**, 1.

[7]   Glendinning, P. and Sparrow, C. (1984) J. Stat. Phys. **35**, 645; Gaspard, P., Kapral, R. and Nicolis, G. (1984) J. Stat. Phys. **35**, 697.

[8]   Guckenheimer, J. ( 1981) in : Lect. Notes in Math. **898**, Springer, Berlin, p.99; Broer, H.W. and Vegter, G. (1984) Ergod. Th. & Dynam. Sys. **4**, 509.

[9]   Gaspard, P. (1987) "Homoclinic Tangencies in Dissipative Dynamical Systems", Ph. D. thesis, Univ. of Brussels (in French).

[10]  Gaspard, P. and Nicolis, G. (1983) J. Stat. Phys. **31**, 499.

[11]  Doedel, E.J. (1981) Cong. Num. **30**, 265.

[12]  Sparrow, C. (1982) "The Lorenz Equations : Bifurcations, Chaos and Strange Attractors", Springer, New York.

[13]  Bryant, P., Jeffries, C. and Nakamura, K. (1987) Nuclear Phys. B (Proc. Suppl.) **2**, 25.

[14]  Knobloch, E. and Weiss, N.O. (1983) Physica D **9**, 379; Knobloch, E., Moore, D.R., Toomre, J. and Weiss, N.O. (1986) J. Fluid Mech. **166**, 409.

[15]  Arnéodo, A., Coullet, P. and Tresser, C. (1980) Phys. Lett. A **79**, 259; (1982) J. Stat. Phys. **27**, 171.

[16]  Hastings, S.P. (1982) SIAM J. Appl. Math. **42**, 247.

[17]  Gunnar, N., Russberg (1989) "CATs, SETs and MAPs : modelling and dynamics of selected nonlinear systems ", Ph. D. thesis, Göteborg University; Wang, X.-J. and Gaspard, P. (1990), in : Spatial inhomogeneities and transient behaviour in chemical kinetics, Eds. P. Gray, G. Nicolis, F. Baras, P. Borckmans and S.K. Scott, Manchester, University Press; Bassett, M.R. and Hudson, J.L. (1988) J. Phys. Chem. **92**, 6963.

[18]  Showalter, K., Noyes, R.M. and Bar-Eli, K. (1978) J. Chem. Phys. **69**, 2514.

[19]  Gaspard, P. (1984) in : "Fluctuations and Sensitivity in Nonequilibrium Systems", Eds. W. Horstemke and D.K. Kondepudi, Springer, Berlin.

[20]  Barkley, D. (1988) J. Chem. Phys. **89**, 5547.

[21]  Argoul, F., Arnéodo, A. and Richetti, P. (1987) Phys. Lett. A **201**, 269.

[22]  Wang, X.-J. and Mou, C.Y. (1985) J. Chem. Phys. **83**, 4554.

[23]  Gray, P., Griffiths, J.F., Hasko, S.M. and Lignola, P.G. (1981) Combustion and Flame **43**, 175; (1981) Proc. R. Soc. Lond. A **374**, 313.

[24]  Decroly, O. and Goldbeter A. (1982) Proc. Natl. Acad. Sci. USA **79**, 6917; Goldbeter, A. and Decroly, O. (1983) Am. J. Physiol. **245**, R478.

[25]  Castets, V., Dulos, E., Boissonade, J. and De Kepper, P. (1990) "Experimental Evidence of a Turing Stationary Structure, preprint, University of Bordeaux.

[26]  Kuramoto, Y. (1984) "Chemical Oscillations, Waves and Turbulence", Springer, Berlin; Sivashinsky, G.I. (1977), Acta Astronaut. **4**, 1177.

[27]  Pomeau, Y., Pumir, A. and Pelce, P. (1984) J. Stat. Phys. **37**, 39.

[28]    Eckmann, J.-P. and Ruelle, D. (1985) Rev. Mod. Phys. **57**, 617.

[29]    Orszag, S.A. (1985) in : "Turbulence and Predictability in Geophysical Fluid Dynamics and Climate Dynamics", North-Holland, Amsterdam, p.107.

[30]    Nicolis, C. and Nicolis, G. unpublished work.

[31]    Boissonade, J., Ouyang, Q., Arnéodo, A., Elezgaray, J., Roux, J.C. and De Kepper, P. (1990) in : "Nonlinear Waves Processes in Excitable Media", Eds. A.V. Holden, M. Markus and H.G. Othmer, Pergamon Press.

[32]    Ouyang, Q. (1989) "Structures de réaction-diffusion dans des systèmes chimiques monophasiques en réacteur ouvert quasi unidimensionnel: Structures spatio-temporelles et structures spatiales stationnaires", Ph. D. thesis, Univ. of Bordeaux (in French).

# EIGENVALUE PROBLEMS WITH THE SYMMETRY OF A GROUP AND BIFURCATIONS

B. WERNER
Institut für Angewandte Mathematik
Universität Hamburg
Bundesstraße 55
D-2000 Hamburg 13
Federal Republic of Germany

ABSTRACT. For parameter dependent nonlinear $\Gamma$-equivariant dynamical systems

$$\dot{x} = g(x, \lambda), \quad g : X \times \mathbb{R} \to X, \quad X = \mathbb{R}^n,$$

along a branch of $\Gamma$-symmetric equilibria $\{(x(s), \lambda(s)) : s \in I\}$ the Jacobians $A(s) := \frac{\partial g}{\partial x}(x(s), \lambda(s))$ share the $\Gamma$-equivariance with $g(., \lambda)$:

$$\gamma A(s) = A(s)\gamma \text{ for all } \gamma \in \Gamma. \tag{0.1}$$

(Here $\Gamma$ is a compact Lie group of orthogonal $n \times n$-matrices).
Since the eigenvalues of $A(s)$ are responsible for steady state and Hopf bifurcation we focus on matrix eigenvalue problems for $\Gamma$-symmetric matrices defined by (1). We show that the block diagonalization of all $\Gamma$-symmetric matrices due to a **symmetry adapted basis** in $X$ w.r.t. $\Gamma$ ([11], [13]) is very useful for the theoretical and the numerical treatment of bifurcation problems with symmetries.

## 1 Introduction

We consider parameter dependent nonlinear $\Gamma$-equivariant dynamical systems

$$\dot{x} = g(x, \lambda), \quad g : X \times \mathbb{R} \to X, \quad X = \mathbb{R}^n,$$

where $g$ satisfies the following equivariance condition

$$g(\gamma x, \lambda) = \gamma g(x, \lambda), \quad x \in X, \lambda \in \mathbb{R}, \gamma \in \Gamma. \tag{1.1}$$

Here $\Gamma$ is a compact Lie group of orthogonal $n \times n$-matrices acting on $X$ and being isomorphic to an abstract group $G$: There is a *real* $n$-dimensional faithful representation $\vartheta$ of $G$ on $\mathbb{R}^n$ with

$$\Gamma = \{\vartheta(s) : s \in G\}. \tag{1.2}$$

(see [6] for the basic background of those problems).
Along a branch of $\Gamma$-*symmetric* equilibria $\{(x(s), \lambda(s)) : s \in I\}$ ($\gamma x(s) = x(s)$ for all $\gamma \in \Gamma$), the Jacobians $A(s) := \frac{\partial g}{\partial x}(x(s), \lambda(s))$ share the $\Gamma$-equivariance with $g(., \lambda)$:

$$\gamma A(s) = A(s)\gamma \text{ for all } \gamma \in \Gamma. \tag{1.3}$$

71

*D. Roose et al. (eds.), Continuation and Bifurcations: Numerical Techniques and Applications, 71–88.*
© 1990 *Kluwer Academic Publishers.*

Since the eigenvalues of $A(s)$ are responsible for steady state and Hopf bifurcation we focus on matrix eigenvalue problems for $\Gamma$-*symmetric* matrices $A$ defined by

$$\gamma A = A\gamma \text{ for all } \gamma \in \Gamma. \tag{1.4}$$

We will make use of the fact that there is a *symmetry adapted basis* in $X$ w.r.t. $\Gamma$ such that all $\Gamma$-symmetric matrices have block diagonal form ([11], [13] for *complex* representations, see Th. 2.3 for a *real* version).

This theory has not yet been used for bifurcations with symmetries and we will show that it is very useful for the theoretical and the numerical treatment of bifurcation problems as well:

- We can classify the eigenvalues according to their *symmetry type* which essentially determines the type of symmetry breaking bifurcation. There is a connection to bifurcation subgroups in [4].

- The bifurcation assumptions concerning genericity in [6] become more obvious. Certain singularities (e.g. of Takens-Bogdanov type) for $\Gamma$-equivariant problems can be described in terms of the movement of eigenvalues with arclength $s$.

- Linearly invariant subspaces of minimal dimension are available for the detection and computation of bifurcation points. The computation of eigenvalues determining stability and indicating bifurcation can be performed less costly using the diagonal blocks of the Jacobians only.

The main group theoretical tools we need are based on representation theory (sec.2). The action of $\Gamma$ on $X$ yields an orthogonal decomposition of $X$ into isotypic components $X_i$ and even smaller subcomponents $Y_i^j$ which are invariant under all $\Gamma$-symmetric matrices $A$.

Though we are motivated by bifurcation problems, eigenvalue problems with symmetries of type (1.3) are of interest on their own. They arise for discrete vibration and stability problems of elastic structures with certain geometrical symmetry - compare [1] for corresponding linear problems. It has been pointed out already in [13] that the block diagonalization due to a symmetry adapted basis allows an efficient way to compute the eigenvalues of $\Gamma$-symmetric matrices.

## 2  Representation Theory

In basic text books ([11], [13]), *complex* representations of a group $G$ on $n$-dimensional *complex* vector spaces are considered. We are interested in *real* representations $\vartheta^{\mathbb{R}}$ on *real* vector spaces $X^{\mathbb{R}}(= \mathbb{R}^n)$. But $\vartheta^{\mathbb{R}}$ induces in a natural way a complex representation $\vartheta^{\mathbb{C}}$ on $X^{\mathbb{C}} := X^{\mathbb{R}} + iX^{\mathbb{R}}(= \mathbb{C}^n)$ by setting

$$\vartheta^{\mathbb{C}}(s)(x + iy) := \vartheta^{\mathbb{R}}(s)x + i\vartheta^{\mathbb{R}}(s)y. \tag{2.1}$$

Now it remains to show how the results in [11], [13] for $\vartheta^{\mathbb{C}}$ can be expressed in terms of $\vartheta^{\mathbb{R}}$.

## 2.1 IRREDUCIBILITY AND ISOTYPIC COMPONENTS

Let $\vartheta$ be a (real or complex) orthogonal (resp. unitary) $n$-dimensional representation of a group $G$ on a (real or complex) vector space $X$. Assume that $\Gamma := \vartheta(G)$ is a compact Lie group in $O(n)$ acting on $X$.

A subspace $Y$ of $X$ is called $\Gamma$-**invariant** iff $\gamma y \in Y$ for all $y \in Y$. A $\Gamma$-invariant subspace $Y$ is called **irreducible** (w.r.t. $\vartheta$) iff the only proper $\Gamma$-invariant subspace of $Y$ is $\{0\}$. In that case the representation $\vartheta_Y$ on $Y$ defined by $\vartheta_Y(s) := \vartheta(s)|_Y$ is called **irreducible** too.

Two irreducible subspaces $Y_1$ and $Y_2$ of $X$ are called $\Gamma$-*isomorphic* iff the representations $\vartheta_{Y_1}$ and $\vartheta_{Y_2}$ are **equivalent** in the sense that after a suitable choice of bases $\vartheta_{Y_1}(s) = \vartheta_{Y_2}(s), s \in G$, for corresponding realizations by matrices.

Hence each irreducible subspace $Y$ of $X$ corresponds to a unique irreducible representation $\vartheta_i$ up to equivalence. We will call $Y$ of **type** $\vartheta_i$.

The **dimension** $n_i$ of an irreducible representation $\vartheta_i$ is defined by the vector space dimension of an irreducible subspace $Y$ of type $\vartheta_i$.

Given an irreducible representation $\vartheta_i$ of $G$ (to be more precise: given an equivalence class of irreducible representations) the corresponding **isotypic component** $X_i$ of $\vartheta_i$ in $X$ is given by the sum of all irreducible subspaces $Y$ of $X$ of type $\vartheta_i$ ($X_i$ might be zero if no subspace of that type exists). There is a unique number $c_i \in \mathbb{N}$, called the **multiplicity** of $\vartheta_i$ in $\vartheta$, such that $X_i$ is the direct sum of $c_i$ irreducible subspaces $X_{i,j}, j = 1, .., c_i$, of type $\vartheta_i$,

$$X_i = \bigoplus_{j=1}^{c_i} X_{i,j}. \tag{2.2}$$

(But note that the $X_{i,j}$ are not uniquely determined.)

For most groups $G$ being relevant in applications, all irreducible representations are known. If $G$ is finite only a finite number of (pairwisely non-equivalent) irreducible representations exists. But even if $G$ is an infinite group, any representation on $X = \mathbb{R}^n$ contains only a finite number $m$ of irreducible representations with multiplicity $c_i > 0$.

One main result in representation theory is

**Theorem 2.1** *There is a unique canonical orthogonal decomposition*

$$X = \bigoplus_{i=1}^{m} X_i. \tag{2.3}$$

For a proof see [11] or [13] ($\mathbb{K} = \mathbb{C}$) or [6] ($\mathbb{K} = \mathbb{R}$).

To indicate the multiplicities $c_i$ of $\vartheta_i$ in the action of $\vartheta$ on $X$, one writes

$$\vartheta = \sum_{i=1}^{m} c_i \vartheta_i. \tag{2.4}$$

In sec.5 we explain these notions by means of an example involving the dihedral group $D_6$. But we should refer to the most common symmetry given by a *reflection symmetry* $\Gamma = \{I, S\}$, where $S^2 = I$. Here there are two isotypic components consisting of symmetric elements $(Sx = x)$ and anti-symmetric elements $(Sx = -x)$. The following theory is trivial in this case and needs no group theory.

For the following we need the notion of the **character** $\chi$ of the representation $\vartheta$:

$$\chi(s) := \text{Trace } \vartheta(s), \quad s \in G.$$

## 2.2 IRREDUCIBLE REPRESENTATIONS OF REAL AND COMPLEX TYPE

Let $\vartheta_i$ be a (real) irreducible representation of $G$ of dimension $n_i$ (hence $\vartheta_i(s), s \in G$, can be understood as real $n_i \times n_i$-matrices). The set

$$C(\vartheta_i) := \{B \in Mat(n_i, \mathbb{R}) : B\vartheta_i(s) = \vartheta_i(s)B \text{ for all } s \in G\}$$

is an algebra over $\mathbb{R}$ isomorphic to the skew field $\mathbb{R}, \mathbb{C}$ or the quaternions $\mathbb{H}$ (see [6], p.41f). Correspondingly we will call $\vartheta_i$ of **real, complex** or of **quaternion type**. The notion of an irreducible representation of real type coincides with that of an *absolutely irreducible representation* in [6].

The dihedral groups $D_k$, the permutation group $S_4$ of an tetrahedron and the infinite groups $O(2)$ and $O(3)$ are examples of groups $G$ having only irreducible representations of real type. For $k > 2$ t he cyclic groups $C_k$ have two-dimensional (real) irreducible representations of complex type - which decompose into two complex one-dimensional complex irreducible representations being complex conjugate to each other, see below.

From now on we will consider only those groups $G$ which have no irreducible representations of quaternionic type. All examples in applications seem to belong to this class.

Given a real representation $\vartheta$ of $G$ on $X = \mathbb{R}^n$ and a corresponding compact Lie group $\Gamma$ acting on $X$ we are interested in the multiplicities $c_i$ of a (real) irreducible representation $\vartheta_i$. We will refer to results in [13] and [11]. To this end the complexifications of $\vartheta$ and $\vartheta_i$ have to be used. While for real types the complexifications are still (complex) irreducible this is not true for complex types. Here the complex representation $\vartheta_i^{\mathbb{C}}$ splits into

$$\vartheta_i^{\mathbb{C}} = \vartheta_{i,1}^{\mathbb{C}} + \vartheta_{i,2}^{\mathbb{C}} \text{ with } \vartheta_{i,2}^{\mathbb{C}} = \overline{\vartheta_{i,1}^{\mathbb{C}}}, \tag{2.5}$$

where $\vartheta_{i,1}^{\mathbb{C}}$ is a $\mathbb{C}$-irreducible representation and the bar denotes complex conjugation. This is the reason why the dimension $n_i$ of a real irreducible representation $\vartheta_i$ of complex type is always even. For reasons which will come clear very soon we indroduce a number $d_i$ by

$$d_i := \begin{cases} 1, & \text{if } \vartheta_i \text{ is of real type} \\ 2, & \text{if } \vartheta_i \text{ is of complex type.} \end{cases}$$

In the following we will assume that $G$ is finite. But all results transfer easily to the infinite case. One only has to replace $\frac{1}{|G|} \sum_{s \in G}$ by the *Haar integral* $\int_G$.

We have

**Theorem 2.2** *Let $\vartheta_i$ be an irreducible representation of real or complex type with multiplicity $c_i > 0$ on $X$ and dimension $n_i$. Then*

$$c_i = \frac{1}{d_i|G|} \sum_{s \in G} \chi_i(s)\chi(s). \tag{2.6}$$

*The corresponding isotypic component $X_i$ is given by the image of the orthogonal projector*

$$P_i := \frac{n_i}{d_i|G|} \sum_{s \in G} \chi_i(s)\vartheta(s). \tag{2.7}$$

$\chi$ *and* $\chi_i$ *are the characters of* $\vartheta$ *and* $\vartheta_i$ *respectively.*

This theorem can be derived from the corresponding "complex" formulas in [11] and [13]. Only the case of an irreducible representation of complex type needs some attention. The complex formulas are here

$$c_{i,1} = \frac{1}{|G|} \sum_{s \in G} \overline{\chi_{i,1}(s)}\chi(s) \quad (= c_{i,2})$$

for the multiplicity of $\vartheta_{i,1}^{\mathbb{C}}$ ($\vartheta_{i,2}^{\mathbb{C}}$) defined in (2.5) and

$$P_{i,1} := \frac{n_{i,1}}{|G|} \sum_{s \in G} \overline{\chi_{i,1}(s)}\vartheta(s) \text{ (with } n_{i,1} = n_i/2)$$

for the projector on the (complex) isotypic components $X_{i,1}$ of $\vartheta_{i,1}^{\mathbb{C}}$. Now $\chi_i(s) = \chi_{i,1}(s) + \overline{\chi_{i,1}(s)}, c_i = c_{i,1}$, and $P_i = P_{i,1} + \overline{P_{i,1}}$.

The (real) isotypic component $X_i$ is given by the real (and imaginary) parts of the vectors in $X_{i,1}$.

(The correspondence between $X_{i,1}$ and $X_i$ is the same as that between the (complex) eigenspace of a non-real eigenvalue of a real matrix and the real space spanned by the real and imaginary parts of the eigenvectors - the *real eigenspace*.)

Note that the introduction of the number $d_i$ characterizing $\vartheta_i$ allows unifying formulas (2.6) and (2.7) for real and complex types of $\vartheta_i$.

## 2.3 BLOCK DIAGONALIZATION OF $\Gamma$-SYMMETRIC MATRICES. SYMMETRY ADAPTED BASIS.

Let us use the notation

$$C(\vartheta) := \{A \in \mathbb{R}^{n,n} : A \text{ is } \Gamma\text{-symmetric}\}.$$

The group $\Gamma$ acting on $X = \mathbb{R}^n$ allows the construction of a symmetry adapted basis in $X$ such that all $A \in C(\vartheta)$ - w.r.t. this basis - are **block diagonal** with a block structure which depends only on $\Gamma$. In other words: there is a decomposition

$$X = \bigoplus_{j=1}^{s} V_j$$

such that $V_j$ is *A-invariant* for *all* $A \in C(\vartheta)$ for $j = 1, .., s$.

It is easy to seen that $V_j := X_j, s := m$, is a suitable choice since the isotypic components are invariant under all $A \in C(\vartheta)$. But there is still a finer block structure:

**Theorem 2.3** *There is a (non-unique) symmetry adapted basis of $X$ such that $A \in C(\vartheta)$ if and only if - w.r.t. this basis - $A$ is a block diagonal matrix with the following properties:*

- *For each irreducible representation $\vartheta_i$ of $G$ with multiplicity $c_i > 0$ on $X$ and dimension $n_i$ there are $n_i/d_i$ identical blocks $A_i \in \mathbb{R}^{d_i c_i, d_i c_i}, i = 1, .., m$.*
- *If $\vartheta_i$ is of complex type then additionally $A_i$ has the "complex" structure*

$$A_i = \begin{pmatrix} a_i & -b_i \\ b_i & a_i \end{pmatrix} \quad \text{with } a_i, b_i \in \mathbb{R}^{c_i, c_i}. \tag{2.8}$$

- *The block structure reflects the decomposition (2.3) and a further decomposition of $X_i$ into $n_i/d_i$ subspaces $Y_i^j$:*

$$X_i = \bigoplus_{j=1}^{n_i/d_i} Y_i^j, \quad \dim Y_i^j = d_i c_i,$$

*where each $Y_i^j$ is A-invariant for all $A \in C(\vartheta)$ and is transformed by $A$ in the same way as described by $A_i$ for $j = 1, .., n_i/d_i$.*

For a proof see [11],p.23f, and [13], p. 100f. Th. 2.3 is only a *real* version of their results. The difference concerns only the $\vartheta_i$-blocks of complex type ($d_i = 2$). Here in (2.8)

$$a_i = \Re A_{i,1}, b_i = \Im A_{i,1},$$

where $A_{i,1}$ is the corresponding complex $\vartheta_{i,1}^{\mathbb{C}}$-block according to (2.5).

For a numerical computation of the eigenvalues of a $\Gamma$-symmetric matrix $A$, one has to know a *symmetry adapted* basis of at least one *symmetry adapted subspace* of the isotypic component $X_i$ for each $i = 1, .., m$.

**Definition 2.4** *A subspace $Y_i$ of $X_i$ is called **symmetry adapted** iff it is invariant under all $\Gamma$-symmetric $A$ and the restriction of $A$ to $Y_i$ is similar to $A_i$ in Th. 2.3 (then $Y_i^1 := Y_i$ is a possible choice in Th.2.3).*

From the "complex" results in [11], [13] we get

**Theorem 2.5** *Let $\vartheta_i$ be an (real) irreducible representation of $G$ with multiplicity $c_i > 0$ in $X$ and dimension $n_i$. Let $r_i(s)$ be the first entry of (the $n_i \times n_i$-matrix) $\vartheta_i(s), s \in G$. (If $\vartheta_i$ is of complex type we assume that the matrices $\vartheta_i(s)$ have already the complex structure as in 2.8 with blocks of size $n_i/d_i$). Then*

$$Q_i := \frac{n_i}{|G|} \sum_{s \in G} r_i(s^{-1})\vartheta(s) \tag{2.9}$$

*is a projector onto a symmetry adapted subspace $Y_i$ of $X_i$.*

For the proof we only have to consider the case that $\vartheta_i$ is of complex type. We get $Q_i = Q_{i,1} + \overline{Q_{i,1}}$, where $Q_{i,1}$ and $Q_{i,2} = \overline{Q_{i,1}}$ are the projectors on the complex symmetry adapted subspaces $Y_{i,1}$ and $\overline{Y_{i,1}}$ corresponding to $\vartheta_{i,1}^{\mathbb{C}}$ and $\overline{\vartheta_{i,1}^{\mathbb{C}}}$ in (2.5) as given in [11] and [13].

If a projector on a subspace $Y$ with given dimension is given, it is an easy task to compute an orthonormal basis of $Y$ numerically (e.g. by a Gram-Schmidt procedure). In our situation the dimension of $Y$ is $c_i d_i$ and can be computed by the formula (2.6).

But often there is an analytical way to construct such a basis by projecting the unit vectors in $X$ onto $Y$.

## 3 The symmetry type of eigenvalues

Now we consider eigenvalues and the corresponding eigenspaces of $\Gamma$- symmetric $A \in C(\vartheta)$. By Th.2.3 it is clear that $\mu$ is an eigenvalue of $A$ iff there is a diagonal block $A_i$ such that $\mu$ is an eigenvalue of $A_i$. Since there is a one-to-one correspondence between $A_i$ and an irreducible representation $\vartheta_i$ the latter can be used to classify eigenvalues of $A$: we call $\mu \in \mathbb{C}$ a $\vartheta_i$-**eigenvalue** of $A$ iff $\mu$ is an eigenvalue of the $\vartheta_i$-block $A_i$. Then we have immediately

**Proposition 3.6** *$\mu$ is a $\vartheta_i$-eigenvalue of a $\Gamma$-symmetric real matrix $A$ if and only if there is a corresponding eigenvector such that its real and imaginary parts belong to the isotypic component $X_i$.*

*The geometrical multiplicity of a $\vartheta_i$-eigenvalue $\mu$ is at least $n_i/d_i$.*

Prop.3.6 implies the well known fact that multi-dimensional irreducible representations of real type ($n_i \geq 2, d_i = 1$) force $\Gamma$-symmetric matrices to have multiple eigenvalues.

For the proof of Prop.3.6 note that $n_i/d_i$ is the number of identical blocks $A_i$ in Th.2.3. Concerning complex eigenvalues one has to consider the *real eigenspace* $E_\mu^{\mathbf{R}}$ of $\mu$ spanned by the real and imaginary parts of the eigenvectors. Now it is easy to be seen that $\mu$ is a $\vartheta_i$-eigenvalue iff $E_\mu^{\mathbf{R}}$ has non-zero vectors in common with the (real) isotypic component $X_i$.

### 3.1 $\Gamma$-SIMPLE EIGENVALUES

**Definition 3.7** *An eigenvalue $\mu$ of a $\Gamma$-symmetric matrix $A$ is called $\Gamma$-**simple** iff the following holds:*

1. *There is one and only one $\vartheta_i$ such that $\mu$ is a $\vartheta_i$-eigenvalue.*

2. *$\mu$ is an algebraically simple eigenvalue of the $\vartheta_i$-block $A_i$ given in Th.2.3.*

*We call $\vartheta_i$ the* **symmetry type** *of the $\Gamma$-simple eigenvalue $\mu$.*

Now we can state the following result.

**Theorem 3.8** *Let $\vartheta_i$ be a real irreducible representation of dimension $n_i$ and multiplicity $c_i > 0$ in $X$. Let $X_i$ be the corresponding isotypic component. Then $\mu$ is a $\Gamma$-simple eigenvalue of symmetry type $\vartheta_i$ of a $\Gamma$-symmetric matrix $A$ if and only if the following conditions hold.*

*If $\mu$ is a real eigenvalue then $\vartheta_i$ is of real type and the eigenspace $E_\mu$ is an $n_i$-dimensional (absolutely) irreducible subspace of $X_i$.*

*If $\mu$ is non-real then*

- *either the real eigenspace $E_\mu^{\mathbf{R}}$ is an $n_i$-dimensional irreducible subspace of $X_i$ and $\vartheta_i$ is of complex type*

- *or the real eigenspace is the direct sum of two $\Gamma$-isomorphic $n_i$-dimensional subspaces of $X_i$ and $\vartheta_i$ is of real type.*

*$\mu$ is semisimple: the algebraical and geometrical multiplicity of any $\Gamma$-simple eigenvalue of symmetry type $\vartheta_i$ is $n_i/d_i$.*

The proof is very easy. Observe that a real eigenvalue of a $\vartheta_i$-block $A_i$ corresponding to an irreducible representation $\vartheta_i$ of complex type has at least geometrical multiplicity two. Hence algebraically simple eigenvalues of those blocks are necessarily non-real. The other properties follow directly from Th.2.3. Note for the multiplicity of $\mu$ that the number of equal blocks $A_i$ is $n_i/d_i$.

If one knows the numbers $n_i, c_i, d_i$ of $\vartheta_i, i = 1, .., m$, then the following a priori information about the numbers of eigenvalues of symmetry type $\vartheta_i$ is known.

**Corollary 3.9** *Let all $\vartheta_i$-eigenvalues of $A \in C(\vartheta)$ be $\Gamma$-simple. Then there are $d_i c_i$ different eigenvalues of symmetry type $\vartheta_i$. If $\vartheta_i$ is of complex type then all eigenvalues of this symmetry type are non-real and appear as $c_i$ pairs of conjugate complex eigenvalues.*

This follows directly from Th.3.8.

### 3.2 GENERICITY OF $\Gamma$-SIMPLE EIGENVALUES

We define $C_S(\vartheta)$ to be the subset of $C(\vartheta)$ of all $\Gamma$-symmetric matrices having only $\Gamma$-simple eigenvalues.

**Theorem 3.10** *The property that a $\Gamma$-symmetric matrix $A$ has only $\Gamma$-simple eigenvalues is generic: $C_S(\vartheta)$ is an open and dense subset of $C(\vartheta)$.*

PROOF: This follows immediately from the following three facts.

1. The property that a real matrix (without any further structure) - as the block $A_i$ corresponding to an irreducible representation of real type - has only algebraically simple eigenvalues, is a generic one.

2. Generically a real matrix of complex structure - as the blocks $A_i$ corresponding to an irreducible representation of complex type - has only algebraically simple non-real eigenvalues.

3. Generically for a matrix with given block diagonal structure (where the blocks are independent of each other), different blocks have different eigenvalues. (The blocks we have to consider here consist of the collection of $n_i/d_i$ identical blocks $A_i$ in Th.2.3.)

∎

## 3.3 PATHS OF Γ-SYMMETRIC MATRICES

Consider a smooth path $A(s), s \in I$, in $C(\vartheta)$, where $I$ is some open interval. We will investigate how the eigenvalues of certain symmetry types vary with $s$. Since algebraically simple eigenvalues of a matrix $A$ depend smoothly on $A$, the following theorem follows directly from Th.2.3 by looking at the $\vartheta_i$-blocks $A_i(s)$ of $A(s)$.

**Theorem 3.11** *Let $\mu_0$ be a Γ-simple eigenvalue of symmetry type $\vartheta_i$ of $A(s_0), s_0 \in I$. Then there is a neighborhood $U$ of $s_0$ in $I$ such that for all $s \in U$ the matrices $A(s)$ have Γ-simple eigenvalues $\mu(s)$ of symmetry type $\vartheta_i$ of $A(s)$ depending smoothly on $s \in U$ and satisfying $\mu(s_0) = \mu_0$.*

Th.3.11 shows that Γ-simple eigenvalues behave like algebraically simple eigenvalues under perturbation.

The next question concerns **critical** parameters $s_0$ for which $A(s_0)$ looses the (generic) property to belong to $C_S(\vartheta)$. Then $A(s_0)$ has no longer only Γ-simple eigenvalues. The generic case will be that the path $A(s)$ crosses certain manifolds of codimension one in $C(\vartheta)$ transversally. Therefore the following three sets are of interest.

1. The set $C_B(\vartheta)$ consists of all matrices $A \in C(\vartheta)$ for which all eigenvalues are Γ-simple with one exception: A block $A_i$ corresponding to an irreducible representation $\vartheta_i$ of real type has one real, algebraically double and geometrically simple eigenvalue $\mu_0$. $\mu_0$ will be considered as an Γ-algebraically double and Γ-geometrically simple $\vartheta_i$-eigenvalue of real type.

2. The set $C_C(\vartheta)$ consists of all matrices $A \in C(\vartheta)$ for which all eigenvalues are Γ-simple with one exception: A block $A_i$ corresponding to an irreducible representation $\vartheta_i$ of complex type has one real, algebraically and geometrically double eigenvalue $\mu_0$. $\mu_0$ will be considered as an Γ-algebraically and Γ-geometrically double $\vartheta_i$-eigenvalue of complex type.

3. The set $C_D(\vartheta)$ consists of all matrices $A \in C(\vartheta)$ for which all eigenvalues are Γ-simple with one exception: Two different blocks $A_i$ and $A_j$ corresponding to irreducible representations $\vartheta_i$ and $\vartheta_j$ of real type have coalescing real algebraically simple eigenvalues $\mu_0$. $\mu_0$ will be considered as an Γ-algebraically and Γ-geometrically double $\vartheta_i$-eigenvalue of real type.

The proof of the codimension one property of these sets and the characterization of a transversality condition is beyond the scope of this paper. But it should be clear that again Th.2.3 is the essential tool.

The transversal crossing of the three manifolds above by the path $A(s)$ is caused by the collision of two Γ-simple eigenvalues of $A(s)$ for $s = s_0$. In numerical experiments we visualized collisions according to the three types above and observed the following behavior which seem to reflect the transversality by which $A(s)$ crosses the manifolds.

1. $C_B(\vartheta)$: Two Γ-simple eigenvalues of the same *real* symmetry type collide at some *real* number $\mu_0$. If the two eigenvalues are real before collision they are non-real (and conjugate complex) to each other after collision - and vice versa. The speed of the eigenvalues at collision is infinite.

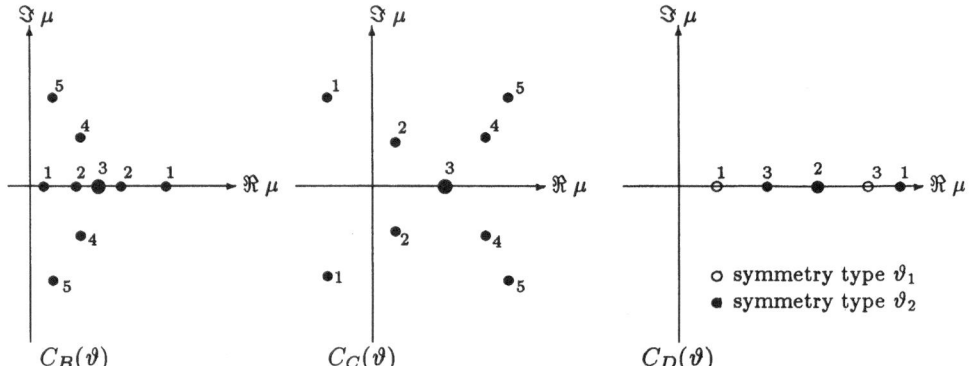

Figure 1: Three different eigenvalue collisions

2. $C_C(\vartheta)$: Two $\Gamma$-simple non-real eigenvalues of the same *complex* symmetry type (complex conjugate to each other) collide at some *real* number $\mu_0$. Before and after collision they are non-real. The speed at collision is finite.

3. $C_D(\vartheta)$: Two $\Gamma$-simple real eigenvalues of different *real* symmetry type collide at some *real* number $\mu_0$. Before and after collision they are both real. The speed is finite. No interaction as in the first case takes place.

The first case $C_B(\vartheta)$ is very much related to a Takens-Bogdanov bifurcation - if the critical eigenvalue $\mu_0$ is zero (see nect section).

The three different types of eigenvalue collisions described by the sets $C_B(\vartheta), C_C(\vartheta)$ and $C_D(\vartheta)$ are shown in Fig.1. The numbers $i$ in the figure refer to different parameter values $s_i, i = 1, 2, \ldots$.

## 4  Bifurcation

For parameter dependend nonlinear $\Gamma$-equivariant dynamical systems

$$\dot{x} = g(x, \lambda), \quad g : X \times \mathbb{R} \to X, \quad X = \mathbb{R}^n,$$

we consider a smooth branch of $\Gamma$-symmetric equilibria $\mathcal{C} := \{(x(s), \lambda(s)) : s \in I\}$, where $I$ is a parameter interval. The $\Gamma$-symmetry of $x(s)$ is described by

$$\gamma x(s) = x(s) \text{ for all } \gamma \in \Gamma.$$

Then the Jacobians $A(s) := \frac{\partial g}{\partial x}(x(s), \lambda(s))$ are $\Gamma$-symmetric. One expects a bifurcation for $s = s_0$ at $(x_0, \lambda_0) = (x(s_0), \lambda(s_0))$ if $A(s_0)$ has an eigenvalue $\mu(s_0)$ with vanishing real part. By Th.3.10 it sounds reasonable to assume that $\mu(s_0)$ is $\Gamma$-simple. Then by Th.3.11 the following definition makes sense:

**Definition 4.12** $(x_0, \lambda_0)$ *is a* $\Gamma$-**simple bifurcation point** *of* **symmetry type** $\vartheta_i$ *if* $\mu(s_0)$ *is a* $\Gamma$-*simple eigenvalue of* $A(s_0)$ *of symmetry type* $\vartheta_i$ *and* $\mu(s)$ *crosses the imaginary axis for* $s = s_0$ *with non-zero speed. If* $\mu(s_0) = 0$ *we call* $(x_0, \lambda_0)$ *a* **static** *bifurcation point and a* **Hopf** *point otherwise.*

Observe that for $\Gamma$-simple static bifurcation points the symmetry type is necessarely real (Th.3.8). This can also be expressed by the fact that the kernel of $A(s_0)$ (which equals the generalized kernel) is absolutely irreducible ([6]). For $\Gamma$-simple Hopf bifurcation points the symmetry type can be real or complex ([6],[3]). Note that our notation for Hopf point contains no resonance conditions.

Now we have to comment the genericity of $\Gamma$-simple bifurcation points. In [6] it is shown that on $\Gamma$-branches generically only $\Gamma$-simple bifurcation points in the sense of Def.4.12 occur. But the proofs in [6] are not complete. We think that Th.3.10 in connection with Th.3.8 helps a lot to understand this genericity statement. But there still remains something to prove in connection with the density part. For static bifurcation points [14] gives a complete proof.

The name $\Gamma$-simple *bifurcation points* could be misleading since it is still not proved in this generality that bifurcation *occurs* at $\Gamma$-simple bifurcation points.

The following arguments help to understand why it is difficult to find a counterexample.

On one hand $\Gamma$-simple Hopf points of complex symmetry type are *simple* Hopf points if the dimension of the irreducible representation $\vartheta_i$ equals $n_i = 2$. In applications there seem to be no groups $G$ with irreducible representations of complex type and of dimension $\geq 4$ (as far as I know).

On the other hand for all irreducible representations of real type I know there always exists at least one *bifurcation subgroup* $\Sigma$ which guarantees bifurcation of solutions with the spatial symmetry $\Sigma$ (see [4] and next section).

## 4.1 BIFURCATION SUBGROUPS

We recall the definition of a *bifurcation subgroup*:

**Definition 4.13** *Let* $\vartheta_i$ *be an irreducible representation of real type of* $G$. *Then a subgroup* $\Sigma$ *of* $\Gamma$ *is called a* **bifurcation subgroup** *for* $\vartheta_i$ *if for one and therefore for any irreducible subspace* $Y$ *of* $X$ *of type* $\vartheta_i$ *the fixed point space*

$$Y^\Sigma := \{y \in Y : \sigma y = y \text{ for all } \sigma \in \Sigma\}$$

*is one-dimensional.*

Now focussing on solutions with at least the symmetry of $\Sigma$ it is easy to show that ($n_i$-multiple) $\Gamma$-simple bifurcation points of real symmetry type $\vartheta_i$ can be reduced to *simple* bifurcation points if $\Sigma$ is a bifurcation subgroup for $\vartheta_i$. This guarantees that for each bifurcation subgroup $\Sigma$ for $\vartheta_i$ there is a branch of (steady state or periodic) solutions with spatial symmetry of $\Sigma$ bifurcating from a $\Gamma$-simple (static or Hopf) bifurcation point of symmetry type $\vartheta_i$ ( [4], [6]). For more details see [4].

Though at least for Hopf points the concept of bifurcation subgroups is not sufficient to describe all bifurcating branches, the symmetry type $\vartheta_i$ of a bifurcation point seems to be the essential information bifurcation theory needs to know about different bifurcations of steady state or periodic solutions ([6], [3]).

Now we want to point out a connection between symmetry adapted subspaces (Def.2.4) and bifurcation subgroups.

**Theorem 4.14** *Let $\Sigma$ be a bifurcation subgroup for an irreducible representation $\vartheta_i$ of $G$ of real type. Let $X_i$ be the corresponding isotypic component. Then $Y_i^\Sigma := X_i \cap X^\Sigma$ is a symmetry adapted subspace.*

$$Q_i^\Sigma := P^\Sigma P_i \tag{4.1}$$

*is the orthogonal projector from $X$ onto $Y_i^\Sigma$. Here $P^\Sigma$ is the orthogonal projector on $X^\Sigma$,*

$$P^\Sigma := \frac{1}{|\Sigma|} \sum_{\sigma \in \Sigma} \sigma.$$

*For $P_i$ see (2.7).*

The proof follows from $\dim X_i \cap X^\Sigma = c_i$ and from the $A$-invariance of $X^\Sigma$ for all $A \in C(\vartheta)$.

We remark that (4.1) is a special case of (2.9) in Th.2.5. If one chooses an orthonormal basis of a $\Gamma$-irreducible subspace $Y$ of $X$ of type $\vartheta_i$ with the first basis vector in $Y^\Sigma$, then the corresponding $n_i \times n_i$-matrices $\vartheta_i(s)$ have a first entry $r_i(s)$ such that $Q_i$ in (2.9) coincides with $Q_i^\Sigma$ in (4.1).

## 4.2 DETECTION AND COMPUTATION OF $\Gamma$-SIMPLE BIFURCATION POINTS

In [4], Th.3.8., subspaces called $X_{\Gamma,\perp}^\Sigma$ and $X_{\Gamma,-}^\Sigma$ have been introduced which turned out to be (linearly) invariant under all $\Gamma$-symmetric $A$. These subspaces had been used for the detection and computation of bifurcation points with bifurcation subgroups $\Sigma$. But now - using the block-diagonalization result in Th.2.3 - we can improve these results.

Following a $\Gamma$-path $\mathcal{C}$ we compute the $\vartheta_i$-blocks $A_i(s)$ of the Jacobians $A(s)$ corresponding to symmetry adapted subspaces $Y_i$ which are in general of lower dimension than the above mentioned subspaces $X_{\Gamma,\perp}^\Sigma$ and $X_{\Gamma,-}^\Sigma$.

The sign change of the determinant of $A_i(s)$ indicates a static bifurcation point of symmetry type $\vartheta_i$. The detection of Hopf points is not that cheap. Computing the eigenvalues of $A_i(s)$ a Hopf point is detected by a pair of conjugate complex eigenvalues crossing the imaginary axis. But every other method for the detection of Hopf points should apply using the blocks $A_i(s)$ instead of $A(s)$.

It is obvious that symmetry adapted subspaces $Y_i$ can also be used for the computation of $\Gamma$-simple bifurcation points of symmetry type $\vartheta_i$ using extended systems (see [4]) with incorporated kernel- or eigenvectors. These vectors contain the essential information for the direction of branching of a $\Sigma$-branch if they lie in $X^\Sigma$ and if $\Sigma$ is a bifurcation subgroup for $\vartheta_i$. Hence we recommend the use of the symmetry adapted subspace $Y_i^\Sigma = X_i \cap X^\Sigma$ in Th.4.14 where $\Sigma$ is any bifurcation subgroup for $\vartheta_i$ (there might be more than one).

But also other methods like the methods based on minimal extended systems ([8]) can utilize the block diagonalization of $A(s)$.

We conclude this subsection by a remark concerning the difference between the subspaces $X_{\Gamma,\perp}^\Sigma$ and $X_{\Gamma,-}^\Sigma$ mentioned above on one hand and the symmetry adapted subspaces $Y_i^\Sigma =$

$X_i \cap X^\Sigma$ (Th.4.14) on the other hand. The last ones are always subspaces of the first ones. For one-dimensional irreducible representations $\vartheta_i$ one can easily show that

$$Y_i^\Sigma = X_{\Gamma,-}^\Sigma = \{x \in X^\Sigma : \gamma x = -x\}$$

where $\gamma$ is any group element in $\Gamma \setminus \Sigma$. Here $x \in Y_i^\Sigma$ is characterized by the symmetry $\sigma x = x$, $\sigma \in \Sigma$ and by the *anti-symmetry* defined by $\gamma x = -x, \gamma \in \Gamma \setminus \Sigma$. For $G = D_4$ also for two-dimensional irreducible representations, $X_{\Gamma,-}^\Sigma$ equals $Y_i^\Sigma$. The advantage of our approach can be demonstrated for $G = D_n$, where $n$ is prime and large. Here the bifurcation subgroups for all two dimensional irreducible representations are given by groups $\Sigma = \{I, S\}$, where $S$ is any reflection. For an $n$-box problem we have $\dim Y_i^\Sigma = m$ in contrast to the much larger number $\dim X_{\Gamma,\perp}^\Sigma = m \cdot n - 2$ (here $m$ is the number of box variables).

## 4.3 NON-$\Gamma$-SIMPLE BIFURCATION POINTS

In section 3.3 we discussed the subsets $C_S(\vartheta), C_B(\vartheta), C_C(\vartheta)$ and $C_D(\vartheta)$ of the space of $\Gamma$-symmetric matrices $C(\vartheta)$. If the eigenvalue $\mu(s_0)$ of the Jacobian $A(s_0) \in C(\vartheta)$ being responsible for bifurcation is **not** $\Gamma$-simple then $A(s_0)$ is not in the generic set $C_S(\vartheta)$.

We shortly concentrate on the case $C_B(\vartheta)$. Here two $\Gamma$-simple real eigenvalues of (real) symmetry type $\vartheta_i$ collide at $\mu = 0$ and split into two non-real (conjugate complex) $\Gamma$-simple eigenvalues of the same symmetry type after collision (or vice versa). This is essentially a Takens-Bogdanov type of bifurcation (see [9]). We call such a bifurcation point according to [5] shortly a B-point of symmetry type $\vartheta_i$.

Takens-Bogdanov bifurcation points are not yet analysed for general symmetries (but see [2] for $\Gamma = O(2)$). By the eigenvalue approach it can easily be understood that B-points can be considered as an *origin of Hopf points* (at least on the $\Gamma$-symmetric branch) (see sec. 4.4).

The other singularities concerning $C_C(\vartheta)$ und $C_D(\vartheta)$ are not yet analysed in this generality - as far as I know. Note that the set $C_C(\vartheta)$ can be generic if the nonlinearity $g$ origins from a potential forcing all eigenvalues of the Jacobians $A(s)$ (being symmetric in the ordinary sense) to be real. As an application one could consider the lattice dome in [10] where the hexagonal symmetry is broken into a rotational $C_6$-symmetry.

Not that the case that more than two $\Gamma$-simple eigenvalues cross the imaginary axis for the same value of the parameter, is another important degeneracy (e.g. Hopf - steady state and Hopf - Hopf interaction, see [6]).

## 4.4 PRE-B-POINTS AND B-POINTS OF SYMMETRY TYPE $\vartheta_i$.

Consider again a $\Gamma$-branch of equilibria and the eigenvalues of the Jacobians $A(s)$. Assume that two $\Gamma$-simple real eigenvalues $\mu_{1,2}(s)$ of (real) symmetry type $\vartheta_i$ collide at $\mu_B$ for $s = s_B$ and split into two non-real (conjugate complex) $\Gamma$-simple eigenvalues after collision (or vice versa). ($\mu_B$ is a $\Gamma$-algebraically double and $\Gamma$-geometrically simple $\vartheta_i$-eigenvalue of $A(s_B)$ and the collision occurs with infinite speed controlled by $\sqrt{|s - s_B|}$). Then we call $y_B := (x(s_B), \lambda(s_B))$ a **pre-B-point** of symmetry type $\vartheta_i$. We call the eigenvalue $\mu_B$ the **collision value** of $y_B$.

B-points are of course special pre B-points with collision value zero.

84

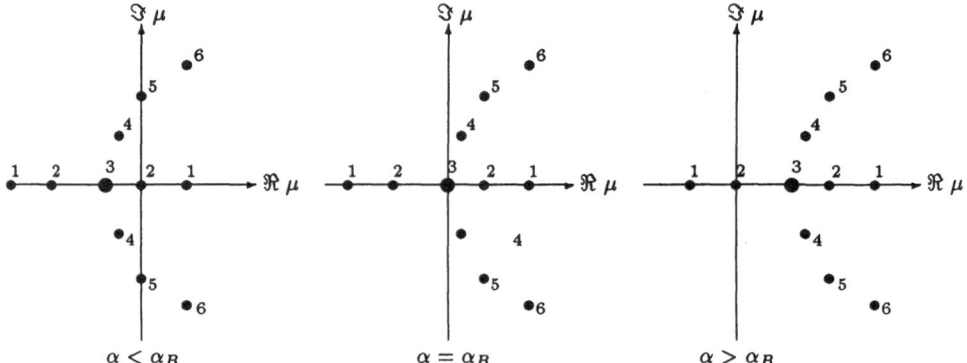

Figure 2: Eigenvalue movement near a B-point

A pre-B-point can be considered as a dynamical bifurcation point since the dynamic changes after bifurcation (a spiraling around the equilibrium vanishes or is borne). It is also denoted as a *node-focus bifurcation point*.

The other fact which makes this point interesting is the following. If the collision value $\mu_B$ of a pre-B-point $y_B$ is rather close to $\mu = 0$, there will be a $\Gamma$-simple static bifurcation point of symmetry type $\vartheta_i$ close to $y_B$ since the speed of $\mu(s)$ is infinite at $s_B$. Moreover a pre-B-point $y_B$ is very often accompanied by a Hopf point (of the same symmetry type) with rather low frequency near $y_B$. In that case Hopf points and static bifurcation points of the same symmetry type $\vartheta_i$ are close together! Note that for the trivial one-dimensional representation $\vartheta_i$ the $\Gamma$-simple static bifurcation point of symmetry type $\vartheta_i$ is a turning point. Of course, this interaction of a pre-B-point, a Hopf point and a turning point is always present in systems without symmetry !

If a second parameter $\alpha$ is involved, pre B-points $y_B(\alpha)$ of symmetry type $\vartheta_i$ and their collision values $\mu_B(\alpha)$ will depend on $\alpha$. Assume that this dependence is smooth and that for some $\alpha = \alpha_B$ one has $\mu_B(\alpha_B) = 0$. Then $y_B$ is a B-point of symmetry type $\vartheta_i$ w.r.t. $\lambda$ for fixed $\alpha = \alpha_B$. Thinking in frames of eigenvalue collisions one can understand that - under the condition that the collision value $\mu_B(\alpha)$ crosses the imaginary axis in zero with nonzero speed - locally either for $\alpha < \alpha_B$ or $\alpha > \alpha_B$ Hopf bifurcation of symmetry type $\vartheta_i$ will generically occur.

This situation is sketched in Fig.2 where for different fixed $\alpha$ six points of two eigenvalue curves $\{\mu(s)\}$ of the same symmetry type are shown (the numbers 1-6 correspond to six parameter values $s_1, s_2, ..., s_6$). For $\alpha < \alpha_B$ there is a static bifurcation (2), a pre B-point (3) and a Hopf point (5). For $\alpha = \alpha_B$ these three bifurcation points coalesce while for $\alpha > \alpha_B$ there is only a static bifurcation (2) and a pre B-point (3).

## 5    Examples and numerical results

Coupled oscillators (cf.[7], [12], [4]) represent nice examples for bifurcation problems with symmetries where dihedral or cyclic groups are involved.

Here we will consider a 6-box Brusselator problem as in [16]. Let us assume that we have six identical chemical reactors ('6 boxes') connected cyclically. The state $z$ and the

Brusselator reaction law $f(z)$ for each box is given by

$$z := (x, y) \in \mathbb{R}^2, \ f(z) := \begin{pmatrix} A - (B+1)x + x^2 y \\ Bx - x^2 y \end{pmatrix}.$$

If $z_j \in \mathbb{R}^2$ is the state vector in box $j$, $j=1,\ldots,6$, then the dynamic of the whole reaction is modelled by the equations

$$\dot{z}_j = f(z_j) + \lambda^{-2} \left[ D_R(z_{j+1} - z_j) + D_L(z_{j-1} - z_j) \right], \ j = 1,\ldots,6, \qquad (5.1)$$

where $z_0 := z_6$, $z_7 := z_1$, and $D_R, D_L \in \mathbb{R}^{2 \times 2}$ are diagonal diffusion matrices determining the clockwise and counterclockwise diffusion (in general $D_R = D_L$).

The bifurcation parameter $\lambda$ controls the size of diffusion of the reactants between the boxes. In comparison with [16] we use $\lambda^{-2}$ instead of $\lambda$.

Set $x := (z_1, \ldots, z_6)$ and let $g(x, \lambda)$ be the right hand side of (5.1). Then $g$ obviously satisfies an equivariance condition (1.3) where $\Gamma$ is isomorphic to the cyclic group $C_6$ in the case of $D_R \neq D_L$ and to the dihedral group $D_6$ (the symmetry group of an hexagon) in the case of $D_R = D_L$. The groups are acting on $X = \mathbb{R}^{12}$.

Each group element $\gamma \in \Gamma$ is represented by a $6 \times 6$-block permutation matrix where each block is the $2 \times 2$- identity matrix.

There is a $\Gamma$-symmetric trivial branch parametrized by $\lambda$ on which $z_j \equiv (A, B/A)$, $j = 1, \ldots 6$.

We are interested in the eigenvalues of the $\Gamma$-symmetric $12 \times 12$-Jacobians $A(\lambda)$, which are responsible for static or Hopf bifurcations on this branch.

If not otherwise noted we have chosen $A = 2$ and $B = 5.9$ in the numerical computations.

## 5.1  $D_6$-SYMMETRY

Here we choose

$$D_R = D_L := \begin{pmatrix} 1 & 0 \\ 0 & 10 \end{pmatrix}.$$

The dihedral group $D_6$ is of order 12 and is generated by the rotation $R$ about the angle $\pi/3$ and by one reflection $S$. Each irreducible representation $\vartheta_i$ is determined by $\vartheta_i(R)$ and $\vartheta_i(S)$. There are six pairwisely non-equivalent irreducible representations of $D_6$ - four are 1-dimensional (called $\vartheta_1, \ldots \vartheta_4$) and two are 2-dimensional ($\vartheta_5, \vartheta_6$). The trivial representation $\vartheta_1$ is given by $R \to 1, S \to 1$, $\vartheta_2$ by $R \to 1, S \to -1$, $\vartheta_3$ by $R \to -1, S \to 1$ and $\vartheta_4$ by $R \to -1, S \to -1$.

The 2-dimensional irreducible representations $\vartheta_5$ and $\vartheta_6$ are defined as follows:

$$\vartheta_5: \quad R \to \begin{pmatrix} \cos\frac{\pi}{3} & -\sin\frac{\pi}{3} \\ \sin\frac{\pi}{3} & \cos\frac{\pi}{3} \end{pmatrix}, \quad S \to \begin{pmatrix} 1 & 0 \\ 0 & -1 \end{pmatrix},$$

$$\vartheta_6: \quad R \to \begin{pmatrix} \cos\frac{2\pi}{3} & -\sin\frac{2\pi}{3} \\ \sin\frac{2\pi}{3} & \cos\frac{2\pi}{3} \end{pmatrix}, \quad S \to \begin{pmatrix} 1 & 0 \\ 0 & -1 \end{pmatrix}.$$

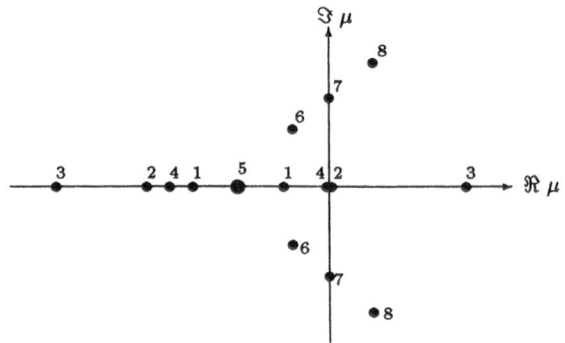

Figure 3: Eigenvalue movements

All irreducible representations are of real type. By means of (2.6) one easily computes the multiplicities

$$c_1 = c_3 = c_5 = c_6 = 2, \quad c_2 = c_4 = 0.$$

Hence we expect - for a given $\lambda$ - two $D_6$-simple eigenvalues of each of the four different symmetry types $\vartheta_1, \vartheta_3, \vartheta_5$ and $\vartheta_6$. We computed these eigenvalues by solving the eigenvalue problem of four $2 \times 2$-matrices using symmetry adapted subspaces as described above.

The movement of all eight eigenvalues in the complex plane has been visualized on a color monitor using different colors for each symmetry type. The movement of the eigenvalues of symmetry type $\vartheta_3, \vartheta_5$ and $\vartheta_6$ are qualitatively the same: for $\lambda = 0$ (infinite diffusion) both eigenvalues are real and negative (number 1 in Fig.3). For increasing $\lambda$ one eigenvalue moves to the left while the other crosses the imaginary axis at $\mu = 0$ (static bifurcation, number 2 in Fig.3). For some larger values of $\lambda$ the "right" (real) eigenvalue and the "left" eigenvalue turn back (3). They approach each other with increasing speed. Before they collide in a collision value of a pre-B-point (5) on the left of the imaginary axis, the "right" eigenvalue changes sign and is thus responsible for a second static bifurcation (4). After collision both eigenvalues bifurcate into the complex plane with infinite speed and cross the imaginary axis (Hopf bifurcation, (7)).

Hence we detected two static and one Hopf bifurcation points of each symmetry type with the exception of the type $\vartheta_1$.

The presence of pre-B-points with collision values close to the imaginary axis indicates that a variation of a second parameter (e.g. of $B$) would lead to a Takens-Bogdanov point. Indeed this is the case if $B$ is decreased.

## 5.2 $C_6$-SYMMETRY

To have only cyclic symmetry the reflectional symmetry has to be destroyed ($D_R \neq D_L$). We choose

$$D_R := \begin{pmatrix} 1 & 0 \\ 0 & 10 \end{pmatrix}, D_L := \begin{pmatrix} 1.2 & 0 \\ 0 & 10.3 \end{pmatrix}.$$

The cyclic group $C_6$ is of order 6 and is generated by the rotation $R$ about the angle $\pi/3$. There are two 1-dimensional irreducible representations $\vartheta_1(R \to 1)$ and $\vartheta_2(R \to -1)$ - both of course of real type. The two (real) 2-dimensional irreducible representations $\vartheta_3$ and $\vartheta_4$ are defined by

$$\vartheta_3: \quad R \to \begin{pmatrix} \cos\frac{\pi}{3} & -\sin\frac{\pi}{3} \\ \sin\frac{\pi}{3} & \cos\frac{\pi}{3} \end{pmatrix}, \quad \vartheta_4: \quad R \to \begin{pmatrix} \cos\frac{2\pi}{3} & -\sin\frac{2\pi}{3} \\ \sin\frac{2\pi}{3} & \cos\frac{2\pi}{3} \end{pmatrix},$$

respectively. They are of complex type.
The multiplicities are

$$c_1 = c_2 = c_3 = c_4 = 2.$$

For a given $\lambda$ we expect two $C_6$-simple eigenvalues of each of the four symmetry types where the two eigenvalues of the complex symmetry types $\vartheta_3$ and $\vartheta_4$ should appear as conjugate complex pairs.
The consequence of the $D_6$-$C_6$-perturbation is that all real $D_6$-simple eigenvalues of 2-dimensional types are broken into pairs of conjugate complex simple eigenvalues of the corresponding 2-dimensional $(C_6)$-types. Hence the corresponding $D_6$-simple static bifurcation points will be perturbed into simple Hopf points while the $D_6$-simple Hopf points will split into two $(C_6$-)simple Hopf points. Indeed we detected now four Hopf points of symmetry types $\vartheta_3$ and $\vartheta_4$ while the two $\vartheta_2$-eigenvalues behave qualitatively like the $\vartheta_3$-eigenvalues in the $D_6$-case.
This observation would justify to call $D_6$-simple static bifurcation points of 2-dimensional symmetry type a *Hopf origin* w.r.t. a perturbation into the group $C_6$. More general a perturbation of $D_n$-equivariant problems into $C_n$-equivariant ones will generically perturb static $D_n$-simple bifurcation points of 2-dimensional symmetry type into $C_n$-simple Hopf points of corresponding symmetry type (see also [3], p.108).

**Acknowledgment:** I thank P. Stork for reading the manuscript and making useful comments.

# References

[1] A. Bossavit. *Symmetry, groups and boundary value problems. A progressive introduction to noncommutative harmonic analysis of partial differential equations in domains with geometrical symmetry.* Computer Meth. in Appl. Mech. and Eng. **56**, 167-215, 1986.

[2] G. Dangelmayr, E. Knobloch. *The Takens-Bogdanov Bifurcation with O(2)-Symmetry.* Phil. Trans. R. Soc. Lond. A 322, 243-279, 1987.

[3] M. Dellnitz. *Hopf-Verzweigung in Systemen mit Symmetrie und deren Numerische Behandlung.* Wissenschaftliche Beiträge aus Europäischen Hochschulen. Verlag an der Lottbek. (Dissertation Universität Hamburg) 1989.

[4] M. Dellnitz, B. Werner. *Computational methods for bifurcation problems with symmetries - with special attention to steady state and Hopf bifurcation points.* J. of Comp. and Appl. Math. **26**, 97-123, 1989. (special issue on **Continuation Techniques and Bifurcation Problems** (H.D. Mittelmann, D. Roose (eds))

[5] B. Fiedler. *Global Hopf bifurcation of two-parameter flows*. Arch. Rat. Mech. and Anal. **94**, 59-81, 1986.

[6] M. Golubitsky, I. Stewart, D. Schaeffer. **Singularities and Groups in Bifurcation Theory**, Vol. 2, Springer 1988.

[7] M. Golubitsky, I. Stewart. *Hopf bifurcation with dihedral group symmetry: coupled nonlinear oscillators*. In **Multiparameter Bifurcation Theory**, M. Golubitsky, J. Guckenheimer (eds.). Contemporary Mathematics **56**, 131-173. Amer. Math. Soc., Providence, 1986.

[8] A. Griewank, G. Reddien. *Characterization and computation of generalized turning points*. SIAM J. Numer. Anal. **21**, 176-185, 1984.

[9] J. Guckenheimer, P. Holmes. **Nonlinear Oscillations, Dynamical Systems and Bifurcations of Vector Fields.** Springer, New-York, 1983.

[10] T. Healey. *A group-theoretic approach to computational bifurcation with symmetry*. Computer Methods in Appl. Mech. and Eng. **67**, 257-295, 1988.

[11] J.P. Serre **Linear representations of finite groups.** Springer 1977.

[12] I. Schreiber, M. Holodniok, M. Kubicek, M. Marek. *Periodic and aperiodic regimes in coupled dissipative chemical oscillators*. J. of Statist. Phys. **43**, 489-518, 1986.

[13] E. Stiefel, A. Fässler **Gruppentheoretische Methoden und ihre Anwendung.** Teubner 1979.

[14] P. Stork. *Statische Verzweigungen mit Symmetrien*. Diplomarbeit. Institut für Angewandte Mathematik der Universität Hamburg 1989.

[15] A. Vanderbauwhede. **Local Bifurcation Theory and Symmetry.** Pitman 1982.

[16] B. Werner. *Computational methods for bifurcation problems with symmetries and applications to steady states of n-box reaction-diffusion models*. In **Numerical Analysis 1987.**, D. F. Griffiths, G. A. Watson (eds.), 279-293, Pitman 1988.

[17] B. Werner, A. Spence. *The computation of symmetry-breaking bifurcation points*. SIAM J. Numer. Anal. **21**, 388-399, 1984.

# STEADY-STATE/STEADY-STATE MODE INTERACTION IN NONLINEAR EQUATIONS WITH Z₂-SYMMETRY

W. WU [1,3],   K.A. CLIFFE [2]   and   A. SPENCE [1]

[1] *School of Mathematical Sciences*
*University of Bath*
*Bath, BA2 7AY*
*United Kingdom*

[2] *Theoretical Physics Division*
*AEA Technology*
*Harwell*
*Oxfordshire, OX11 0RA*
*United Kingdom*

[3] On leave from:
*Department of Mathematics*
*University of Jilin*
*Changchun*
*China*

ABSTRACT. This paper is concerned with an example of steady-state/steady-state mode interaction in two parameter nonlinear problems satisfying a $Z_2$-symmetry (equivariance) condition. Specifically we analyze the solution structure near a point where a path of symmetry-breaking bifurcations and a path of fold points intersect. The treatment is such that numerical information is obtained which could prove useful when switching from one path to another.

## 1. Introduction

Consider the nonlinear equation with two parameters

$$f(x,\lambda,\alpha)=0, \quad f:X\times R^2 \to X, \tag{1.1}$$

where $X$ is a Banach space and $f$ is $C^4$ on $X\times R$. We are interested in solutions near *singular* points of (1.1), i.e., points $(x_0,\lambda_0,\alpha_0)$ where the linearization of $f$ is a Fredholm operator of index zero satisfying

$$\text{Null}\,(f_x^0)= \text{span}\,(\varphi_0), \quad \varphi_0 \in X\backslash\{0\}, \tag{1.2a}$$

$$\text{Range}\,(f_x^0)=\{x\in X, \ \psi_0 x=0\}, \quad \psi_0 \in X'\backslash\{0\} \tag{1.2b}$$

where $f_x^0:=f_x(x_0,\lambda_0,\alpha_0)$ and $X'$ is the dual space of $X$. (Note that throughout this paper we write $f_x^0$ for $f_x(x_0,\lambda_0,\alpha_0)$, $f_\lambda^0$ for $f_\lambda(x_0,\lambda_0,\alpha_0)$ etc.) Equation (1.1) is often studied as a first step towards the understanding of the evolution equation

$$x_t+f(x,\lambda,\alpha)=0,$$

89

*D. Roose et al. (eds.), Continuation and Bifurcations: Numerical Techniques and Applications, 89–103.*

since a singular solution point $(x_0,\lambda_0,\alpha_0)$ of (1.1) may play an important role in the determination of the boundary of the stable and the unstable "steady-state" solutions of the evolution equation (see Guckenheimer and Holmes [6] and Seydel [10]). A natural system to use to determine a singular point is the following extended system (Moore and Spence [8], Seydel [11], Werner and Spence [13]):

$$F(y,\alpha):=\begin{bmatrix} f(x,\lambda,\alpha) \\ f_x(x,\lambda,\alpha)\varphi \\ l\varphi-1 \end{bmatrix}=0, \tag{1.3}$$

where $l\in X'$ is some scaling function and

$$y:=(x,\varphi,\lambda)\in Y:=X\times X\times\mathbf{R},$$

$$F:Y\times\mathbf{R}\to Y.$$

One of our main assumptions, which arises in many applications (see [1], [4], [5], [10], [13]), is that $f$ satisfies a $Z_2$-symmetry relation (equivariance condition): there exists an $s\in L(X)$ such that

$$s\neq I, \quad s^2=I, \quad sf(x,\lambda,\alpha)=f(sx,\lambda,\alpha) \quad \forall (x,\lambda,\alpha)\in X\times\mathbf{R}^2. \tag{1.4}$$

In this paper we use system (1.3) to provide an analysis of a steady-state/steady-state mode interaction (Golubitsky *et al.* [5]) which occurs when a path of fold point bifurcations intersects with a path of symmetry-breaking bifurcations. This paper can be regarded as a sequel to Cliffe and Spence [1] in that it extends their analysis near a quartic symmetry-breaking bifurcation point and provides a systematic computational approach for jumping on to the path of fold points.

First we provide some background theory and definitions. It is well-known that (1.4) induces the splitting

$$X=X_s\oplus X_a$$

where $X_s:=\{x\in X, sx=x\}$ and $X_a:=\{x\in X, sx=-x\}$ are the *symmetric* and the *antisymmetric* subspaces of $X$ respectively, and that if $x_0\in X_s$ then eigenvectors of $f_x(x_0,\lambda_0,\alpha_0)$ lie either in $X_s$ or in $X_a$.

There are two cases to describe depending on whether or not symmetry is broken. First we consider the *symmetry−breaking* case, that is,

$$x_0\in X_s, \quad \varphi_0\in X_a. \tag{1.5}$$

If we introduce an "antisymmetric" subspace of $X'$ as follows

$$X'_a:=\{\psi\in X', \quad \psi x=0 \ \forall x\in X_s\},$$

then (1.2) and (1.5) imply that

$$\psi_0\in X'_a. \tag{1.6}$$

We may now introduce the following expressions to be used later in the nondegeneracy conditions:

$$b_\lambda:=\psi_0 f^0_{xx}v_\lambda\varphi_0+\psi_0 f^0_{x\lambda}\varphi_0, \tag{1.7}$$

$$b_z:=\psi_0 f^0_{xxx}\varphi^3_0+3\psi_0 f^0_{xx}\varphi_0 v_z, \tag{1.8}$$

where $v_\lambda$ and $v_z$ are defined by

$$f_x^0 v_\lambda + f_\lambda^0 = 0, \quad v_\lambda \in X_s, \tag{1.9}$$

$$f_x^0 v_z + f_{xx}^0 \varphi_0^2 = 0, \quad v_z \in X_s. \tag{1.10}$$

**Definition 1.1.**
Let $(x_0, \lambda_0, \alpha_0)$ be a solution point of (1.1) satisfying (1.2) and (1.5), then we call it a
  (i) *symmetry−breaking bifurcation point (S−point)* if $b_\lambda \neq 0$;
  (ii) *quadratic S−point* if $b_\lambda \neq 0$ and $b_z \neq 0$;
  (iii) *high order S−point* if $b_\lambda \neq 0$ and $b_z = 0$;
  (iv) *quartic S−point* if $b_\lambda \neq 0$, $b_z = 0$ and $b_{zz} \neq 0$, where $b_{zz}$ is a constant defined implicitly by (3.15) in Theorem 3.4.

It is clear from the discussion in [4, Chapter VI] that in the two parameter problem (1.1) we would expect to encounter at most quartic $S$-points. (Note that since dim Null$(f_x^0)=1$ all bifurcation points are "simple" and so we could write "simple quadratic $S$-point" etc. However since we never consider a 2 dimensional null space of $f_x^0$, we drop the term "simple".)

Now let us turn to the case when symmetry is not broken. We introduce the expressions

$$c_\lambda = \psi_0 f_\lambda,$$

$$c_{xx} = \psi_0 f_{xx} \varphi_0 \varphi_0$$

and make the following definition:

**Definition 1.2.**
Let $(x_0, \lambda_0, \alpha_0)$ be a solution point of (1.1) satisfying (1.2), then we call it a
  (i) *fold point* if $c_\lambda \neq 0$;
  (ii) *quadratic fold point* if $c_\lambda \neq 0$, $c_{xx} \neq 0$.

Schematic solution diagrams of (1.1) for fixed $x_0$ near $S$-points are given in Figure 3.2 of [1].

Note that both $S$-points and fold points satisfy (1.3) which suggests that the system (1.3) can be used to analyze the solution structure of (1.1) near the point where the paths intersect. This paper contains a theoretical analysis of the behavior of the solutions of (1.3) near a high order $S$-point and, as indicated above, we find an interaction of a quadratic $S$-point path and a fold point path in $(x, \lambda, \alpha) \in X \times \mathbf{R}^2$ space subject to an appropriate "unfolding" condition. This result is of course well-known as an example of a steady-state/steady-state mode interaction (Golubitsky *et.al.* [5], Chapter XIX), but our approach is nevertheless interesting for two main reasons. First, at a practical level, it provides some computational information which enables the path of fold points to be easily computed. This is important when dealing with large systems arising, say, from discretization of partial differential equations where accurate starting values to compute the path of fold points are essential (see Section 5). Second, of theoretical benefit to us, the analysis aids our understanding of paths of $S$-points in the whole space $Y = X \times X \times \mathbf{R}$ rather than in $Y_s := X_s \times X_a \times \mathbf{R}$. This is crucial in the treatment of (more interesting) symmetry-breaking Takens-Bogdanov points. Our approach relies on the fact that $F$ given by (1.3) will inherit a symmetry relation from $f$ provided we choose $l$ appropriately (see (2.1)). As we shall see, along a path of $S$-points of $F(y, \alpha) = 0$, $F_y$ has always precisely one antisymmetric null vector but this *does not* induce bifurcation at a quadratic $S$-point. However bifurcation in $F(y, \alpha) = 0$ does occur at a high order $S$-point, subject to the unfolding condition, resulting in a path of fold points.

The outline of this paper is as follows. In Section 2 we show how to choose $l$ in (1.3) such that $F$ will inherit a symmetry relation from $f$. We also confirm (Theorem 2.3) that in $Y_s \times \mathbf{R}$ there exists precisely one solution path of (1.3) at any $S$-point subject to the unfolding condition. Section 3 contains the main results. First we show in Theorem 3.3 that the solution path of (1.3) mentioned above is the only one in the whole space $Y \times \mathbf{R}$ at a quadratic $S$-point. Theorem 3.4 states that generically at a high order $S$-point there exists precisely one other solution path of (1.3) in $Y \times \mathbf{R}$, which corresponds to (conjugate) fold point solutions, and also that if the high order $S$-point is a quartic $S$-point then the paths of fold points are quadratic fold points. Some tedious calculations are contained in Section 4. Section 5 is devoted to a brief discussion of how to jump on to the path of fold points near a high order $S$-point which arises in a finite element discretization of the Navier-Stokes equations for steady axisymmetric flow in the Taylor problem.

To end this section, we remark that when using an extended system to analyze bifurcation phenomena it is often possible, and indeed desirable, to confine the analysis to an extended symmetry space (for example $X_s \times X_a \times \mathbf{R}$, see Werner and Spence [13]). However in our approach to the analysis near a high order $S$-point with its associated fold points, we need to consider the system $F(y,\alpha)=0$ in the whole space $Y=X \times X \times \mathbf{R}$.

## 2. Bifurcation analysis in the symmetric space $Y_s$

In this section we define a symmetry on $Y=X \times X \times \mathbf{R}$ to ensure that $F$ satisfies a symmetry relation of the same type as (1.4), and then rework some fairly standard theory (cf. Werner and Spence [13]).

The first step is to make precise the requirements on $l$, the scaling function in (1.3). We require that, for the given $\varphi_0 \in X_a$, $l$ satisfies

$$ls=-l, \quad l\varphi_0-1=0. \tag{2.1}$$

This condition is not restrictive. In fact, for any $\tilde{l} \in X'$ satisfying $\tilde{l}\varphi_0 \neq 0$ we may make the choice

$$l:=\frac{\tilde{l}-\tilde{l}s}{2\tilde{l}\varphi_0},$$

which clearly satisfies (2.1).

Our first result is about the symmetry on $F$ inherited from that on $f$.

**Lemma 2.1.**
Let $y=(x,\varphi,\lambda) \in Y$ and define a linear operator $S$ on $Y$ by

$$Sy:=(sx,-s\varphi,\lambda) \tag{2.2}$$

where $s$ satisfies (1.4). If (2.1) holds then

$$S \neq I, \quad S^2=I, \quad SF(y,\alpha)=F(Sy,\alpha) \quad \forall (y,\alpha) \in Y \times \mathbf{R}. \tag{2.3}$$

Accordingly, we may split $Y$ into

$$Y=Y_s \oplus Y_a,$$

where $Y_s:=X_s \times X_a \times \mathbf{R}$ and $Y_a:=X_a \times X_s \times \{0\}$ are the symmetric and the antisymmetric subspaces of $Y$ respectively. As before we may define $Y'_a$ by

$$Y'_a := \{\Psi \in Y', \ \Psi y = 0, \ \forall y \in Y_s\}.$$

The following lemma can be proved by direct calculation.

**Lemma 2.2.**
Let $(x_0, \lambda_0, \alpha_0)$ be a $S$-point and assume that (2.1) holds then

$$\text{Null}(F_y^0) = \text{span}(\Phi_0), \quad \Phi_0 = (\varphi_0, v_z, 0) \in Y_a, \tag{2.4a}$$

$$\text{Range}(F_y^0) = \{y \in Y, \ \Psi_0 y = 0\}, \quad \Psi_0 = (\psi_0, 0, 0) \in Y'_a. \tag{2.4b}$$

An immediate consequence of this lemma is

**Theorem 2.3.**
Under the same assumption as for Lemma 2.2, there exists locally in $Y_s \times \mathbf{R}$ precisely one solution path of (1.3) passing through $(y_0, \alpha_0) = (x_0, \varphi_0, \lambda_0, \alpha_0)$. It can be parameterized as $(y(\alpha), \alpha) = (x(\alpha), \varphi(\alpha), \lambda(\alpha), \alpha)$ with tangent vector $(V_\alpha, 1)$ at $(y_0, \alpha_0)$ where $V_\alpha \in Y_s$ is defined by $F_y^0 V_\alpha + F_\alpha^0 = 0$.

**Proof.**
Note that $y_0 \in Y_s$ and that $Y_s$ is an invariant subspace in the sense that $F_y^0 \eta \in Y_s \ \forall \eta \in Y_s$. By Lemma 2.2, the restriction $F_y^0|_{Y_s}$ is nonsingular, and so the Implicit Function Theorem implies the existence and uniqueness of $y(\alpha)$ for given $\alpha$ near $\alpha_0$. The last statement of the theorem follows from evaluating the following equation at $\alpha = \alpha_0$,

$$\frac{d}{d\alpha} F(y(\alpha), \alpha) = F_y \dot{y} + F_\alpha = 0, \quad \dot{y} \in Y_s.$$

It is clear that if $(x_0, \lambda_0, \alpha_0)$ is a quadratic $S$-point then all the points $(x(\alpha), \lambda(\alpha), \alpha)$ near $(x_0, \lambda_0, \alpha_0)$ are quadratic $S$-points. But the situation is not so straightforward at a high order $S$-point and further analysis is needed. It is well known from the discussion in Golubitsky and Schaeffer [4, Chapter VI] that an unfolding condition is necessary in order for $f(x, \lambda, \alpha)$ to be a universal unfolding of $f(x_0, \lambda_0, \alpha_0)$. The following theory interprets this unfolding condition as a "crossing condition" on $b_z$ at $\alpha_0$.

Let $w \in X_a$ be some constant vector such that

$$\psi_0 w - 1 = 0. \tag{2.5}$$

It is clear that there exists a unique smooth function $\psi(\alpha): \mathbf{R} \to X'_a$ satisfying

$$\psi(\alpha) f_x^\alpha = 0, \quad \psi(\alpha) w - 1 = 0, \quad \psi(\alpha_0) = \psi_0, \tag{2.6}$$

where we use $f_x^\alpha$ to denote $f_x(x(\alpha), \lambda(\alpha), \alpha)$ etc. Let us define (cf. (1.8))

$$b(\alpha) := \psi(\alpha) f_{xxx}^\alpha (\varphi(\alpha))^3 + 3\psi(\alpha) f_{xx}^\alpha \varphi(\alpha) v(\alpha), \tag{2.7}$$

where $v(\alpha)$ is given by (cf.(1.10))

$$f_x^\alpha v(\alpha) + f_{xx}^\alpha (\varphi(\alpha))^2 = 0, \quad v(\alpha) \in X_s. \tag{2.8}$$

Then $b(\alpha_0) = b_z$, $v(\alpha_0) = v_z$.
Let $b_z = 0$ and $V_\alpha = (V_1, V_2, V_3) \in X_s \times X_a \times \mathbf{R}$ where $V_\alpha$ is defined in Theorem 2.3. We define

$$b_{z\alpha} := \psi_0 \{(f_{xxx}^0 V_1 + f_{xx\lambda}^0 V_3 + f_{xx\alpha}^0) \varphi_0^3 + 3(f_{xx}^0 V_1 + f_{xx\lambda}^0 V_3 + f_{xx\alpha}^0) \varphi_0 v_z \tag{2.9}$$

$$+(f_{xx}^0 V_1 + f_{x\lambda}^0 V_3 + f_{x\alpha}^0)v_\varphi + 3f_{xxx}^0 \varphi_0^2 V_2 + 3f_{xx}^0 v_z V_2 + 3f_{xx}^0 \varphi_0 v_m\},$$

where $v_\varphi$ and $v_m$ are given by

$$f_x^0 v_\varphi + f_{xxx}^0 \varphi_0^3 + 3f_{xx}^0 \varphi_0 v_z = 0, \quad lv_\varphi = 0, \quad v_\varphi \in X_a, \tag{2.10}$$

$$f_x^0 v_m + f_{xxx}^0 \varphi_0^2 V_1 + f_{xx\lambda}^0 \varphi_0^2 V_3 + f_{xx\alpha}^0 \varphi_0^2 + f_{xx}^0 V_1 v_z \tag{2.11}$$

$$+ f_{x\lambda}^0 V_3 v_z + f_{x\alpha}^0 v_z + 2f_{xx}^0 V_2 \varphi_0 = 0, \quad v_m \in X_s.$$

The condition $b_{z\alpha} \neq 0$ corresponds to the existence of the $\alpha_1 u$ term in the universal unfolding of $(u^k - \lambda)x$ in Proposition 3.4(a) of [4, p259]. The next Lemma produces the crossing condition on $b_z$ at $\alpha_0$.

**Lemma 2.4.**
a). At a high order $S$-point $(x_0, \lambda_0, \alpha_0)$,

$$b_{z\alpha} = \left. \frac{db(\alpha)}{d\alpha} \right|_{\alpha=\alpha_0}, \tag{2.12}$$

and hence,
b). If $b_{z\alpha} \neq 0$, then $b(\alpha)$ crosses zero with nonzero velocity as $\alpha$ passes through $\alpha_0$.

**Proof.**

Write $\dot{v}_0 := \left. \frac{dv(\alpha)}{d\alpha} \right|_{\alpha=\alpha_0}$ etc. Note

$$V_1 = \dot{x}_0, \quad V_2 = \dot{\varphi}_0, \quad V_3 = \dot{\lambda}_0, \quad v_m = \dot{v}_0,$$

$$\dot{\psi}_0 f_x^0 + \psi_0 f_{xx}^0 V_1 + \psi_0 f_{x\lambda}^0 V_3 + \psi_0 f_{x\alpha}^0 = 0, \quad \dot{\psi}_0 w = 0.$$

Differentiating (2.7) and Evaluating it at $\alpha=\alpha_0$ yields (2.12). The result (b) is an obvious corollary of (a).

The result of Lemma 2.4 b) ensures that, provided $b_{z\alpha} \neq 0$, a reliable numerical technique for the detection of a high order $S$-point is to search for a sign change in $b(\alpha)$. A final Corollary is:

**Corollary 2.5.**
At a high order $S$-point $(x_0, \lambda_0, \alpha_0)$, if $b_{z\alpha} \neq 0$ then all points $(x(\alpha), \lambda(\alpha), \alpha)$ near but not equal to $(x_0, \lambda_0, \alpha_0)$ on the solution path mentioned in Theorem 2.3 are quadratic $S$-points.

### 3. Bifurcation analysis in the whole space $Y \times R$

In this section, we use a Liapunov-Schmidt reduction procedure to investigate the bifurcation of solutions of $F(y, \alpha) = 0$ from a high order $S$-point of $f(x, \lambda, \alpha) = 0$ in the whole space $Y \times R$. Our main results are given by Theorems 3.3 and 3.4.

Let us define

$$G(u, \xi, \eta) := (I - W\Psi_0)F(y_0 + \xi\Phi_0 + u, \alpha_0 + \eta), \tag{3.1}$$

$$G: Y_0 \times R_2 \to Y_1,$$

where $W=(w,0,0)^T$ (see (2.5)), $\Phi_0$ is given by (2.4a) and $Y_0$, $Y_1$ are subspaces of $Y$ defined by

$$Y_0:=\{y=(x,\varphi,\lambda)\in Y, \ lx=0\},$$

$$Y_1:=\{y\in Y, \ \Psi_0 y=0\}.$$

Note that $\Psi_0 W=1$ and

$$Y=Y_0\oplus \text{span}\,\{\Phi_0\}=Y_1\oplus \text{span}\,\{W\}. \tag{3.2}$$

Since $G(0,0,0)=0$ and $G_u^0:=G_u(0,0,0)=F_y^0:Y_0\to Y_1$ is an isomorphism, there exists a unique smooth function $u=u(\xi,\eta)\in Y_0$ such that

$$G(u(\xi,\eta),\xi,\eta)=0 \ \ \forall(\xi,\eta) \ \text{near} \ (0,0). \tag{3.3}$$

As is standard, let us introduce the reduced (bifurcation) function

$$g(\xi,\eta):=\Psi_0 F(y_0+\xi\Phi_0+u(\xi,\eta), \ \alpha_0+\eta),$$

and the bifurcation analysis of $F(y,\alpha)=0$ near $(y_0,\alpha_0)$ reduces to the bifurcation analysis of

$$g(\xi,\eta)=0 \tag{3.4}$$

near $(\xi,\eta)=(0,0)$.

We claim that $g(\xi,\eta)$ is an odd function of $\xi$ as a result of the symmetry relation of $F$, that is:

**Lemma 3.1.**

$$g(-\xi,\eta)=-g(\xi,\eta) \ \ \forall(\xi,\eta) \ \text{near} \ (0,0). \tag{3.5}$$

**Proof.**
Note that

$$\Psi_0 Sy=-\Psi_0 y \ \ \ \forall y\in Y, \tag{3.6}$$

and from (3.3) and (3.6) it follows that

$$G(Su(\xi,\eta),-\xi,\eta)=SG(u(\xi,\eta),\xi,\eta)=0.$$

Then the uniqueness of $u(\xi,\eta)$ implies

$$Su(\xi,\eta)=u(-\xi,\eta). \tag{3.7}$$

Thus, (3.5) holds since

$$g(-\xi,\eta)=\Psi_0 F(y_0-\xi\Phi_0+u(-\xi,\eta),\alpha_0+\eta)=\Psi_0 F(S(y_0+\xi\Phi_0+u(\xi,\eta)),\alpha_0+\eta)$$

$$=-\Psi_0 F(y_0+\xi\Phi_0+u(\xi,\eta),\alpha_0+\eta)=-g(\xi,\eta).$$

It is well-known (see Golubitsky and Shaeffer [4, p249]) that the property (3.5) restricts the form of $g(\xi,\eta)$ to

$$g(\xi,\eta)=\xi q(\xi^2,\eta)$$

for some smooth function $q:\mathbf{R}^2\to\mathbf{R}$. The following lemma reveals a more profound structure of $g(\xi,\eta)$, which will play a central role in our analysis.

**Lemma 3.2.**
Let $(x_0, \lambda_0, \alpha_0)$ be an $S$-point. Then there exists a smooth function $p: \mathbf{R}^2 \to \mathbf{R}$ such that

$$g(\xi, \eta) = \xi^3 p(\xi^2, \eta). \tag{3.8}$$

**Proof.**
From (3.5) it follows that

$$g(0, \eta) = 0, \quad g_{\xi\xi}(0, \eta) = 0, \quad \forall \eta \text{ near } 0. \tag{3.9}$$

We claim that

$$g_\xi(0, \eta) = 0 \quad \forall \eta \quad \text{near } 0, \tag{3.10}$$

and so (3.8) follows from (3.9), (3.10) using the Taylor expansion at $\xi = 0$ with respect to $\xi$ only. Now, all that remains is to verify (3.10). Observe that the solution path $(y(\alpha), \alpha)$ in Theorem 2.3 now can be written as $(y_0 + u(0, \eta), \alpha_0 + \eta)$. Let

$$\Phi(\eta) := (\varphi(\alpha_0 + \eta), v(\alpha_0 + \eta), 0),$$

where $\varphi(\alpha)$ is given in Theorem 2.3 and $v(\alpha)$ by (2.8). Then

$$F_y(y_0 + u(0, \eta), \alpha_0 + \eta)\Phi(\eta) = 0. \tag{3.11}$$

Recall (3.2), note $l\varphi(\alpha) - 1 = 0$ and split $\Phi(\eta)$ into

$$\Phi(\eta) = \Phi_0 + \theta(\eta), \quad \theta(\eta) \in Y_0. \tag{3.12}$$

Notice that, by (3.3),

$$G_\xi(u(0, \eta), 0, \eta) = (I - W\Psi_0)F_y(y_0 + u(0, \eta), \alpha_0 + \eta)(\Phi_0 + u_\xi(0, \eta)) = 0, \tag{3.13}$$

and since $(I - W\Psi_0)F_y(y_0 + u(0, \eta), \alpha_0 + \eta): Y_0 \to Y_1$ is an isomorphism for $\eta$ near 0 we conclude by (3.11), (3.12) and (3.13) that

$$u_\xi(0, \eta) = \theta(\eta).$$

Finally, (3.10) follows from

$$g_\xi(0, \eta) = \Psi_0 F_y(y_0 + u(0, \eta), \alpha_0 + \eta)(\Phi_0 + u_\xi(0, \eta))$$
$$= \Psi_0 F_y(y_0 + u(0, \eta), \alpha_0 + \eta)\Phi(\eta) = 0.$$

Now, we are ready to state the main results of the bifurcation analysis in the following two theorems.

**Theorem 3.3.**
Let $(x_0, \lambda_0, \alpha_0)$ be a quadratic $S$-point. Then the solution path $(y(\alpha), \alpha)$ mentioned in Theorem 2.3 is locally the only one in the whole space $(y, \alpha) \in Y \times \mathbf{R}$.

**Proof.**
This is an immediate consequence of the fact that for all quadratic $S$-points

$$p(0, 0) = -\frac{1}{3}b_z \neq 0, \tag{3.14}$$

where $p$ is given in Lemma 3.2. The proof of this equality is postponed till the next section.

**Theorem 3.4.**
Assume that $(x_0,\lambda_0,\alpha_0)$ is a high order S-point, i.e. $b_z=0$, and that the unfolding condition $b_{z\alpha}\neq 0$ holds.
a) Locally in $Y\times R$ there exist two solution paths of (1.3) passing though $(x_0,\lambda_0,\alpha_0)$. Except at $(x_0,\lambda_0,\alpha_0)$, the first path consists of quadratic S-points in $X_s\times X_a\times R$ as given by Theorem 2.3. The second path consists of fold points not in $X_s\times X_a\times R$ which has a parameterization of the form $(y(\xi),\alpha(\xi^2))$ with a tangent $(\Phi_0,0)$ at $(y(0),\alpha(0))=(y_0,\alpha_0)$. This path is conjugate in the sense that if $(y(\xi),\alpha(\xi^2))$ is a solution then so is $(y(-\xi),\alpha(\xi^2))$. In $(x,\lambda,\alpha)$ space the bifurcation diagram of (1.3) looks like that for an S-point.
b) If in addition it is known that $p(\xi^2,\eta)$ defined in Lemma 3.2 satisfies

$$b_{zz}:=p_{\xi\xi}(0,0)\neq 0 \tag{3.15}$$

then the bifurcation diagram of (1.3) in $(x,\lambda,\alpha)$ space looks like that for a *quadratic S*−point and the fold points on the asymmetric branches are *quadratic*.

**Proof.**
a) The first step in the proof is to define a new nonlinear equation

$$h(\xi,\eta)=0 \tag{3.16}$$

with

$$h(\xi,\eta)=\begin{cases}\xi^{-2}g(\xi,\eta)=\xi p(\xi^2,\eta), & \xi\neq 0,\\ 0, & \xi=0.\end{cases} \tag{3.17}$$

Clearly (3.4) is equivalent to (3.16) in the sense that if $(\xi,\eta)$ solves (3.4) then $(\xi,\eta)$ solves (3.16) and vice versa. Now $h(\xi,\eta)$ clearly satisfies (1.4) with $s\xi=-\xi$ and $h_\xi(0,0)=p(0,0)=0$, since $b_z=0$ at a high order S-point (cf. (3.14)). Our claim (to be confirmed in the next section) is that

$$p_\eta(0,0)=\frac{1}{6}g_{\xi\xi\xi\eta}(0,0)=-\frac{1}{3}b_{z\alpha}\neq 0. \tag{3.18}$$

Thus, since

$$b_\eta:=h_{\xi\eta}(0,0)=p_\eta(0,0)$$

(cf. (1.7)) we have that

$$b_{z\alpha}\neq 0 \Rightarrow b_\eta\neq 0 \Rightarrow (0,0)\text{ is an }S-\text{point of }(3.16).$$

Thus the existence of asymmetric ($\xi\neq 0$) solutions near $\eta=0$ of $h(\xi,\eta)=0$, and hence of $g(\xi,\eta)=0$, is ensured with these solutions having the parameterization $(\xi,\eta(\xi^2))$. The conjugate nature of this solution is clear since if $(\xi,\eta(\xi^2))$ is a solution so is $(-\xi,\eta(\xi^2))$. Returning to (1.3) we observe that its solutions can be written in the form

$$(y(\xi),\alpha(\xi^2))=(y_0+\xi\Phi_0+u(\xi,\eta(\xi^2)),\ \alpha_0+\eta(\xi^2)). \tag{3.19}$$

Again we see that if $(y(\xi),\alpha(\xi^2))$ is a solution, so is $(y(-\xi),\alpha(\xi^2))$.
It is thus immediately clear that $\dfrac{d\alpha}{d\xi}=0$ at $\xi=0$ and hence by

$$\frac{d}{d\xi}F(y(\xi),\alpha(\xi))|_{\xi=0}=F_y^0\dot{y}_0+F_\alpha^0\dot{\alpha}_0=0$$

we have $\dot{y}_0=\Phi_0$ with a suitable scaling, and so the tangent vector can be written as

$$(\dot{y}_0,\dot{\alpha}_0)=(\Phi_0,0).$$

Now, let us introduce $\psi(\xi)\in X'$ along the path $(y(\xi),\alpha(\xi))$ as the unique solution of the following system $(f_x^\xi:=f_x(x(\xi),\lambda(\xi),\alpha(\xi))$ etc.)

$$\psi(\xi)f_x^\xi=0, \tag{3.21a}$$

$$\psi(\xi)w-1=0, \tag{3.21b}$$

where $w$ is given by (2.5). Note that $\psi(0)=\psi_0$ and $\psi(0)f_\lambda^0=0$. We can show that $(x(\xi),\lambda(\xi),\alpha(\xi))$, $\xi\neq0$, are all fold points by verifying that

$$\frac{d}{d\xi}\left.(\psi(\xi)f_\lambda^\xi)\right|_{\xi=0}\neq0 \tag{3.22}$$

and hence

$$\psi(\xi)f_\lambda^\xi\neq0 \quad \forall\,\xi\neq0. \tag{3.23}$$

To prove (3.22), we notice that along the asymmetric solution path

$$\dot{x}_0=\varphi_0, \quad \dot{\varphi}_0=v_z, \quad \dot{\lambda}_0=0, \quad \dot{\alpha}_0=0, \tag{3.24}$$

$$\dot{\psi}_0 f_x^0+\psi_0 f_{xx}^0\varphi_0+\psi_0 f_{x\lambda}^0\dot{\lambda}_0+\psi_0 f_{x\alpha}^0\dot{\alpha}_0=\dot{\psi}_0 f_x^0+\psi_0 f_{xx}^0\varphi_0=0. \tag{3.25}$$

Hence,

$$\frac{d}{d\xi}(\psi(\xi)f_\lambda^\xi)\bigg|_{\xi=0}=\dot{\psi}_0 f_\lambda^0+\psi_0 f_{x\lambda}^0\varphi_0=-\dot{\psi}_0 f_x^0 v_\lambda+\psi_0 f_{x\lambda}^0\varphi_0$$

$$=\psi_0 f_{xx}^0\varphi_0 v_\lambda+\psi_0 f_{x\lambda}^0\varphi_0=b_\lambda\neq0.$$

This completes the proof of (a).

b) If in addition (3.15) holds then $p(\xi^2,\eta)=a\xi^2+\eta+h.o.t$ with $a\neq0$, and so $h(\xi,\eta)=a\xi^3+\eta\xi$ which gives the canonical form for quadratic $S$-point. The details for the fact of that the fold points are quadratic are omitted but this may be inferred from the results in [4] Chapter VI.

By Theorems 3.4(a) and 2.3, at a high order $S$-point we can interpret the nondegeneracy conditions $b_\lambda\neq0$ as a 'crossing condition' on $\psi(\xi)f_\lambda^\xi$ along the asymmetric solution path, and $b_{z\alpha}\neq0$ as a crossing condition on $b(\alpha)$ along the symmetric solution path.

In computations, we may use (1.3) to follow an $S$-point path by varying $\alpha$ and detect a high order $S$-point by a change of sign of $b(\alpha)$. A quartic $S$-point can be located precisely by an extended system presented in Cliffe and Spence [1]. Then we may jump to the fold point path by employing the tangent vector given in Theorem 3.4 and using any suitable continuation algorithm (see Deuflehard et al. [2], Doedel [3] and Section 5).

Finally in this section we show that in $(\lambda,\alpha)$ space the paths of bifurcation points are tangential at the high order $S$-point, which is observed in numerical results in the Taylor problem (see figure 2.1 in Cliffe and Spence [1]). For completeness we state the result as a theorem (cf. Corollary 8 in Spence et al. [12]).

### Theorem 3.5.
At a high order $S$-point $(x_0,\lambda_0,\alpha_0)$, the projections in $(\lambda,\alpha)$ space of the two solution paths given in Theorem 3.4 are tangential at $(\lambda_0,\alpha_0)$.

**Proof.**

For the $S$-point path $(y(\eta), \alpha(\eta))$ (recall Theorem 2.3) we have

$$\dot{\alpha}_0 = 1, \quad \dot{\lambda}_0 = (\dot{y}_0)_3 = V_3 = -\frac{b_\alpha}{b_\lambda},$$

where we write $\dot{\alpha}_0 := \frac{d\alpha}{d\eta}\big|_{\eta=0}$ etc. and

$$b_\alpha := \psi_0(f^0_{xx}v_\alpha + f^0_{x\alpha})\varphi_0,$$

with

$$f^0_x v_\alpha + f^0_\alpha = 0, \quad v_\alpha \in X_s.$$

Hence

$$\frac{d\lambda}{d\alpha}\bigg|_{\eta=0} = \frac{\dot{\lambda}_0}{\dot{\alpha}_0} = -\frac{b_\alpha}{b_\lambda}. \tag{3.26}$$

On the other hand, at $(x_0, \lambda_0, \alpha_0)$ the fold point path $(y(\xi), \alpha(\xi))$ has a tangent vector $(\Phi_0, 0) = (\varphi_0, v_z, 0, 0)$, and the third and fourth components imply that (with $\dot{\lambda}_0 := \frac{d\lambda}{d\xi}\big|_{\xi=0}$ etc.)

$$\dot{\lambda}_0 = \dot{\alpha}_0 = 0. \tag{3.27}$$

We have to go further by noting that

$$\frac{d^2}{d\xi^2} F(y(\xi), \alpha(\xi))\big|_{\xi=0} = F^0_{yy}\dot{y}_0^2 + 2F^0_{y\alpha}\dot{y}_0\dot{\alpha}_0 + F^0_{\alpha\alpha}\dot{\alpha}_0^2$$

$$+ F^0_y \ddot{y}_0 + F^0_\alpha \ddot{\alpha}_0 = F^0_{yy}\Phi_0^2 + F^0_\alpha \ddot{\alpha}_0 + F^0_y \ddot{y}_0 = 0$$

and hence

$$\ddot{\lambda}_0 = (\ddot{y}_0)_3 = -\frac{b_\alpha}{b_\lambda}\ddot{\alpha}_0. \tag{3.28}$$

Now (3.27) together with (3.28) implies

$$\frac{d\lambda}{d\alpha}\bigg|_{\xi=0} = -\frac{b_\alpha}{b_\lambda}. \tag{3.29}$$

Comparing (3.26) and (3.29) yields the desired conclusion.

## 4. Some details of the calculations.

In this section, we provide the details of calculations leading to (3.14) and (3.18). As we have seen, these two equalities are key points in the proofs.

Recall that by definition $u(\xi, \eta) \in Y_0$ for all $(\xi, \eta)$ near $(0,0)$. Note that by (3.7)

$$Su_\xi(\xi, \eta) = -u_\xi(-\xi, \eta),$$

$$Su_\eta(\xi, \eta) = u_\eta(-\xi, \eta)$$

etc. First, with the notation $u^0 := u(0,0)$ we can show

$$u^0, \ u^0_\eta, \ u^0_{\xi\xi}, \ u^0_{\xi\xi\eta} \in Y_s \cap Y_0 = Y_s, \tag{4.1}$$

$$u_\xi^0, \; u_{\xi\eta}^0, \; u_{\xi\xi\xi}^0 \in Y_a \cap Y_0. \tag{4.2}$$

The following lemma will make our calculations easier.

**Lemma 4.1.**
Let $A, B \subset Y$ satisfy

$$F_y^0 A + B = 0, \tag{4.3}$$

then

$$\Psi_0 F_{yy}^0 \Phi_0 A = -\psi_0 B_2, \tag{4.4}$$

where $B := (B_1, B_2, B_3) \in X \times X \times \mathbf{R} = Y$.

**Proof.**
Multiply the second equation of the linear system (4.3) by $\psi_0$ and compare it with (4.4).

**Proof of (3.14).**
Note that

$$6p(0,0) = g_{\xi\xi\xi}^0 = \Psi_0 F_{yyy}^0 (\Phi_0 + u_\xi^0)^3 + 3\Psi_0 F_{yy}^0 (\Phi_0 + u_\xi^0) u_{\xi\xi}^0, \tag{4.5}$$

though we need to specify $u_\xi^0$ and $u_{\xi\xi}^0$. For brevity let us write

$$H(\xi,\eta) := G(u(\xi,\eta),\xi,\eta) = 0. \tag{4.6}$$

If we differentiate $H(\xi,\eta)$ with respect to $\xi$ and evaluate it at $(\xi,\eta) = (0,0)$, then we have

$$H_\xi^0 = F_y^0 u_\xi^0 = 0. \tag{4.7}$$

This together with (4.2) implies that

$$u_\xi^0 = 0. \tag{4.8}$$

Also note that $u_{\xi\xi}^0$ satisfies

$$H_{\xi\xi}^0 = F_{yy}^0 \Phi_0^2 + F_y^0 u_{\xi\xi}^0 = 0. \tag{4.9}$$

Now, (3.14) follows from (4.5), (4.8), (4.9) and Lemma 4.1:

$$6p(0,0) = g_{\xi\xi\xi}^0 = \Psi_0 F_{yyy}^0 \Phi_0^3 + 3\Psi_0 F_{yy}^0 \Phi_0 u_{\xi\xi}^0 = \psi_0 f_{xxx}^0 \varphi_0^3 + 3\psi_0 (F_{yy}^0 \Phi_0 \Phi_0)_2$$

$$= \psi_0 f_{xxx}^0 \varphi_0^3 - 3\psi_0 (f_{xx}^0 \varphi_0^3 + 2f_{xx}^0 \varphi_0 v_z) = -2\psi_0 f_{xxx}^0 \varphi_0^3 - 6\psi_0 f_{xx}^0 \varphi_0 v_z = -2b_z,$$

where here and below we use the subscript notation $B = (B_1, B_2, B_3)$ for any $B \in Y$ as in Lemma 4.1.

**Proof of (3.18).**
Now we assume

$$b_z = 0. \tag{4.10}$$

First we note that

$$6p_\eta(0,0) = g_{\xi\xi\xi\eta}^0 = \Psi_0 \{ F_{yyyy}^0 \Phi_0^3 u_\eta^0 + F_{yyy\alpha}^0 \Phi_0^3 + 3F_{yyy}^0 \Phi_0 u_{\xi\xi}^0 u_\eta^0 + 3F_{yy\alpha}^0 \Phi_0 u_{\xi\xi}^0 \tag{4.11}$$

$$+ 3F_{yyy}^0 \Phi_0^2 u_{\xi\eta}^0 + 3F_{yy}^0 u_{\xi\eta}^0 u_{\xi\xi}^0 + 3F_{yy}^0 \Phi_0 u_{\xi\xi\eta}^0 + F_{yy}^0 u_\eta^0 u_{\xi\xi\xi}^0 + F_{y\alpha}^0 u_{\xi\xi\xi}^0 \}.$$

We need to specify $u_\eta^0$, $u_{\xi\xi}^0$ etc. From

$$H_\eta^0 = F_y^0 u_\eta^0 + F_\alpha^0 = 0$$

it follows that

$$u_\eta^0 = V_\alpha, \tag{4.12}$$

where $V_\alpha = (V_1, V_2, V_3)$ is defined in Theorem 2.3. Solving

$$H_{\xi\eta}^0 = F_{yy}^0 \Phi_0 u_\eta^0 + F_{y\alpha}^0 \Phi_0 + F_y^0 u_{\xi\eta}^0 = 0, \quad u_{\xi\eta}^0 \in Y_a \cap Y_0,$$

gives

$$u_{\xi\eta}^0 = (V_2, v_m, 0), \tag{4.13}$$

where $v_m$ is given by (2.11). Noting (4.9) and (4.10) we have that

$$u_{\xi\xi}^0 = (v_z, v_\varphi, 0), \tag{4.14}$$

where $v_\varphi$ is defined by (2.10). Similarly, from (4.13) and

$$H_{\xi\xi\xi}^0 = F_{yyy}^0 \Phi_0^3 + 3F_{yy}^0 \Phi_0 u_{\xi\xi}^0 + F_y^0 u_{\xi\xi\xi}^0 = 0, \quad u_{\xi\xi\xi}^0 \in Y_a \cap Y_0$$

we have

$$(u_{\xi\xi\xi}^0)_1 = v_\varphi. \tag{4.15}$$

Now, we are ready to calculate the right hand side of (4.11). By virtue of (4.12)-(4.15), direct calculations lead to the following equalities:

$$\Psi_0(F_{yyyy}^0 \Phi_0^3 u_\eta^0 + F_{yy\alpha}^0 \Phi_0^3) = \psi_0(f_{xxx}^0 V_1 + f_{xx\lambda}^0 V_3 + f_{xx\alpha}^0) \varphi_0^3, \tag{4.16}$$

$$\Psi_0(F_{yyy}^0 \Phi_0 u_{\xi\xi}^0 u_\eta^0 + F_{yy\alpha}^0 \Phi_0 u_{\xi\xi}^0) = \psi_0(f_{xx}^0 V_1 + f_{xx\lambda}^0 V_3 + f_{xx\alpha}^0) \varphi_0 v_z, \tag{4.17}$$

$$\Psi_0(F_{yyy}^0 \Phi_0^2 u_{\xi\eta}^0 + F_{yy}^0 u_{\xi\eta}^0 u_{\xi\xi}^0) = \psi_0(f_{xx}^0 \varphi_0^2 + f_{xx}^0 v_z) V_2, \tag{4.18}$$

$$\Psi_0(F_{yy}^0 u_\eta^0 u_{\xi\xi\xi}^0 + F_{y\alpha}^0 u_{\xi\xi\xi}^0) = \psi_0(f_x^0 V_1 + f_{x\lambda}^0 V_3 + f_{x\alpha}^0) v_\varphi. \tag{4.19}$$

To find $\Psi_0 F_{yy}^0 \Phi_0 u_{\xi\xi\eta}^0$ we note that

$$H_{\xi\xi\eta}^0 = F_{yyy}^0 \Phi_0^2 u_\eta^0 + F_{yy\alpha}^0 \Phi_0^2 + F_{yy}^0 u_{\xi\xi}^0 u_\eta^0 + F_{y\alpha}^0 u_{\xi\xi}^0 + 2F_{yy}^0 \Phi_0 u_{\xi\eta}^0 + F_y^0 u_{\xi\xi\eta}^0 = 0.$$

Hence, by Lemma 4.2,

$$\Psi_0 F_{yy}^0 \Phi_0 u_{\xi\xi\eta}^0 = -\psi_0(F_{yyy}^0 \Phi_0^2 u_\eta^0 + F_{yy\alpha}^0 \Phi_0^2 + F_{yy}^0 u_{\xi\xi}^0 u_\eta^0 + F_{y\alpha}^0 u_{\xi\xi}^0 + 2F_{yy}^0 \Phi_0 u_{\xi\eta}^0)_2 \tag{4.20}$$

$$= -\psi_0 \{(f_{xxx}^0 V_1 + f_{xx\lambda}^0 V_3 + f_{xx\alpha}^0) \varphi_0^3 + 3(f_{xx}^0 V_1 + f_{x\lambda}^0 V_3 + f_{x\alpha}^0) \varphi_0 v_z$$

$$+ (f_{xx}^0 V_1 + f_{x\lambda}^0 V_3 + f_{x\alpha}^0) v_\varphi + 3f_{xx}^0 \varphi_0^2 V_2 + 3f_{xx}^0 v_z V_2 + 2f_{xx}^0 \varphi_0 v_m \}.$$

Finally, substituting (4.16)-(4.20) into (4.11) gives

$$6p_\eta(0,0) = g_{\xi\xi\xi\eta}^0 = -2b_{z\alpha}.$$

This completes the proof.

## 5. Numerical Results

We end this paper with a brief section about the usefulness of the information given by Theorem 3.4 about the tangent direction of the path of fold points. Many continuation algorithms use tangent information and we illustrate this with a brief look at the pseudo-arc length approach (Keller [7], Doedel [3]) where the path of fold points is parametered by $s$. The

102

equations to be solved to jump from $(y_0, \alpha_0) \in Y_s \times \mathbf{R}$ onto the asymmetric fold path are

$$F(y(s), \alpha(s)) = 0,$$

$$(dy/ds)_{s=0}(y(s) - y_0) + (d\alpha/ds)_{s=0}(\alpha(s) - \alpha_0) - s = 0.$$

It is easily shown that $(dy/ds, d\alpha/ds)_{s=0}$ is tangent to the curve at $(y_0, \alpha_0)$ and hence from Theorem 3.4 we may take $(dy/ds, d\alpha/ds)_{s=0} = (\Phi_0, 0)$. The usual Euler-Newton technique may now be employed.

To illustrate the effectiveness of this approach we present some numerical results for a finite element discretization of the Navier-Stokes equations arising in the Finite Taylor problem (see Cliffe and Spence [1] for details and further references). Table 1 below shows the norms of the corrections in the Newton method with $s=0.1$, using an Euler predict step from the quartic point shown in [1, Figure 2.1]. The quadratic convergence is evident, showing the importance of a step in the tangent direction.

The calculations were done in grid with around 1500 degrees of freedom. The finite element program was run on a CRAY2 computer.

TABLE 1

| Iteration | 1 | 2 | 3 | 4 |
|---|---|---|---|---|
| $\|$ Newton Correction $\|_\infty$ | 1.51 | $3.31 \times 10^{-1}$ | $4.42 \times 10^{-3}$ | $2.43 \times 10^{-7}$ |

References

[1]   Cliffe, K.A. and Spence, A. (1984) The calculation of high order singularities in the finite Taylor problem. In *Numerical Methods for Bifurcation Problems* (eds. T. Küpper, H.D. Mittelmann, H.Weber), *ISNM 70, Birkhäuser, Basel*, 129-144.
[2]   Deuflehard, P., Fiedler, B. and Kunkel, P. (1987) Efficient numerical path following beyond critical points. *SIAM J. Numer. Analysis*, 24, 912-927.
[3]   Doedel, E.J. (1981) AUTO: a program for the automatic bifurcation analysis of autonomous systems, *Congressus Numerantium*, 30, 265-284.
[4]   Golubitsky, M. and Schaeffer, D.G. (1986) *Singularities and Groups in Bifurcation Theory, Vol.1. Appl. Math. Sci.* 51, *Springer, New York.*
[5]   Golubitsky, M., Stewart, I.N. and Schaeffer, D.G. (1988) *Singularities and Groups in Bifurcation Theory, Vol. 2. Appl. Math. Sci.* 69, *Springer, New York.*
[6]   Guckenheimer, J. and Holmes, P. (1983) *Nonlinear Oscillations, Dynamical Systems and Bifurcations of Vector Fields. Springer, New York.*
[7]   Keller, H.B. (1977) Numerical solution of bifurcation and nonlinear eigenvalue problems, in Rabinowtz, P.H. ed. *Applications of Bifurcation Theory, Academic Press, New York*, 359-384.
[8]   Moore, G. and Spence, A., (1980) The calculation of turning points of nonlinear equations. *SIAM J. Numer. Analysis*, 17, 567-576.
[9]   Rheinboldt, W.C. (1986) *Numerical Analysis of Parameterized Nonlinear Equations. Wiley-Interscience, New York.*
[10]  Seydel, R. (1988) *From Equilibrium to Chaos. Elsevier, New York.*
[11]  Seydel, R. (1979) Numerical computation of branch points in ordinary differential equations, *Numer. Math.* 32, 51-68.

[12] Spence, A., Aston, P.J. and Wu, W. (1989) Bifurcation and stability analysis in nonlinear equations using symmetry-breaking in extended systems, To appear in *Numerical Analysis, Dundee, Longman,* eds. Griffiths, D.F. and Watson, G.A.
[13] Werner, B. and Spence, A. (1984) The computation of symmetry-breaking bifurcation points. *SIAM J. Numer. Anal.* 21, 388-399.

# SYMBOLIC COMPUTATION AND BIFURCATION METHODS

E. FREIRE, E. GAMERO, and E. PONCE
*Department of Applied Mathematics (University of Sevilla)*
*Escuela Técnica Superior de Ingenieros Industriales*
*Avda. Reina Mercedes s/n*
*41012-SEVILLA, Spain*

ABSTRACT. From the numerical point of view, continuation methods constitute a widely used tool in the qualitative analysis of dynamical systems and, specifically, for the calculation of bifurcation points. However, analytical methods are frequently needed to characterize the bifurcation, being also a valuable guide for the numerical approach. As in the effective application of bifurcation methods the hand calculation of very long expressions is normally required, computer algebra implementations avoiding the intermediate expression swell problem are needed. Using the normal form approach and exploiting the possibilities of Lie transforms, several algorithms well suited to symbolic computation have been developed. In particular, algorithms implementing the center manifold reduction and the computation of Hopf bifurcation are reported. As an example, the above algorithms are applied to the Fitzhugh nerve equations, as studied by Golubitsky & Langford in a celebrated paper. A coefficient of its normal form never computed is explicitly shown, what implies an unfounded statement in the quoted work.

## 1. Introduction

For a given dynamical system, the appearance of a nonhyperbolic equilibrium point for certain values of the parameters can lead to a bifurcation phenomenon. Then, to determine direction, kind, stability and number of eventually emanating branches is a question of major interest. For that, analytical methods have been developed. Such bifurcation methods are a valuable guide in the use of numerical methods, giving us not only initial guesses but determining the type of branches to be searched for. In the effective application of bifurcation methods, the hand calculation (as opposed to numerical evaluation) of very long expressions is normally required. The situation is even more embarrasing if, e. g., we are dealing with systems in presence of symmetries; then, a higher degree of accuracy is frequently needed. These tasks are consequently tedious and very error-prone, thereby being good candidates for computer algebra implementations.

However, a direct translation of the standard procedures over a computer algebra system often leads to intermediate expression swells, in such a way that the computer facilities can be exhausted without success. To overcome these difficulties and to optimize the computational effort, specific algorithms well suited to symbolic computation must be derived.

*D. Roose et al. (eds.), Continuation and Bifurcations: Numerical Techniques and Applications, 105–122.*
© 1990 *Kluwer Academic Publishers.*

In most cases a two-step analysis is needed. Firstly, we have to do a dimensional reduction to get rid of the hyperbolic directions and to facilitate the subsequent analysis. Afterwards, several changes of variables can be useful in order to characterize in a easier way the present bifurcation.

In this paper, using the normal form theory and exploiting the possibilities of Lie transforms, a computer algebra unified approach for the two steps just mentioned is derived. In particular, the Hopf bifurcation case by means of a center manifold reduction is considered but work is in progress to take into account — with the same ideas— other classical bifurcations.

Using MACSYMA, Rand et al. ([12],[13],[14]) have introduced computer algebra in bifurcation methods. For the case of Hopf bifurcation, they use the normal form approach, but their procedure is not optimized because they do not take full advantage of transformation theory leading up to normal forms.

The paper is organized as follows. In section 2 the use of the normal form theory as applied to a dimensional reduction by means of the center manifold calculation is presented. The next section is devoted to the derivation of the normal form in the Hopf bifurcation case and, in section 4, the results of the corresponding REDUCE programs [8], when applied to the Fitzhugh nerve equations, are shown.

## 2. Bifurcations, normal forms and center manifold reduction

As it is well known, the presence of nonhyperbolic equilibrium points is a necessary condition for the appearance of bifurcation phenomena. So, consider the system

$$\dot{u} = Au + f(u,v)$$
$$\dot{v} = Bv + g(u,v) \qquad (2.1)$$

where $u \in \mathbb{R}^n$, $v \in \mathbb{R}^m$, and $A, B$ are constant matrices such that all the eigenvalues of $A$ have zero real parts while all the eigenvalues of $B$ have negative real parts. This is the most important case in the applications. The functions $f$ and $g$ are $C^r$ with $f(0,0) = 0$, $Df(0,0) = 0$, $g(0,0) = 0$, $Dg(0,0) = 0$. The origin is obviously a nonhyperbolic equilibrium. In a more general context, this situation may be obtained for a characteristic value of a certain parameter (the bifurcation parameter) which, for sake of simplicity, we will not explicitly write.

To study whether we are effectively dealing with a bifurcation, and to determine the subsequent equilibrium structure, a deeper analysis is needed. For instance, in the study of periodic oscillations of nonlinear systems, intimately related with the Hopf bifurcation, several authors ([5], [7], [9], [10], [15]) have derived formulae characterizing the bifurcation (direction, stability and amplitudes of bifurcating periodic orbits, ...).

To facilitate the analysis, it is more convenient to get rid of the $m$ hyperbolic directions, thereby concentrating ourselves on the nonhyperbolic variables. For that, although there are other possibilities as, for example, the application of Lyapunov-Schmidt theory (see e.g. [5]), the center manifold theory is a useful technique, as we will see in the following.

Returning to the system (2.1), there exists a local invariant manifold: $v = h(u)$ with $h(0) = 0$, $Dh(0) = 0$ and $h$ is $C^r$; it is the so-named center manifold [1]. The flow in this manifold is governed by the equation

$$\dot{u} = Au + f(u, h(u)) \tag{2.2}$$

which constitutes the so-named reduced ($n$-dimensional) system. It contains all the necessary information to determine the asymptotic behavior for the flow near the origin of the $(n + m)$-dimensional system (2.1). As the center manifold is invariant for the flow, the following equation must be satisfied

$$Dh(u)\{Au + f(u, h(u))\} - Bh(u) - g(u, h(u)) = 0 \tag{2.3}$$

In practice we consider polynomial approximations to $h$ and its computation proceeds as follows. Let $V(k, n, m)$ denote the linear space of all $m$-vector functions of the $n$-vector $u$ which are homogeneous polynomials in $u$ of degree $k$. Thus, we can suppose

$$h(u) = \sum_{k \geq 2} h_k(u), \quad h_k \in V(k, n, m) \tag{2.4}$$

and our objective will be to compute $h_k(u)$ up to $k_{\max}$, a certain degree, getting an approximation to (2.2).

If we define

$$\begin{aligned} L(h(u)) &= Dh(u)Au - Bh(u) \\ N(h(u)) &= g(u, h(u)) - Dh(u)f(u, h(u)) \end{aligned} \tag{2.5}$$

then (2.3) can be rewritten as $L(h(u)) = N(h(u))$. Note that $L$ is a linear operator and $L(V(k, n, m)) \subset V(k, n, m)$ for all $k$. So, it is required that

$$L_k(h_k(u)) = n_k(u) \tag{2.6}$$

where $L_k$ is $L$ restricted to $V(k, n, m)$ and $n_k(u) \in V(k, n, m)$ represents the $k$-degree terms of Taylor expansion of $N$ —for that it suffices to consider the approximation up to $k - 1$ degree of $h$—. The equation (2.6) constitutes a linear system to be solved in $V(k, n, m)$ whose dimension is $m \cdot \binom{k+n-1}{k}$.

In applications, (2.1) can be a large system (the value of $m + n$ is high); further, one can consider linear degeneracies of codimension greater than one (high value of $n$). In other cases, as in presence of symmetries, we deal with high-codimension nonlinear degeneracies, forcing a growth in the order of necessary accuracy (high value of $k_{max}$). In sum, the linear system (2.6) might be a very large system and so its computer algebra resolution would be effectively impossible unless a careful insight is provided.

The setting of (2.6) itself involves computational complexities. It should be noticed that a direct substitution of previous terms of $h$ in the Taylor expansion of $N$ to obtain $n_k$ produces not only $k$-degree terms but lower and higher ones which are not required and consequently the computational effort would not be optimized. Essentially, this technique is used in some previous works (e.g. [14]), but —as indicated in [4]— an approach well suited to symbolic computation can be derived, following an iterative scheme and making good use of the previous steps. So, it is possible to minimize the number of operations and the memory requirements.

In the study of behavior near a degenerate equilibrium of a dynamical system is of great interest to use certain coordinate changes by means of which it is possible to "simplify" its differential equation, so obtaining the so-called normal forms. These forms are simpler than the initial system to the effect that nonlinear terms which are not essential have been removed.

The coordinate transformations yielding normal forms can be used for center manifolds calculations (see Chow & Hale [2]). Let us make the following near-identity transformation in (2.1):

$$\begin{pmatrix} u \\ v \end{pmatrix} = \begin{pmatrix} \bar{u} \\ \bar{v} \end{pmatrix} + \begin{pmatrix} 0 \\ \bar{h}(\bar{u}) \end{pmatrix} \qquad (2.7)$$

where $\bar{u} \in \mathbb{R}^n, \bar{v} \in \mathbb{R}^m$ and $\bar{h}(0) = 0, D\bar{h}(0) = 0$. The new differential equations are:

$$\begin{aligned} \dot{\bar{u}} &= A\bar{u} + \bar{f}(\bar{u}, \bar{v}) \\ \dot{\bar{v}} &= B\bar{v} + \bar{g}(\bar{u}, \bar{v}) \end{aligned} \qquad (2.8)$$

where

$$\begin{aligned} \bar{f}(\bar{u}, \bar{v}) &= f(\bar{u}, \bar{v} + \bar{h}(\bar{u})) \\ \bar{g}(\bar{u}, \bar{v}) &= - \{D\bar{h}(\bar{u})A\bar{u} - B\bar{h}(\bar{u})\} + \\ &\quad + \{g(\bar{u}, \bar{v} + \bar{h}(\bar{u})) - D\bar{h}(\bar{u})f(\bar{u}, \bar{v} + \bar{h}(\bar{u}))\} \end{aligned} \qquad (2.9)$$

If we choose now $\bar{h}(\bar{u})$ in such a way that for (2.8) $\bar{v} = 0$ is an invariant hyperplane, we get $\bar{g}(\bar{u}, 0) = 0$ and therefore, we deduce that $\bar{h}(\bar{u})$ must satisfy the equation (2.3) corresponding to center manifolds; from now, we identify $\bar{h}$ and $h$. Furthermore the system

$$\dot{\bar{u}} = A\bar{u} + \bar{f}(\bar{u}, 0)$$

turns out to be the reduced system. So, the center manifold computation for (2.1) is equivalent to calculate the transformation (2.7) leading to (2.8) with the above invariance condition. From a geometrical point of view the role of this coordinate transformation is to flat the center manifold.

In Meyer & Schmidt [11] and Chow & Hale [2] an approach is presented to the transformation theory leading to normal forms using Lie transforms. They arrive to a recursive algorithm to obtain the transformed equations from original ones. We now give a review of the ideas behind their algorithm and how to use them in our problem.

Suppose the following formal expansions:

$$f(u,v) = \sum_{k \geq 2} f_k(u,v), \quad f_k \in V(k, n+m, n)$$

$$g(u,v) = \sum_{k \geq 2} g_k(u,v), \quad g_k \in V(k, n+m, m)$$

$$\bar{f}(\bar{u}, \bar{v}) = \sum_{k \geq 2} \bar{f}_k(\bar{u}, \bar{v}), \quad \bar{f}_k \in V(k, n+m, n)$$

$$\bar{g}(\bar{u}, \bar{v}) = \sum_{k \geq 2} \bar{g}_k(\bar{u}, \bar{v}), \quad \bar{g}_k \in V(k, n+m, m)$$

(2.10)

Taking into account (2.9) and (2.5) it must be concluded that

$$\bar{g}_k(\bar{u}, 0) = -L_k(h_k(\bar{u})) + n_k(\bar{u}), \quad k \geq 2 \tag{2.11}$$

Thus, in the above notation our objective is to obtain $h_k$, $\bar{f}_k$, so that (2.11) vanishes.

If $u = \epsilon X$, $v = \epsilon Y$, $\epsilon \in \mathbb{R}$ in (1.1), then

$$\dot{X} = AX + \sum_{k \geq 1} F_k(X, Y)\epsilon^k / k!$$

$$\dot{Y} = BY + \sum_{k \geq 1} G_k(X, Y)\epsilon^k / k!$$

(2.12)

where

$$F_k(X, Y) = k!\, f_{k+1}(X, Y)$$
$$G_k(X, Y) = k!\, g_{k+1}(X, Y), \quad k \geq 1 \tag{2.13}$$

and they are homogeneous polynomials in $(X, Y)$ of degree $k + 1$. Also define $F_0(X, Y) = AX$ and $G_0(X, Y) = BY$.

Now consider a transformation of variables:

$$\begin{pmatrix} X \\ Y \end{pmatrix} = \begin{pmatrix} \tilde{X} \\ \tilde{Y} \end{pmatrix} + \begin{pmatrix} 0 \\ H(\tilde{X},\epsilon) \end{pmatrix} = \begin{pmatrix} \tilde{X} \\ \tilde{Y} + \sum_{k\geq 1} H_k(\tilde{X})\epsilon^k/k! \end{pmatrix} \tag{2.14}$$

where the $H_k$ are homogeneous in $\tilde{X}$ of degre $k+1$. Then the differential equations for $(\tilde{X}, \tilde{Y})$ are

$$\dot{\tilde{X}} = A\tilde{X} + \sum_{k\geq 1} \tilde{F}_k(\tilde{X},\tilde{Y})\epsilon^k/k!$$

$$\dot{\tilde{Y}} = B\tilde{Y} + \sum_{k\geq 1} \tilde{G}_k(\tilde{X},\tilde{Y})\epsilon^k/k! \tag{2.15}$$

where the $\tilde{F}_k, \tilde{G}_k$ are homogeneous polynomials in $(\tilde{X}, \tilde{Y})$ of degree $k+1$. Consequently, the changes of variables $u = \epsilon X$, $v = \epsilon Y$; $\bar{u} = \epsilon\tilde{X}$, $\bar{v} = \epsilon\tilde{Y}$ and (2.14) yield the system (2.8) provided that

$$H_k(\tilde{X}) = k!\, h_{k+1}(\tilde{X}), \quad k \geq 1 \tag{2.16}$$

and so we obtain

$$\tilde{F}_k(\tilde{X},\tilde{Y}) = k!\, \bar{f}_{k+1}(\tilde{X},\tilde{Y})$$

$$\tilde{G}_k(\tilde{X},\tilde{Y}) = k!\, \bar{g}_{k+1}(\tilde{X},\tilde{Y}) \tag{2.17}$$

In fact, transforming (2.12) by the changes defined by (2.14) is equivalent to transforming (2.1) by the changes of the form (2.7). The reason justifying the above set of transformations is that the $\tilde{F}_k$, $\tilde{G}_k$ can be recursively computed from $F_i$, $G_i$, $H_i$, $i \leq k$; and so, the relations (2.13), (2.16) and (2.17) enable us to calculate recursively $\bar{f}_k$, $\bar{g}_k$.

We now introduce the following notation:

$$\begin{pmatrix} R(X,Y) \\ S(X,Y) \end{pmatrix} \times \begin{pmatrix} 0 \\ T(X) \end{pmatrix} = \begin{pmatrix} \dfrac{\partial R(X,Y)}{\partial Y}T(X) \\ \dfrac{\partial S(X,Y)}{\partial Y}T(X) - \dfrac{\partial T(X)}{\partial X}R(X,Y) \end{pmatrix} \tag{2.18}$$

Notice that this convention is related to the Lie bracket operator when applied to the two particular functions above.

If we define the sequence

$$\begin{pmatrix} F_l^i \\ G_l^i \end{pmatrix}, \quad l,i = 0,1,2,\ldots$$

by the recursive relations:

$$F_l^0 = F_l, \quad G_l^0 = G_l, \quad l = 0, 1, 2, \ldots$$

$$\begin{pmatrix} F_l^i \\ G_l^i \end{pmatrix} = \begin{pmatrix} F_{l+1}^{i-1} \\ G_{l+1}^{i-1} \end{pmatrix} + \sum_{j=0}^{l} \binom{l}{j} \begin{pmatrix} F_{l-j}^{i-1} \\ G_{l-j}^{i-1} \end{pmatrix} \times \begin{pmatrix} 0 \\ H_{j+1} \end{pmatrix} \quad \begin{array}{l} l = 0, 1, 2, \ldots \\ i = 1, 2, 3, \ldots \end{array} \quad (2.19)$$

then it can be proved ([2], [11]):

$$\begin{pmatrix} \tilde{F}_k \\ \hat{G}_k \end{pmatrix} = \begin{pmatrix} F_0^k \\ G_0^k \end{pmatrix}, \quad k = 1, 2, \ldots \qquad (2.20)$$

We remark that the computations (2.19) can be accomplished by considering the so-called Lie triangle:

$$Z_0^0$$

$$Z_1^0 \quad Z_0^1$$

$$Z_2^0 \quad Z_1^1 \quad Z_0^2 \qquad \qquad \text{where } Z_l^i = \begin{pmatrix} F_l^i \\ G_l^i \end{pmatrix},$$

$$Z_3^0 \quad Z_2^1 \quad Z_1^2 \quad Z_0^3$$

$$\vdots \quad \vdots \quad \vdots \quad \vdots \quad \ddots$$

and each element can be calculated by using the elements in the column one step to the left and up. From (2.20) the searched elements are $Z_0^k$, which are on the diagonal of the Lie triangle. Note that in each row the terms involved have always same degree.

Recall that our objective is to obtain $h_k$, $\bar{f}_k$, $k \geq 2$, and now, since (2.16), (2.17), this is equivalent to compute $H_k$, $\tilde{F}_k$, $k \geq 1$. From (2.17), the condition $\bar{g}(\bar{x}, 0) = 0$ becomes $\hat{G}_k(\tilde{X}, 0) = 0$, $k \geq 1$, and then we can write (see 2.11 and 2.16):

$$\hat{G}_k(\tilde{X}, 0) = G_0^k(\tilde{X}, 0) = k! \left\{ -L_{k+1}(\frac{H_k(\tilde{X})}{k!}) + n_{k+1}(\tilde{X}) \right\} = 0, \quad k \geq 1 \qquad (2.21)$$

We recognize in (2.21) the equation satisfying the $k$-approximation of the center manifold, which is obtained in a recursive way as the second component of element $Z_0^k$ on the diagonal of Lie triangle. Further, the first component of $Z_0^k$ is precisely $\tilde{F}_k$ —see (2.20)— which leads us to the reduced system.

We can rewrite (2.21) as

$$L_{k+1}(H_k(\tilde{X})) = N_{k+1}(\tilde{X}), \quad k \geq 1 \tag{2.22}$$

where $N_{k+1}(\cdot) = k!\, n_{k+1}(\cdot)$. A key observation in order to obtain $H_k$ is that we can organize the calculations separately computing $L_{k+1}$ on one hand and $N_{k+1}$ on the other hand. As $L_{k+1}$ is a linear operator, once a polynomial basis is choosen, its representation is almost a straightforward task (see [4]).

We now perform some adaptations which permits us to achieve $N_{k+1}$. For that we set

$$Z^i_{k-i} = W^i_{k-i} + \begin{pmatrix} 0 \\ -L_{k+1}(H_k(\tilde{X})) \end{pmatrix}, \quad k \geq 1, \quad 1 \leq i \leq k \tag{2.23}$$

and then it can be strictly proved that a recursive relation analogous to (2.19) holds for the $W$'s. In fact the last term in the summation leading to $Z^1_{k-1}$ (i.e. with $j = k - 1$) becomes

$$Z^0_0 \times \begin{pmatrix} 0 \\ H_k(\tilde{X}) \end{pmatrix} = \begin{pmatrix} 0 \\ -L_{k+1}(H_k(\tilde{X})) \end{pmatrix} \tag{2.24}$$

and then,

$$W^1_{k-1} = \begin{pmatrix} F_k \\ G_k \end{pmatrix} + \sum_{j=0}^{k-2} \binom{k-1}{j} \begin{pmatrix} F_{k-j-1} \\ G_{k-j-1} \end{pmatrix} \times \begin{pmatrix} 0 \\ H_{j+1} \end{pmatrix} \tag{2.25}$$

Furthermore, taking into account that

$$\begin{pmatrix} 0 \\ S(X) \end{pmatrix} \times \begin{pmatrix} 0 \\ T(X) \end{pmatrix} = \begin{pmatrix} 0 \\ 0 \end{pmatrix}$$

we obtain

$$Z^i_l \times \begin{pmatrix} 0 \\ H_{j+1} \end{pmatrix} = W^i_l \times \begin{pmatrix} 0 \\ H_{j+1} \end{pmatrix} \tag{2.26}$$

and therefore

$$W^i_{k-i} = W^{i-1}_{k-i+1} + \sum_{j=0}^{k-i} \binom{k-i}{j} W^{i-1}_{k-i-j} \times \begin{pmatrix} 0 \\ H_{j+1} \end{pmatrix}, \quad 2 \leq i \leq k \tag{2.27}$$

for all $k \geq 2$. With this notation, we construct a similar triangle without the first column:

$$W_0^1$$

$$W_1^1 \quad W_0^2$$

$$W_2^1 \quad W_1^2 \quad W_0^3$$

$$W_3^1 \quad W_2^2 \quad W_1^3 \quad W_0^4$$

$$\vdots \quad\quad \vdots \quad\quad \vdots \quad\quad \vdots \quad\quad \ddots$$

Note that (2.21) together with (2.23) implies that the first $n$ components of $W_0^k$ and $Z_0^k$ are the same, giving us $\tilde{F}_k$, and the last $m$ components now provide us $N_{k+1}$. This strategy along with the determination of $L_{k+1}$ for each $k$ allows us the setting of the linear system (2.22). For that, selecting an appropriate polynomial basis can be useful, working so with vectorial representations rather than polynomials.

As the main objective is the reduced system, it should be noticed that by computing the first $n$ components of the next row in the above triangle up to $W_0^{k+1}$, the $(k+1)$-approximation to the reduced equation is obtained. For more details about the implementation of this step see [4].

## 3. The Hopf bifurcation case

Once the reduced system is obtained, we only deal with the nonhyperbolic variables getting a valuable simplification of the problem. Suppose a Hopf bifurcation is present; we are interested in obtaining a good characterization of the bifurcation and for that several changes of variables can help us.

So, consider the system

$$\begin{aligned}
\dot{x} &= f^1(x, y, \mu) \\
\dot{y} &= f^2(x, y, \mu)
\end{aligned} \tag{3.1}$$

corresponding to (2.2) with an isolated equilibrium point at the origin whose jacobian matrix for this point has the canonical form

$$A(\mu) = \begin{bmatrix} \alpha(\mu) & -\omega(\mu) \\ \omega(\mu) & \alpha(\mu) \end{bmatrix}$$

where $\mu$ is the bifurcation parameter and $f^1, f^2$ are smooth. For $\mu = 0$ it is verified $\alpha(0) = 0$, $\omega(0) = \omega_0 > 0$ and $\alpha'(0) \neq 0$. The appearance of bifurcating periodic orbits for the system is named a Hopf bifurcation [10].

As it is outlined in the following, in view of the hypothesis $\alpha'(0) \neq 0$, to characterize this bifurcation (number and stability of bifurcating periodic solutions) it is enough to consider the system at $\mu = 0$ :

$$\dot{x} = -\omega_0 y + \sum_{k \geq 2} f_k^1(x, y)$$

$$\dot{y} = \omega_0 x + \sum_{k \geq 2} f_k^2(x, y) \tag{3.2}$$

where formal expansions are assumed for $f^1(x, y, 0)$, $f^2(x, y, 0)$ and $f_k^1, f_k^2 \in V(k)$, the linear space of all homogeneous polynomials in $x, y$ of degree $k$.

It is possible to transform (3.2) by means of succesive near-identity transformations into the normal form ([6], [7]):

$$\dot{x} = -\omega_0 y + \sum_{j \geq 1} \{a_j(x^2 + y^2)^j x - b_j(x^2 + y^2)^j y\}$$

$$\dot{y} = \omega_0 x + \sum_{j \geq 1} \{a_j(x^2 + y^2)^j y + b_j(x^2 + y^2)^j x\} \tag{3.3}$$

which is expressed in polar coordinates as

$$\dot{r} = r \sum_{j \geq 1} a_j r^{2j}$$

$$\dot{\theta} = \omega_0 + \sum_{j \geq 1} b_j r^{2j} \tag{3.4}$$

The parameterized system (3.1) can be brought to the following form

$$\dot{r} = r(\alpha'(0)\mu + \sum_{j \geq 1} a_j r^{2j})$$

$$\dot{\theta} = \omega_0 + \sum_{j \geq 1} b_j r^{2j} \tag{3.5}$$

For $|r| \ll 1$, the dominant term of the $\theta$-equation is $\omega_0$ and so the bifurcating behavior is determined by the $r$-equation. Thus, $a_j$ are essential for the characterization of Hopf bifurcation ([5], [7], [9]).

Here follows a brief description about the derivation of normal form (3.3). It is assumed that the normal form is already computed to the $k - 1$ step:

$$\begin{pmatrix} \dot{x} \\ \dot{y} \end{pmatrix} = A(0) \begin{pmatrix} x \\ y \end{pmatrix} + J_{2k-1}(x, y) + R_{2k}(x, y) + \cdots \tag{3.6}$$

where

$$J_{2k-1}(x, y) = \sum_{j=1}^{k-1} (x^2 + y^2)^j \begin{pmatrix} a_j & -b_j \\ b_j & a_j \end{pmatrix} \begin{pmatrix} x \\ y \end{pmatrix}$$

and $R_{2k} \in V(2k,2)$, the linear space of 2-vector homogeneous polynomials of $2k$-degree. If the near identity transformation

$$\begin{pmatrix} x \\ y \end{pmatrix} = \begin{pmatrix} \tilde{x} \\ \tilde{y} \end{pmatrix} + P_{2k}(\tilde{x}, \tilde{y}), \qquad \text{where } P_{2k} \in V(2k,2) \tag{3.7}$$

is performed the following is obtained:

$$\begin{pmatrix} \dot{\tilde{x}} \\ \dot{\tilde{y}} \end{pmatrix} = A(0) \begin{pmatrix} \tilde{x} \\ \tilde{y} \end{pmatrix} + J_{2k-1}(\tilde{x}, \tilde{y}) + \{R_{2k}(\tilde{x}, \tilde{y}) - L_{2k} P_{2k}(\tilde{x}, \tilde{y})\} + \cdots \tag{3.8}$$

where the linear operator $L_{2k} : V(2k,2) \to V(2k,2)$ defined by

$$L_{2k} P_{2k}(\tilde{x}, \tilde{y}) = DP_{2k}(\tilde{x}, \tilde{y}) A(0) \begin{pmatrix} \tilde{x} \\ \tilde{y} \end{pmatrix} - A(0) P_{2k}(\tilde{x}, \tilde{y})$$

has been introduced.

It turns out ([2], [6]) that $L_{2k}$ is injective and so it is possible to solve uniquely

$$L_{2k} P_{2k} = R_{2k} \tag{3.9}$$

and therefore to remove $2k$-degree terms.

Dropping the tildes, the normal form obtained is

$$\begin{pmatrix} \dot{x} \\ \dot{y} \end{pmatrix} = A(0) \begin{pmatrix} x \\ y \end{pmatrix} + J_{2k-1}(x, y) + R_{2k+1}(x, y) + \cdots \tag{3.10}$$

where $R_{2k+1} \in V(2k+1, 2)$. Now, to complete the $k$-step, a new near-identity transformation must be applied:

$$\begin{pmatrix} x \\ y \end{pmatrix} = \begin{pmatrix} \tilde{x} \\ \tilde{y} \end{pmatrix} + P_{2k+1}(\tilde{x}, \tilde{y}), \qquad P_{2k+1} \in V(2k+1, 2) \tag{3.11}$$

transforming (2.10) into

$$\begin{pmatrix} \dot{\tilde{x}} \\ \dot{\tilde{y}} \end{pmatrix} = A(0) \begin{pmatrix} \tilde{x} \\ \tilde{y} \end{pmatrix} + J_{2k-1}(\tilde{x}, \tilde{y}) + \{R_{2k+1}(\tilde{x}, \tilde{y}) - L_{2k+1} P_{2k+1}(\tilde{x}, \tilde{y})\} + \cdots \tag{3.12}$$

The linear operator $L_{2k+1}$ is no longer injective. Let $V^r(2k+1, 2)$ denote the range of $L_{2k+1}$, and $V^c(2k+1, 2)$ its kernel; it then holds that:

a) $V(2k+1, 2) = V^r(2k+1, 2) \oplus V^c(2k+1, 2)$

b) $V^c(2k+1, 2) = \mathrm{span}\{(x^2 + y^2)^k \begin{pmatrix} x \\ y \end{pmatrix}, (x^2 + y^2)^k \begin{pmatrix} -y \\ x \end{pmatrix}\}$

c) $L_{2k+1} V^r(2k+1, 2) = V^r(2k+1, 2)$

$$\tag{3.13}$$

According to (3.13.a), one can decompose $R_{2k+1} = R^r_{2k+1} + R^c_{2k+1}$ (the superscripts denote range and kernel components respectively); also, it is possible to choose uniquely $P_{2k+1} \in V^r(2k+1,2)$ such that

$$L_{2k+1}P_{2k+1} = R^r_{2k+1} \tag{3.14}$$

and so the $(2k+1)$-degree terms of (3.12) become $R^c_{2k+1}$. Now, dropping the tildes, the system (3.12) takes the form:

$$\begin{pmatrix} \dot{x} \\ \dot{y} \end{pmatrix} = A(0) \begin{pmatrix} x \\ y \end{pmatrix} + J_{2k+1}(x,y) + \cdots \tag{3.15}$$

and the step is completed.

It should be noticed that the obtained normal form is equivariant under arbitrary rotations, which is a consequence of the symmetry of the linear part $(-\omega_0 y, \omega_0 x)^t$ with respect to the rotation group. As it has been shown above, this normal form has a simple representation in polar coordinates (see 3.4).

Summarizing, to compute $a_k, b_k$, one might:

(1) Calculate $R_{2k}$ in (3.6) which represents the $2k$-degree terms produced by the previous transformation of the original system;

(2) Solve the $4k$-dimensional linear equation (3.9) so obtaining $P_{2k}$;

(3) Calculate $R_{2k+1}$ in (3.10) taking into account the previous transformations and the corresponding one to $P_{2k}$;

(4) Decompose $R_{2k+1}$ according to (3.13.a);

(5) Solve the $(4k+2)$-dimensional linear equation (3.14) to obtain $P_{2k+1}$.

A direct translation of this computational scheme using MACSYMA can be found in [13]. The authors of this work perform (2) and (5) in real coordinates; as it will be seen below, use of complex coordinates results in both halving of the dimension and a simpler structure of the matrix representation of linear operators $L_{2k}, L_{2k+1}$, making the projection involved in (4) also easier. In the quoted work the calculations of $R_{2k}, R_{2k+1}$ are performed by direct substitution of previous transformations; moreover, no profit is made from the corresponding computations in the previous steps. To summarize, it seems that proceeding in this way the computational effort is not being optimized.

As it has been mentioned, to increase the efficiency in the Hopf bifurcation computation, it is natural to introduce complex variables [9]. But what is more relevant is the possibility of using Lie transforms in the theory of normal forms ([2], [11]), which leads to a recursive way to obtain the transformed equations from the original ones. With these ideas in mind, an efficient procedure for the symbolic computation of Hopf bifurcation normal forms can be derived.

Thus, making $z = x + iy$, $\bar{z} = x - iy$ in (3.2), where the bars now denote conjugation, one obtains

$$
\begin{aligned}
\dot{z} &= \omega_0 i z + \sum_{k \geq 2} F_k(z, \bar{z}) \\
\dot{\bar{z}} &= -\omega_0 i \bar{z} + \sum_{k \geq 2} \overline{F_k(z, \bar{z})}
\end{aligned}
\tag{3.16}
$$

where

$$
F_k(z, \bar{z}) = f_k^1(\frac{z + \bar{z}}{2}, \frac{z - \bar{z}}{2i}) + i f_k^2(\frac{z + \bar{z}}{2}, \frac{z - \bar{z}}{2i}), \quad k \geq 2
$$

and it will be later assumed that $F_1(z, \bar{z}) = \omega_0 i z$. Note that $F_k \in \mathcal{V}(k)$, the linear space of complex homogeneous polynomials in $z, \bar{z}$ of degree $k$.

Now, consider the near-identity transformation

$$
z = w + \sum_{k \geq 2} u_k(w, \bar{w})
\tag{3.17}
$$

where $u_k \in \mathcal{V}(k)$. Conjugation of (3.17) provides the transformation to be considered for $\bar{z}$. This change of variables yields:

$$
\dot{w} = \omega_0 i w + \sum_{k \geq 1} W_k(w, \bar{w})/k!
\tag{3.18}
$$

where $W_k \in \mathcal{V}(k + 1)$. It is clear that it suffices to work only with the equations and transformations involving the variables $z, w$, since the conjugation operation produces the corresponding ones for $\bar{z}, \bar{w}$.

The key observation is that $W_k$ can be obtained by recursive expressions as follows: For $k = 0, 1, \ldots$ ; $l = 1, 2, \ldots, k$, let the following be defined

$$
W_k^0 = k! \, F_{k+1}
$$

$$
W_{k-l}^l = W_{k-l+1}^{l-1} + \sum_{j=0}^{k-l} \binom{k - l}{j} W_{k-l-j}^{l-1} \odot U_j
\tag{3.19}
$$

where $U_j \in \mathcal{V}(j + 2)$ are related to the transformation (3.17). Also, a $\odot$-operation for a pair $W(w, \bar{w})$ and $U(w, \bar{w})$ has been introduced such that:

$$
W \odot U = \frac{\partial W}{\partial w} U + \frac{\partial W}{\partial \bar{w}} \bar{U} - \frac{\partial U}{\partial w} W - \frac{\partial U}{\partial \bar{w}} \bar{W}
\tag{3.20}
$$

From this it is obvious that $W_{k-l}^l \in \mathcal{V}(k+1)$ for all $l$. With these definitions it can be proved that

$$W_k = W_0^k, \quad k = 1, 2, \ldots \tag{3.21}$$

It must be remarked that the computations in (3.19) can be accomplished by considering again a Lie triangle. The elements $W_0^k$ that are searched for appear on the diagonal of the Lie triangle:

$$W_0^0$$

$$W_1^0 \quad W_0^1$$

$$W_2^0 \quad W_1^1 \quad W_0^2$$

$$W_3^0 \quad W_2^1 \quad W_1^2 \quad W_0^3$$

$$\vdots \quad \vdots \quad \vdots \quad \vdots \quad \ddots$$

where each element can be calculated by using the entries in the column one step to the left and up.

In order to simplify $W_k$ —so putting (3.18) in normal form— an appropriate choice of $U_k$ must be made. It is easy to verify —see (3.8), (3.10)— that

$$W_k = W_0^k = \mathcal{R}_k - \mathcal{L}_k U_{k-1}, \quad k \geq 1 \tag{3.22}$$

where $\mathcal{L}_k U = -W_0^0 \odot U$, $U \in \mathcal{V}(k+1)$. $\mathcal{L}_k : \mathcal{V}(k+1) \to \mathcal{V}(k+1)$ is a linear operator and $\mathcal{R}_k \in \mathcal{V}(k+1)$, which depends on the $U_j$, $0 \leq j \leq k-2$ and so, if these are known, then $\mathcal{R}_k$ is known.

Since $\mathcal{L}_k$ is nonsingular for $k$ odd, it is possible to choose $U_{k-1} \in \mathcal{V}(k+1)$ such that $W_k$ vanishes in these cases (cf. 3.9).

If $k$ is even, the linear operator $\mathcal{L}_k$ is singular. Let $\mathcal{V}^r(k+1)$ denote its range and $\mathcal{V}^c(k+1)$ its kernel. It then holds (see 3.13) that:

a) $\mathcal{V}(k+1) = \mathcal{V}^r(k+1) \oplus \mathcal{V}^c(k+1)$

b) $\mathcal{V}^c(k+1) = \mathrm{span}\{(w\overline{w})^{k/2} w\}$ \hfill (3.23)

c) $\mathcal{L}_k(\mathcal{V}^r(k+1)) = \mathcal{V}^r(k+1)$

According to (3.23) it is now possible to choose $U_{k-1} \in \mathcal{V}^r(k+1)$ uniquely such that $W_k \in \mathcal{V}^c(k+1)$. In other words, if this is done likewise for $k = 2m$, then $W_k(w, \overline{w}) = d_m(w\overline{w})^m w$, $m \geq 1$, $d_m \in \mathbb{C}$. Therefore, it has been shown that (3.18) adopts the following normal form:

$$\dot{w} = \omega_0 i w + \sum_{m \geq 1} \frac{d_m}{(2m)!}(w\overline{w})^m w \tag{3.24}$$

Comparing (3.24) with the real normal form (3.3), it is concluded that

$$\frac{d_m}{(2m)!} = a_m + ib_m, \quad m \geq 1 \tag{3.25}$$

Using this computing process, several improvements are achieved. On the one hand the dimension of the equations to be solved is halved and the corresponding operators take a diagonal form. On the other, only the operations needed are performed, lowering the possibility of exhausting the computing facilities with unnecessary terms that must later be truncated (see [13]). For computational details and examples about the sketched algorithm see [3].

## 4. A case study: the Fitzhugh nerve equations

In [5], M. Golubitsky and W. F. Langford carried out a deep study of degenerate Hopf bifurcation using Lyapunov-Schmidt method and singularity theory and obtained a polynomial form of the equation characterizing the Hopf bifurcation. Furthermore, explicit formulae for the coefficients of that polynomial normal form were given (see [5, §5]). In section 6 of that paper, the Fitzhugh nerve equation is analysed, and the appearance of a codimension three degenerate Hopf bifurcation is assured.

With the above approach we have studied the Fitzhugh nerve equation and the results obtained disagree with the achieved ones in [5]. Now the results of our algorithms when applied to Fitzhugh equation are presented.

As we have shown, the Poincaré normal form for the Hopf bifurcation is expressed in polar coordinates as

$$\dot{r} = r(a_1 r^2 + a_2 r^4 + \cdots) \tag{4.1}$$

while in [5] the bifurcation equation is written as:

$$x(a_{10}x^2 + a_{20}x^4 + \cdots) = 0 \tag{4.2}$$

where the coefficients $a_{i0}$ are related to the corresponding ones in (4.1). In fact, as it is easily shown, the computations can be arranged in such a way that

$$a_1 = -a_{10}, \quad a_2 = -a_{20}, \quad \ldots \tag{4.3}$$

Explicit formulae for $a_{10}$, $a_{20}$ are presented in section 5 of [5] but they can lead, even for low dimensional systems, to very large calculations. In [5, §6], two simple biochemical models (two-dimensional differential equations) are analyzed. In the following, we restrict ourselves to the second example corresponding to the Fitzhugh nerve equations.

According to [5, page 411], these equations can be written as:

$$\frac{dX}{dt} = \lambda + Y + X - \frac{1}{3}X^3$$
$$\frac{dY}{dt} = -\rho(X + \beta Y) \tag{4.4}$$

where the parameters satisfy $\lambda \in \mathbb{R}$ and $0 < \beta, \rho < 1$. $\lambda$ is taken as the bifurcation parameter and it suffices to consider $\lambda \geq 0$.

Under these conditions, a unique steady state $(X_0, Y_0)$ exists for every $\lambda$, and it is given by

$$Y_0 = -X_0/\beta$$
$$\lambda = \frac{1}{3}X_0^3 + \frac{1-\beta}{\beta}X_0 \tag{4.5}$$

Now, by changing variables $X = X_0 + u_1, Y = Y_0 + u_2$, the following equation is obtained:

$$\frac{du}{dt} = Au + \begin{pmatrix} -1 \\ 0 \end{pmatrix} h(u) \tag{4.6}$$

where

$$A = \begin{pmatrix} 1 - X_0^2 & 1 \\ -\rho & -\rho\beta \end{pmatrix}$$
$$h(u) = X_0 u_1^2 + \frac{1}{3}u_1^3 \tag{4.7}$$

The Hopf condition $trA = 0$ along with $\lambda \geq 0$ lead to $X_0^c = \sqrt{1 - \rho\beta}$. Now, we make a new change of variables:

$$u_1 = x$$
$$u_2 = -\rho\beta x - \omega_0 y \tag{4.8}$$

where $\omega_0 = \{\rho(1 - \rho\beta^2)\}^{\frac{1}{2}}$ and $\pm\omega_0 i$ are the eigenvalues of $A$ at Hopf bifurcation. The new equations for the critical parameter value $X_0^c$ are

$$\begin{pmatrix} \dot{x} \\ \dot{y} \end{pmatrix} = \begin{pmatrix} 0 & -\omega_0 \\ \omega_0 & 0 \end{pmatrix} \begin{pmatrix} x \\ y \end{pmatrix} + f(x) \begin{pmatrix} 1 \\ r_0 \end{pmatrix} \tag{4.9}$$

where

$$f(x) = -X_0^c x^2 - x^3/3$$
$$r_0 = -\frac{\rho\beta}{\omega_0} \tag{4.10}$$

The algorithm reported above, when applied to the above equations, produces:

$$a_1 = -\frac{\rho}{8\omega^2}\{\rho\beta^2 - 2\beta + 1\}$$

$$a_2 = -\frac{\rho^2}{288\omega^4}\{(\rho\beta^2 - 2\beta + 1)\frac{20 - 23\rho\beta + 3\rho^2\beta^3}{\omega^2} + 20(1 - \rho\beta)\beta^2\} \quad (4.11)$$

Note that $a_1$ is in full accordance with the coefficient $a_{10}$ shown in [5, Eq. (6.25), page 412].

The condition $a_1 = 0$ leads to $\rho_c = (2\beta - 1)/\beta^2$ and Golubitsky & Langford argue that "preliminary calculations indicate that $a_{20}$ has a unique simple zero $\beta_0 \in (\frac{1}{2}, 1)$". From this, the occurrence of a codimension three bifurcation with bifurcation parameters $\lambda$, $\rho$ and $\beta$ is established; the resulting bifurcation diagrams would be those shown in figure 4.5 of [5]. The authors, now, emphasize the consequence for the Fitzhugh equations: "the model could exhibit three concentric limit cycles, two of which are stable giving hysteresis".

As they do not show the expression of $a_{20}$, we cannot make any comparison. From our expression for $a_2$, making $\rho = \rho_c$, it is easily obtained:

$$a_2(\beta) = -\frac{5\beta}{288(1 - \beta)} \quad (4.12)$$

and this expression has no zero for $\beta \in (0, 1)$!

Therefore, the resulting bifurcation diagrams are those shown in figure 4.3 p. 396 of [5], and so, we are, indeed, dealing with a codimension two bifurcation with bifurcation parameters $\lambda$ and $\rho$. What this implies is that the Fitzhugh nerve equations could exhibit two concentric limit cycles, one stable and another unstable.

The mistake pointed out shows the help that can be obtained with the above approach in generating bifurcation formulae.

## 5. Conclusions

In order to be an effective help in the qualitative and quantitative analysis of dynamical systems, good implementations of bifurcation methods on a computer are needed. The standard procedures, when translated into a computer algebra system, give rise to unoptimized algorithms; they are expensive and very time-consuming and they not always succeed because of eventual intermediate expression swells.

In this paper, we have presented an unified computer algebra approach which overcomes those limitations for the dimensional reduction and the Hopf bifurcation case. Work is in progress to obtain analogous algorithms for other types of bifurcations, looking for a general analytical tool in bifurcation analysis.

## Acknowledgements

E. Ponce wishes to thank Arun V. Holden and John Brindley from CNLS at Leeds (UK) for their hospitality and their help while preparing the manuscript, and to the Consejería de Educación de la Junta de Andalucía for partial economic support.

## REFERENCES

1. Carr J. (1981) Applications of Centre Manifold Theory, Springer, Nueva York.
2. Chow S. and Hale J. K. (1982) Methods of Bifurcation Theory, Springer, New York.
3. Freire E., Gamero E., and Ponce E. (1989), An Algorithm for Symbolic Computation of Hopf Bifurcation, in E. Kaltofen and S.M. Watt (eds.), Computers & Mathematics, Springer, New York, pp 109–118.
4. Freire E., Gamero E., Ponce E. and Franquelo L.G. (1989), An Algorithm for Symbolic Computation of Center Manifolds, in P. Gianni (ed.), Symbolic and Algebraic Computation, Lecture Notes in Computer Science, vol 358, Springer, New York, pp 218–230.
5. Golubitsky M. and Langford W.F. (1981) Classification and Unfoldings of Degenerate Hopf Bifurcations, J. Diff. Eqns. 41, 375–415.
6. Guckenheimer J. and Holmes P. (1986) Nonlinear Oscillations, Dynamical Systems, and Bifurcations of Vector Fields, Appl. Math. Sci. 42, Springer, New York.
7. Hassard B.D., Kazarinoff N.D. and Wang Y.-H. (1980) Theory and Applications of the Hopf Bifurcation, Cambridge University Press, Cambridge.
8. Hearn A.C. (ed.) (1985) *REDUCE 3.2*, The Rand Corporation, Santa Monica.
9. Hsü I.D. and Kazarinoff N.D. (1976) An Applicable Hopf Bifurcation Formula and Instability of Small Periodic Solutions of the Field-Noyes Model, J. Math. Anal. Appl. 55, 61–89.
10. Marsden J.E. and McCracken M. (1976) The Hopf Bifurcation and its Applications, Springer, New York.
11. Meyer K. R. and Schmidt D. S. (1977) Entrainment Domains, Funkcialaj Ekvacioj 20, 171–192.
12. Rand R. H. (1985) Derivation of the Hopf Bifurcation Formula using Lindstedt's Perturbation Method and MACSYMA, in R. Pavelle (ed.), Applications of Computer Algebra, Kluwer Academic Publishers, Boston.
13. Rand R. H. and Armbruster D. (1987) Perturbation Method, Bifurcation Theory, and Computer Algebra, Appl. Math. Sci. 65, Springer, New York.
14. Rand R. H. and Keith W. L. (1985) Normal Form and Center Manifold Calculations on MACSYMA, in R. Pavelle (ed.), Applications of Computer Algebra, Kluwer Academic Publishers, Boston.
15. Takens F. (1973) Unfoldings of certain Singularities of Vector Fields: Generalized Hopf Bifurcations, J. Diff. Eqns. 14, 476–493.

# BIFURCATION ANALYSIS:
# A COMBINED NUMERICAL AND ANALYTICAL
# APPROACH

**M. Kleczka**

Institut B für Mechanik, Universität Stuttgart
Pfaffenwaldring 9, D-7000 Stuttgart 80

**W. Kleczka, E. Kreuzer**

Meerestechnik II, Universität Hamburg-Harburg
Eißendorfer Str. 42, D-2100 Hamburg 90

**ABSTRACT.** This paper proposes a generally applicable approach for the bifurcation analysis of nonlinear dissipative systems, described by smooth, ordinary and autonomous differential equations. The analysis is done in two stages. The first stage is the brute force calculation of Lyapunov exponents. The second stage is the more sophisticated partially analytical investigation of selected points of interest. Therefore we use a parametrized Taylor series approximation of the Poincaré map, where the analytical derivations can be performed by using computer algebra. The approximation enables the determination of the type of bifurcation, the iterative calculation of bifurcation points and stability analysis at the critical value by application of center manifold theory. The proposed approach is demonstrated by application to the Duffing oscillator in the version of Ueda.

## 1. Introduction

The only way to come to a systematic analysis of nonlinear dynamical engineering systems is the development of computer supported tools. The first tools applicable not only to a specific system, but to a class of systems have been purely numerical, e. g. Kreuzer [5]. The following methods have proven to be generally applicable and useful:

- Cell mapping approach, e. g. Hsu [2].

- Calculation of Lyapunov exponents, e. g. Kleczka [3].

- Calculation of fixed points and periodic solutions, e. g. Wilmers [10].

These tools are not designed to provide an overall insight in the system under consideration, even if the experienced expert can draw conclusions from such kind of numerical results. With these tools the engineer is able to speed up the analysis, as

123

*D. Roose et al. (eds.), Continuation and Bifurcations: Numerical Techniques and Applications, 123–137.*
© 1990 *Kluwer Academic Publishers.*

the time consuming generation and critical examination of endless serpentine lines of trajectories can be omitted. On the background of the exponential development of computing power and forthcoming parallel computers these numerical tools promise to be very powerful in the near future, even for higher dimensional practical engineering problems.

The next step in the development of tools is the application of analytical concepts. By means of computer algebra system, such as MACSYMA [6], tedious paper and pencil calculations can be avoided, e. g. W. Kleczka [4]. In this domain the most appreciable approaches are:

- Center manifold formalism, Carr [1] , Armbruster [7].

- Local approximation of the Poincaré map, e. g. Troger [8].

These methods are relevant and applicable for a wide range of nonlinear dynamical systems. They allow a detailed investigation that affords an insight into the dynamical structure of the system. Only a systematic and widely automated approach makes a reliable (errorfree) application of these methods to more complicated engineering systems possible.

An appropriate strategy to solve nonlinear dynamical problems in engineering systems seems to be the use of generally applicable, proven tools. And the most promising approach seems to be the complementary application of numerical and analytical tools.

## 2. Local approximation of the Poincaré map

The discretization of time by using a Poincaré surface of section is a well–known and very useful method for simplification of the analysis of a continuous dynamical system. The dimension of state space is reduced by one; instead of a time continuous differential equation a 'simple' point mapping has to be treated:

$$\dot{x} = f(x) \quad \longrightarrow \quad \xi_{k+1} = g(\xi_k) . \tag{1}$$

Unfortunately, in general the corresponding point mapping cannot be derived in an analytical form; the Poincaré map can be determined only by numerical integration of the underlying continuous system.

Because of the lack of an analytical mapping, analytical tools working on point mappings cannot be applied. A global analytical Poincaré map not beeing available, we content ourselves with a local approximation, a power series expansion in the neighborhood of a periodic solution, resp. a fixed point, Fig 1.

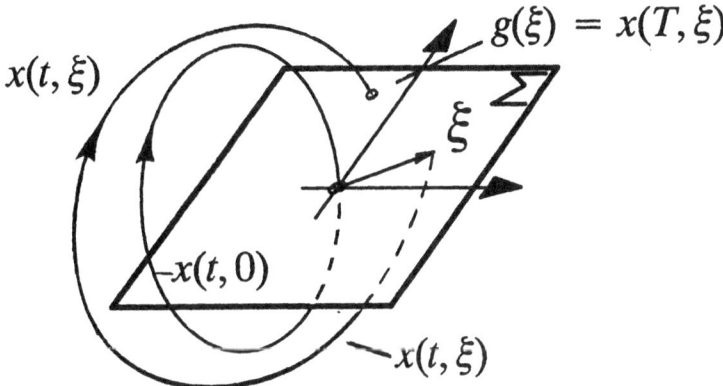

Figure 1: Poincaré map: periodic solution and its neighborhood.

Without loss of generality the coordinate system is chosen such that

$$x(t = 0, \boldsymbol{\xi} = 0) = x(T, 0) = 0 , \qquad (2)$$

respectively

$$g(\boldsymbol{\xi} = 0) = 0 . \qquad (3)$$

A local approximation of the Poincaré map, $g(\boldsymbol{\xi})$, is defined by the Taylor series expansion

$$g(\boldsymbol{\xi}) = g' \cdot \boldsymbol{\xi} + \frac{1}{2} g'' \{\boldsymbol{\xi}, \boldsymbol{\xi}\} + \frac{1}{6} g''' \{\boldsymbol{\xi}, \boldsymbol{\xi}, \boldsymbol{\xi}\} \cdots \qquad (4)$$

where $\{\boldsymbol{\xi}, \cdots, \boldsymbol{\xi}\}$ denotes the outer product. If $x(t, \boldsymbol{\xi})$ denotes the solution of

$$\dot{x} = f(x) \quad , \quad x(t = 0) = \boldsymbol{\xi} , \qquad (5)$$

the Poincaré map is defined by

$$g(\boldsymbol{\xi}) = x(T, \boldsymbol{\xi}) . \qquad (6)$$

The unknown derivative $g'$ can be expressed as

$$g' = g'(T,0) = \left.\frac{\partial x(T,\xi)}{\partial \xi}\right|_{\xi=0} = \left.\frac{\partial x(t,\xi)}{\partial \xi}\right|_{\substack{t=T \\ \xi=0}} \cdot \tag{7}$$

The connection to the differential equation

$$\dot{x} = f(x) \tag{8}$$

is made by differentiation with respect to $\xi$ :

$$\frac{d}{dt}\left(\frac{\partial x(t,\xi)}{\partial \xi}\right) = \frac{\partial f(x(t,\xi))}{\partial \xi} = \frac{\partial f(x)}{\partial x}\left(\frac{\partial x(t,\xi)}{\partial \xi}\right) \cdot \tag{9}$$

For a $n$–dimensional dynamical system this is a system of $n \times n$ first order differential equations. With the initial conditions

$$g'(t=0,\xi=0) = \frac{\partial x(t=0,\xi=0)}{\partial \xi} = \begin{bmatrix} 1 & & 0 \\ & \ddots & \\ 0 & & 1 \end{bmatrix}, \tag{10}$$

numerical integration up to $t=T$ leads to

$$g' = g'(t=T,\xi=0) = \frac{\partial x}{\partial \xi}(t=T,\xi) \cdot \tag{11}$$

Further derivations of system's differential equation (8) supply ordinary differential equations for $g'', g''', \cdots$

A more compact and even more simple formulation to derive the equations for the sought–after terms can be given in index notation. If we define

$$\widehat{\phantom{a}_i} = \frac{\partial}{\partial \xi_i} \quad , \quad _{,i} = \frac{\partial}{\partial x_i} \quad , \tag{12}$$

first order terms are given by

$$\dot{x}_{i,\widehat{\jmath}} = f_{i,\widehat{\jmath}} \cdot$$

Application of the chain rule yields

$$\dot{x}_{i,\hat{j}} = f_{i,a} x_{a,\hat{j}} \ . \tag{13}$$

For second order terms, Eqn. (13) has to be differentiated again with respect to $\xi_i$,

$$\dot{x}_{i,\widehat{jk}} = \left( f_{i,a} x_{a,\hat{j}} \right)_{\hat{k}} = f_{i,a\hat{k}} x_{a,\hat{j}} + f_{i,a} x_{a,\widehat{jk}} \ ,$$

$$\dot{x}_{i,\widehat{jk}} = f_{i,ab} x_{a,\hat{k}} x_{a,\hat{j}} + f_{i,a} x_{a,\widehat{jk}} \ . \tag{14}$$

The next steps can be performed equivalently. For third order terms, the result is

$$
\begin{aligned}
\dot{x}_{i,\widehat{jkl}} = \ & f_{i,abc} \, x_{c,\hat{l}} \, x_{b,\hat{k}} \, x_{a,\hat{j}} + \\
& f_{i,ab} \, x_{b,\widehat{kl}} \, x_{a,\hat{j}} + f_{i,ab} \, x_{b,\hat{k}} \, x_{a,\widehat{jl}} + \\
& f_{i,ab} \, x_{b,\hat{l}} \, x_{a,\widehat{jk}} + f_{i,a} \, x_{a,\widehat{jkl}} \ .
\end{aligned}
\tag{15}
$$

The only non–vanishing initial conditions are

$$x_{i,\hat{j}}(t = 0) = \delta_{ij} \ . \tag{16}$$

An important point is, that this systematic index formulation can be fully automated. A MACSYMA–routine could be written, that produces the whole set of differential equations (8) , (13) , (14) , (15) up to an approximation order of $p = 3$ for systems of arbitrary dimension $n$. Table 1 gives the correlation between $p, n$ and the number of coupled differential equations to be derived and solved.

Table 1: Number of differential equations

|       | n=2 | n=3 | n=4  |
|-------|-----|-----|------|
| p=2   | 14  | 39  | 84   |
| p=3   | 30  | 120 | 340  |
| p=4   | 62  | 363 | 1364 |

An approximation of the Poincaré map is not very useful, if the influence of parameters cannot be studied. Therefore, parameters have to be included in a symbolic way. This is possible by a simple extension of state space. The parameters $\lambda_i$ under variation (control parameters) can be interpreted as state variables,

$$\dot{x} = f(x, \lambda) ,$$
$$\dot{\lambda} = 0 .$$

(17)

The parameter dependence, for instance bifurcations with respect to parameter fluctuations, can be studied by means of a local Taylor series approximation of the Poincaré map of the extended system (17).

## 3. Application to the Duffing oscillator

We are now going to apply the outlined strategy to the well–known Duffing oscillator in the version of Ueda [9],

$$\ddot{x} + \delta \dot{x} + x^3 = a \cdot \cos(t) .$$

(18)

The parameters under variation are the parameter of the linear damping $\delta$ and the amplitude of excitation $a$.

The first step is the overview analysis by means of calculation of Lyapunov exponents $\sigma_i$ for

$$5 \leq a \leq 15 ,$$
$$0.02 \leq \delta \leq 0.2 .$$

(19)

Fig. 2 reveals a broad spectrum of dynamics: bifurcations ($\sigma_1 = 0$), regular ($\sigma_1 < 0$) and chaotic motion ($\sigma_1 > 0$), all kinds of coexistence phenomena.

The famous Ueda chart Fig. 3 is in nearly perfect agreement with the numerically generated chart, Fig. 2.

Arbitrarily we have chosen to analyze the system in more detail along $a = 7.5$, varying $\delta$. Figure 4 gives the spectrum of Lyapunov exponents and again periodic and chaotic behavior, which coexist in a certain range of $\delta$, is found.

In the magnification of the periodic range, Fig. 5 , one Lyapunov exponent approaching the value of zero indicates the occurrence of at least two bifurcations.

A detailed investigation of the bifurcations has to answer the following questions:

- What type of bifurcation occurs?

- What are the exact values for the bifurcation points?

- What is the stability at the critical point?

To come to solutions, we use a more analytical approach that is based on a local approximation of the Poincaré map. To include the control parameter $\delta$, the extended system

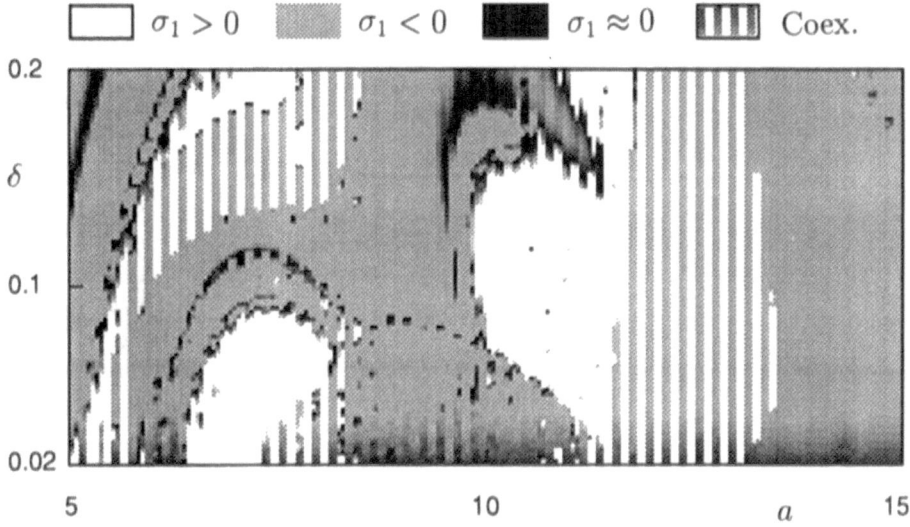

Figure 2: Numerically generated stability chart for the Duffing oscillator.

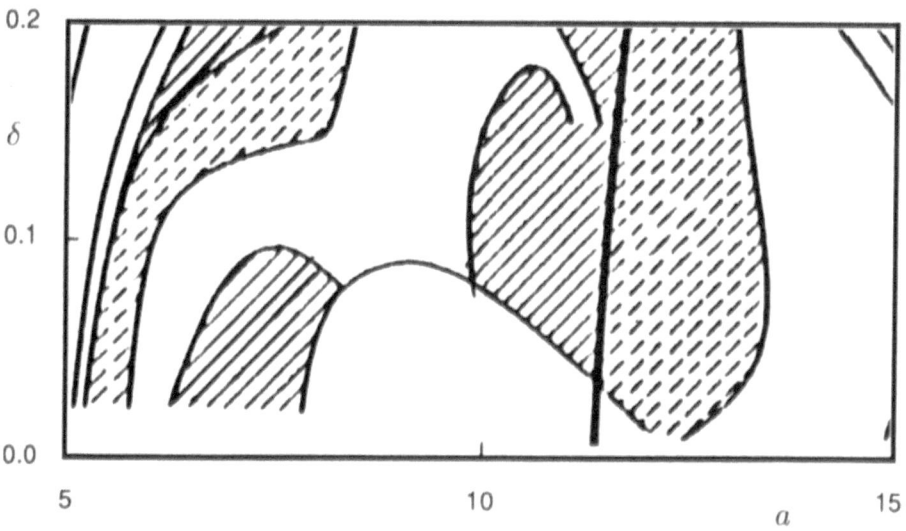

Figure 3: Stability chart of Ueda generated mainly by analog simulations.

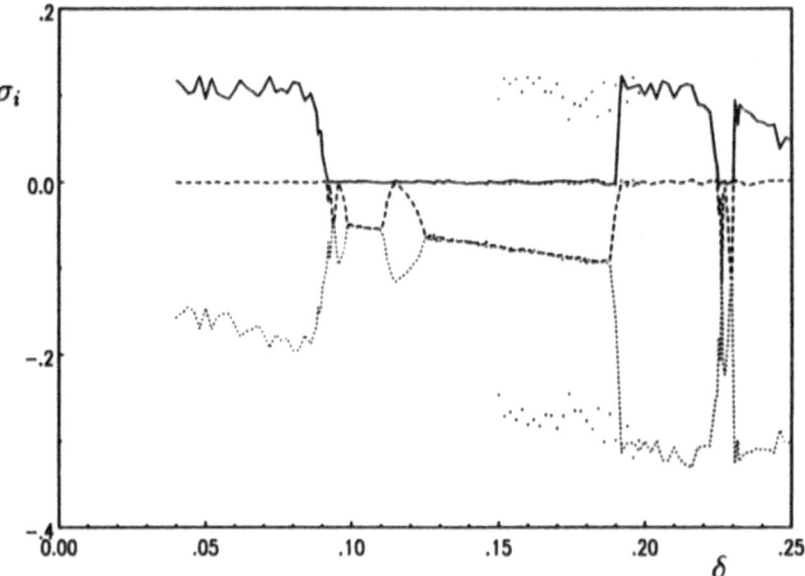

Figure 4: Spectrum of Lyapunov exponents for $a = 7.5$.

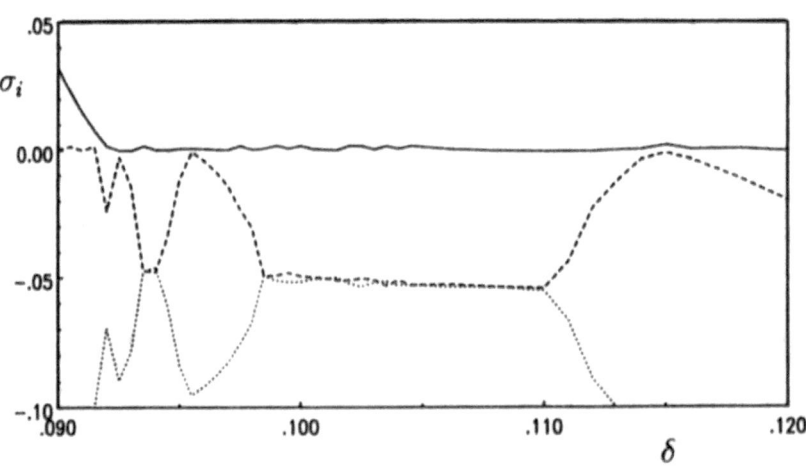

Figure 5: Lyapunov exponents indicating bifurcations.

$$\begin{bmatrix} \dot{x}_1 \\ \dot{x}_2 \\ \dot{\delta} \end{bmatrix} = \begin{bmatrix} x_2 \\ -x_1^3 + \delta x_2 + a\cos(t) \\ 0 \end{bmatrix} \tag{20}$$

will be considered. A MACSYMA–procedure generates the set of 120 coupled ordinary differential equations that provides by numerical integration the coefficients for the Taylor series expansion of the Poincaré map up to terms of order $p = 3$. These can be condensed, as terms appear multiply in the outer products. This condensation and ordering is also performed by a MACSYMA–procedure and finally $3 \times 19$ coefficients remain. We define the vector of relative coordinates with respect to the fixed point $[x_{10}, x_{20}, \delta_0]^T$,

$$\xi = \begin{bmatrix} x \\ y \\ \varepsilon \end{bmatrix} = \begin{bmatrix} x_1 - x_{10} \\ x_2 - x_{20} \\ \delta - \delta_0 \end{bmatrix}. \tag{21}$$

The expansion terms up to order three compose the $(19 \times 1)$ vector

$$\zeta = \left[ x, y, \varepsilon \left| x^2, xy, x\varepsilon, y^2, y\varepsilon, \varepsilon^2 \right. \right.$$

$$\left. x^3, x^2 y, x^2 \varepsilon, xy^2, xy\varepsilon, x\varepsilon^2, y^3, y^2\varepsilon, y\varepsilon^2, \varepsilon^3 \right]^T, \tag{22}$$

while the coefficients are gathered in three $(19 \times 1)$ vectors,

$$\begin{aligned} a^T &= [\ a_1 \quad a_2 \quad a_3 \quad a_4 \quad \cdots \quad a_{18} \quad a_{19}\ ], \\ b^T &= [\ b_1 \quad b_2 \quad b_3 \quad b_4 \quad \cdots \quad b_{18} \quad b_{19}\ ], \\ c^T &= [\ 0 \quad 0 \quad 1 \quad 0 \quad \cdots \quad 0 \quad 0\ ]. \end{aligned} \tag{23}$$

Now the local approximation of the Poincaré map is given by

$$\xi_{k+1} = \begin{bmatrix} a^T \\ b^T \\ c^T \end{bmatrix} \zeta_k. \tag{24}$$

It is appropriate to reduce Eqn. (24) by the third component that only provides

$$\varepsilon_{k+1} = \varepsilon_k,$$

and to interpret $\varepsilon$ no longer as a state variable, but as a parameter. For the analysis of the non–critical case a linear approximation is sufficient, so from Eqn. (24) we derive a linear parametrized approximation of the Poincaré map,

$$\underbrace{\begin{bmatrix} x_{k+1} \\ y_{k+1} \end{bmatrix}}_{\boldsymbol{\eta}_{k+1}} = \underbrace{\begin{bmatrix} a_1 + a_6\varepsilon + a_{15}\varepsilon^2 & a_2 + a_8\varepsilon + a_{18}\varepsilon^2 \\ b_1 + b_6\varepsilon + b_{15}\varepsilon^2 & b_2 + b_8\varepsilon + b_{18}\varepsilon^2 \end{bmatrix}}_{K(\varepsilon)} \underbrace{\begin{bmatrix} x_k \\ y_k \end{bmatrix}}_{\boldsymbol{\eta}_k} + \underbrace{\begin{bmatrix} a_3 \\ b_3 \end{bmatrix}}_{\boldsymbol{\kappa}} \varepsilon \ .$$

$$(25)$$

Now we come to the point: Eqn. (25) enables the analytical stability analysis with respect to the parameter $\varepsilon$.

We define the following subsidiary terms:

$$
\begin{aligned}
M &= a_1 + b_2 \ , \\
N &= a_6 + b_8 \ , \\
O &= a_{15} + b_{18} \ , \\
P &= a_1 b_2 - a_2 b_1 \ , \\
Q &= a_1 b_8 + a_6 b_2 - a_2 b_6 - a_8 b_1 \ , \\
R &= a_1 b_{18} + a_6 b_8 + a_{15} b_2 - a_2 b_{15} - a_8 b_6 - a_{18} b_1 \ , \\
S &= a_6 b_{18} + a_{15} b_8 - a_8 b_{15} - a_{18} b_6 \ , \\
T &= a_{15} b_{18} - a_{18} b_{15} \ .
\end{aligned}
$$

The characteristic equation of $K(\varepsilon)$ is given by

$$\lambda^2 - (M + N\varepsilon + O\varepsilon^2)\lambda + (P + Q\varepsilon + R\varepsilon^2 + S\varepsilon^3 + T\varepsilon^4) = 0. \qquad (26)$$

If the system is sufficiently close to the bifurcation point, i. e. the absolute value of at least one eigenvalue $\lambda$ of $K(\varepsilon = 0)$ is close to one, then the type of the expected bifurcation is known:

a) $\lambda \approx -1 \quad \Rightarrow \quad$ Flip ,

b) $\lambda \approx +1 \quad \Rightarrow \quad$ Pitchfork ,

c) $\lambda = a \pm ib, b \neq 0, |a \pm ib| \approx 1 \quad \Rightarrow \quad$ Hopf .

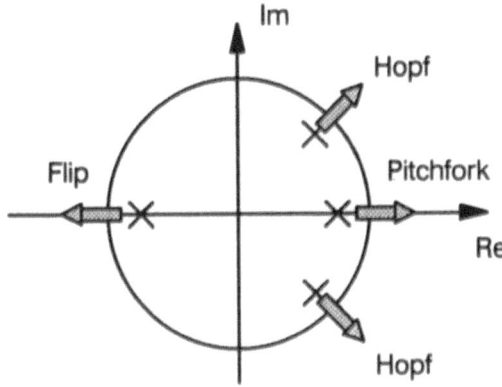

Figure 6: Possible situations close to bifurcations of codimension 1.

Then the bifurcation point can be estimated by setting $\lambda$ in Eqn. (26) to the expected critical value, neglecting third and higher order terms of $\varepsilon$. For $\lambda^* = \pm 1$ the approximate bifurcation parameter $\varepsilon^*$ is given by

$$(R \mp O)\varepsilon^{*2} + (Q \mp N)\varepsilon^* + (1 + P \mp M) = 0 . \tag{27}$$

The approximate value for the bifurcation parameter is given by the one solution of Eqn. (27), which is closer to the linear approximation

$$\varepsilon^*_{lin} = -\frac{1 + P \mp M}{Q \mp N} . \tag{28}$$

Furthermore, from Eqn. (25) an estimation for the fixed point $\bar{\eta}$ in dependence of $\varepsilon$ is gained:

$$\bar{\eta} = K \, \bar{\eta} + \kappa \, \varepsilon ,$$

$$\bar{\eta} = -\varepsilon [K - I]^{-1} \kappa . \tag{29}$$

Now the following procedure seems to be appropriate:

1. Select the parameter in the vicinity of a bifurcation point.

2. Calculate the fixed point of the Poincaré map.

3. Derive a parametrized approximation of the Poincaré map.

4. Calculate the eigenvalues of the linearized map.

5. Determine the type of bifurcation to be expected.

6. Calculate an estimation of the bifurcation point in parameter space and state space.

7. Correct the estimated bifurcation point in state space.

8. Continue iteratively, until the bifurcation point is reached within a certain tolerance.

The sequence of iterations for the bifurcation at $\delta \approx 0.96$ is given in Table 2. It is remarkable, that the iteration for the starting point at $\delta = 0.1$ gives a very good approximation even if the eigenvalues are conjugate complex[1].

Table 2: Sequence of iterations for the flip bifurcation

| Iteration | $\delta$ | $\varepsilon^*$ | $\lambda$ |
|-----------|----------|-----------------|-----------|
| 0 | 0.100000 | -0.0032 | $-0.29 \pm 0.26i$ |
| 1 | 0.096800 | -0.00082 | -0.790 |
| 2 | 0.095980 | -0.00028 | -0.928 |
| 3 | 0.095700 | -0.00011 | -0.974 |
| 4 | 0.095590 | -0.000037 | -0.990 |
| 5 | 0.095553 | -0.000015 | -0.996 |
| 6 | 0.095538 | -0.000006 | -0.998 |
| 7 | **0.095532** | | -0.999 |

The question of stability at the critical point can be answered by the center manifold theory. Therefore, the approximated point mapping has to be transformed into normal form, where the linear parts are decoupled. Then the center manifold formalism can be applied. Both tasks are performed by MACSYMA–routines [7]. For the flip bifurcation at $\delta \approx 0.96$ the center manifold in normal coordinates $(\nu, \mu)$ comes out to be

$$\nu(\mu) = 16.04\mu^2 - 261.04\mu^3 + 4412\mu^4 - 65707\mu^5 + O[\mu^6]. \tag{30}$$

---

[1]The expected type of bifurcation is a flip bifurcation, as the real part of the eigenvalue is negative and a Hopf bifurcation from a periodic solution is not possible for this type of system.

The dynamics on the center manifold is described by

$$\mu_{k+1} = -1.0\mu_k - 27.6\mu_k^2 + 26.65\mu_k^3 + 1243\mu_k^4 - 51991\mu_k^5 + O[\mu_k^6] \, . \tag{31}$$

To analyze the stability of the bifurcation point for the period doubling flip bifurcation the double mapping

$$\mu_{k+2} = 1.0\mu_k - 1577\mu_k^3 + O[\mu_k^4] \, . \tag{32}$$

has to be taken into account. As for flip bifurcation the quadratic term vanishes, the stability depends on the sign of the cubic term. For the above case, the negative sign indicates a stable, supercritical flip bifurcation, Fig. 7

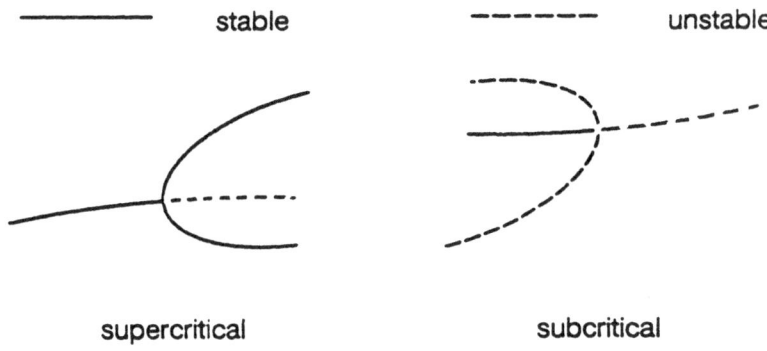

Figure 7: Stable, supercritical and unstable, subcritical bifurcations.

The analysis of the second bifurcation at $\delta \approx 0.115$ gives a similar picture. As the real part of the eigenvalue for the starting point $\delta = 0.13$ is positiv, a pitchfork bifurcation is supposed to occur. Indeed the iteration converges very fast to the point of pitchfork bifurcation, Table 3.

The center manifold analysis at the critical point yields

$$\nu(\mu) = 1.235\mu^2 + 5.265\mu^3 + 16.96\mu^4 + 79.75\mu^5 + O[\mu^6]. \tag{33}$$

The dynamics on the center manifold is described by

$$\mu_{k+1} = 1.0\mu_k - 10.5\mu_k^3 - 9.4\mu_k^4 - 83\mu_k^5 + O[\mu_k^6]. \tag{34}$$

Table 3: Sequence of iterations for the pitchfork bifurcation

| Iteration | $\delta$ | $\varepsilon^*$ | $\lambda$ |
|---|---|---|---|
| 0 | 0.13000 | -0.008 | $0.19 \pm 0.22i$ |
| 1 | 0.12200 | -0.004 | 0.552 |
| 2 | 0.11800 | -0.002 | 0.799 |
| 3 | 0.11600 | -0.0008 | 0.919 |
| 4 | 0.11520 | -0.0003 | 0.967 |
| 5 | 0.11490 | -0.00014 | 0.985 |
| 6 | 0.11476 | -0.00006 | 0.994 |
| 7 | 0.11470 | -0.00003 | 0.997 |
| 8 | **0.11467** | | 0.999 |

The quadratic portion vanishes at the critical point. The stability is governed by the cubic term where the negative sign guarantees stable, supercritical behavior.

## 4. Conclusions

For the given example a complementary numerical and analytical approach has proven to be successful. In the first stage the purely numerical analysis by means of calculation of Lyapunov exponents provides for an overview on the dynamics of the system in a broad range of parameters. Even if brute force this method is very valuable, as it requires no a priori information on system's dynamics, and it is very efficient, as it is perfectly suited for vector and parallel processors. The detailed investigation of specific points of interest relies on a local closed–form approximation of the Poincaré map. For the analytical part of the generation of the approximation, a computer algebra package has been applied successfully. It has been outlined how parameters can be included in a symbolic way. The bifurcation analysis has been performed on the parametrized approximation of the Poincaré map. The most important results have been the iterative calculation of bifurcation points and the stability analysis of the critical point by means of center manifold theory.

# References

[1] Carr, J.:
'*Applications of Centre Manifold Theory*'. New York/...: Springer-Verlag, 1981.

[2] Hsu, C. S.:
'*A Theory of Cell–to–Cell Mapping Dynamical Systems*'. J. of Appl. Mech. 47, 1980, 931–939.

[3] Kleczka, M.:
'*Zur Berechnung der Ljapunov-Exponenten und deren Bedeutung*'. Stuttgart: Universität Stuttgart, Institut B für Mechanik, Studienarbeit STUD-16, 1985.

[4] Kleczka, W.:
'*Einsatz von Computer–Algebra zur Analyse nichtlinearer dynamischer Probleme*'. Stuttgart: Universität Stuttgart, Institut B für Mechanik, Studienarbeit STUD-50, 1989.

[5] Kreuzer, E.:
'*Numerische Untersuchung nichtlinearer dynamischer Systeme*'. Berlin/...: Springer-Verlag, 1987.

[6] MACSYMA-Group of SYMBOLICS, Inc.:
'*MACSYMA Reference Manual, Version 11*'. 11 Cambridge Center, Cambridge, MA 02142 (1986).

[7] Rand, R. H.; Armbruster, D.:
'*Perturbation Methods, Bifurcation Theory and Computer Algebra*'. New York/...: Springer-Verlag, 1987.

[8] Troger, H.; Lindtner, E.; Steindl, A.:
'*Generic One-parameter Bifurcations in the Motion of a Simple Robot*'. Wien: Technische Universität Wien, 1988.

[9] Ueda, Y.:
'*Steady motions exhibited by Duffing's equation: A picture book of regular and stochastic motions*'. In: 'New Approaches to Nonlinear Problems in Dynamics' (Editor: P. J. Holmes), SIAM, Philadelphia, 1980, 311-322.

[10] Wilmers, C.:
'*Verzweigungsphänomene in mechanischen Oszillatoren*'. Stuttgart: Universität Stuttgart, Institut B für Mechanik, Diplomarbeit DIPL-24, 1988.

# INTRODUCTION TO THE NUMERICAL SOLUTION OF SYMMETRY-BREAKING BIFURCATION PROBLEMS.

P. J. ASTON
*Department of Mathematics*
*University of Surrey,*
*Guildford,*
*Surrey, GU2 5XH.*
*UK.*

ABSTRACT. This paper gives an introduction to systematic methods for obtaining reliable numerical solutions of symmetry-breaking bifurcation problems. The symmetry properties of a nonlinear equation are described by an equivariance property on which all the theory is based. Fixed point spaces are described since they are invariant under the nonlinear operator. The isotypic decomposition is carefully developed as the isotypic components are invariant under the linearisation of the nonlinear operator and this is employed in the efficient detection and computation of bifurcation points. The theory is illustrated throughout with an example and numerical results are presented.

## 1. Introduction

Our aim is to give a brief introduction to the rapidly expanding field of symmetry-breaking bifurcation theory, concentrating on numerical solution techniques. This topic uses the elegant and well established theory of groups and group representations as a framework for understanding and analysing the branching behaviour of the solutions of non-linear, parameter dependent equations which occur throughout applied mathematics. We try to keep the theory to a minimum and refer the reader to Ledermann (1973) and Miller (1972) for an introduction to group theory and group representation theory, respectively. We also refer the reader to Vanderbauwhede (1982) and Golubitsky, Stewart and Schaeffer (1988) (which we will refer to as GSS) which cover many more aspects of this topic than are contained in this introduction. The approach of GSS is to assume that a Lyapunov-Schmidt reduction has been performed at a bifurcation point and they then do a very careful study of the bifurcation equations, taking account of the symmetry. A more computational approach is given by Dellnitz and Werner (1989).

Our approach is to consider an equation which has certain symmetry properties and then to investigate how the symmetry may be utilised for the efficient numerical solution of the problem. For the sake of clarity, we illustrate the theory with an example in finite dimensions, although the analysis does extend to infinite dimensional problems (see Aston (1989)). We take as our example, the Kuramoto-Sivashinsky equation (Scovel, Kevrekidis and Nicolaenko (1988))

$$u_t + 4u^{(iv)} + \lambda(u'' + uu') = 0 \tag{1.1}$$

139

*D. Roose et al. (eds.), Continuation and Bifurcations: Numerical Techniques and Applications,* 139–152.
© 1990 *Kluwer Academic Publishers.*

where $u$ is $2\pi$-periodic and of zero mean. This equation models several different physical systems and has many non-trivial steady-state and time-dependent solutions including chaotic regimes (see Hyman and Nicolaenko (1986)) but we shall restrict attention to steady-state solutions. Bifurcation from the trivial solution $u = 0$ occurs at $\lambda = 64$ resulting in a primary branch of solutions which are all odd with period $\frac{\pi}{2}$. We shall consider the problem of determining all the secondary bifurcations which occur from this branch of solutions. For the sake of simplicity, we do not consider the infinite dimensional problem (1.1) but apply a numerical method resulting in an approximate finite-dimensional problem. We employ a spectral Galerkin method where we approximate $u$ by

$$u(x) \equiv \sum_{k=1}^{N} a_k \sin kx + b_k \cos kx \in \mathcal{F}$$

where $\mathcal{F} = \mathrm{sp}\{\sin kx, \cos kx\}_{k=1}^{N}$. (Note that there is no constant term since $u$ has zero mean.) Substituting this approximation into (1.1) and taking the inner product with the $2N$ basis functions of the function space $\mathcal{F}$ gives the finite-dimensional problem

$$g(U, \lambda) = 0, \qquad g : \mathbf{R}^{2N} \times \mathbf{R} \rightarrow \mathbf{R}^{2N} \tag{1.2}$$

where $U = [a_1, b_1, a_2, b_2, \ldots, a_N, b_N]^T \in \mathbf{R}^{2N}$.

## 2. Equivariance Property of the Equation

We consider a nonlinear, parameter-dependent equation

$$g(x, \lambda) = 0, \qquad g : X \times \mathbf{R} \rightarrow X \tag{2.1}$$

where $X$ is a finite-dimensional Hilbert space. Many equations have the *equivariance property*

$$\gamma g(x, \lambda) = g(\gamma x, \lambda) \tag{2.2}$$

where $\gamma$ is an invertible linear transformation on $X$ e.g. rotations, translations, reflections etc. Clearly, if $\gamma_1$ and $\gamma_2$ both satisfy (2.2), then so does the product $\gamma_1 \gamma_2$. Also, if $\gamma$ satisfies (2.2), then so does $\gamma^{-1}$. Finally, the identity transformation on $X$ satisfies (2.2). Thus, $\Gamma = \{\gamma \text{ satisfying } (2.2)\}$ forms a *group*. Strictly speaking, $\Gamma$ is a group of linear transformations on $X$ which is isomorphic to an abstract group. However, we generally associate the name of the abstract group with the group of transformations. Also, we assume that $\Gamma$ is a *compact Lie group* which we will not define precisely (see Miller (1972)). This includes all finite groups and a particular class of continuous groups including the orthogonal groups $0(n)$ of all orthogonal linear transformations on $\mathbf{R}^n$.

The whole of the theory of symmetry-breaking bifurcation is based on the equivariance condition (2.2). In particular, we will show in Section 6 that differentiating (2.2) leads to many useful results. One immediate consequence of the equivariance condition (2.2) is that if $(x, \lambda)$ is a solution of (2.1), then $(\gamma x, \lambda)$ is also a solution for all $\gamma \in \Gamma$. These are called *conjugate solutions*.

Example : The *KS* equation (1.1) is equivariant with respect to the transformations

$$R_\alpha u(s) = u(s+\alpha), \qquad \alpha \in [0,2\pi) \tag{2.3a}$$

$$S\, u(s) = - u(-s). \tag{2.3b}$$

We also define

$$SR_\alpha u(s) = -u(-s - \alpha) \tag{2.3c}$$

for reasons which will become apparent later (see Section 4). We illustrate the equivariance property (2.2) for the reflection $S$. Now

$$(S u)' = \frac{d}{ds}\left[-u(-s)\right] = u'(-s) = -S u'.$$

It follows from this that $(Su)'' = Su''$ and $(Su)^{(iv)} = Su^{(iv)}$. Also

$$Su(Su)' = -u(-s)u'(-s) = S(uu').$$

Thus,

$$S(u^{(iv)} + \lambda(u'' + uu')) = (Su)^{(iv)} + \lambda((Su)'' + Su(Su)')$$

as required. The translational equivariance is proved similarly and is more straightforward since translation commutes with differentiation.

The group generated by the "rotations" (or translations) $R_\alpha$, $\alpha \in [0,2\pi)$ and the reflection $S$ is called 0(2), the group of all orthogonal linear transformations of the plane (and also the symmetry group of the circle).

It can be shown that the finite-dimensional system (1.2) inherits the equivariance properties of the full equation, only the linear transformations are now defined in terms of invertible matrix representations of $R_\alpha$ and $S$ which "act" on $\mathbf{R}^{2n}$ .

## 3. Fixed Point Spaces

Fixed point spaces play an important role in the theory since they are subspaces of $X$ which are invariant under $g$ (see Lemma 6.4). They are used to define the symmetry of a branch of solutions (see Section 6).

We define a *subgroup* of $\Gamma$ to be a subset of $\Gamma$ which is also a group and which is *closed* in $\Gamma$. If $\Sigma$ is a subgroup of $\Gamma$, then the fixed point subspace $X^\Sigma$ of $X$ is defined by

$$X^\Sigma = \{x \in X : \sigma x = x \ \forall \sigma \in \ \Sigma\}.$$

This is the subspace of all the elements of $X$ which have the symmetry of the subgroup $\Sigma$.

Example: Consider the dihedral group $D_n$ which consists of the "rotations" $R_{\frac{2\pi m}{n}}$ and the reflections $SR_{\frac{2\pi m}{n}}$, $m = 0, 1, ..., n-1$. The group is generated by $R_{\frac{\pi}{2}}$ and $S$. From the definitions (2.3), we conclude that

$$S u = u \Rightarrow - u(-s) = u(s) \Rightarrow u \text{ is an odd function}$$

$$R_{\frac{2\pi}{n}} u = u \Rightarrow u\left(s + \frac{2\pi}{n}\right) = u(s) \Rightarrow u \text{ has period } \frac{2\pi}{n}.$$

Thus,

$$\mathcal{F}^{D_n} = \{u \in \mathcal{F} : u \text{ is odd and has period } \frac{2\pi}{n}\}$$

and so a basis for $\mathcal{F}^{D_n}$ consists of the functions $\sin kns$, $k \in \mathbf{Z}^+$ such that $kn \leq N$.
  Note that the primary branch of solutions of the $KS$ equation which we are considering has $u \in \mathcal{F}^{D_4}$.


## 4. Invariant Subspaces

Invariant subspaces can be decomposed into isotypic components (see Section 5) which feature prominently in the bifurcation analysis of Section 6. Thus, we now consider invariant and irreducible subspaces which are defined as follows:

(i)  A subspace $W$ of $X$ is $\Gamma$-*invariant* if

$$\gamma w \in W \qquad \forall \gamma \in \Gamma, w \in W.$$

(ii) If $W$ is $\Gamma$-invariant and has no proper, $\Gamma$-invariant subspaces, then it is called $\Gamma$-*irreducible*.

(If there is no ambiguity with regard to the group $\Gamma$, we drop the prefix and refer to subspaces simply as invariant etc.) Thus the irreducible subspaces are the "smallest" non-trivial invariant subspaces. There is a matrix representation of $\Gamma$ on invariant subspaces which is itself called irreducible if the subspace is irreducible and the dimension of the representation is defined to be the dimension of the space on which it is defined.

Example: Consider the action of 0(2) on $\mathcal{F}$ defined by (2.3).

(i)  $W_1 = \mathrm{sp}\{\sin ks\}$ is *not* an invariant subspace of $\mathcal{F}$ since

$$R_\alpha \sin ks = \sin k(s+\alpha)$$
$$= \cos k\alpha \underline{\sin ks} + \sin k\alpha \underline{\cos ks}$$

which is not contained in $W_1$ for all $\alpha \in [0,2\pi)$.

(ii) $W_2 = \text{sp}\{\sin ks, \cos ks\}$ *is* an invariant subspace of $\mathcal{F}$ with matrix representation $T$ given by

$$T(R_\alpha) = \begin{pmatrix} \cos k\alpha & \sin k\alpha \\ -\sin k\alpha & \cos k\alpha \end{pmatrix}, \qquad T(S) = \begin{pmatrix} 1 & 0 \\ 0 & -1 \end{pmatrix}$$

The definition (2.3c) of $SR_\alpha$ ensures that $T$ satisfies the homomorphism property

$$T(S)T(R_\alpha) = T(SR_\alpha) .$$

We observe that $W_2$ is also an irreducible subspace of $\mathcal{F}$ and so $T$ is an irreducible representation of $0(2)$ (of dimension 2).

## 5. Group Theoretic Decomposition of $X$

An implicit assumption of the equivariance condition (2.2) is that $X$ is $\Gamma$-invariant. We now describe two decompositions which apply to any invariant space. The first, Theorem 5.1, is a well known result from group representation theory (see Ledermann (1976)). The second, the isotypic decomposition, is well known for complex spaces (Miller (1972)) but seems to have been largely ignored until recently for real spaces. The isotypic components are invariant under $g_x(x,\lambda)$ (Theorem 6.3) and thus play an important role in the bifurcation analysis of Section 6.

THEOREM 5.1

If $X$ is $\Gamma$-invariant, then it can be decomposed as

$$X = W_1 \oplus W_2 \oplus \ ... \ \oplus \ W_l$$

where each $W_i$ is a $\Gamma$-irreducible subspace of $X$. (This decomposition is not unique.)

On each irreducible subspace $W_i$, there is an irreducible (matrix) representation of $\Gamma$. Collecting together all the irreducible subspaces whose representations are "the same" (or more precisely, *equivalent*, that is, the same up to a change of basis) gives the *isotypic decomposition* of $X$,

$$X = V_1 \oplus V_2 \oplus \ ... \ \oplus \ V_L$$

where $V_i = W_{i_1} \oplus W_{i_2} \oplus \ldots \oplus W_{i_{m_i}}$ and the representations on $W_{ij}$ are "the same" for $j = 1, \ldots, m_i$. The projections onto the isotypic components $V_i$ are known (see Werner (1990)) and this decomposition is unique. Thus, there is one isotypic component associated with each irreducible representation of $\Gamma$. The isotypic components also have the important property that *every* irreducible subspace of $V_i$ has "the same" representation. Further, the isotypic components are mutually orthogonal.

Example: Consider the subgroup $D_4$ of $0(2)$ generated by $R_{\frac{\pi}{2}}$ and $S$ and let $W_1 = \mathrm{sp}\{\sin 4s\}$. Clearly

$$R_{\frac{\pi}{2}} \sin 4s = \sin 4s$$
$$S \sin 4s = \sin 4s$$

using the definitions (2.3), and so $W_1$ is invariant and must therefore be irreducible since it has no proper subspaces. Similarly,

$$R_{\frac{\pi}{2}} \cos 4s = \cos 4s$$
$$S \cos 4s = -\cos 4s$$

and so $W_2 = \mathrm{sp}\{\cos 4s\}$ is also invariant and thus irreducible. Now $\mathrm{sp}\{\sin s\}$ is *not* invariant, but $W_3 = \mathrm{sp}\{\sin s, \cos s\}$ is invariant and irreducible and

$$R_{\frac{\pi}{2}}\begin{pmatrix} \sin s \\ \cos s \end{pmatrix} = \begin{pmatrix} \cos \frac{\pi}{2} & \sin \frac{\pi}{2} \\ -\sin \frac{\pi}{2} & \cos \frac{\pi}{2} \end{pmatrix}\begin{pmatrix} \sin s \\ \cos s \end{pmatrix}, \quad S\begin{pmatrix} \sin s \\ \cos s \end{pmatrix} = \begin{pmatrix} 1 & 0 \\ 0 & -1 \end{pmatrix}\begin{pmatrix} \sin s \\ \cos s \end{pmatrix}.$$

In this way, we can identify all the irreducible subspaces of $\mathcal{F}$ and their corresponding irreducible representations. These, together with the associated isotypic components of $\mathcal{F}$, with $N = 16$ are as follows:

(i) $\quad R_{\frac{\pi}{2}} = [1], \quad S = [1] : V_1 = \mathrm{sp}\{\sin 4ks\}_{k=1}^4 \qquad (= \mathcal{F}^{D_4})$

(ii) $\quad R_{\frac{\pi}{2}} = [1], \quad S = [-1] : V_2 = \mathrm{sp}\{\cos 4ks\}_{k=1}^4$

(iii) $\quad R_{\frac{\pi}{2}} = [-1], \quad S = [1] : V_3 = \mathrm{sp}\{\sin (4k-2)s\}_{k=1}^4$

(iv) $\quad R_{\frac{\pi}{2}} = [-1], \quad S = [-1] : V_4 = \mathrm{sp}\{\cos (4k - 2)s\}_{k=1}^4$

(v) $\quad R_{\frac{\pi}{2}} = \begin{pmatrix} 0 & 1 \\ -1 & 0 \end{pmatrix}, \; S = \begin{pmatrix} 1 & 0 \\ 0 & -1 \end{pmatrix} : V_5 = \mathrm{sp}\{\sin (4k-3)s, \; \sin (4k-1)s,$

$$\cos (4k-3)s, \; \cos (4k-1)s\}_{k=1}^4$$

## 6. Application to Bifurcation Theory

We now redefine $\Gamma$ to be the symmetry group of the (primary) branch of solutions of (2.1) under consideration, i.e. there is a branch of solutions contained in $X^\Gamma \times \mathbf{R}$. Our aim is to give conditions for bifurcation to occur such that the secondary branches are contained in $X^\Sigma \times \mathbf{R}$ for some subgroup $\Sigma (\neq \Gamma)$ of $\Gamma$. Such a bifurcation is called symmetry-breaking since the secondary branches "have less symmetry" than the primary branch. We also describe how such bifurcation points can be detected and computed in an efficient way.

An important result which applies to many practical situations is the Equivariant Branching Lemma (EBL) of Cicogna (1981) and Vanderbauwhede (1982).

EQUIVARIANT BRANCHING LEMMA

If $x_0 \in X^\Gamma$, $\mathcal{N}_0 := \mathrm{Null}\,(g_x(x_0, \lambda_0))$ is non-trivial and

(i) $\mathcal{N}_0 \cap X^\Gamma = \{0\}$,

(ii) there exists a subgroup $\Sigma$ of $\Gamma$ such that

$$\dim(\mathcal{N}_0 \cap X^\Sigma) = 1,$$

(iii) a non-degeneracy condition holds,

then there exists a secondary branch of solutions which bifurcates at $(x_0, \lambda_0)$, contained in $X^\Sigma \times \mathbf{R}$.

The EBL gives sufficient conditions for the occurrence of bifurcation although bifurcation can occur with the secondary branch contained in $X^\Sigma \times \mathbf{R}$ when $\dim(\mathcal{N}_0 \cap X^\Sigma) > 1$ (see for example, Lauterbach (1986)). This result is essentially bifurcation from a simple eigenvalue in the appropriate fixed point space.

Condition (ii) of the EBL may seem unnecessary, since it would not be unreasonable to assume that generically, $\mathcal{N}_0$ would be one-dimensional at a singular point. However, due to the symmetry of the problem, this is not the case as we now show . Differentiating the equivariance condition (2.2) with respect to $x$ gives

$$\gamma g_x(x, \lambda)\phi = g_x(\gamma x, \lambda)\gamma \phi.$$

Now if $x \in X^\Gamma$, then $\gamma x = x$ for all $\gamma \in \Gamma$ and so

$$\gamma g_x(x, \lambda)\phi = g_x(x, \lambda)\gamma \phi.$$

Thus, if $\phi \in \mathcal{N}_0$, then $\gamma \phi \in \mathcal{N}_0$ also. This result can be restated as follows.

LEMMA 6.1

If $x_0 \in X^\Gamma$, then $\mathcal{N}_0$ is $\Gamma$-invariant.

Recall that the "smallest" invariant subspaces of $X$ are the irreducible ones. Many groups give rise to irreducible subspaces of dimension greater than one and so multiple zero eigenvalues *can* occur generically. However, we have the following result (GSS, p. 82).

THEOREM 6.2

If $x_0 \in X^\Gamma$, and $\mathcal{N}_0$ is non-trivial, then generically it is $\Gamma$-irreducible.

Thus, generically, $\mathcal{N}_0$ will be the smallest possible invariant subspace i.e. irreducible. We now show how the isotypic decomposition of $X$ plays an important role.

THEOREM 6.3

If $x \in X^\Gamma$, then $g_x(x,\lambda) : V_i \to V_i$ where the $V_i$ are the $\Gamma$-isotypic components of $X$.

This result is proved by showing that $g_x(x, \lambda)$ commutes with the projection operators onto the isotypic components $V_i$ (see Werner (1990)). It follows from this that $g_x(x,\lambda)$ can be decomposed into "block diagonal" form as

$$g_x(x, \lambda) = \text{diag}(g_x^i(x, \lambda))$$

where $g_x^i(x, \lambda) := g_x(x, \lambda)|_{V_i} : V_i \to V_i$. In some cases, if the irreducible representation associated with $V_i$ has dimension greater than one, then $g_x^i(x, \lambda)$ can be decomposed further (see Werner (1990)).

It is thus possible for different "blocks" of $g_x(x,\lambda)$ to become singular as $\lambda$ is varied. Now every irreducible subspace of an isotypic component has "the same" irreducible representation (for an appropriate choice of basis). Thus, if $\mathcal{N}_0 \subset V_i$ is irreducible, then the precise form of the irreducible representation on $\mathcal{N}_0$ is known. Thus, it is possible to determine *a priori* whether or not a subgroup $\Sigma$ of $\Gamma$ exists such that $\dim(\mathcal{N}_0 \cap X^\Sigma) = 1$ when $\mathcal{N}_0 \subset V_i$ (assuming the generic condition of irreducibility, see Theorem 6.2). If such a subgroup exists, it is easily proved that $g_x(x, \lambda) : X^\Sigma \to X^\Sigma$ also. Hence, $g_x(x,\lambda) : V_i \cap X^\Sigma \to V_i \cap X^\Sigma$ and so the "block" of $g_x(x,\lambda)$ restricted to $V_i \cap X^\Sigma$ will (generically) have a simple zero eigenvalue at a singular point, which can thus be detected by a sign change of the determinant of this "block". Moreover, since the "blocks" of $g_x(x,\lambda)$ depend only on the solution $(x, \lambda) \in X^\Gamma \times \mathbf{R}$ and are independent of each other, the determinants of all the relevant "blocks" associated with different isotypic components can be computed in parallel, thus increasing the computational efficiency still further.

We note that the fixed point space $X^\Gamma$ is a $\Gamma$-isotypic component of $X$ ($V_1$ say) associated with the trivial one-dimensional irreducible representation $\gamma = [1]$ for all $\gamma \in \Gamma$ (see the example of Section 5). Thus, if $g_x^1(x, \lambda)$ becomes singular, then generically, the null space will be one-dimensional and there will be a left eigenvector $\psi$ which is also in $X^\Gamma$. Now it follows from differentiating the equivariance condition (2.2) with respect to $\lambda$ that $g_\lambda(x, \lambda) \in X^\Gamma$ since $x \in X^\Gamma$ and so generically $< \psi, g_\lambda(x, \lambda) > \neq 0$ at the singular point which thus corresponds to a turning point. However, if any other "block" of $g_x(x, \lambda)$, associated with $V_i$ say, becomes singular, then there are left eigenvector(s) $\psi_j$ which are in $V_i$. In this case $< \psi_j, g_\lambda(x, \lambda) > = 0$ due to the orthogonality of the isotypic components and so the singular point will be a bifurcation point. Thus, generically, if symmetry is not broken, bifurcation will not occur. We can thus label branches of solutions of (2.1) by their symmetry group. Note that condition (i) of the EBL simply excludes the possibility of the singular point being a turning point.

Once a bifurcation point has been detected with $\mathcal{N}_0 \subset V_i$, it can be solved for directly using the extended system

$$
G(y) := \begin{pmatrix} g(x, \lambda) \\ g_x(x, \lambda)\phi \\ < \ell, \phi > -1 \end{pmatrix} = 0, \qquad G : Y \rightarrow Y \tag{6.1}
$$

$$
y = (x, \phi, \lambda) \in Y : = X^\Gamma \times (V_i \cap X^\Sigma) \times \mathbf{R}
$$

where $\ell \in V_i \cap X^\Sigma$ is chosen appropriately to normalise $\phi$. If the conditions of the EBL hold at $(x_0, \lambda_0)$, then $y_0 = (x_0, \phi_0, \lambda_0)$, where $\phi_0 \in \mathcal{N}_0 \cap V_i \cap X^\Sigma$, is an isolated solution of (6.1). This is proved in exactly the same way as the corresponding result of Werner and Spence (1984).

Having determined the fixed point space $\Sigma$ of a branch of solutions, we can utilise the following result, which is derived directly from the equivariance condition (2.2).

LEMMA 6.4

If $x \in X^\Sigma$, then $g(x, \lambda) \in X^\Sigma$ also, that is

$$
g : X^\Sigma \times \mathbf{R} \rightarrow X^\Sigma.
$$

Thus, the branch of solutions in $X^\Sigma \times \mathbf{R}$ can be determined by solving the reduced problem $g_\Sigma(x, \lambda) = 0$, $g_\Sigma : = g|_{X^\Sigma \times \mathbf{R}}$.

Finally, we make the following remarks:

(i) There is a simple test to determine whether the bifurcation is either a symmetric pitchfork with conjugate secondary branches or is non-symmetric with non-conjugate secondary branches (see Dellnitz and Werner (1989)).

(ii) Irreducible representations can be divided into different classes. One such class is that of absolutely irreducible representations (see GSS, p.40). Generically a "block" of $g_x(x,\lambda)$ associated with an isotypic component corresponding to a non-absolutely irreducible representation will never become singular and so bifurcation is generically associated only with those isotypic components corresponding to absolutely irreducible representations (GSS, p82).

(iii) Hopf bifurcation can be treated in a similar way and can be associated generically with both absolutely irreducible and non-absolutely irreducible representations. (see GSS, Dellnitz and Werner (1989)).

(iv) Bifurcations from a branch of solutions contained in $X^\Gamma \times \mathbf{R}$ can be unfolded by perturbations (arising from a naive numerical approximation) which do not preserve the $\Gamma$-equivariance and so the application of a numerical method, ignoring the symmetry of the problem, could give rise to very misleading results. Thus, it is important that the systematic approach outlined here is used to obtain accurate numerical results.

Example: The isotypic component $V_1$ is associated only with turning points and so we do not consider this one. If $\mathcal{N}_0 \subset V_i$, $i = 2, 3, 4$ is irreducible, then it is one-dimensional and so we have only to identify the fixed point space in which an irreducible subspace lies. This will also be the fixed point subspace of the isotypic component since the representation on each irreducible subspace is "the same". Thus, we have that $V_2 \subset \mathcal{F}^{\mathbf{Z}_4}$, where $\mathbf{Z}_4$ is the cyclic group generated by $R_{\frac{\pi}{2}}$, $V_3 \subset \mathcal{F}^{D_2}$ where $D_2$ is generated by $R_\pi$ and $S$, and $V_4 \subset \mathcal{F}^{\bar{D}_2}$ where $\bar{D}_2$ is generated by $R_\pi$ and $SR_{\frac{\pi}{2}}$.

If $\mathcal{N}_0 \subset V_5$ is irreducible, then there exists $\phi_1, \phi_2 \in V_5$ such that $\mathcal{N}_0 = \mathrm{sp}\{\phi_1, \phi_2\}$ and

$$R_{\frac{\pi}{2}}\begin{pmatrix} \phi_1 \\ \phi_2 \end{pmatrix} = \begin{pmatrix} 0 & 1 \\ -1 & 0 \end{pmatrix}\begin{pmatrix} \phi_1 \\ \phi_2 \end{pmatrix}, \qquad S\begin{pmatrix} \phi_1 \\ \phi_2 \end{pmatrix} = \begin{pmatrix} 1 & 0 \\ 0 & -1 \end{pmatrix}\begin{pmatrix} \phi_1 \\ \phi_2 \end{pmatrix}$$

Clearly, since $S\phi_1 = \phi_1$, we have

$$\phi_1 = \sum_{k=1}^{N/4} a_{4k-3} \sin(4k-3)s + a_{4k-1}\sin(4k-1)s$$

and as $R_{\frac{\pi}{2}}\phi_1 = \phi_2$, we obtain

$$\phi_2 = \sum_{k=1}^{N/4} a_{4k-3}\cos(4k-3)s - a_{4k-1}\cos(4k-1)s.$$

Thus, $\mathcal{N}_0 \cap X^{\mathbf{Z}_2} = \mathrm{sp}\{\phi_1\}$ where $\mathbf{Z}_2 = \{I, S\}$. Also $\mathcal{N}_0 \cap X^{\tilde{\mathbf{Z}}_2} = \mathrm{sp}\{\phi_1 + \phi_2\}$ where $\tilde{\mathbf{Z}}_2 = \{I, SR_{\frac{\pi}{2}}\}$ and so this case gives rise to a *multiple bifurcation point* since 2 distinct branches of solutions bifurcate.

A simple calculation confirms that all these bifurcations are symmetric pitchforks (see Dellnitz and Werner (1989)) and so in Fig. 6.1 we plot only one of the 2 conjugate secondary banches at each bifurcation point. The branches are labelled by their symmetry group. The region near the multiple bifurcation point is blown up in Fig. 6.2. Also, due to the underlying $O(2)$ symmetry, the bifurcation associated with $V_2$ will not be a steady–state bifurcation but a bifurcation to rotating (or travelling) waves (see Aston, Spence and Wu (1989, 1990)).

Finally, for each solution branch, the reduced problem arising from Lemma 6.4 consists of solving only for the non-zero coefficients of $U \in \mathbf{R}^{2N}$.

150

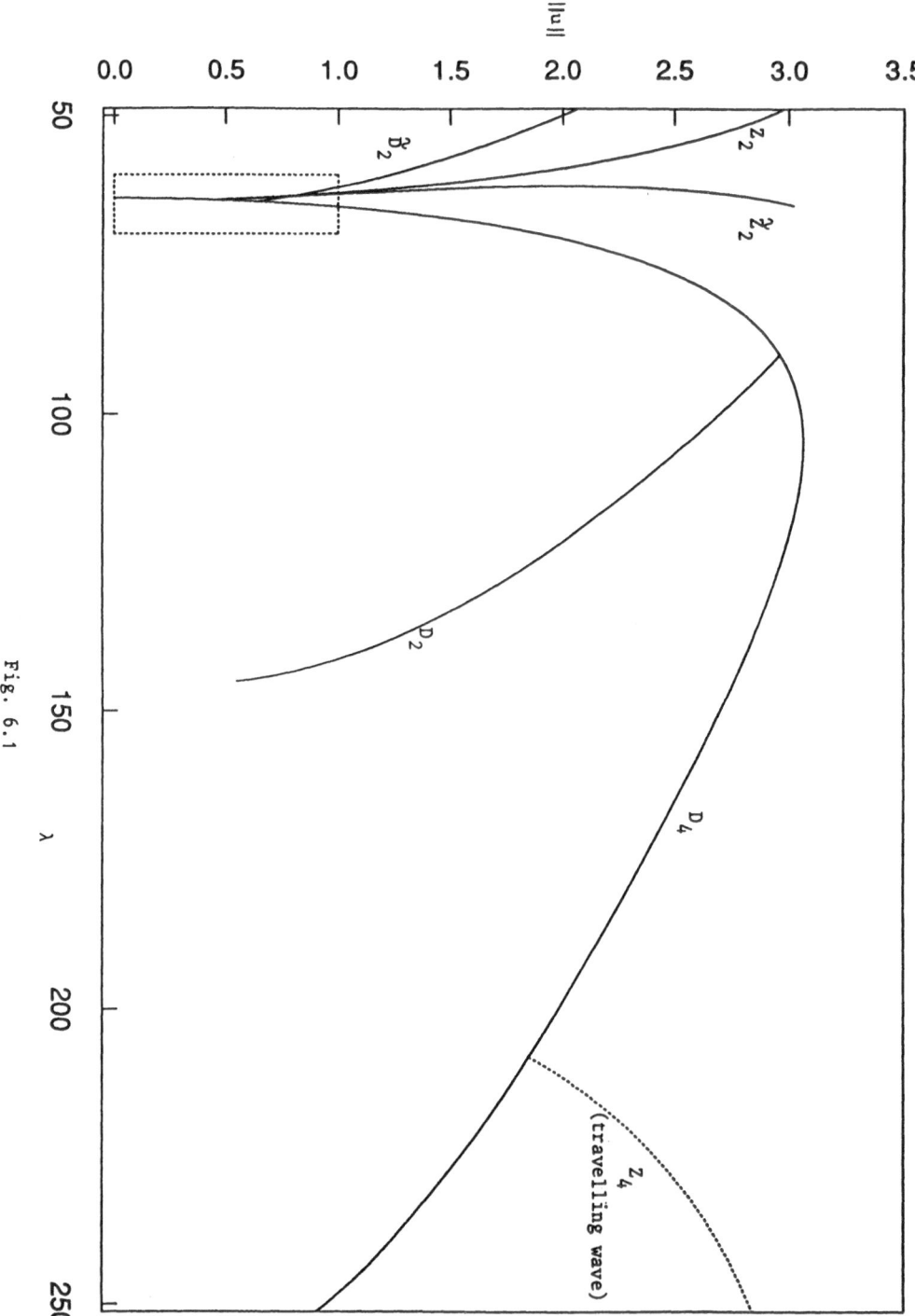

Fig. 6.1

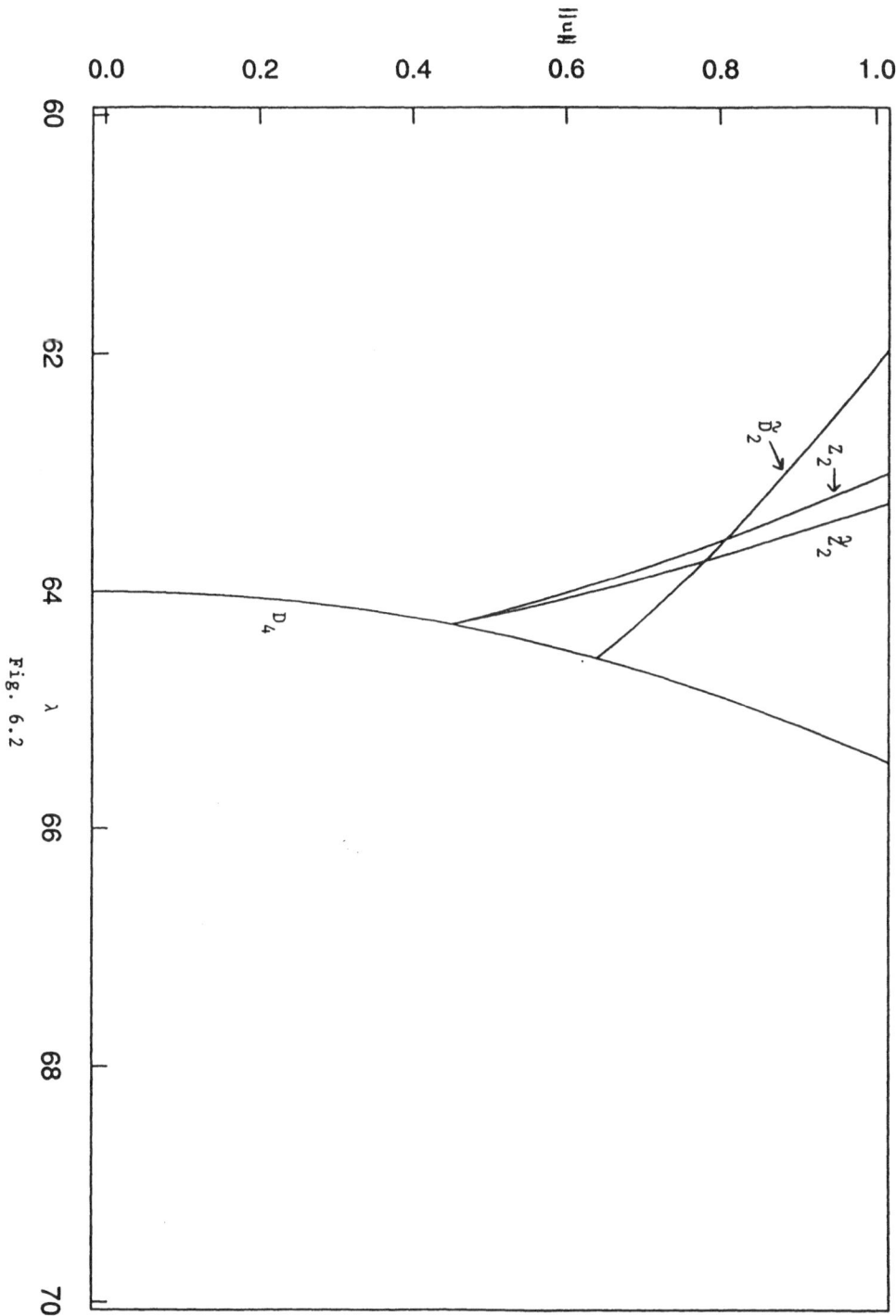

Fig. 6.2

## REFERENCES

Aston, P. J. (1990). Analysis and computation of symmetry-breaking bifurcation and scaling laws using group theoretic methods. To appear.

Aston, P. J., Spence, A. and Wu, W. (1990). Bifurcation to rotating waves in equations with 0(2)-symmetry. Submitted to *SIAM J. Appl. Math.*

Aston, P. J., Spence, A. and Wu, W. (1990). Bifurcation to rotating waves from non-trivial steady-states. These proceedings.

Cicogna, G. (1981). Symmetry breakdown from bifurcation. *Lettre al Nuovo Cimento* **31**, 600-602.

Dellnitz, M. and Werner, B. (1989). Computational methods for bifurcation problems with symmetries - with special attention to steady state and Hopf bifurcation points. *J. Comp. Appl. Math.* **26**, 97-123.

Golubitsky, M., Stewart, I. and Schaeffer, D. G. (1988). *Singularities and Groups in Bifurcation Theory, Vol. 2.* Appl. Math. Sci. **69**, Springer, New York.

Hyman, J. M. and Nicolaenko, B. (1986). The Kuramoto-Sivashinsky equation: a bridge between pde's and dynamical systems. *Physica D,* **18**, 113-126.

Lauterbach, R. (1986). An example of symmetry-breaking with submaximal isotropy. In Multiparameter Bifurcation Theory, eds. Golubitsky, M. and Guckenheimer, J., *Contemp. Math.* **56**, 217-222, AMS, Providence.

Ledermann, W. (1973). *Introduction to Group Theory,* Longman, London.

Ledermann, W. (1977). *Introduction to Group Characters*, Cambridge University Press, Cambridge.

Miller, W. (1972). *Symmetry Groups and their Applications*, Academic Press, London.

Scovel, J. C., Kevrekidis, I. G. and Nicolaenko, B. (1988). Scaling laws and the prediction of bifurcations in systems modelling pattern formation. *Phys. Letts. A* **130**, 73-80.

Vanderbauwhede, A. (1982). *Local Bifurcation and Symmetry,* Research Notes in Mathematics **75**, Pitman, London.

Werner, B. (1990). Eigenvalue problems with the symmetry of a group and bifurcations. These proceedings.

Werner, B. and Spence, A. (1984). The computation of symmetry breaking bifurcation points. *SIAM J. Num. Anal.* **21**, 388-399.

# A COMPUTATIONAL METHOD AND PATH FOLLOWING FOR PERIODIC SOLUTIONS WITH SYMMETRY

M. DELLNITZ
Institut für Angewandte Mathematik
Universität Hamburg
Bundesstraße 55
D-2000 Hamburg 13
Federal Republic of Germany

ABSTRACT. In this paper a Galerkin method is used for the computation of periodic solutions of autonomous ODEs with underlying symmetry. It is shown how the spatial and temporal symmetry of periodic solutions can be utilized to make their numerical computation more efficient. The method is illustrated for the example of a four-cell-Brusselator with tetrahedral symmetry.

## 1 Introduction

In applications often periodic solutions of parameter dependent ordinary differential equations of the form

$$\dot{x} = g(x, \lambda), \qquad x \in X := \mathbb{R}^n, \quad \lambda \in \mathbb{R}, \tag{1.1}$$

have to be computed. The main aim of this paper is to describe a numerical method, which can be used to do this efficiently. Thereby we will always assume that the underlying dynamical system modelled by (1.1) possesses symmetry. It is well known that in this case *multiple* Hopf bifurcations can occur generically in (1.1) (cf. GOLUBITSKY-STEWART-SCHAEFFER [9]) and, accordingly, we have to expect the existence of (a lot of) different branches of periodic solutions bifurcating in corresponding Hopf bifurcation points.

Therefore an appropriate program for the computation of periodic solutions of (1.1) is given by the following three steps:

1. First of all compute a (multiple) Hopf bifurcation point in the system (1.1).

2. Obtain an overview about the different types of branches of periodic solutions bifurcating at this point. Jump onto a branch which is of interest.

3. Compute this branch of periodic solutions by path following.

For all three steps the underlying symmetry can be utilized to enable an efficient numerical treatment. With respect to the first step see, for instance, DELLNITZ-WERNER [4]. In this paper we will focus on the second and the third step.

In order to obtain an overview of the different branches of periodic solutions bifurcating in a multiple Hopf bifurcation point we will use the *Equivariant Hopf Theorem* of GOLUBITSKY-STEWART-SCHAEFFER [9]. For this we will give a brief description of the phenomenon of *symmetry-breaking Hopf bifurcation* in Section 2.1. Thereby we will avoid group theoretical

*D. Roose et al. (eds.), Continuation and Bifurcations: Numerical Techniques and Applications*, 153–167.
© 1990 *Kluwer Academic Publishers.*

notions as much as possible. Nevertheless, some results from group representation theory have to be mentioned.

In order to show the usefulness of the Equivariant Hopf Theorem in practice and to show how it can be applied to a given problem, we will study the example of a *four-cell-Brusselator* with tetrahedral symmetry (cf. Section 2.2).

In Section 3 we will deal with the third step of the above mentioned program. We will use a Galerkin method based on Fourier expansions for the numerical computation of periodic solutions of (1.1) (see eg ZHENG [17]). In comparison with other methods, in particular the (multiple) shooting method (cf. KUBIČEK-MAREK [10]), this approach provides several advantages:

- The stability of the periodic solution that has to be computed does not affect the convergence or even the applicability of this method.

- The spatial <u>and</u> temporal symmetry of a periodic solution of (1.1) can be used to reduce the effort for its numerical computation extremly.

- The numerical computation of a branch of <u>periodic</u> solutions of (1.1) requires the computation of a corresponding branch of <u>steady-state</u> solutions of a certain <u>static</u> bifurcation problem. Therefore all the analytical and numerical techniques and results for static bifurcation problems with symmetries can directly be applied to this situation.

Finally, the applicability of the method will also be illustrated for the above mentioned example of the four-cell-Brusselator with tetrahedral symmetry (see Section 3).

## 2 Hopf Bifurcation with Symmetry

In this section we briefly describe the phenomenon of symmetry-breaking Hopf bifurcation. The section is devided into two parts: In the first part we will state the *"Equivariant Hopf Theorem"*, the symmetry adapted generalization of the *"Classical Hopf Theorem"* (cf. eg AMANN [1], MARSDEN-MCCRACKEN [11]). This equivariant version is due to GOLUBITSKY-STEWART [7] and – in a more general setting – due to GOLUBITSKY-STEWART-SCHAEFFER [9].

In the second part we will use this Theorem in order to proof the existence of branches of periodic solutions with certain types of symmetry, bifurcating in a *multiple* Hopf bifurcation point. For this we will analyse in detail the example of a multiple Hopf bifurcation in a four-cell-Brusselator with the symmetry of a tetrahedron.

### 2.1 THE HOPF THEOREM WITH SYMMETRY

In the following we will essentially use notations and definitions from the book [9] of GOLUBITSKY-STEWART-SCHAEFFER .

We consider a parameter dependent system of ODEs

$$\dot{x} = g(x, \lambda), \tag{2.1}$$

where $x \in X := \mathbb{R}^n$, and $\lambda \in \mathbb{R}$ is the parameter. We will assume $g$ to be sufficiently smooth.

According to symmetry, the right hand side $g$ of (2.1) satisfies the following *equivariance condition* :

$$\gamma g(x, \lambda) = g(\gamma x, \lambda) \qquad \forall \gamma \in \Gamma, \quad x \in X, \tag{2.2}$$

where $\Gamma \subset \mathbf{O}(n)$ is a compact Lie group. This reflects the fact, that the underlying dynamical system possesses a certain $\Gamma$-symmetry (examples for such systems are, for instance, given in FIEDLER [6], GOLUBITSKY-STEWART [8], STEINDL-TROGER [13] and also in this article).

Since (2.1) is autonomous, not only the $\Gamma$-symmetry but also an $S^1$-symmetry is relevant for the analytical treatment of symmetry-breaking Hopf bifurcation (cf. GOLUBITSKY-STEWART [7]). For explanation, we introduce a period scaling parameter $\nu \neq 0$ and with $t \to \nu t$ we obtain from (2.1) the system

$$\dot{y} = \frac{1}{\nu} g(y, \lambda). \tag{2.3}$$

Then $2\pi$-periodic solutions of (2.3) correspond to $\frac{2\pi}{\nu}$-periodic solutions of (2.1). Now we define an action of the group $\Gamma \times S^1$ on the space $\mathcal{C}_{2\pi}$ of continuous $2\pi$-periodic mappings $\mathbb{R} \to X$ by

$$(\gamma, \theta)v(t) := \gamma v(t + \theta), \quad (\gamma, \theta) \in \Gamma \times S^1, \quad v \in \mathcal{C}_{2\pi}. \tag{2.4}$$

From (2.2) it is immediately seen that the operator

$$\Phi : \mathcal{C}_{2\pi}^1 \times \mathbb{R} \times \mathbb{R} \longrightarrow \mathcal{C}_{2\pi}$$

defined by

$$\Phi(y, \lambda, \nu) := \nu \dot{y} - g(y, \lambda) \tag{2.5}$$

satisfies the equivariance condition

$$(\gamma, \theta)\Phi(y, \lambda, \nu) = \Phi((\gamma, \theta)y, \lambda, \nu) \quad \forall (\gamma, \theta) \in \Gamma \times S^1, \quad y \in \mathcal{C}_{2\pi}^1.$$

Solutions $(u, \lambda_0, \nu_0)$ of the equation $\Phi = 0$ correspond to $\frac{2\pi}{\nu_0}$-periodic solutions of (2.1) for $\lambda = \lambda_0$. This indicates that for the analytical treatment of Hopf bifurcation in systems with $\Gamma$-symmetry, $\Gamma \times S^1$ is the appropriate group of symmetries.

We will always characterize the symmetry of a $2\pi$-periodic mapping $v(t)$ by its *isotropy subgroup* $\Sigma_v$,

$$\Sigma_v := \{(\gamma, \theta) \in \Gamma \times S^1 : (\gamma, \theta)v(t) = v(t)\} \subset \Gamma \times S^1. \tag{2.6}$$

Examples for isotropy subgroups and corresponding $2\pi$-periodic mappings will be given in the next subsection.

Now let us assume that $(x_0, \lambda_0) \in X \times \mathbb{R}$ is a steady-state solution of the system (2.1), where $x_0$ is $\Gamma$-symmetric ($\Sigma_{x_0} = \Gamma \times S^1$) and the Jacobian $g_x(x_0, \lambda_0)$ has a purely imaginary eigenvalue $i\omega, \omega \neq 0$. For the formulation of the Equivariant Hopf Theorem, we have to introduce three nondegeneracy assumptions (H1)-(H3). The first one is well known from the classical result too. We assume

$$\sigma(g_x(x_0, \lambda_0)) \cap i\mathbb{R} = \{i\omega, -i\omega\}. \tag{H1}$$

Here $\sigma(g_x(x_0, \lambda_0))$ denotes the spectrum of the Jacobian $g_x(x_0, \lambda_0)$. Note that $i\omega$ is not necessarily a simple eigenvalue of $g_x(x_0, \lambda_0)$.

By (2.2) the *real eigenspace*

$$N_\omega := \{z \in X : (g_x(x_0, \lambda_0)^2 + \omega^2 I)z = 0\} \tag{2.7}$$

is invariant under the action of $\Gamma$. Furthermore $N_\omega$ is even invariant under an action of $\Gamma \times S^1$, which is defined by (cf. GOLUBITSKY-STEWART-SCHAEFFER [9], Chap. XVI, Lemma 3.2)

$$(\gamma, \theta)z := \gamma e^{-\theta J} z, \quad (\gamma, \theta) \in \Gamma \times S^1, \quad z \in N_\omega. \tag{2.8}$$

Thereby $J : N_\omega \to N_\omega$ is the restriction of $g_x(x_0, \lambda_0)$ on $N_\omega$. We will give an explicit description of this action in the example below.

The second assumption essentially concerns the structure of $N_\omega$:

$$\left\{ \begin{array}{l} i\omega \text{ is a semisimple eigenvalue of } g_x(x_0, \lambda_0) \\ \text{and } N_\omega \text{ is } \Gamma\text{-simple.} \end{array} \right\} \tag{H2}$$

Recall that a subspace $W$ of $X$ is called $\Gamma$-*simple* if it is either the direct sum $V_1 \oplus V_2$ of two absolutely $\Gamma$-irreducible isomorphic subspaces $V_1, V_2$ or $W$ itself is nonabsolutely $\Gamma$-irreducible (cf. GOLUBITSKY-STEWART-SCHAEFFER [9], Chap. XVI, Def. 1.3). With respect to the genericity of this assumption see also GOLUBITSKY-STEWART-SCHAEFFER [9] or WERNER [16].

Finally, let $\mu(\lambda) \pm i\rho(\lambda)$ be the associated pair of complex-conjugate eigenvalues of the Jacobian $g_x(x(\lambda), \lambda)$ along the branch of steady-state solutions through $(x_0, \lambda_0)$ with $\mu(\lambda_0) = 0$, $\rho(\lambda_0) = \omega$. Then we assume that this pair crosses the imaginary axis with nonzero speed:

$$\frac{d\mu}{d\lambda}(\lambda_0) \neq 0. \tag{H3}$$

**Theorem 2.1** (Equivariant Hopf Theorem): *Let the system of ODEs (2.1) satisfy (H1), (H2) and (H3). Suppose that for a subgroup $\Sigma \subset \Gamma \times S^1$ the fixed point space*

$$N_\omega^\Sigma := \{v \in N_\omega : (\sigma, \varphi)v = v \quad \forall(\sigma, \varphi) \in \Sigma\}$$

*(cf. (2.8)) satisfies the condition*

$$\dim N_\omega^\Sigma = 2.$$

*Then there exists a unique branch of small-amplitude periodic solutions to (2.1) with period near $2\pi/\omega$, having $\Sigma$ as their group of symmetries (isotropy subgroup).*

PROOF: See GOLUBITSKY-STEWART-SCHAEFFER [9], Chap. XVI. ∎

In the following we will call the point $(x_0, \lambda_0)$, which fulfils (H1)-(H3), a *symmetry-breaking Hopf bifurcation point*.

## 2.2  EXAMPLE: FOUR-CELL-BRUSSELATOR WITH TETRAHEDRAL SYMMETRY

Now we will show how Theorem 2.1 can be applied to a given problem with underlying symmetry, in order to proof the existence of bifurcating branches of periodic solutions with certain symmetry types in a symmetry-breaking Hopf bifurcation point.

The problem we will look at is the following: For $j = 1, 3, 5, 7$ (mod 8, $x_0 := x_8$) consider the equations

$$\begin{aligned} \dot{x}_j &= A - (B+1)x_j + x_j^2 x_{j+1} + \lambda(-3x_j + x_{j+2} + x_{j+4} + x_{j+6}), \\ \dot{x}_{j+1} &= Bx_j - x_j^2 x_{j+1} + 10\lambda(-3x_{j+1} + x_{j+3} + x_{j+5} + x_{j+7}), \end{aligned} \tag{2.9}$$

$A, B \in \mathbb{R}$ positive constants.

These are the celebrated *Brusselator equations* (cf. PRIGOGINE-LEFEVER [12]), where in this case four cells – with corresponding concentrations $x_j, x_{j+1}$ ($j = 1, 3, 5, 7$) in every cell – are identically coupled with each other. The parameter $\lambda$ measures the strength of the coupling between the cells. We can visualize this type of symmetry in the way that we put the four cells at the vortices of a tetrahedron (see Figure 1).

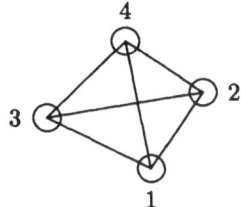

Figure 1: Schematic representation of the four-cell-Brusselator with tetrahedral symmetry

Therefore the system (2.9) is $\Gamma$-symmetric, where $\Gamma \subset \mathbf{O}(8)$ is isomorphic to the symmetry group $T = S_4$ of the tetrahedron. This group consists of $4! = 24$ elements corresponding to all possible permutations of the numbers (cells) $1, 2, 3, 4$. Accordingly, we will denote the elements of $\Gamma$ by $\gamma_{ijkl}$. For instance, the matrix

$$\gamma_{3124} := \begin{pmatrix} 0 & 0 & I & 0 \\ I & 0 & 0 & 0 \\ 0 & I & 0 & 0 \\ 0 & 0 & 0 & I \end{pmatrix} \in \Gamma, \quad I \in \mathbb{R}^{2,2},$$

corresponds to a cyclic permutation of the cells $1, 2, 3$, while cell 4 remains at the same place.

Looking at the *bifurcation graph* of this problem (cf. DELLNITZ-WERNER [4]) one knows that there is exactly one type of $\Gamma$-symmetry-breaking Hopf bifurcation, which generically can occur in (2.9). This is a multiple Hopf bifurcation, which is related to a 3-dimensional absolutely irreducible representation of $T$. Setting $A = 2.0, B = 5.9$ and using the methods described in DELLNITZ-WERNER [4], it is possible to compute such a Hopf bifurcation point in the system (2.9), namely

$$(x_0, \lambda_0) = (2.0, 2.95, 2.0, 2.95, 2.0, 2.95, 2.0, 2.95, 0.02045).$$

The frequency is $\omega = 0.62058$.

The underlying representation is given – up to equivalence – in Table 1 (see also CORN-WELL [2], Table D.3, Appendix D). From this we exactly know, how the group $\Gamma$ is acting on the real eigenspace $N_\omega$: $N_\omega$ splits into the direct sum of two isomorphic absolutely irreducible 3-dimensional subspaces and $\Gamma$ is acting on each of these spaces like described in Table 1.

For completeness we also describe for this case how $S^1$ is acting on $N_\omega$ (cf. GOLU-BITSKY-STEWART [7]). Let $z = v_1 + v_2 \in V_1 \oplus V_2 = N_\omega$ (see (H2)). Then $(I, \theta)z = (\cos\theta - \sin\theta)v_1 + (\sin\theta + \cos\theta)v_2$. This can be proved using (2.8).

For the application of Theorem 2.1 we have to compute isotropy subgroups $\Sigma$ of $\Gamma \times S^1$ which fulfil the condition

$$\dim N_\omega^\Sigma = 2. \tag{2.10}$$

For this we will make use of the fact that all proper isotropy subgroups $\Sigma$ of $\Gamma \times S^1$, which are candidates for satisfying (2.10), must be *twisted subgroups* (GOLUBITSKY-STEWART-SCHAEFFER [9], Proposition 7.2, Chapter XVI).

$$\gamma_{1234} \leftrightarrow \begin{pmatrix} 1 & 0 & 0 \\ 0 & 1 & 0 \\ 0 & 0 & 1 \end{pmatrix}, \qquad \gamma_{3124} \leftrightarrow \begin{pmatrix} 0 & -1 & 0 \\ 0 & 0 & -1 \\ 1 & 0 & 0 \end{pmatrix}, \qquad \gamma_{2314} \leftrightarrow \begin{pmatrix} 0 & 0 & 1 \\ -1 & 0 & 0 \\ 0 & -1 & 0 \end{pmatrix},$$

$$\gamma_{4132} \leftrightarrow \begin{pmatrix} 0 & 0 & -1 \\ 1 & 0 & 0 \\ 0 & -1 & 0 \end{pmatrix}, \qquad \gamma_{2431} \leftrightarrow \begin{pmatrix} 0 & 1 & 0 \\ 0 & 0 & -1 \\ -1 & 0 & 0 \end{pmatrix}, \qquad \gamma_{4213} \leftrightarrow \begin{pmatrix} 0 & 1 & 0 \\ 0 & 0 & 1 \\ 1 & 0 & 0 \end{pmatrix},$$

$$\gamma_{3241} \leftrightarrow \begin{pmatrix} 0 & 0 & 1 \\ 1 & 0 & 0 \\ 0 & 1 & 0 \end{pmatrix}, \qquad \gamma_{1423} \leftrightarrow \begin{pmatrix} 0 & 0 & -1 \\ -1 & 0 & 0 \\ 0 & 1 & 0 \end{pmatrix}, \qquad \gamma_{1342} \leftrightarrow \begin{pmatrix} 0 & -1 & 0 \\ 0 & 0 & 1 \\ -1 & 0 & 0 \end{pmatrix},$$

$$\gamma_{2143} \leftrightarrow \begin{pmatrix} 1 & 0 & 0 \\ 0 & -1 & 0 \\ 0 & 0 & -1 \end{pmatrix}, \qquad \gamma_{3412} \leftrightarrow \begin{pmatrix} -1 & 0 & 0 \\ 0 & 1 & 0 \\ 0 & 0 & -1 \end{pmatrix}, \qquad \gamma_{4321} \leftrightarrow \begin{pmatrix} -1 & 0 & 0 \\ 0 & -1 & 0 \\ 0 & 0 & 1 \end{pmatrix},$$

$$\gamma_{2134} \leftrightarrow \begin{pmatrix} 1 & 0 & 0 \\ 0 & 0 & -1 \\ 0 & -1 & 0 \end{pmatrix}, \qquad \gamma_{1324} \leftrightarrow \begin{pmatrix} 0 & -1 & 0 \\ -1 & 0 & 0 \\ 0 & 0 & 1 \end{pmatrix}, \qquad \gamma_{1243} \leftrightarrow \begin{pmatrix} 1 & 0 & 0 \\ 0 & 0 & 1 \\ 0 & 1 & 0 \end{pmatrix},$$

$$\gamma_{3214} \leftrightarrow \begin{pmatrix} 0 & 0 & 1 \\ 0 & 1 & 0 \\ 1 & 0 & 0 \end{pmatrix}, \qquad \gamma_{4231} \leftrightarrow \begin{pmatrix} 0 & 1 & 0 \\ 1 & 0 & 0 \\ 0 & 0 & 1 \end{pmatrix}, \qquad \gamma_{1432} \leftrightarrow \begin{pmatrix} 0 & 0 & -1 \\ 0 & 1 & 0 \\ -1 & 0 & 0 \end{pmatrix},$$

$$\gamma_{2341} \leftrightarrow \begin{pmatrix} 0 & 0 & 1 \\ 0 & -1 & 0 \\ -1 & 0 & 0 \end{pmatrix}, \qquad \gamma_{4123} \leftrightarrow \begin{pmatrix} 0 & 0 & -1 \\ 0 & -1 & 0 \\ 1 & 0 & 0 \end{pmatrix}, \qquad \gamma_{4312} \leftrightarrow \begin{pmatrix} -1 & 0 & 0 \\ 0 & 0 & 1 \\ 0 & -1 & 0 \end{pmatrix},$$

$$\gamma_{3421} \leftrightarrow \begin{pmatrix} -1 & 0 & 0 \\ 0 & 0 & -1 \\ 0 & 1 & 0 \end{pmatrix}, \qquad \gamma_{3142} \leftrightarrow \begin{pmatrix} 0 & -1 & 0 \\ 1 & 0 & 0 \\ 0 & 0 & -1 \end{pmatrix}, \qquad \gamma_{2413} \leftrightarrow \begin{pmatrix} 0 & 1 & 0 \\ -1 & 0 & 0 \\ 0 & 0 & -1 \end{pmatrix}$$

Table 1: Underlying 3-dimensional absolutely irreducible representation of $T$

**Definition 2.2** Let $H \subset \Gamma$ be a subgroup and let $\Theta : H \to S^1$ be a group homomorphism. We call
$$H^\Theta := \{(h, \Theta(h)) \in \Gamma \times S^1 : h \in H\}$$
a *twisted subgroup* of $\Gamma \times S^1$ and $\Theta$ the corresponding *twist*.

For twisted subgroups $H^\Theta$ of $\Gamma \times S^1$ acting on $N_\omega$ the following equation holds (see again GOLUBITSKY-STEWART-SCHAEFFER [9], Chapter XVI):

$$\dim N_\omega^{H^\Theta} = 2 \frac{1}{|H|} \sum_{h \in H} \text{Tr}(h) \cos \Theta(h). \qquad (2.11)$$

Here $\text{Tr}(h)$ denotes the trace of the matrix in Table 1 associated with the element $h \in H \subset \Gamma$.

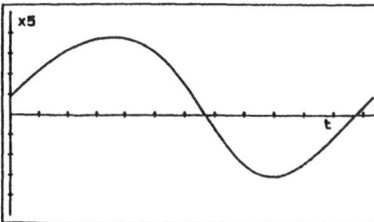
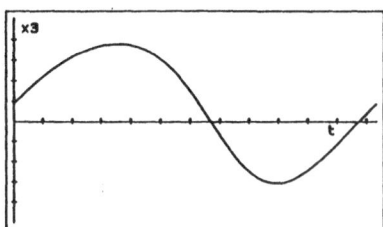
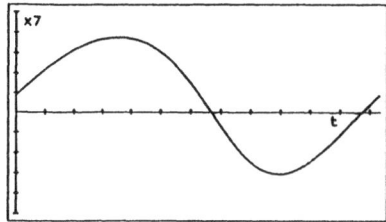
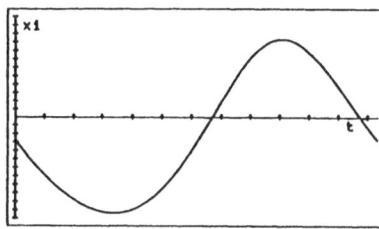

Figure 2: $\Sigma_1$-symmetric periodic solution of (2.9); $\lambda = 0.0209$, $\tau = 12.323$

Using (2.11) it is an easy task to verify that the following twisted subgroups of $\Gamma \times S^1$ satisfy the condition on the dimension in (2.10):

$$
\begin{aligned}
\Sigma_1 &= \{(\gamma_{1234}, 0), (\gamma_{1423}, 0), (\gamma_{1342}, 0), \\
&\quad (\gamma_{1243}, 0), (\gamma_{1324}, 0), (\gamma_{1432}, 0)\} \\
\Sigma_2 &= \{(\gamma_{1234}, 0), (\gamma_{3412}, 0), (\gamma_{3214}, 0), (\gamma_{1432}, 0), \\
&\quad (\gamma_{2143}, \pi), (\gamma_{4123}, \pi), (\gamma_{2341}, \pi), (\gamma_{4321}, \pi)\} \\
\Sigma_3 &= \{(\gamma_{1234}, 0), (\gamma_{2413}, \tfrac{3}{2}\pi), (\gamma_{4321}, \pi), (\gamma_{3142}, \tfrac{1}{2}\pi)\} \\
\Sigma_4 &= \{(\gamma_{1234}, 0), (\gamma_{1243}, 0), (\gamma_{2134}, \pi), (\gamma_{2143}, \pi)\} \\
\Sigma_5 &= \{(\gamma_{1234}, 0), (\gamma_{4132}, \tfrac{2}{3}\pi), (\gamma_{2431}, \tfrac{4}{3}\pi)\}
\end{aligned}
\tag{2.12}
$$

Therefore we know that in the Hopf bifurcation point $(x_0, \lambda_0)$ at least 5 branches of periodic solutions with symmetry types (isotropy subgroups) $\Sigma_1$-$\Sigma_5$ are bifurcating. Numerically computed corresponding periodic solutions of (2.9) with period $\tau$ are shown in Figures 2 – 6. The numerical method which has been used will be described in the next section.

Of course, for every isotropy subgroup $\Sigma_j$ $(j = 1, \ldots, 5)$ there is a number of *conjugate* isotropy subgroups, which also satisfy (2.10). But those conjugate subgroups essentially describe the same type of symmetry. For instance, the subgroup

$$
\Sigma_4' := \{(\gamma_{1234}, 0), (\gamma_{2134}, 0), (\gamma_{1243}, \pi), (\gamma_{2143}, \pi)\}
$$

is conjugate to $\Sigma_4$ and the only difference for a corresponding periodic solution is that the concentrations $x_1, x_7$ and $x_3, x_5$ in Figure 5 are interchanged.

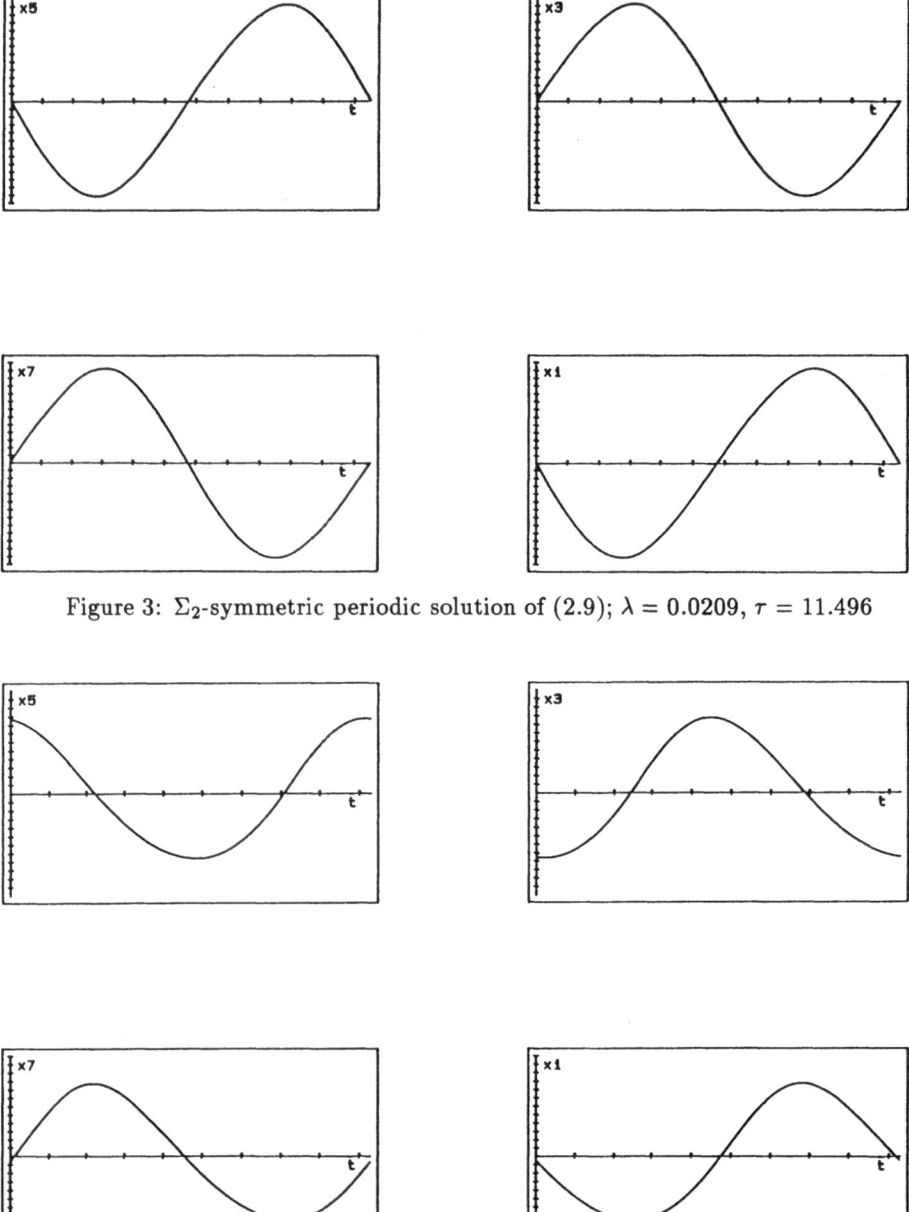

Figure 3: $\Sigma_2$-symmetric periodic solution of (2.9); $\lambda = 0.0209$, $\tau = 11.496$

Figure 4: $\Sigma_3$-symmetric periodic solution of (2.9); $\lambda = 0.0200$, $\tau = 9.273$

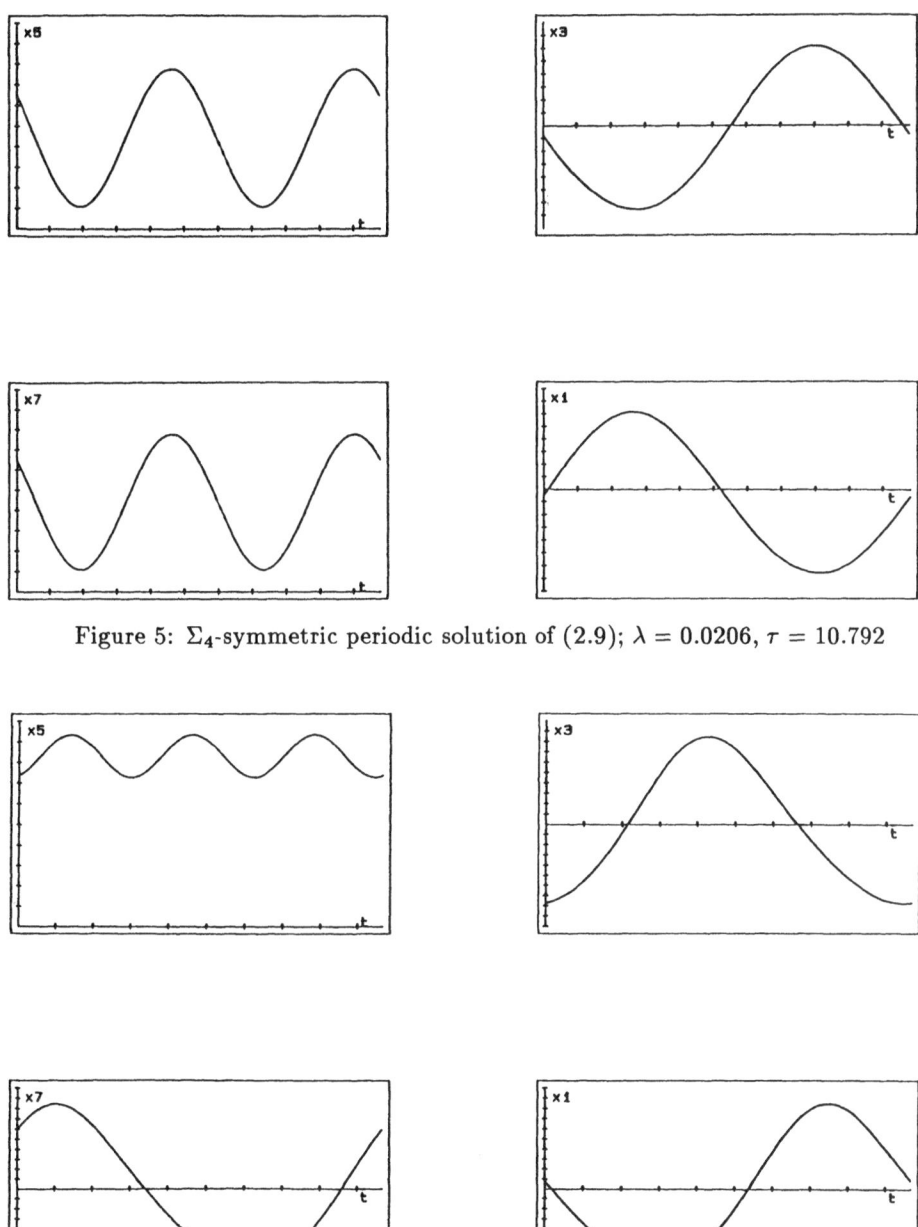

Figure 5: $\Sigma_4$-symmetric periodic solution of (2.9); $\lambda = 0.0206$, $\tau = 10.792$

Figure 6: $\Sigma_5$-symmetric periodic solution of (2.9); $\lambda = 0.0202$, $\tau = 9.663$

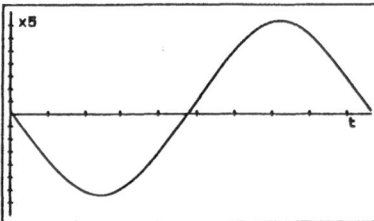
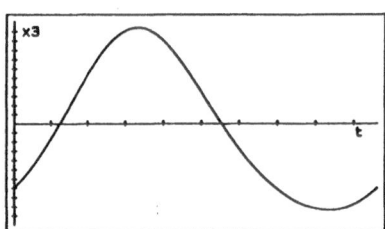

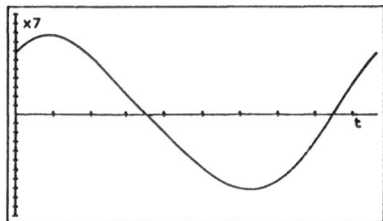
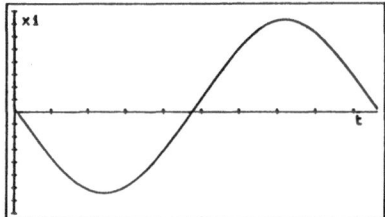

Figure 7: $\tilde{\Sigma}_6$-symmetric periodic solution of (2.9); $\lambda = 0.0202$, $\tau = 9.607$

**Remark 2.3**  a) The subgroups $\tilde{\Sigma}_1, \tilde{\Sigma}_2$ of $\Gamma$, consisting of the purely spatial parts of $\Sigma_1, \Sigma_2$ (the elements $(h,0)$ of $\Sigma_1, \Sigma_2$), are *bifurcation subgroups* of $\Gamma$ related to the absolutely irreducible representation in Table 1 (cf. DELLNITZ-WERNER [4]). The reason for this is that the subgroups

$$\tilde{\Sigma}_1 = \Sigma_1,$$
$$\tilde{\Sigma}_2 = \{(\gamma_{1234}, 0), (\gamma_{3412}, 0), (\gamma_{3214}, 0), (\gamma_{1432}, 0)\}$$

already satisfy the condition (2.10). Consequently, for a <u>static</u> bifurcation point with the underlying absolutely irreducible representation given in Table 1, at least branches of steady-state solutions with this type of symmetry have to bifurcate. We know that for the four-cell-Brusselator in (2.9) such a bifurcation point and these bifurcating branches exist. (See also the (numerical) results for another problem with this type of symmetry in EILBECK-LOMDAHL-SCOTT [5] (Fig. 3, p. 325).)

b) We have also found numerically a branch of periodic solutions bifurcating in the symmetry-breaking Hopf bifurcation point $(x_0, \lambda_0)$ with *submaximal symmetry* $\tilde{\Sigma}_6$, where

$$\tilde{\Sigma}_6 = \{(\gamma_{1234}, 0), (\gamma_{3214}, 0)\}.$$

One of these solutions is shown in Figure 7.
The submaximal character of this symmetry is immediately seen by comparison of this solution with that in Figure 2. It possesses less spatial symmetry and no additional temporal symmetry. In other words, $\tilde{\Sigma}_6$ is a proper subgroup of a subgroup of $\Gamma$ being conjugate to $\Sigma_1$ and $\Sigma_1$ itself is a maximal isotropy subgroup of $\Gamma \times S^1$, since it is a bifurcation subgroup of $\Gamma$ (see also DELLNITZ [3]).

## 3 Computation of Periodic Solutions

In this section we will show how the symmetry of periodic solutions can be utilized in order to obtain an efficient method for their numerical computation.

The method we will use is the following (cf. ZHENG [17]). Again we consider the system of ODEs (see (2.3))

$$\dot{y} = \frac{1}{\nu} g(y, \lambda), \tag{3.1}$$

where $g$ is $\Gamma$-equivariant. Our aim is to compute $2\pi$-periodic solutions $x(t, \nu, \lambda)$ of this equation.

Let

$$x(t) := \sum_{j=-\infty}^{\infty} c_j e^{ijt}, \quad c_j \in \mathbb{C}^n, \quad c_{-j} = \bar{c}_j, \tag{3.2}$$

be the Fourier expansion of $x \in C_{2\pi}$. Inserting (3.2) into (3.1) yields that $x(t)$ is a $2\pi$-periodic solution of (3.1) if and only if

$$\mathcal{F}(\mathbf{c}, \nu, \lambda) = 0, \tag{3.3}$$

where – identifying $\mathbb{C}$ with $\mathbb{R}^2$ – $\mathcal{F}$ is a mapping $\mathcal{L} \times \mathbb{R} \times \mathbb{R} \to \mathcal{L}$ ($\mathcal{L} := (l^2)^{2n}$, the space of sequences of Fourier vectors $\mathbf{c} := (c_j)_{j=0,1,2,...}$). In components, $\mathcal{F}$ can be written as

$$\mathcal{F}(\mathbf{c}, \nu, \lambda) = (d_j(\mathbf{c}, \nu, \lambda) - ijc_j)_{j=0,1,2,...}, \tag{3.4}$$

where

$$\sum_{j=-\infty}^{\infty} d_j(\mathbf{c}, \nu, \lambda) e^{ijt} \tag{3.5}$$

denotes the Fourier expansion of $\frac{1}{\nu} g(x(t), \lambda)$.

For the numerical treatment we truncate the Fourier expansions in (3.2), (3.5) in order to obtain an approximation

$$\sum_{j=-k+1}^{k-1} c_j e^{ijt} \approx x(t)$$

of a $2\pi$-periodic solution of (3.1). Therefore we have to solve the equation ($\mathbf{c}_f :=$ $(c_j)_{j=0,1,...,k-1}$)

$$\mathcal{F}_f(\mathbf{c}_f, \nu, \lambda) := (d_j(\mathbf{c}_f, \nu, \lambda) - ijc_j)_{j=0,1,...,k-1} = 0 \tag{3.6}$$

for the numerical computation of branches of periodic solutions of (3.1). The argument $\mathbf{c}_f$ of $d_j$ has to be interpreted as an element of $\mathcal{L}$ (it has to be filled up with zeros). Observe that (3.6) is a finite dimensional static bifurcation problem.

Now $\mathcal{F}_f$ is a mapping $Y \times \mathbb{R} \times \mathbb{R} \to Y$, $Y := X^{2k}$, and therefore we have to fix one component of $(\mathbf{c}_f, \nu)$ in order to be able to apply path following methods to the finite dimensional static bifurcation problem (3.6). An appropriate condition is obtained using the phase-shift equivariance of (3.1). We additionally force the vector $\mathbf{c}_f$ to fulfil

$$\Im(c_m^q) = 0, \tag{3.7}$$

where $\Im(c_m^q)$ denotes the imaginary part of the $q$-th component of $c_m$, $m \in \{1, \ldots, k-1\}$, $q \in \{1, \ldots, n\}$. This does not affect the possibility of computing every $2\pi$-periodic solution of (3.1), since

$$x(t + \vartheta) = \sum_{j=-\infty}^{\infty} (e^{ij\vartheta} c_j) e^{ijt}$$

is a $2\pi$-periodic solution of (3.1) iff $x(t)$ is one.

Now we will show how the symmetry of a periodic solution – given by its isotropy subgroup $\Sigma \subset \Gamma \times S^1$ – can be used to make the method described above more efficient. At first we define an action of the group $\Gamma \times S^1$ on $\mathcal{L}$ by

$$(\gamma, \theta)\mathbf{c} := (\gamma c_0, \gamma e^{i\theta} c_1, \gamma e^{i2\theta} c_2, \ldots) = (\gamma e^{ij\theta} c_j)_{j=0,1,2,\ldots} \tag{3.8}$$

and similarly on $Y$ by

$$(\gamma, \theta)\mathbf{c}_f := (\gamma c_0, \gamma e^{i\theta} c_1, \ldots, \gamma e^{i(k-1)\theta} c_{k-1}) = (\gamma e^{ij\theta} c_j)_{j=0,1,\ldots,k-1} . \tag{3.9}$$

**Proposition 3.1** *The mappings $\mathcal{F}$ and $\mathcal{F}_f$ are $\Gamma \times S^1$-equivariant:*

$$\left. \begin{array}{rcl} (\gamma, \theta)\mathcal{F}(\mathbf{c}, \nu, \lambda) & = & \mathcal{F}((\gamma, \theta)\mathbf{c}, \nu, \lambda) \\ (\gamma, \theta)\mathcal{F}_f(\mathbf{c}_f, \nu, \lambda) & = & \mathcal{F}_f((\gamma, \theta)\mathbf{c}_f, \nu, \lambda) \end{array} \right\} \quad \forall (\gamma, \theta) \in \Gamma \times S^1. \tag{3.10}$$

*Thereby the actions of $\Gamma \times S^1$ are defined in (3.8) and (3.9).*

PROOF: Let $(\gamma, \theta)$ be an element of $\Gamma \times S^1$. By the $\Gamma$-equivariance of $g$ we have (using the notation from above)

$$\begin{aligned} \sum_{j=-\infty}^{\infty} \gamma d_j(\mathbf{c}, \nu, \lambda) e^{ijt} & = \frac{1}{\nu} \gamma g(x(t), \lambda) = \\ & = \frac{1}{\nu} g(\gamma x(t), \lambda) = \\ & = \sum_{j=-\infty}^{\infty} d_j((\gamma, 0)\mathbf{c}, \nu, \lambda) e^{ijt} \end{aligned}$$

and

$$\begin{aligned} \sum_{j=-\infty}^{\infty} e^{ij\theta} d_j(\mathbf{c}, \nu, \lambda) e^{ijt} & = \frac{1}{\nu} g(x(t+\theta), \lambda) = \\ & = \sum_{j=-\infty}^{\infty} d_j((I, \theta)\mathbf{c}, \nu, \lambda) e^{ijt} . \end{aligned}$$

Therefore, by the uniqueness of the Fourier expansion, for all $(\gamma, \theta) \in \Gamma \times S^1$ we have

$$(\gamma, \theta)\mathbf{d}(\mathbf{c}, \nu, \lambda) = \mathbf{d}((\gamma, \theta)\mathbf{c}, \nu, \lambda).$$

From (3.4) the equivariance of $\mathcal{F}$ with respect to $\Gamma \times S^1$ immediately follows.

Defining the embedding $E_k : Y \to \mathcal{L}$ and the projection $P_k : \mathcal{L} \to Y$ by

$$E_k \mathbf{c}_f := (c_0, c_1, \ldots, c_{k-1}, 0, 0, \ldots), \quad P_k \mathbf{c} := (c_j)_{j=0,1,\ldots,k-1}$$

one easily sees that $E_k, P_k$ are also $\Gamma \times S^1$-equivariant with respect to the different actions of $\Gamma \times S^1$ in (3.8), (3.9).

Since $\mathcal{F}_f(\mathbf{c}_f, \nu, \lambda) = P_k \mathcal{F}(E_k \mathbf{c}_f, \nu, \lambda)$ for some $k$, also the equivariance of $\mathcal{F}_f$ immediately follows. ∎

In the next proposition the connection between symmetry types of functions in $\mathcal{C}_{2\pi}$ and of vectors in $\mathcal{L}$ is described.

**Proposition 3.2** *Let $\Sigma_u \subset \Gamma \times S^1$ be the isotropy subgroup of the $2\pi$-periodic function $u \in C_{2\pi}$ and let $\sum_{j=-\infty}^{\infty} c_j e^{ijt}$ be its Fourier expansion. Then*

$$(\sigma, \varphi)\mathbf{c} = \mathbf{c} \qquad \forall (\sigma, \varphi) \in \Sigma_u. \tag{3.11}$$

PROOF: We have for all $(\sigma, \varphi) \in \Sigma_u$

$$\sum_{j=-\infty}^{\infty} c_j e^{ijt} = u(t) = \sigma u(t + \varphi) = \sum_{j=-\infty}^{\infty} (\sigma e^{ij\varphi} c_j) e^{ijt}.$$

Now the equation (3.11) is immediately obtained using the fact that the Fourier expansion of a periodic function is unique. ∎

Let $u(t) = \sum_{j=-\infty}^{\infty} c_j e^{ijt}$ be a $2\pi$-periodic function with isotropy subgroup $\Sigma_u$. Then, by Proposition 3.2, $\mathbf{c} \in \mathcal{L}$ is an element of

$$\mathcal{L}^{\Sigma_u} := \{\mathbf{b} \in \mathcal{L} : (\sigma, \varphi)\mathbf{b} = \mathbf{b} \quad \forall (\sigma, \varphi) \in \Sigma_u\} \subset \mathcal{L},$$

the fixed point space of $\Sigma_u$ in the infinite dimensional space $\mathcal{L}$. Therefore – in order to compute $u$ approximately – we only have to compute solutions $(\mathbf{c}_f, \nu, \lambda)$ of (3.6), where $\mathbf{c}_f$ is an element of

$$Y^{\Sigma_u} := \{\mathbf{b}_f \in Y : (\sigma, \varphi)\mathbf{b}_f = \mathbf{b}_f \quad \forall (\sigma, \varphi) \in \Sigma_u\} \subset Y,$$

the fixed point space of $\Sigma_u$ in the finite dimensional space $Y$.

Since fixed point spaces are invariant spaces for equivariant mappings (cf. VANDERBAU-WHEDE [14]), by Proposition 3.1 we only have to consider the *reduced system*

$$\mathcal{F}_f^{\Sigma_u}(\mathbf{c}_f^{\Sigma_u}, \nu, \lambda) = 0 \tag{3.12}$$

instead of (3.6), where

$$\mathbf{c}_f^{\Sigma_u} \in Y^{\Sigma_u}$$

and

$$\mathcal{F}_f^{\Sigma_u} : Y^{\Sigma_u} \times \mathbb{R}^2 \to Y^{\Sigma_u}$$

is the restriction of $\mathcal{F}_f$ to $Y^{\Sigma_u}$. Observe that the system (3.12) in general has a much lower dimension than the original system (3.6).

With respect to the description of the explicit numerical implementation of reduced systems in $m$-cell-problems see WERNER [15]. We will indicate the usefulness of this reduction process by an example, namely the one of the last section.

**Example 3.3** *Four-cell-Brusselator with tetrahedral symmetry:*
In this example we will compute the fixed point spaces $\mathcal{L}^{\Sigma_j}$ for the subgroups $\Sigma_1, \Sigma_2, \Sigma_5$ in the list (2.12). We begin by rewriting (3.11) in components as

$$\sigma e^{ij\varphi} c_j = c_j \quad \forall (\sigma, \varphi) \in \Sigma_u, \quad j \in \mathbb{N}_0, \tag{3.13}$$

where $\Sigma_u$ is the isotropy subgroup of $u(t) = \sum_{j=-\infty}^{\infty} c_j e^{ijt}$.

- $\Sigma_u = \Sigma_1$: In this case there are only purely spatial symmetries in $\Sigma_u$ and (3.13) yields that

$$c_j = (a_j, b_j, b_j, b_j)^T, \quad j \in \mathbb{N}_0,$$

where $a_j, b_j \in \mathbb{C}^2$. Therefore the system (3.12) has half the dimension of the system (3.6).

- $\underline{\Sigma_u = \Sigma_2}$: Here we obtain from (3.13) that

$$
\left.\begin{array}{rcl}
c_{2j} & = & (a_j, a_j, a_j, a_j)^T \\
c_{2j+1} & = & (b_j, -b_j, b_j, -b_j)^T
\end{array}\right\} \quad j \in \mathbb{N}_0, \quad a_j, b_j \in \mathbb{C}^2,
$$

and the reduced system (3.12) only has a fourth of the dimension of the original one in (3.6).

- $\underline{\Sigma_u = \Sigma_5}$: Condition (3.13) yields

$$
\left.\begin{array}{rcl}
c_{3j} & = & (b_j, b_j, a_j, b_j)^T \\
c_{3j+1} & = & (c_j, e^{i2/3\pi} c_j, 0, e^{i4/3\pi} c_j)^T \\
c_{3j+2} & = & (d_j, e^{i4/3\pi} d_j, 0, e^{i2/3\pi} d_j)^T
\end{array}\right\} \quad j \in \mathbb{N}_0, \quad a_j, b_j, c_j, d_j \in \mathbb{C}^2.
$$

Even in this case, where the symmetry group $\Sigma_u$ has only order three, the dimension of the system (3.6) is reduced to a third in (3.12).

See again Figures 2, 3 and 6 for the corresponding numerically computed solutions of (3.1).

# References

[1] H. Amann. **Gewöhnliche Differentialgleichungen.** de Gruyter 1983.

[2] J. F. Cornwell. **Group Theory in Physics.** Vol. 1, Academic Press 1984.

[3] M. Dellnitz. **Hopf-Verzweigung in Systemen mit Symmetrie und deren Numerische Behandlung,** Wissenschaftliche Beiträge aus Europäischen Hochschulen, Reihe 11, Band 1, Verlag an der Lottbek, Ammersbek 1989.

[4] M. Dellnitz, B. Werner. *Computational methods for Bifurcation problems with symmetries – with special attention to steady state and Hopf bifurcation points.* Journal of Computational and Applied Mathematics **26**, 97-123, 1989.

[5] J.C. Eilbeck, P.S. Lomdahl, A.C. Scott. *The discrete self-trapping equation.* Physica 16D, 318-338, 1985.

[6] B. Fiedler. *Global Bifurcation of Periodic Solutions with Symmetry.* Lecture Notes in Mathematics, **1309**, Springer 1988.

[7] M. Golubitsky, I. Stewart. *Hopf bifurcation in the presence of symmetry.* Arch. Rat. Mech. Anal., **87**, 107-165, 1985.

[8] M. Golubitsky, I. Stewart. *Hopf bifurcation with dihedral group symmetry: coupled nonlinear oscillators.* In **Multiparameter Bifurcation Theory**, M. Golubitsky, J. Guckenheimer (eds.). Contemporary Mathematics **56**, 131-173. Amer. Math. Soc., Providence, 1986.

[9] M. Golubitsky, I. Stewart, D. Schaeffer. **Singularities and Groups in Bifurcation Theory.** Vol. 2, Springer 1988.

[10] M. Kubíček, M. Marek. **Computational Methods in Bifurcation Theory and Dissipative Structures.** Springer 1983.

[11] J. E. Marsden, M. McCracken. **The Hopf Bifurcation and Its Applications.** Springer 1976.

[12] I. Prigogine, R. Lefever. *Symmetry breaking instabilities in dissipative systems II.* Journal of Chem. Physics **48**, No. 4, 1695-1700, 1968.

[13] A. Steindl, H. Troger. *Bifurcation of the equilibrium of a spherical double pendulum at a multiple eigenvalue.* International Series of Numerical Mathematics **79**, 277-287, Birkhäuser 1987.

[14] A. Vanderbauwhede. **Local Bifurcation Theory and Symmetry.** Pitman 1982.

[15] B. Werner. *Computational methods for bifurcation problems with symmetries and applications to steady states of n-box reaction-diffusion models.* In **1987 Dundee Conference on Numerical Analysis.**, D. F. Griffiths, G. A. Watson (eds.). Pitman 1988.

[16] B. Werner. *Eigenvalue problems with the symmetry of a group and bifurcations.* This volume, 1990.

[17] Q. Zheng. *Ein Algorithmus zur Berechnung nichtlinearer Schwingungen bei Algebro-Differentialgleichungen.* Hamburger Beiträge zur Angewandten Mathematik, Reihe D, **3**, 1988.

# GLOBAL BIFURCATIONS AND THEIR NUMERICAL COMPUTATION

W.-J. BEYN
Fakultät für Mathematik
Universität Konstanz
Postfach 55 60
D-7750 Konstanz

ABSTRACT. Global bifurcations in dynamical systems often occur from homoclinic or heteroclinic orbits. The best known effect is the termination of a branch of periodic orbits at a homoclinic orbit. In this paper we extend our numerical approach to connecting orbits and the error analysis developed in [1]. The basic nondegeneracy condition is characterized by a geometric transversality condition. Further, the analysis of the error obtained by truncating to a finite interval is generalized in order to include periodic boundary conditions and to explain the superconvergence phenomenon with respect to the parameter as observed in [1].

## 1. INTRODUCTION

Global changes in the asymptotic regime of a parametrized dynamical system

$$\dot{x} = f(x,\lambda), \quad f \in C^1(\mathbb{R}^{m+p}, \mathbb{R}^m) \tag{1.1}$$

are often related to the appearance or disappearance of connecting orbits. Here we call a solution $(\bar{x}(t)(t\in\mathbb{R}), \bar{\lambda})$ of (1.1) a *connecting orbit pair* (COP) if

$$\bar{x}(t) \to \bar{x}_\pm \text{ as } t \to \pm\infty \quad \text{and} \quad f(\bar{x}_\pm, \bar{\lambda}) = 0. \tag{1.2}$$

For the homoclinic case (i.e. $\bar{x}_- = \bar{x}_+$) the most important effect is the birth of periodic orbits or more complicated invariant sets (see e.g. Shil'nikov [12], Sparrow [13], Guckenheimer & Holmes [9], Glendinning [8] and section 3 below). Moreover, both homoclinic and heteroclinic ($\bar{x}_+ \neq \bar{x}_-$) orbits occur when determining the shape and speed of traveling waves in parabolic systems, (see e.g. Fife [6]).

In Beyn [1] we have introduced the notion of *nondegenerate connecting orbit pairs*. these turn out to be regular solutions of the infinite boundary value problem (1.1), (1.2) if a suitable phase condition is added:

*D. Roose et al. (eds.), Continuation and Bifurcations: Numerical Techniques and Applications, 169–181.*
© 1990 *Kluwer Academic Publishers.*

$$\Psi(x,\lambda) = 0 \tag{1.3}$$

In this paper we characterize this nondegeneracy by the transversal intersection of certain stable and unstable manifolds.

Moreover, we extend our numerical approach from [ 1 ], which was based on earlier work of de Hoog & Weiß [ 3 ], Keller & Lentini [10]. Here we consider approximating finite boundary value problems of the following general type

$$\dot{x} = f(x,\lambda) \text{ on } J = [T_-, T_+] \tag{1.4a}$$

$$B(x(T_-), x(T_+), \lambda) = 0 \tag{1.4b}$$

$$\Psi_J(x,\lambda) \qquad = 0 \tag{1.4c}$$

We assume (1.4c) to be a scalar condition and (1.4b) to be a set of $m + p - 1$ so called *asymptotic boundary conditions*. The whole system (1.4) is then a boundary value problem of dimension $m + p$ for the unknowns $(x,\lambda)$. The general form (1.4b) includes the most efficient projection conditions (see [ 1 ]) as well as the periodic b.c. which are convenient for the homoclinic one-parameter case. For the solutions of (1.4) and (1.1) - (1.3) we present a detailed error analysis. This will explain the superconvergence phenomenon for the parameter as observed in [ 1 ] and it will give us a theorem on the bifurcation of periodic orbits from a homoclinic orbit.

A related approach to connecting orbits has been developed by Friedman & Doedel [ 5 ],[ 7 ]. The main difference in their approach is that the stationary points $\bar{x}_\pm$ as well as certain eigenvalues and eigenvectors of the linearizations $f_x(\bar{x}_\pm, \bar{\lambda})$ are introduced as new unknowns into the system. This simplifies the implementation of the boundary conditions but also increases the dimension of the system and requires some a-priori knowledge about the structure of the spectrum. Other differences relate to the integral phase condition and the use of weighted Banach spaces in [ 5 ],[ 7 ] whereas here we employ the theory of exponential dichotomies Coppel [ 2 ], Palmer [11].

## 2. A WELL-POSED PROBLEM FOR CONNECTING ORBITS

The theory of exponential dichotomies [ 2 ],[11] turns out to be a useful tool when dealing with linearizations at connecting orbits. Therefore we start with some results on linear differential operators

$$Lx = \dot{x} - A(t)x, \quad x \in C^1(J, \mathbf{R}^m), \quad A \in C^0(J, \mathbf{R}^{m,m}) \tag{2.1}$$

where $J$ is an open interval $(T_-, T_+)$. We include the cases where

$T_- = -\infty$ or $T_+ = +\infty$ or both.

L is said to have an *exponential dichotomy* on J, if there is a fundamental solution matrix $Y(t)$, $t \in J$, a projector P in $\mathbf{R}^m$ and constants $K, \alpha_1, \alpha_2 > 0$ such that for all $t, s \in J$

$$\|Y(t)PY(s)^{-1}\| \le K \exp(-\alpha_1(t-s)) \quad \forall t \ge s \qquad (2.2a)$$

$$\|Y(t)(I-P)Y(s)^{-1}\| \le K \exp(-\alpha_2(s-t)) \quad \forall s \ge t \qquad (2.2b)$$

For later estimates it is useful to introduce the *critical exponents* $(\alpha_s, \alpha_u)$ of L given by

$$\alpha_s = \sup\{\alpha_1 > 0: \text{ there exist } P, Y, \| \ \|, \alpha_2, K \text{ with } (2.2)\}$$

$$\alpha_u = \sup\{\alpha_2 > 0: \text{ there exist } P, Y, \| \ \|, \alpha_1, K \text{ with } (2.2)\}$$

It is easy to show (see [2], p. 16) that for any given $\| \ \|, Y(t)$ and any $\varepsilon > 0$ there exists a projector P and a constant K such that (2.2a) and (2.2b) hold with $\alpha_1 = \alpha_s - \varepsilon, \alpha_2 = \alpha_u - \varepsilon$. For the constant coefficient case $Lx = \dot{x} - Ax$ with a hyperbolic matrix A, we may take $J = \mathbf{R}$ and find the critical exponents

$$\alpha_s = \text{Min}\{-\text{Re}\lambda: \lambda \text{ is an eigenvalue of A with } \text{Re}\lambda < 0\} \qquad (2.3a)$$

$$\alpha_u = \text{Min}\{\text{Re}\lambda: \lambda \text{ is an eigenvalue of A with } \text{Re}\lambda > 0\} \qquad (2.3b)$$

where $P = P_s$ resp. $I-P = P_u$ are the projectors onto the stable resp. unstable subspaces of A.

A close inspection of the roughness theorem [2] and Lemma 3.4 in [11] also shows that the exponential dichotomy on an interval $(t_0, \infty)$ as well as the critical exponents are invariant under perturbations of $A(t)$ which vanish as $t \to \infty$. Combining these results gives us the first part of the following Lemma.

Lemma 2.1
For a linear operator $Lx = \dot{x} - A(t)x$ assume that $A(t) \to A_\pm$ as $t \to \pm\infty$ with hyperbolic matrices $A_\pm$ and let $\alpha_{\pm s}$, $\alpha_{\pm u}$ be the corresponding spectral bounds (2.3). Then, for any $t_0 \in \mathbf{R}$, L has an exponential dichotomy on both $(-\infty, t_0)$ and $(t_0, \infty)$ with critical exponents $(\alpha_{-s}, \alpha_{-u})$ and $(\alpha_{+s}, \alpha_{+u})$. Moreover, if $P_-$ and $P_+$ are projectors for which the dichotomy estimates (2.2) hold on $(-\infty, t_0)$ and $(t_0, \infty)$ then

$$Y(t)P_{\pm}Y(t)^{-1} \to P_{\pm s} \text{ as } t \to \pm\infty \tag{2.4}$$

where $P_{\pm s}$ denotes the projector onto the stable subspace of $A_{\pm}$. Finally, the adjoint operator $L*x = \dot{x} + A^T(t)x$ also has exponential dichotomies on $(-\infty, t_0)$ and $(t_0, \infty)$ with projectors $I - P_-^T$ and $I - P_+^T$. The critical exponents are $(\alpha_{-u}, \alpha_{-s})$ and $(\alpha_{+u}, \alpha_{+s})$.

The last statement follows immediately from (2.2) and the fact that $Y^{-1T}(t)$ is a fundamental matrix of $L*$.

In our next step we introduce the Banach spaces
$$X^k(J) = \{x \in C^k(J, \mathbb{R}^m) : x^{(j)}(t) \text{ converges as } t \to T_{\pm} \text{ for } j = 0, \ldots, k\}$$
$$\|x\|_k = \sum_{j=0}^{k} \sup\{\|x^{(j)}(t)\| : t \in J\}$$
The solutions of linear inhomogenous initial value problems in these spaces are described in the following Lemma (see [2] and [1], Appendix for a proof).

Lemma 2.2

Under the assumptions of Lemma 2.1 the general solution $x \in X^1(t_0, \infty)$ of $Lx = r \in X^0(t_0, \infty)$ is given by $x(t) = Y(t)Y(t_0)^{-1}\xi + (G_+r)(t)$ where
$$\xi \in R(P_+) \text{ and } (G_+r)(t) = \int_{t_0}^{t} Y(t)PY(s)^{-1}r(s)ds - \int_{t}^{\infty} Y(t)(I-P)Y(s)^{-1}r(s)ds$$

Finally we recall from [1] that a COP $(\bar{x}, \bar{\lambda}) \in X^1(\mathbb{R}) \times \mathbb{R}^p$ of the system (1.1) is called *nondegenerate* if the following conditions hold

$$A_{\pm} := f_x(\bar{x}_{\pm}, \bar{\lambda}) \text{ is hyperbolic with stable dimension } m_{\pm s} \tag{2.5}$$

$$p = m_{-s} - m_{+s} + 1 \tag{2.6}$$

the only solutions $(y, \mu) \in X^1(\mathbb{R}) \times \mathbb{R}^p$ of the variational system $\dot{y} = f_x(\bar{x}, \bar{\lambda})y + f_\lambda(\bar{x}, \bar{\lambda})\mu$ are $y = c\dot{\bar{x}}(c \in \mathbb{R}), \mu = 0$ $\tag{2.7}$

A nondegenerate COP $(\bar{x}, \bar{\lambda})$ was shown in [1] to be a regular solution of the operator equation

$$\dot{x} - f(x, \lambda) = 0, \quad \Psi(x, \lambda) = 0 \tag{2.8}$$

provided the phase condition $\Psi \in C^1(X^0(\mathbb{R}) \times \mathbb{R}^p, \mathbb{R})$ satisfies

$$\Psi(\bar{x}, \bar{\lambda}) = 0, \quad \Psi_x(\bar{x}, \bar{\lambda})\dot{\bar{x}} \neq 0 \tag{2.9}$$

The following Theorem gives a geometrical equivalent of the nondegeneracy condition (2.7). First we apply the implicit function theorem to obtain locally unique stationary points $x_{\pm}(\lambda)$ of (1.1) such that

$$x_\pm(\bar\lambda) = \bar x_\pm.$$

## Theorem 2.3

Let $(\bar x, \bar\lambda) \in X^1(\mathbf{R}) \times \mathbf{R}^p$ be a COP of the system (1.1) which satisfies (2.5) and (2.6). Then $(\bar x, \bar\lambda)$ is nondegenerate if and only if the immersed stable and unstable manifolds

$$M_{+s} = \{(x,\lambda) : \|\lambda - \bar\lambda\| < \varepsilon, \ \Phi(t,x,\lambda) \to x_+(\lambda) \text{ as } t \to \infty\}$$

$$M_{-u} = \{(x,\lambda) : \|\lambda - \bar\lambda\| < \varepsilon, \ \Phi(t,x,\lambda) \to x_-(\lambda) \text{ as } t \to -\infty\}$$

(2.10)

intersect transversely at any point $(\bar x(t_0), \bar\lambda)$, $t_0 \in \mathbf{R}$, of the COP.

Remark: In (2.10) we have just given the set definition of $M_{+s}$, $M_{-u}$ by using the t-flow $\Phi(t,\cdot,\lambda)$ of (1.1). The manifold structure will be made precise in the following proof. For a two-dimensional saddle-saddle connection the transverse intersection is illustrated in Figure 1.

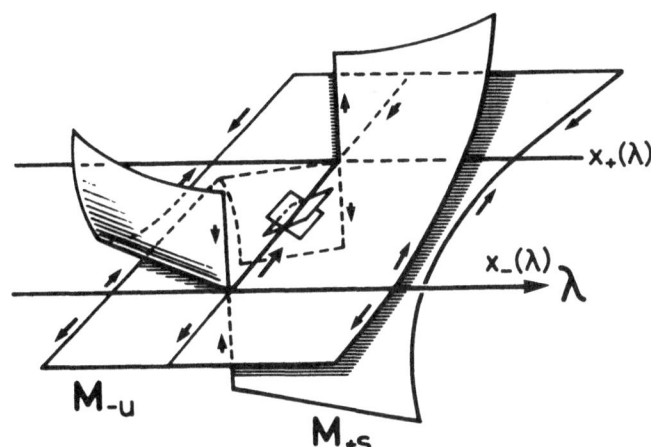

Figure 1. Illustration of transversal intersection for the two-dimensional heteroclinic case.

Proof: We fix some $t_0 \in \mathbf{R}$ and apply Lemma 2.1 to the linearization $Lx = \dot x - f_x(\bar x, \bar\lambda)x$. The parametrization of $M_{+s}$ near $(\bar x(t_0), \bar\lambda)$ will be obtained from the initial value problem

$$\dot y = f(y,\lambda) \text{ in } [t_0, \infty), \quad P_+(y(t_0) - \bar x(t_0)) = \xi \in R(P_+)$$

(2.11)

In fact, this may be written as an operator equation for
$(y,\xi,\lambda) \in X^1(t_0,\infty) \times R(P_+) \times \mathbf{R}^p$. Using Lemma 2.2 we can invoke the
implicit function theorem to obtain solutions $y(\cdot,\xi,\lambda) \in X^1(t_0,\infty)$ with
$\xi$ and $\lambda - \bar\lambda$ close to zero such that $y(t,0,\bar\lambda) = \bar{x}(t)$. $M_{+s}$ is then locally
parametrized by $(y(t_0,\xi,\lambda),\lambda)$ with tangent space

$$T_1 := T_{(\bar{x}(t_0),\bar\lambda)}M_{+s} = \{(\xi + z_+(t_0)\lambda,\lambda) : \xi \in R(P_+), \lambda \in \mathbf{R}^p\}$$

where $z_+(t) = (G_+ f_\lambda(\bar{x},\bar\lambda))(t)$. Similarly, we find

$$T_2 := T_{(\bar{x}(t_0),\bar\lambda)}M_{-u} = \{(\eta + z_-(t_0)\lambda,\lambda) : \eta \in N(P_-), \lambda \in \mathbf{R}^p\}$$

where $z_-(t) = (G_- f_\lambda(\bar{x},\bar\lambda))(t) = \int_{-\infty}^{t} Y(t)P_- Y(s)^{-1} f_\lambda(\bar{x}(s),\bar\lambda)ds$

$$- \int_{t}^{t_0} Y(t)(I - P_-)Y(s)^{-1} f_\lambda(\bar{x}(s),\bar\lambda)ds.$$

From $\dot{\bar{x}} \in N(L)$ we conclude $\dot{\bar{x}}(t_0) \in R(P_+) \cap N(P_-)$, hence
$(\dot{\bar{x}}(t_0),0) \in T_1 \cap T_2$ and $\dim (T_1 \cap T_2) \geq 1$. From (2.6) we obtain
$\dim (T_1 + T_2) = \dim(T_1) + \dim(T_2) - \dim(T_1 \cap T_2) = m_{+s} + p + m - m_{-s} + p$
$-\dim(T_1 \cap T_2) = m + p + 1 - \dim(T_1 \cap T_2) \leq m + p$. Thus the transversality
condition $T_1 + T_2 = \mathbf{R}^{m+p}$ is equivalent to

$$T_1 \cap T_2 = \text{span}\{(\dot{\bar{x}}(t_0),0)\} \tag{2.12}$$

Let us now assume that $(\bar{x},\bar\lambda)$ is nondegenerate and consider an element

$$(\xi + z_+(t_0)\mu,\mu) = (\eta + z_-(t_0)\mu,\mu) \in T_1 \cap T_2 \tag{2.13}$$

where $\xi \in R(P_+)$, $\eta \in N(P_-)$. Then the function

$$y(t) = \begin{cases} z_+(t)\mu + Y(t)Y(t_0)^{-1}\xi , & t \geq t_0 \\ z_-(t)\mu + Y(t)Y(t_0)^{-1}\eta , & t < t_0 \end{cases} \tag{2.14}$$

is continuous at $t_0$ and satisfies $Ly = f_\lambda(\bar{x},\bar\lambda)\mu$. Hence $\mu = 0$ and $y = c\dot{\bar{x}}$
for some $c \in \mathbf{R}$ by (2.7) and we have $(\xi + z_+(t_0)\mu,\mu) = (c\dot{\bar{x}}(t_0),0)$.
Conversely, assume (2.12) and let $y \in X^1(t_0,\infty)$, $\mu \in \mathbf{R}$ satisfy
$Ly = f_\lambda(\bar{x},\bar\lambda)\mu$. Then by Lemma 2.2 and the analogous result on $(-\infty,t_0)$ we
may write $y$ as in (2.14) for some $\xi \in R(P_+)$, $\eta \in N(P_-)$. The continuity of
$y$ now yields (2.13) and we find $\mu = 0$, $y(t_0) = c\dot{\bar{x}}(t_0)$ from our assumption.
Finally, (2.14) gives us $y(t) = c\, Y(t)Y(t_0)^{-1}\dot{\bar{x}}(t_0) = c\dot{\bar{x}}(t)$ for all $t \in \mathbf{R}$

□

## 3. THE APPROXIMATION ERROR

We consider numerical approximations of a COP obtained by solving the finite boundary value problem (1.4). Condition (1.4c) is a finite phase condition, such as

$$\Psi_J(x,\lambda) = \int_{T_-}^{T_+} \dot{x}_0(t)^T (x(t) - x_0(t))dt, \tag{3.1}$$

where $x_0$ is an initial approximation (e.g. given by the last solution on a branch of COP's). For more general integral conditions see [5], [7].

A very efficient type of asymptotic boundary conditions (1.4b) are the so called *projection boundary conditions* which have been analyzed in detail in [1],[3]. They are defined by

$$B(x(T_-), x(T_+),\lambda) = B_-(\lambda)(x(T_-) - x_-(\lambda)), B_+(\lambda)(x(T_+) - x_+(\lambda))) \tag{3.2}$$

where the rows of $B_-(\lambda) \in \mathbf{R}^{m-s,m}$ resp. $B_+(\lambda) \in \mathbf{R}^{m+u,m}$ form a basis of the stable subspace of $f_x^T(x_-(\lambda),\lambda)$ resp. the unstable subspace of $f_x^T(x_+(\lambda),\lambda)$. Notice that $m_{+u} + m_{-s} = m + p - 1$ follows from (2.6) and that the projection conditions force the endpoints $x(T_\pm)$ to lie in the linearized stable and unstable manifolds of $x_\pm(\lambda)$. Of course, (3.2) requires to compute the stationary solutions $x_\pm(\lambda)$ as well as smooth bases of stable and unstable subspaces (see [1] for more details on the implementation).

In the homoclinic case ($p = 1$) a simple alternative are periodic boundary conditions

$$B(x(T_-), x(T_+),\lambda) = x(T_+) - x(T_-) \tag{3.3}$$

and this has been extensively used by Doedel & Kernevez [4]. However, the periodicity condition introduces larger errors than the projection conditions.

For an illustration of these errors we consider the following two-dimensional system from [1]

$$\dot{x}_1 = x_2, \quad \dot{x}_2 = x_1 - x_1^2 + \lambda x_2 + \mu x_1 x_2 \tag{3.4}$$

For fixed $\mu$ this system has at some $\bar{\lambda} = \bar{\lambda}(\mu)$ a homoclinic orbit based at the origin. For $\mu = 0.5$ and the simple phase condition $x_2(0) = 0$, table 1 shows the errors

$$e_x(T) = \sup\{\|\bar{x}(t) - x_T(t)\| : |t| \le T\}, \quad e_\lambda(T) = |\bar{\lambda} - \lambda_T|$$

where $x_T, \lambda_T$ denotes the solution of (1.4) on $[-T,T]$. Actually, the solution on $[-15,15]$ was taken as exact solution and all finite boundary value problems were solved at high accuracy ($\sim 10^{-15}$) so that the error arising from truncation to a finite interval becomes dominant. Anticipating an error behaviour $O(e^{-\delta T})$ we also display the terms $\delta_{x,\lambda}(T) = \ln e_{x,\lambda}(T-1) - \ln e_{x,\lambda}(T)$.

| T | $e_x(T)$ | $\delta_x(T)$ | $e_\lambda(T)$ | $\delta_\lambda(T)$ | |
|---|---|---|---|---|---|
| 5 | 1.01 E-3 | 1.59 | 3.24 E-5 | 2.57 | |
| 6 | 2.02 E-4 | 1.61 | 2.13 E-6 | 2.72 | projection |
| 7 | 4.03 E-5 | 1.61 | 1.31 E-7 | 2.79 | b.c. |
| 8 | 8.03 E-6 | 1.61 | 7.79 E-9 | 2.82 | |
| 9 | 1.60 E-6 | 1.62 | 4.57 E-10 | 2.84 | |
| 5 | 1.02 E-1 | 0.863 | 5.15 E-3 | 1.45 | |
| 6 | 4.53 E-2 | 0.813 | 1.11 E-3 | 1.53 | periodic |
| 7 | 2.02 E-2 | 0.806 | 2.30 E-4 | 1.58 | b.c. |
| 8 | 9.03 E-3 | 0.806 | 4.66 E-5 | 1.60 | |
| 9 | 4.03 E-3 | 0.807 | 9.33 E-6 | 1.61 | |

Table 1

We note that $\bar{\lambda} = -0.429505849$ and that $f_x(0,\bar{\lambda})$ has eigenvalues $-\alpha_s = -1.237552425$, $\alpha_u = 0.808046576$. Table 1 then suggests that the exponents for the x-error are $2\alpha_u$ for the projection conditions and $\alpha_u$ for the periodic conditions. The $\lambda$-error, however, shows a super-convergence with exponents $2\alpha_u + \alpha_s$ and $2\alpha_u$ respectively. These effects will be explained by the following approximation theorem which is a generalization of the corresponding result [1], Theorem 3.2. We will write $J \to R$ for $J = (T_-, T_+)$, $T_\pm \to \pm\infty$ and we use the phrase 'J sufficiently large' correspondingly.

Theorem 3.1
Let $(\bar{x}, \bar{\lambda})$ be a nondegenerate COP of (1.1) with endpoints $\bar{x}_\pm$ and assume $f \in C^2(R^{m+p}, R^m)$. Further assume

A1: $B \in C^2(\mathbb{R}^{2m+p}, \mathbb{R}^{m+p-1})$, $B(\bar{x}_-,\bar{x}_+,\bar{\lambda}) = 0$ and the

matrix $(\bar{B}_- C_{-s}, \bar{B}_+ C_{+u}) \in \mathbb{R}^{m+p-1,m+p-1}$ is nonsingular.

Here $\bar{B}_\pm = \frac{\partial B}{\partial x(T_\pm)}(\bar{x}_-,\bar{x}_+,\bar{\lambda})$ and the columns of the matrices

$C_{+u} \in \mathbb{R}^{m,m+u}$ resp. $C_{-s} \in \mathbb{R}^{m,m-s}$ form a basis of the unstable

resp. stable subspace of $f_x(\bar{x}_\pm,\bar{\lambda})$.

A2: $\Psi_J \in C^2(X^0(J) \times \mathbb{R}^p, \mathbb{R})$, $\Psi_J(\bar{x}_{|J},\bar{\lambda}) \to 0$ as $J \to \mathbb{R}$,

$|\Psi_{J,x}(\bar{x}_{|J},\bar{\lambda})\dot{\bar{x}}_{|J}| \geq g > 0$ for all large J and the derivatives

$\Psi_J', \Psi_J''$ are bounded uniformly in J in some tube $K_\delta$, $\delta > 0$

(see (3.5)).

Then there exists a constant $\rho \leq \delta$ such that the boundary value
problem (1.4) has a unique solution $(x_J, \lambda_J)$ in

$$K_\rho = \{(x,\lambda) \in X^1(J) \times \mathbb{R}^p : \|x-\bar{x}_{|J}\|_1 + \|\lambda - \bar{\lambda}\| \leq \rho\} \tag{3.5}$$

for sufficiently large J. Moreover, there exist $\tau_J \to 0$ as $J \to \mathbb{R}$ and to
any $\varepsilon > 0$ a constant $C_\varepsilon$ such that the following estimates hold with
$d = 1$ and $\bar{y}_J(t) = \bar{x}(t + \tau_J)$

$$\|\bar{y}_J - x_J\|_1 \leq C_\varepsilon \{\exp(-(d\alpha_{+s} - \varepsilon)T_+) + \exp(-(d\alpha_{-u} - \varepsilon)|T_-|)\} \tag{3.6}$$

$$\|\bar{\lambda} - \lambda_J\| \leq C_\varepsilon \{\exp(-(\delta_+ - \varepsilon)T_+) + \exp(-(\delta_- - \varepsilon)|T_-|)\}, \tag{3.7}$$

$$\delta_+ = \text{Min}(2d\alpha_{+s}, d\alpha_{+s} + \alpha_{+u}), \delta_- = \text{Min}(2d\alpha_{-u}, d\alpha_{-u} + \alpha_{-s}).$$

Here $\alpha_{\pm s}$ and $\alpha_{\pm u}$ are the spectral bounds for $f_x(\bar{x}_\pm,\bar{\lambda})$ as in (2.3).
Finally, for the projection conditions the above estimates hold with
$d = 2$.

Remark: The phase shift $\tau_J$ is constructed in such a way that
$\Psi_J(\bar{y}_J,\bar{\lambda}) = 0$ holds. If we replace $\bar{y}_J$ by $\bar{x}_{|J}$ in (3.6) then error terms
depending on $\Psi_J$ will appear (see [1]).

Before proceeding to the proof we notice some important consequences
of Theorem 3.1. Assumption A1 is satisfied for the projection conditions
if f is in $C^3$ and A2 is a rather mild though technical assumption on $\Psi_J$

which is satisfied in standard cases. For our example above we have $\alpha_u < \alpha_s < 2\alpha_u$, thus (3.6) gives the exponent $2\alpha_u$ and (3.7) the exponent $2\alpha_u + \alpha_s$.

Similarly, A1 is automatically satisfied in the homoclinic case (p=1) for periodic boundary conditions and we obtain the exponents $\alpha_u$ and $2\alpha_u$ for our example as in Table 1. In the periodic case, Theorem 3.1 also yields a general result on bifurcation of periodic orbits from a homoclinic orbit. We take $J = (-T,T)$ and the simple phase condition

$$\Psi_J(x,\lambda) = \dot{\bar{x}}(0)^T(x(0) - \bar{x}(0)) \tag{3.8}$$

Then A2 is satisfied and we also have $\tau_J = 0$ since $\Psi_J(\bar{x}|_{[-T,T]}, \bar{\lambda}) = 0$

## Corollary 3.2

Let $(\bar{x}, \bar{\lambda})$ be a nondegenerate homoclinic orbit pair of a one-parameter dynamical system (1.1) with f in $C^2$. Then there exists a $T_0 > 0$ and a branch of 2T-periodic orbits $x_T \in X^1(-T,T), \lambda_T \in \mathbb{R}$ $(T \geq T_0)$ which after a suitable choice of phase satisfy the estimates

$$\|\bar{x}|_{[-T,T]} - x_T\|_1 \leq C_\varepsilon \exp(-(\text{Min}(\alpha_s,\alpha_u) - \varepsilon)T)$$

$$|\bar{\lambda} - \lambda_T| \leq C_\varepsilon \exp(-(2\text{Min}(\alpha_s,\alpha_u) - \varepsilon)T)$$

where $\alpha_s, \alpha_u$ are the spectral bounds for $f_x(\bar{x}_+, \bar{\lambda})$ according to (2.3).

Even in the two-dimensional case this result is more general than the standard global bifurcation theorems. For example, the saddle connection bifurcation in [9], Ch. 6 requires a nonvanishing trace in addition to our nondegeneracy condition (which is formulated there in a geometric way similar to Theorem 2.3).

Proof of Theorem 3.1. We merely sketch the basic steps of the proof which is very similar to that of Theorem 3.2 in Beyn [1]. Let us write (1.4) as an operator equation

$$F_J(x,\lambda) = 0 \tag{3.9}$$

where $F_J : X^1(J) \times \mathbb{R}^p \to X^0(J) \times \mathbb{R}^{m+p}$ is defined by the left hand sides of (1.4). The unique solvability of (3.9) in some $K_\rho$ follows from a local contraction theorem (see [1], Lemma 3.1) provided we have a uniform bound for the inverse of the Frechet derivative $F_J'(\bar{x}|_J, \bar{\lambda})$. For that purpose we set

$$Lx = \dot{x} - f_x(\bar{x}_{|J}, \bar{\lambda})x, \quad \bar{B}_{\pm} = \frac{\partial B}{\partial x(T_{\pm})} (\bar{x}(T_-), \bar{x}(T_+), \bar{\lambda})$$

$$\bar{B}_{\lambda} = \frac{\partial B}{\partial \lambda} (\bar{x}(T_-), \bar{x}(T_+), \bar{\lambda}), \quad \bar{f}_{\lambda} = f_{\lambda}(\bar{x}_{|J}, \bar{\lambda})$$

and consider the variational equations

$$Ly - \bar{f}_{\lambda}\mu = r \in X^0(J) \tag{3.10a}$$

$$\bar{B}_- y(T_-) + \bar{B}_+ y(T_+) + \bar{B}_{\lambda}\mu = b \in \mathbb{R}^{m+p-1} \tag{3.10b}$$

$$\Psi'_J(\bar{x}_{|J}, \bar{\lambda})(y, \mu) = \beta \in \mathbb{R} \tag{3.10c}$$

For these we show an estimate uniformly in J

$$\begin{pmatrix} \|y\|_1 \\ \|\mu\| \end{pmatrix} \leq C_{\varepsilon} \begin{pmatrix} 1 & 1 \\ 1 & \gamma(\varepsilon, J) \end{pmatrix} \begin{pmatrix} \|r\|_0 \\ \|b\| + |\beta| \end{pmatrix} \tag{3.11}$$

where $\gamma(\varepsilon, J) = \exp(-(\alpha_{+u} - \varepsilon) T_+) + \exp(-(\alpha_{-s} - \varepsilon)|T_-|)$.

First we take a matrix function $\Phi = (\varphi_1, \ldots, \varphi_p)$ such that the columns $\varphi_i \in X^1(\mathbb{R})$ form a basis of $N(L^*)$(see [1] Prop. 2.4). By Lemma 2.1 these satisfy an estimate

$$\|\varphi_i(t)\| \leq C_{\varepsilon} \begin{cases} \exp(-(\alpha_{-s} - \varepsilon)t), & t \leq 0 \\ \exp(-(\alpha_{+u} - \varepsilon)t), & t \geq 0 \end{cases} \tag{3.12}$$

We multiply (3.10a) by $\Phi^T(t)$, integrate over J and find by partial integration

$$\Phi^T(t)y(t) \Big|_{T_-}^{T_+} - \int_{T_-}^{T_+} \Phi^T(t)\bar{f}_{\lambda}(t)dt\mu = \int_{T_-}^{T_+} \Phi^T(t)r(t)dt$$

Using (3.12) and the nondegeneracy of $(\bar{x}, \bar{\lambda})$([1], Prop. 2.4) this gives us the estimate

$$\|\mu\| \leq C_{\varepsilon} (\|r\|_0 + \gamma(\varepsilon, J) \|y\|_0) \tag{3.13}$$

which is the key to the superconvergence phenomenon. In much the same way as in ([1], Appendix) we use A1 to find an estimate $\|y\|_1 \leq C(\|r\|_0 + \|b\| + |\beta|)$. Combining this with (3.13) yields (3.11).

In the next step we replace $\gamma(\varepsilon, J)$ by 1 in (3.11) and then obtain an inequality

$$\|\bar{y}_J - x_J\|_1 + \|\bar{\lambda} - \lambda_J\| \leq C_{\varepsilon}\|B(\bar{y}_J(T_-), \bar{y}_J(T_+), \bar{\lambda})\| =: \sigma(\varepsilon, J) \tag{3.14}$$

as in ([1], Theorem 3.2). Using A1 and the exponential approach of the connecting orbit towards the stationary points we find

$$\sigma(\varepsilon,J) = O(\exp(-(d\alpha_{+s} - \varepsilon)T_+) + \exp(-(d\alpha_{-u} - \varepsilon)|T_-|)) \tag{3.15}$$

which proves (3.6).

For the refined estimate (3.7) we first notice that (3.11) also holds with $\bar{y}_J$ in place of $\bar{x}_{|J}$. We then apply (3.11) with $y = \bar{y}_J - x_J$, $\mu = \bar{\lambda} - \lambda_J$ and find for the right hand sides by the $C^2$-smoothness of $f, B$ and $\psi_J$

$$\|r\|_0 = \|f(\bar{y}_J,\bar{\lambda}) - f(x_J,\lambda_J) - f_x(\bar{y}_J,\bar{\lambda})(\bar{y}_J - x_J) - f_\lambda(\bar{y}_J,\bar{\lambda})(\bar{\lambda} - \lambda_J)\|_0$$

$$\leq C_\varepsilon(\|\bar{y}_J - x_J\|_0 + \|\bar{\lambda} - \lambda_J\|)^2 \leq C_\varepsilon\sigma(\varepsilon,J)(\|\bar{y}_J - x_J\|_0 + \|\bar{\lambda} - \lambda_J\|)$$

$$\|b\| \leq \|B(\bar{y}_J(T_+),\bar{y}_J(T_-),\bar{\lambda})\| + C_\varepsilon(\|\bar{y}_J - x_J\|_0 + \|\bar{\lambda} - \lambda_J\|)^2$$

$$\leq \sigma(\varepsilon,J)(1 + C_\varepsilon(\|\bar{y}_J - x_J\|_0 + \|\bar{\lambda} - \lambda_J\|))$$

$$|\beta| \leq C_\varepsilon\sigma(\varepsilon,J)(\|\bar{y}_J - x_J\|_0 + \|\bar{\lambda} - \lambda_J\|).$$

Inserting this into (3.11) and taking the terms involving $\|\bar{y}_J - x_J\|_0 + \|\bar{\lambda} - \lambda_J\|$ to the left we end up with

$$\begin{pmatrix} \|\bar{y}_J - x_J\|_1 \\ \|\bar{\lambda} - \lambda_J\| \end{pmatrix} \leq C_\varepsilon \begin{pmatrix} 1 & 1 \\ 1 & \sigma(\varepsilon,J) + \gamma(\varepsilon,J) \end{pmatrix} \begin{pmatrix} 0 \\ \sigma(\varepsilon,J) \end{pmatrix} = \begin{pmatrix} \sigma(\varepsilon,J) \\ (\sigma^2 + \gamma)(\varepsilon,J) \end{pmatrix}$$

Now the exponential terms for $\sigma$ and $\gamma$ yield the final estimate

□

We remark that further numerical examples with connecting orbits in spaces of dimension greater than two appear in [1]. There we have employed the projection boundary conditions (3.2) and the integral phase condition (3.1). Moreover, in that paper we developed an adaptive strategy for choosing the finite interval which partially was based on qualitative error estimates as in Theorem 3.1 and partially on further numerical observations of the error behaviour at the boundary.

References

[ 1] Beyn, W.-J. (1989) 'The numerical computation of connecting orbits in dynamical systems', to appear in IMA J. Numer. Anal.

[ 2] Coppel, W.A. (1978) Dichotomies in Stability Theory. Lecture Notes in Mathematics 629, Springer, New York.

[ 3] De Hoog, F.R., Weiss, R. (1980) 'An approximation theory for boundary value problems on infinite intervals'. Computing 24, 227 - 239.

[ 4] Doedel, E.J., Kernevez, J.P. (1986) 'AUTO: Software for contin-uation problems in ordinary differential equations', Appl. Math. Technical Report, California Institute of Technology.

[ 5] Doedel, E.J., Friedman, M.J. (1989) 'Numerical computation of heteroclinic orbits', to appear in J. Comp. Appl. Math.

[ 6] Fife, P.C. (1979) Mathematical Aspects of Reacting and Diffusing Systems. Lecture Notes in Biomathematics 28, Springer, New York.

[ 7] Friedman, M.J., Doedel, E.J. (1989) 'Numerical computation and con-tinuation of invariant manifolds connecting fixed points', preprint.

[ 8] Glendinning, P. (1988) 'Global bifurcations in Flows' in T. Bedford, J. Swift (eds.), New Directions in Dynamical Systems, London Math. Soc. Lecture Note Series 127, Cambridge University Press, pp. 120 - 149.

[ 9] Guckenheimer, J., Holmes, Ph. (1983) Nonlinear oscillations, dynamical systems, and bifurcations of vector fields, Appl. Math. Sci. Vol. 42, Springer, New York.

[10] Lentini, M., Keller, H.B. (1980) 'Boundary value problems over semi-infinite intervals and their numerical solution', SIAM J. Numer. Anal. 17, 577 - 604.

[11] Palmer, K.J. (1984) 'Exponential dichotomies and transversal homoclinic points', J. Differential Equations 55, 225 - 256

[12] Shil'nikov, L.P. (1965) 'A case of the existence of a denumerable set of periodic motions', Sov. Math. Dokl. 6, pp. 163 - 166.

[13] Sparrow, C. (1982) The Lorenz Equations: Bifurcations, Chaos and Strange Attractors, Springer, New York.

# COMPUTATION OF INVARIANT MANIFOLD BIFURCATIONS

Yu. A. KUZNETSOV
*Research Computing Centre*
*USSR Academy of Sciences*
*Pushchino Moscow Region*
*142292 USSR*

ABSTRACT. Numerical methods for computation of homoclinic trajectories in ordinary differential equations depending upon parameters are presented. The situation is considered when the trajectory is formed by an intersection of the stable and the unstable manifolds of a saddle equilibrium point with only one positive eigenvalue. Explicit formulae are derived for quadratic expansions of the manifolds. Methods for computation of bifurcation parameter values for which a homoclinic trajectory exists are discussed. Procedures are presented for the computation of the orientability of the manifolds at the bifurcation parameter values. Some applications are given.

## 1. Introduction

Many applications involve a problem of finding out trajectories homoclinic or heteroclinic to equilibria of a saddle type in a system of nonlinear differential equations depending upon parameters:

$$\dot{x} = A(\alpha)x + F(x; \alpha), \tag{1}$$

where $x \in \mathbb{R}^n$, $\alpha \in \mathbb{R}^m$ and right hand side of (1) is sufficiently smooth.
The problem usually arises in two general situations. The first situation appears when we are studying a limit cycle in the model equations and analyzing the mechanisms of its origination and destruction. It is well known due to Silnikov [6 ,7] that existence of a trajectory homoclinic to the saddle at some parameter values leads to the appearance of a limit cycle for close parameter values.
The second situation appears when we are studying stationary propagating waves in some nonlinear equations with partial derivatives. Profiles of this waves usually correspond to homoclinic or heteroclinic trajectories of some ordinary differential equations.
In what follows we will discuss computational problems appearing under homoclinic trajectory considerations. We will restrict ourselves to the case when system (1) has a saddle equilibrium point $x = 0$ at the origin.

*D. Roose et al. (eds.), Continuation and Bifurcations: Numerical Techniques and Applications, 183–195.*

We suppose that linearization matrix $A(\alpha)$ has for all $\alpha$ exactly one positive eigenvalue $\lambda(\alpha)$, while all other eigenvalues have negative real parts. In this case the saddle has two smooth invariant manifolds: $(n-1)$ - dimensional stable manifold $W^s(0)$ formed by all incoming trajectories and one-dimensional unstable manifold $W^u(0)$ formed by two outgoing trajectories $\Gamma^1(0)$ and $\Gamma^2(0)$. Suppose the system depends upon only one parameter $\alpha$ $(m = 1)$. For some parameter values one of the outgoing trajectories (usually referred as *separatrices*) may return to the saddle 0 forming a *separatrix loop*. This loop is a homoclinic trajectory and it belongs to the intersection of the stable and the unstable invariant manifolds.

The appearance of a separatrix loop is a bifurcation of the global phase portrait of system (1). This bifurcation is of codimension one which may be generally treated only numerically.

The numerical analysis of separatrix loop bifurcations may be divided into several stages. We have to find invariant manifolds in the vicinity of the saddle, compute the global part of the unstable manifold and find bifurcation parameter value and analyze (if necessary) topology of the invariant manifolds near the loop.

The following paragraphs contain the description of all this stages and some applications of the presented methods.

## 2. Expansions of invariant manifolds near a saddle point

Let us rewrite system (1) for a fixed value of parameter $\alpha$ in the following form:

$$\dot{y} = \lambda y + G(y,z)$$
$$\dot{z} = Bz + H(y,z), \tag{2}$$

where $y \in \mathbb{R}^1$, $z \in \mathbb{R}^{n-1}$, $\lambda > 0$ , all eigenvalues of matrix $B$ have negative real parts, $G$ and $H$ are smooth nonlinear functions.

The stable manifold $W^s(0)$ and the unstable manifold $W^u(0)$ of saddle 0 in system (2) can be presented in a neighborhood of 0 as graphs of smooth functions $Y$ and $Z$:

$$W^s(0) = \{(y,z) : y = Y(z)\},$$
$$W^u(0) = \{(y,z) : z = Z(y)\}.$$

Suppose function $G$ has a representation

$$G(y,z) = <z,Dz> + O(y^2, y\|z\|, \|z\|^3),$$

where $< \cdot , \cdot >$ denotes the standard scalar product in $\mathbb{R}^{n-1}$, $\|\cdot\| = < \cdot , \cdot >^{1/2}$ and matrix $D$ is a symmetric $(n-1) \times (n-1)$ matrix: $D^+ = D$, while function $H$ may be represented in a form

$$H(y,z) = Ry^2 + O(y^3, y\|z\|, \|z\|^3),$$

where $R \in \mathbb{R}^{n-1}$.

Let us find quadratic terms in the expansion of $Y$ at the origin:

$$Y(z) = <z, Cz> + O(\|z\|^3).$$

Here $C$ is unknown symmetric matrix, $C^+ = C$. This matrix satisfies an equation of the following form:

$$(B^+ - \lambda)C + CB = D,$$

and can be found from this equation.

A quadratic term in the expansion of function $Z$:

$$Z(y) = Qy^2 + O(y^3),$$

is determined by vector $Q$, which can be found from a linear equation of the form

$$(2\lambda - B)Q = R.$$

The next terms in the expansions of $Y$ and $Z$ can be computed in the same way. Each step of a procedure reduces to solving linear equations for the unknown coefficients of the expansions [4].

## 3.  Computation of a bifurcation point

Let us rewrite system (1) depending upon parameter $\alpha$ in the same form as in the previous section

$$\dot{y} = \lambda(\alpha)y + G(y,z; \alpha)$$
$$\dot{z} = B(\alpha)z + H(y,z;\alpha)$$

(3)

with the same assumptions concerning $\lambda, B, G, H$ for all considered values of parameter $\alpha$. The problem now is to find a bifurcation parameter value $\alpha = \alpha^*$ for which system (3) has a separatrix loop, e.g. $\Gamma^u(0)$ returns into saddle 0 when $t$ approaches $+\infty$.

### 3.1. "SPLIT FUNCTION" METHOD [5]

The first method for computation of $\alpha^*$ is based on the construction of a scalar function $f(\alpha)$ with the following property

$$f(\alpha^*) = 0.$$

Any function having this property is called a *split function*, because it may be considered as a measure of a distance between the invariant manifolds. We construct this function by solving a single initial value problem for (3).

Let us introduce a cylinder $\|z\| = \delta$ in $\mathbb{R}^n$, with $\delta > 0$. Consider initial value problem for (3) with the following initial conditions:

$$y^{(0)} = \varepsilon$$

$$z^{(0)} = Z(\varepsilon; \alpha),$$

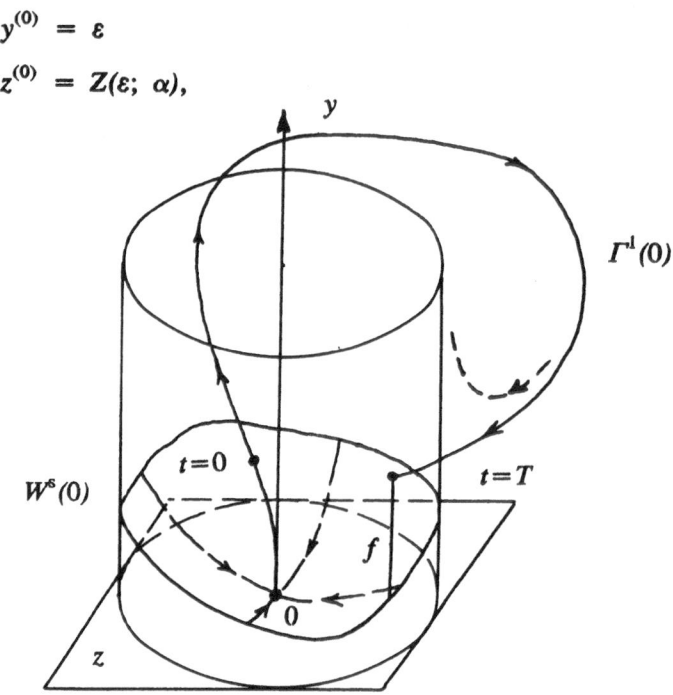

Figure 1.   "Split function" computation.

where $\varepsilon > 0$ is a small number and function $Z(y; \cdot)$ is defined in the previous section for fixed parameter values. In the other words, point $(y^{(0)}, z^{(0)})$ belongs to separatrix $\Gamma^1(0)$ near saddle $0$ (see Fig.1). Let us solve the problem on an interval $t \in [0,T]$, where $T > 0$ is a return time value determined by the following condition:

$$\|z(T)\| = \delta.$$

We assume that for $t^{(1)} = T$ the point $(y^{(1)}, z^{(1)}) = (y(T), z(T))$ of intersection of $\Gamma^1(0)$ and the cylinder is sufficiently close to the saddle $0$ (see Fig.1).

Now we can chose as the split function the distance $f$ from point $(y^{(1)}, z^{(1)})$ to the stable manifold $W^s(0)$ measured in the $y$-direction:

$$f(\alpha) = y^{(1)} - Y(z^{(1)}; \alpha).$$

It is obvious that $f(\alpha^*) = 0$.

In computations we may approximate the invariant manifolds by quadratic terms of their expansions:

$$z^{(0)} \approx Q(\alpha)\varepsilon^2,$$

$$f(\alpha) \approx y^{(1)} - <z^{(1)}, C(\alpha)z^{(1)}>,$$

and we have to apply one of the available numerical methods to solve the initial value problem.

Having a good initial guess for $\alpha^*$, we can find this value with a desired accuracy by Newton iterations for nonlinear scalar equation $f(\alpha) = 0$.

## 3.2. BVP METHOD

The described method has a limit of applicability. If $\lambda$ is relatively large then the initial value problem solution is numerically unstable near invariant manifold $W^s(0)$ due to strong trajectory divergence (see Fig.1). Thus, a numerically obtained solution for the separatrix may be very different from the exact one and may never intersect the cylinder $\|z\| = \delta$ in the vicinity of saddle 0. In the later case the split function is undefined.

To overcome this difficulty a boundary value problem approach is useful [4].

Let us introduce a new time variable $\tau = Tt$, where $T$ is an unknown return time. The variable $\tau$ belongs to interval [0,1]. Consider a so called *extended system* of $(n+2)$ equations:

$$\dot{y} = T (\lambda(\alpha)y + G(y,z; \alpha))$$

$$\dot{z} = T (B(\alpha)z + H(y,z; \alpha))$$

$$\dot{T} = 0 \tag{4}$$

$$\dot{\alpha} = 0,$$

with $\tau \in [0,1]$ as time.

Impose the following boundary conditions for system (4):

$$\tau = 0 : \quad y^{(0)} = \varepsilon$$

$$z^{(0)} = Z(\varepsilon; \alpha),$$

$$\tau = 1 : \quad \|z^{(1)}\| = \delta \tag{5}$$

$$y^{(1)} = Y(z^{(1)}; \alpha).$$

We have $n$ conditions at $\tau = 0$ and two conditions at $\tau = 1$. The main condition is the second one which places the return point $(y^{(1)}, z^{(1)})$

*exactly* on the stable manifold. Solving equations (4) with boundary conditions (5) we obtain bifurcation parameter value $\alpha = \alpha^*$ as well as return time value and the global part of the unstable manifold $W^u(0)$.

In real computations we may approximate the invariant manifolds by described quadratic expansions and have to use one of available codes to solve the boundary value problem.

## 4. Computation of a saddle separatrix loop orientation

Suppose that for bifurcation parameter value $\alpha = \alpha^*$ an eigenvalue $\mu(\alpha)$ of $A(\alpha)$ nearest to the imaginary axis is real and simple. Assume also that separatrix $\Gamma^1(0)$ returns to the saddle along eigenvector $V: A(\alpha)V = \mu V$. This is a general case. In this case topology of the stable invariant manifold is important in the separatrix loop bifurcation analysis [6].

Let us fix a small neighbourhood of the loop. The closure of a part of manifold $W^s(0)$ containing $\Gamma^1(0)$ and lying in the neighbourhood is a nonsmooth $(n - 1)$-dimensional manifold which may be either orientable or nonorientable (see Fig.2). In the first case the loop is called *orientable* while in the second one it is *nonorientable*.

The loop orientation determines orientations of invariant manifolds $W^s(L)$ and $W^u(L)$ of a limit cycle $L = L(\alpha)$ appearing under small parameter variations in system (3). It determines also whether the cycle exists for $\alpha < \alpha^*$ or for $\alpha > \alpha^*$.

So, it is very important to develop methods for computation of a loop orientation.

Let us write system (3) in the following form:

$$\dot{y} = \lambda(\alpha)y + G\ (y,u,v;\ \alpha)$$

$$\dot{u} = \mu(\alpha)u + H_1(y,u,v;\ \alpha) \tag{6}$$

$$\dot{v} = E(\alpha)v + H_2(y,u,v;\ \alpha),$$

where $y,u \in \mathbb{R}^1$, $v \in \mathbb{R}^{n-2}$. Here $u$ is the coordinate in $V$-direction.

### 4.1. INITIAL VALUE PROBLEM METHOD

Introduce two $(n-1)$-dimensional planes

$$\Pi^1 = \{(y,u,v) : y = \varepsilon\},\ \Pi^2 = \{(y,u,v) : u = \delta\}.$$

We may use $(u,v)$ as coordinates in $\Pi^1$ and $(y,v)$ as coordinates in $\Pi^2$. It is more convenient to introduce new coordinates in the planes, $(u^1,v^1)$ in $\Pi^1$ and $(y^2,v^2)$ in $\Pi^2$, so that intersections of $W^s(0)$ and these planes would have equations $u^1 = 0$ and $y^2 = 0$ respectively, while $\Gamma^1(0)$ goes through origins of these new coordinate systems. We assume also that new coordinates have the same positive directions as the original ones (see Fig.3).

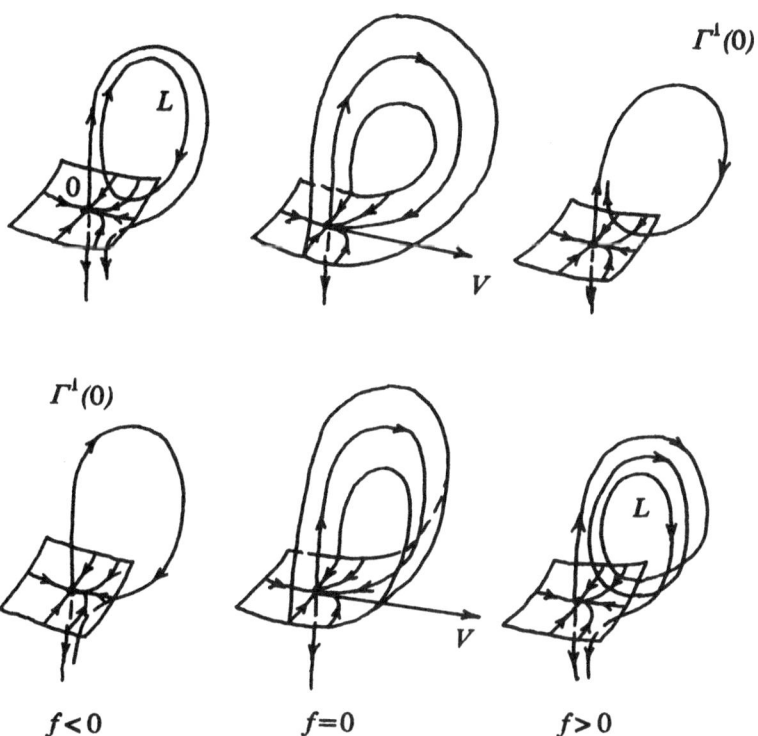

$f < 0$       $f = 0$       $f > 0$

Figure 2.    Orientable and nonorientable loop bifurcations in $\mathbb{R}^3$.

System of differential equations (6) defines a map along trajectories

$$\Phi : \Pi^1 \Rightarrow \Pi^2,$$

or

$$y^2 = \Phi_1(u^1, v^1)$$
$$v^2 = \Phi_2(u^1, v^1).$$

Loop orientation is determined by the following expression:

$$\Delta = sign \; (\Phi_1(\gamma, 0)),$$

190

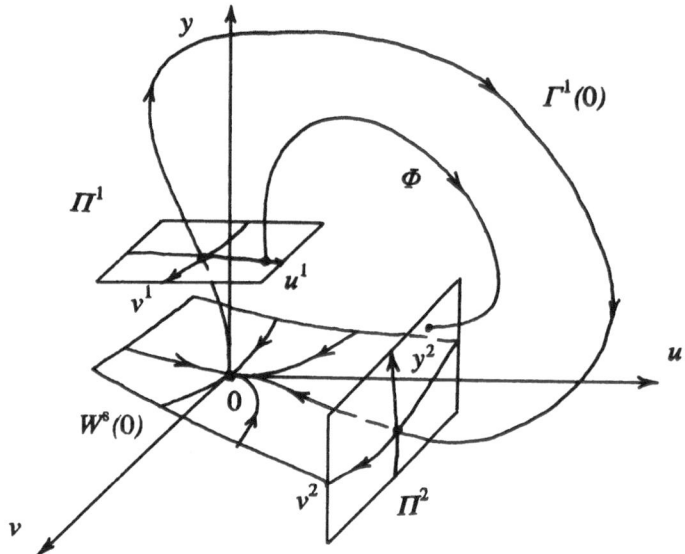

Figure 3.   Map $\Phi$ definition.

where $\gamma$ is a small positive number. The loop is orientable if $\Delta = 1$ and nonorientable if $\Delta = -1$.

In real computations we may choose $\varepsilon$ and $\delta$ small enough to use original coordinates $(y,u,v)$ in the planes $\Pi^1$ and $\Pi^2$. One can use also expansions of the invariant manifolds near the saddle.

## 4.2. TANGENT PLANE CONTINUATION METHOD

At each point $x(t)$ on the separatrix loop a tangent plane $\Pi^T(t)$ to the stable invariant manifold $W^s(0)$ is defined. We know limit positions of this plane for $t$ goes to $\pm\infty$ :

$$\Pi^T(-\infty) = \{(y,u,v) : u = 0\ \},$$

$$\Pi^T(+\infty) = \{(y,u,v) : y = 0\ \}.$$

Let $n^- = (0, 1, 0)$ be a normal vector to plane $\Pi^T(-\infty)$. This vector as well as the tangent plane can be continued along separatrix loop. For $t = +\infty$ we get a normal vector to plane $\Pi^T(+\infty)$:

$$n^+ = (\Delta, 0, 0),$$

where $\Delta = \pm 1$. The value of $\Delta$ determines the loop orientation.

In computations one may start with $n^-$ and a point near the saddle on $\Gamma^1(0)$, continue this vector numerically up to large $t$ values, and then evaluate the sign of its first coordinate.

## 5.  Applications

In this section we present two applications of the formulated methods to the ecological modeling. Two generalizations of the following very simple model of a forest age structure dynamics will be described (see details in [1, 2]). In some scaled variables the model has the form:

$$\dot{x} = \rho y - (y - 1)^2 x - sx$$
$$\dot{y} = x - hy,$$

(7)

where $x$ and $y$ are tree numbers (within an area) of "young" and "old" age classes respectively, and $\rho, s, h$ are parameters. It is assumed that there exists an optimal value of "old" tree density under which the recruitment of "young" trees is greatest.

## 5.1. A MODEL OF FOREST-PEST INTERACTION[1]

Let us introduce an insect pest into the model (7):

$$\dot{x} = \rho y - (y - 1)^2 x - sx - xz$$
$$\dot{y} = x - hy$$
$$\dot{z} = -\varepsilon z + bxz.$$

(8)

Here $z$ is the insect density, $\varepsilon$ is an insect mortality parameter and terms with $xz$ represent the forest-pest interaction.

A parametric portrait of system (8) in $(\rho, h)$-plane for fixed values of $s, \varepsilon$ and $b$ is shown in Fig.4 , while corresponding phase portraits are presented in Fig.5. Curve $P$ corresponds to the existence of a heteroclinic trajectory going from saddle $E_2$ to saddle $E_1$. Due to invariance of plane $z = 0$, the appearance of this heteroclinic trajectory means the presence of a heteroclinic cycle and existence of a limit cycle for close parameter values.

"Split function" approach was applied in our numerical computations. The code CURVE [3] was used for continuation of bifurcation curve $P$ defined by equation $f(\rho, h) = 0$. The same method is used in the second example.

## 5.2. A MODEL OF TRAVELLING FOREST BOUNDARY[2]

The model

$$x_t = \rho y - (y - 1)^2 x - sx + x_{\xi\xi}$$
$$y_t = x - hy$$

(9)

was used for qualitative description of a spatially distributed forest with local dynamics governed by system (7). Here $\xi$ is a space variable

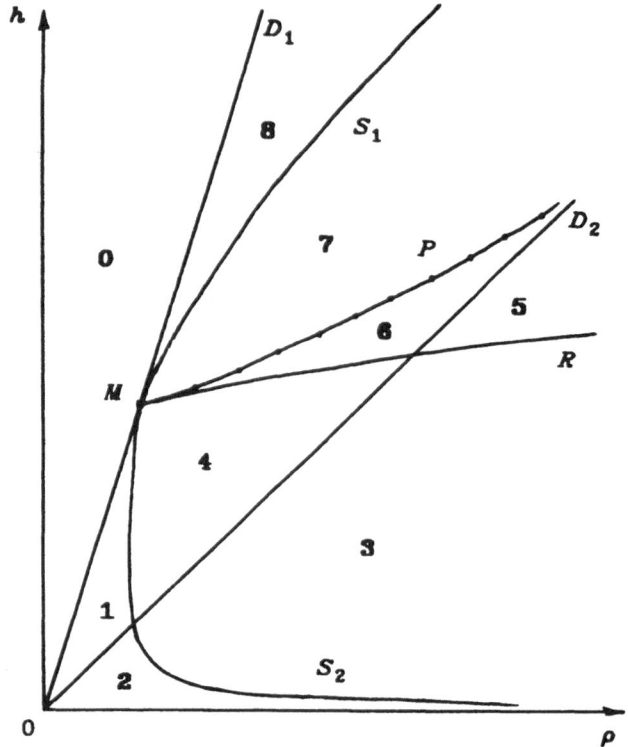

Figure 4. Parametric portrait of system (8).

and indices denote corresponding partial derivatives.

Let us consider a problem of existence of nonhomogeneous solutions of (9) which are travelling waves propagating with a constant speed:

$$x(\zeta,t) = U(\xi + ct), \quad y(\zeta,t) = V(\xi + ct),$$

where $c$ is the propagation speed. These solutions satisfy the following equations:

$$\dot{U} = W$$

$$\dot{W} = cW - \rho V + (V - 1)^2 U + sU \qquad (10)$$

$$\dot{V} = (U - hV)/c,$$

where $\zeta = \xi + ct$ plays the role of "time".

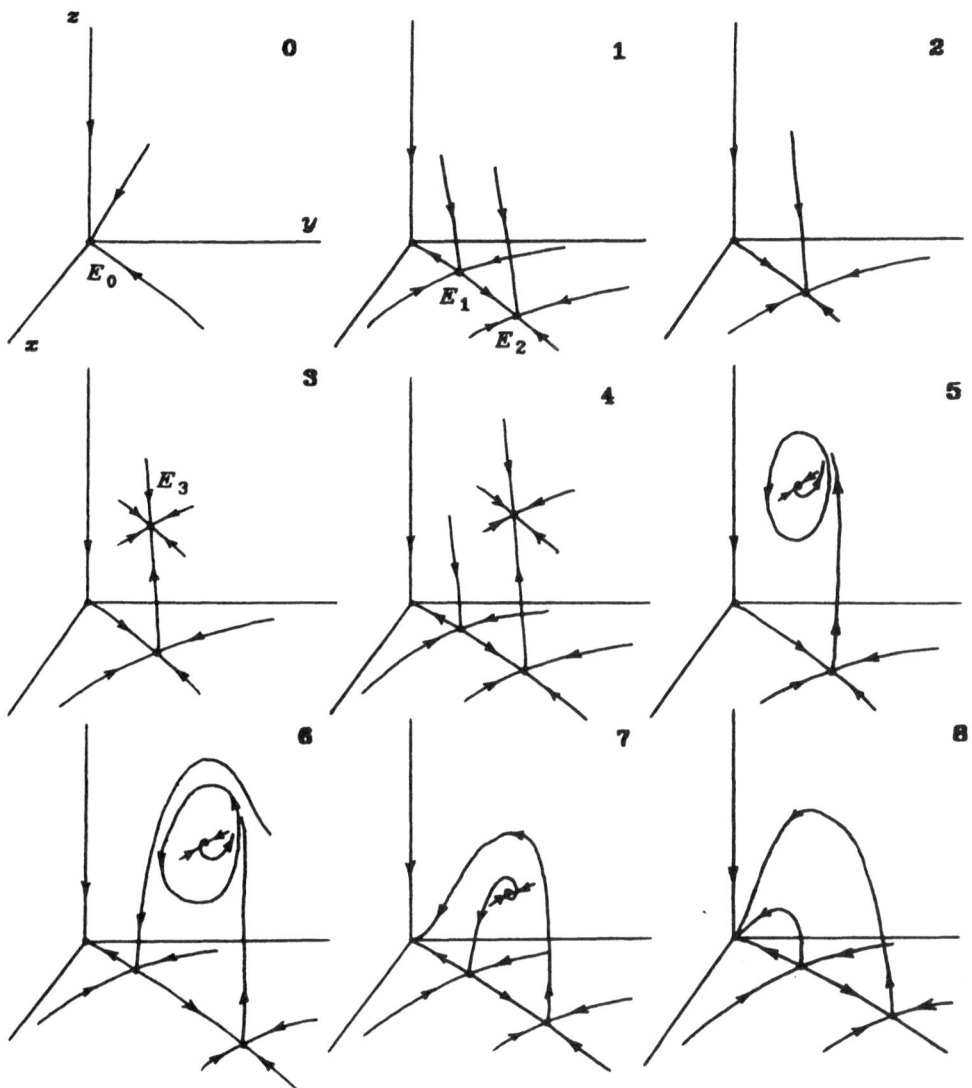

Figure 5.   Phase portraits of system (8).

194

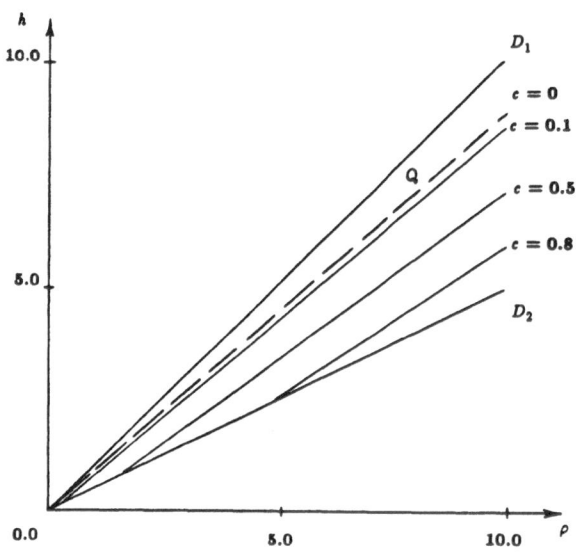

Figure 6.    Isolines of constant speed propagating front in model (9).

  There are parameter values for which system (10) has a separatrix going from one saddle to the other. For these parameter values system (9) has a travelling wave front corresponding to a propagating forest boundary. Several computed curves $c = const$ are presented in Fig.6.    With an unexpected accuracy the curves may be approximated by straight lines. A hypothesis is that they are really straight lines for model (9).

## 6.   References

1. Antonovsky, M.Ya., Clark, W., Kuznetsov, Yu.A. (1987) 'The influence of pests on forest age structure dynamics: The simplest mathematical models'. WP-87-70. International Institute for Applied Systems Analysis, Laxenburg, Austria.

2.   Antonovsky, M.Ya., Aponina, E.A., Kuznetsov, Yu.A. (1989) 'Spatial-temporal structure of mixed-age forest boundary: The simplest mathematical model'. WP-89-54. International Institute for Applied Systems Analysis, Laxenburg, Austria.

3.  Balabaev, N.K., Lunevskaya, L.V. (1978) 'Computation of a curve in

$n$-dimensional space'. FORTRAN Software Series, 1., Research Computing Centre, USSR Academy of Sciences, Pushchino (in Russian).

4. Hassard, B.D. (1980) 'Computation of invariant manifolds', in Ph.Holmes (ed.), New Approaches to Nonlinear Problems in Dynamics. SIAM, Philadelphia, 27-42.

5. Kuznetsov, Yu. A. (1983) 'One-dimensional invariant manifolds in ordinary differential equations depending upon parameters', FORTRAN Software Series, 8., Research Computing Centre, USSR Academy of Sciences, Pushchino (in Russian).

6. Silnikov, L.P.(1968) 'On the appearance of a periodic motions from the homoclinic trajectory to a saddle type equilibrium' , Matem. Sbornik, 77(119), N 3, 464-472 (in Russian).

7. Silnikov, L.P. (1970) 'On the structure of an extended neighbourhood of a structurally stable equilibrium of a saddle-focus type', Matem. Sbornik, 81(123), N 1, 92-113 (in Russian).

# A METHOD FOR HOMOCLINIC AND HETEROCLINIC CONTINU-ATION IN TWO AND THREE DIMENSIONS

A.J. RODRÍGUEZ-LUIS, E. FREIRE and E. PONCE
*Department of Applied Mathematics (University of Sevilla)*
*Escuela Técnica Superior de Ingenieros Industriales*
*Avda. Reina Mercedes s/n*
*41012–SEVILLA, Spain*

ABSTRACT. A numerical method for the detection and continuation of homoclinic and heteroclinic orbits is developed for the case of biparametric dynamical systems in two and three dimensions. We formulate a continuation problem for which the regularity conditions are studied. The numerical method is applied to several systems, some of them well–known.

## 1. Introduction

As is well known, homoclinic and heteroclinic orbits are global behaviour of major interest, for example, in the understanding of the appearance of chaotic phenomena. So it seems important to develop numerical methods for the detection and continuation of such kinds of orbits.

Much bifurcation behaviour is often related to the existence of homoclinic or heteroclinic connections [7], [8], [15]. For uniparametric dynamical systems, homoclinic orbits generically arise for an isolated parameter value, and the corresponding localization problem is formulated in terms of the zero–calculation of an application. In the two–parameter case, the problem is typically to obtain a homoclinic curve in the parameter plane; such a curve appears as the solution of a continuation problem. The idea behind the method we propose is geometrical; a different method but also in the geometrical spirit is due to Y. Kuznetsov [11]. Another approach, proposed by W. Beyn [2], truncates the boundary value problem to a finite interval.

In the next section the method is explained and some theoretical discussion is presented, giving sufficient conditions to assure the regularity of the continuation problem formulated.

In the third section we consider several two–parameter dynamical systems both in two and three dimensions, determining their curves of homoclinic or heteroclinic connections.

*D. Roose et al. (eds.), Continuation and Bifurcations: Numerical Techniques and Applications, 197–210.*
© 1990 *Kluwer Academic Publishers.*

## 2. Homoclinic Continuation in Three Dimensions

First let us consider the uniparametric autonomous system

$$\dot{x} = f(x,\mu) \quad x \in \mathbb{R}^3,\ \mu \in \mathbb{R}. \tag{2.1}$$

Let us suppose that the origin, $x = 0$, is a hyperbolic equilibrium ( $f(0,\mu) = 0$ for all $\mu$ ) and, in fact, a saddle point; without loosing generality, let us also suppose the linearization has the form

$$D_x f(0,\mu) = \begin{pmatrix} A(\mu) & 0 \\ 0 & B(\mu) \end{pmatrix} \tag{2.2}$$

where the $2 \times 2$ matrix $A(\mu)$ has its spectrum in the left–half plane and $B(\mu)$ is positive. In this case, the stable (unstable) manifold $W^s_\mu$ ( $W^u_\mu$ ) of the origin can be locally expressed as $x_3 = h^s(x_1,x_2,\mu)$ ( $x_1 = h^u_1(x_3,\mu)$, $x_2 = h^u_2(x_3,\mu)$ ).

Let us assume that for a parameter value $\mu_0$ the system has a homoclinic orbit $\gamma_0$, $\gamma_0 \subset W^s_{\mu_0} \cap W^u_{\mu_0}$. Let $S$ be a transversal section to the homoclinic orbit in the point $q_0 \in \gamma_0$; let us suppose, for instance, $S \subset \{x \in \mathbb{R}^3 : x_3 = c\}$, i.e., $S$ is contained in a plane (see fig. 1).

Let $p^s_0$, $p^u_0$ be two points belonging to $\gamma_0$, close to the origin, $\|p^s_0\|^2 = \|p^u_0\|^2 = \epsilon$, such that $p^s_0 \in W^s_{\mu_0}$ and $p^u_0 \in W^u_{\mu_0}$. For $\mu$ near $\mu_0$, the orbit of $p^s$, when $\|p^s\|^2 = \epsilon$ and $p^s$ close to $p^s_0$, will intersect $S$ in $q^s$ close to $q_0$; in the same way, the orbit of $p^u$, when $\|p^u\|^2 = \epsilon$ and $p^u$ close to $p^u_0$, will intersect $S$ in $q^u$ close to $q_0$ (see fig. 1).

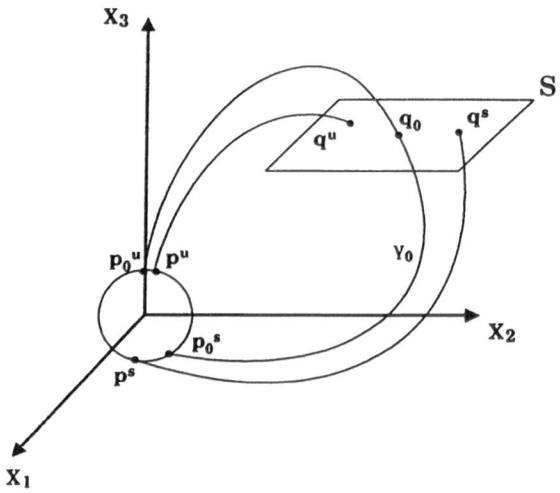

Figure 1. The Geometry of the application $F(p^s, p^u, \mu)$.

If $\phi(x, \mu, t)$ denotes the flow solution, we can write:

$$q_{1,2}^s = \phi_{1,2}\left(p^s, \mu, \tau^s(p^s, \mu)\right), \qquad q_{1,2}^u = \phi_{1,2}\left(p^u, \mu, \tau^u(p^u, \mu)\right) \qquad (2.3)$$

where $\tau^s(p^s, \mu)$ and $\tau^u(p^u, \mu)$ are given by

$$
\begin{array}{ll}
\phi_3\left(p^s, \mu, \tau^s(p^s, \mu)\right) = c, & \tau^s(p_0^s, \mu_0) = \tau_0^s \\
\phi_3\left(p^u, \mu, \tau^u(p^u, \mu)\right) = c, & \tau^u(p_0^u, \mu_0) = \tau_0^u
\end{array}
\qquad (2.4)
$$

with $\phi\left(q_0, \mu_0, \tau_0^s\right) = p_0^s$, $\phi\left(q_0, \mu_0, \tau_0^u\right) = p_0^u$.

We will define $F : U \subset \mathbb{R}^3 \times \mathbb{R}^3 \times \mathbb{R} \longrightarrow \mathbb{R}^7$, where $U$ is a neighbourhood of $(p_0^s, p_0^s, \mu_0)$, as:

$$
F(p^s, p^u, \mu) = \begin{pmatrix}
q_1^s - q_1^u \\
q_2^s - q_2^u \\
p_3^s - h^s(p_1^s, p_2^s, \mu) \\
p_1^u - h_1^u(p_3^u, \mu) \\
p_2^u - h_2^u(p_3^u, \mu) \\
\|p^s\|^2 - \epsilon \\
\|p^u\|^2 - \epsilon
\end{pmatrix}
\qquad (2.5)
$$

It is clear that $F(p_0^s, p_0^u, \mu_0) = 0$.

We will now give conditions on the homoclinic orbit $\gamma_0$ in such a way that the matrix $DF(p_0^s, p_0^u, \mu_0)$ is nonsingular, and thus the zero of $F$ will be regular (and, in particular, isolated).

Let us introduce the variational equation

$$\dot{y} = D_x f(\varphi_0(t), \mu_0) y + D_\mu f(\varphi_0(t), \mu_0) \lambda \qquad (2.6)$$

where $\lambda \in \mathbb{R}$ and $\varphi_0(t) = \phi(q_0, \mu_0, t)$ is a parameterization of the homoclinic orbit $\gamma_0$. According to Beyn [2] the homoclinic orbit $\gamma_0$ is termed **nondegenerate** if the unique bounded solutions of the above variational system are $\lambda = 0$, $y = k\dot{\varphi}$, where $k \in \mathbb{R}$.

Under the nondegeneracy condition we will show the nonsingular character of the matrix $DF(p_0^s, p_0^u, \mu_0) = L_0$; for this we will solve

$$
L_0 \begin{pmatrix} v^s \\ v^u \\ \lambda \end{pmatrix} = 0
\qquad (2.7)
$$

where $v^s \in \mathbb{R}^3$, $v^u \in \mathbb{R}^3$, $\lambda \in \mathbb{R}$. We shall see that $v^s = 0$, $v^u = 0$, $\lambda = 0$ is the unique solution.

From the first two scalar equations of (2.7) we have:

$$R_0 \left[ Y^{-1}(-\tau_0^s)v^s - Y^{-1}(-\tau_0^u)v^u + \lambda \int_{-\tau_0^s}^0 Y^{-1} D_\mu f - \lambda \int_{-\tau_0^u}^0 Y^{-1} D_\mu f \right] = 0 \quad (2.8)$$

where

$$R_0 = \begin{pmatrix} 1 & 0 & \dfrac{-f_1(q_0, \mu_0)}{f_3(q_0, \mu_0)} \\ 0 & 1 & \dfrac{-f_2(q_0, \mu_0)}{f_3(q_0, \mu_0)} \end{pmatrix} \qquad (2.9)$$

and $Y(t)$ is the principal fundamental matrix of the homogeneous part of the variational equation; we note that the transversal character of the section $S$ at $q_0$ guarantees $f_3(q_0, \mu_0) \neq 0$.

The equation (2.8) is equivalent to

$$Y^{-1}(-\tau_0^s)\bar{v}^s - Y^{-1}(-\tau_0^u)\bar{v}^u + \lambda \int_{-\tau_0^s}^0 Y^{-1}D_\mu f - \lambda \int_{-\tau_0^u}^0 Y^{-1}D_\mu f = 0 \qquad (2.10)$$

where $v^s = \bar{v}^s + k_1 f(p_0^s, \mu_0)$, $v^u = \bar{v}^u + k_2 f(p_0^u, \mu_0)$, $k_1, k_2 \in \mathbb{R}$.

Now we introduce the following solution of the variational equation (2.6):

$$z(t) = Y(t)v + \lambda Y(t) \int_0^t Y^{-1}D_\mu f \qquad (2.11)$$

We choose the initial condition $z(0) = v$ as

$$v = Y^{-1}(-\tau_0^s)\bar{v}^s + \lambda \int_{-\tau_0^s}^0 Y^{-1}D_\mu f = Y^{-1}(-\tau_0^u)\bar{v}^u + \lambda \int_{-\tau_0^u}^0 Y^{-1}D_\mu f \qquad (2.12)$$

where we have used (2.10). It is clear that $z(-\tau_0^s) = \bar{v}^s$ and $z(-\tau_0^u) = \bar{v}^u$. Let us study the behaviour of $z(t)$ as $t \longrightarrow +\infty$, $t \longrightarrow -\infty$.

It is possible to build a fundamental set of solutions, $\{\varphi^1, \varphi^2, \varphi^3\}$, for the homogeneous variational equation [9], satisfying the following conditions:

$$\begin{aligned} &\varphi^1(t) = \dot{\varphi}_0(t) \\ &\varphi^2(t) \text{ such that } \varphi^2(0) \in T_{q_0}W_{\mu_0}^s \\ &\varphi^3(t) \text{ such that } \varphi^3(0) \notin T_{q_0}W_{\mu_0}^s \end{aligned} \qquad (2.13)$$

where $T_{q_0}W_{\mu_0}^s$ denotes the tangent space to the stable manifold $W_{\mu_0}^s$ at $q_0$. Furthermore the above fundamental set of solutions verifies:

$$\lim_{t \to \pm\infty} \varphi^1(t) = 0$$

$$\lim_{t \to +\infty} \varphi^2(t) = 0 , \quad \varphi^2(t) \text{ unbounded as } t \longrightarrow -\infty \qquad (2.14)$$

$$\varphi^3(t) \text{ unbounded as } t \longrightarrow \pm\infty$$

Let $\psi(t)$ be a solution of the adjoint equation of the homogeneous variational equation such that

$$\lim_{t \to \pm\infty} \psi(t) = 0 \qquad (2.15)$$

Then we obtain

$$\psi^T(t) \cdot \varphi^1(t) = 0, \quad \psi^T(t) \cdot \varphi^2(t) = 0, \quad \text{for all } t \in \mathbb{R} \tag{2.16}$$

We will now show that there exists $\lim z(t)$ as $t \longrightarrow +\infty$. For this we analyze the "likely unbounded parts" in the expression (2.11) which defines $z(t)$. Firstly we write:

$$z(0) = v = \alpha_1 \varphi^1(0) + \alpha_2 \varphi^2(0) + \alpha_3 \varphi^3(0) \tag{2.17}$$

Multiplying by $\psi^T(0)$ we obtain for the "likely unbounded part" of $Y(t)v$, that is, $\alpha_3 Y(t)\varphi^3(0)$:

$$\alpha_3 \psi^T(0)\varphi^3(0) = \psi^T(-\tau_0^s)\bar{v}^s + \lambda \int_{-\tau_0^s}^0 \psi^T D_\mu f =$$
$$= \psi^T(-\tau_0^s)\bar{v}^s + \lambda \psi^T(0) \int_{-\tau_0^s}^0 Y^{-1} D_\mu f \tag{2.18}$$

Secondly we consider the other term in (2.11):

$$\lambda Y(t) \int_0^t Y^{-1} D_\mu f = \lambda Y(t) \left[ \int_0^t P Y^{-1} D_\mu f - \int_t^\infty (I - P) D_\mu f \right] + Y(t)u$$
$$\text{with } u = \lambda \int_0^\infty (I - P) Y^{-1} D_\mu f \tag{2.19}$$

where we have used the exponential dichotomy on $[0, +\infty)$ of the homogeneous variational equation, and hence $P$ denotes the corresponding projection [2], [3], [12].

The first term on the right hand side of (2.19) has a limit as $t \longrightarrow +\infty$. Now we write:

$$u = \beta_1 \varphi^1(0) + \beta_2 \varphi^2(0) + \beta_3 \varphi^3(0) \tag{2.20}$$

and multiplying by $\psi^T(0)$ we obtain for the "likely unbounded part" of $Y(t)u$, $\beta_3 Y(t)\varphi^3(0)$:

$$\beta_3 \psi^T(0)\varphi^3(0) = \lambda \psi^T(0) \int_0^\infty (I - P) Y^{-1} D_\mu f \tag{2.21}$$

Adding (2.18) and (2.21) we obtain:

$$(\alpha_3 + \beta_3) \psi^T(0)\varphi^3(0) = \lambda \left[ \frac{\partial h^s}{\partial \mu} + \psi^T(0) \int_{-\tau_0^s}^\infty (I - P) Y^{-1} D_\mu f \right] \tag{2.22}$$

where we have used the third scalar equation of (2.7):

$$0 = \left( -\frac{\partial h^s}{\partial p_1^s}, -\frac{\partial h^s}{\partial p_2^s}, 1 \right) \cdot \bar{v}^s - \lambda \frac{\partial h^s}{\partial \mu} = \psi^T(-\tau_0^s) \cdot \bar{v}^s - \lambda \frac{\partial h^s}{\partial \mu} \tag{2.23}$$

A careful computation using the properties of the local stable manifold $h^s(p_1^s, p_2^s, \mu)$ leads to the expression:

$$\frac{\partial h^s}{\partial \mu} = -\psi^T(0) \int_{-\tau_0^s}^{\infty} (I - P)Y^{-1}D_\mu f \qquad (2.24)$$

and so, from (2.22) we can deduce the existence of $\lim z(t)$ as $t \longrightarrow \infty$. Reasoning in an analogous way it can be shown that there exists $\lim z(t)$ as $t \longrightarrow -\infty$. Therefore, in particular, $z(t)$ is a bounded solution of the variational equation (2.11) and applying the hypothesis of nondegeneracy we conclude that $\lambda = 0$ and $z(t) = k\varphi^1(t) = k\dot\varphi_0(t)$, $k \in \mathbb{R}$. Hence

$$\begin{aligned}
\bar{v}^s &= z(-\tau_0^s) = k\dot\varphi_0(-\tau_0^s) = kf(p_0^s, \mu_0) \\
\bar{v}^u &= z(-\tau_0^u) = k\dot\varphi_0(-\tau_0^u) = kf(p_0^u, \mu_0)
\end{aligned} \qquad (2.25)$$

and so

$$\begin{aligned}
v^s &= (k + k_1)f(p_0^s, \mu_0) \\
v^u &= (k + k_2)f(p_0^u, \mu_0)
\end{aligned} \qquad (2.26)$$

Now from (2.26) and the last two scalar equations of (2.7) we have $v^s = v^u = 0$.

Therefore we have proved the following result: if the homoclinic orbit $\gamma_0$ is nongenerate then $(p_0^s, p_0^u, \mu_0)$ is a regular zero of $F$.

Let us now consider the biparametric system

$$\dot{x} = f(x, \mu), \quad x \in \mathbb{R}^3, \quad \mu = (\mu_1, \mu_2) \in \mathbb{R}^2 \qquad (2.27)$$

We will also assume that for a parameter value $\mu_0 = (\mu_1^0, \mu_2^0)$ the system (2.27) has a homoclinic orbit $\gamma_0$. Then, proceeding as above we have a continuation problem. We will define, in an analogous way as was done in (2.5), $F : V \subset \mathbb{R}^3 \times \mathbb{R}^3 \times \mathbb{R}^2 \longrightarrow \mathbb{R}^7$, where $V$ is a neighbourhood of $(p_0^s, p_0^u, \mu_1^0, \mu_2^0)$; it is clear that $F(p_0^s, p_0^u, \mu_1^0, \mu_2^0) = 0$.

In this case, the homoclinic continuation is equivalent to the continuation of the solutions of

$$F(p^s, p^u, \mu_1, \mu_2) = 0 \qquad (2.28)$$

Now, the nondegenerate character of $\gamma_0$ with respect to either $\mu_1$ or $\mu_2$ leads to the nonsingularity of $D_{w_i}F(p_0^s, p_0^u, \mu_1^0, \mu_2^0)$ where $w_i = (p^s, p^u, \mu_i)$, $i = 1$ or $2$. Therefore we obtain, using the implicit function theorem, a regular curve of homoclinic orbits in the parameter plane $(\mu_1, \mu_2)$.

In order to carry out the numerical computation we need to approximate the stable and unstable manifolds. This approximation can be simply linear, i. e., the tangent spaces in the hyperbolic equilibrium, or, for more accurate results we could use higher order approximations computed, for example, with an analogous procedure as proposed in [5], [6].

For the resolution of the continuation problems introduced we have used version 6.0 of PITCON [13].

The method explained above can be used analogously for heteroclinic continuation problems. In the case of bidimensional systems the method is obviously easier to apply than in the three–dimensional case.

## 3. Examples

In this section we present numerical results obtained by means of the proposed continuation method. We will start with several two–dimensional systems.

Consider the two–parameter system that provides an unfolding of the Arnold–Takens–Bogdanov bifurcation:

$$\dot{x} = y \qquad \dot{y} = \mu_1 + \mu_2 y + ax^2 + bxy , \qquad \mu_1 \leq 0, \ \mu_2 \geq 0 \qquad (3.1)$$

for the case $a = b = 1$. This has two equilibria $(\pm\sqrt{-\mu_1}, 0)$. The equilibrium on the right, $(\sqrt{-\mu_1}, 0)$, is a saddle point; while the other equilibrium exhibits a subcritical Hopf bifurcation on the semiparabola $\mu_1 = -\mu_2^2$. The approximation of the homoclinic curve in a neighbourhood of the origin, $(\mu_1, \mu_2) = (0,0)$, using the Melnikov method, is given by the semiparabola $\mu_1 = -(49/25)\mu_2^2$ [10]. In figure 2 the dotted line represents this approximation while the solid line corresponds to the homoclinic bifurcation curve obtained from the continuation method.

The second system we will analyse has, as (3.1), the linear part of the form $\dot{x} = y$, $\dot{y} = 0$, and corresponds to the Arnold–Takens–Bogdanov bifurcation with "cubic" symmetry:

$$\dot{x} = y \qquad \dot{y} = \mu_1 x + \mu_2 y + a_3 x^3 + b_3 x^2 y , \qquad \mu_1, \ \mu_2 \geq 0 \qquad (3.2)$$

for the case $a_3 = b_3 = -1$. It has three equilibria, $(0,0)$ and $(\pm\sqrt{\mu_1}, 0)$. The origin is a saddle point while the other two exhibit a subcritical Hopf bifurcation when $\mu_1 = \mu_2$. The linear approximation of the curve of the homoclinic orbits, provided by the Melnikov method, predicts that this curve is tangential to $\mu_2 = (4/5)\mu_1$ at the origin [10] (see fig. 3).

We will now present a Van der Pol type oscillator given by:

$$\dot{x} = y \qquad \dot{y} = \alpha\beta x(1 - x^2) + y(\beta + 1 - \alpha - 3\beta x^2) , \qquad \alpha, \ \beta > 0 \qquad (3.3)$$

with three equilibria: $(0,0)$ and $(\pm 1, 0)$. The origin is always a saddle point and the other two equilibria change their stability by means of a subcritical Hopf bifurcation. The solid line (see fig. 4), that represents the curve of homoclinic bifurcation, has at the point $\alpha = 1$, $\beta = 0$, a slope of $-(4/5)$. In figure 5 we see two such homoclinic connections when $\alpha = 0.2$ and $\alpha = 0.9$. It is interesting to remark that this system also has a saddle–node bifurcation of periodic orbits. Moreover, for small $\alpha$–values, relaxation oscillations appear.

Let us consider an example of heteroclinic continuation in a bidimensional system that appears [2], [4] in the study of traveling waves of the form $u(z,t) = x(z - \lambda t)$

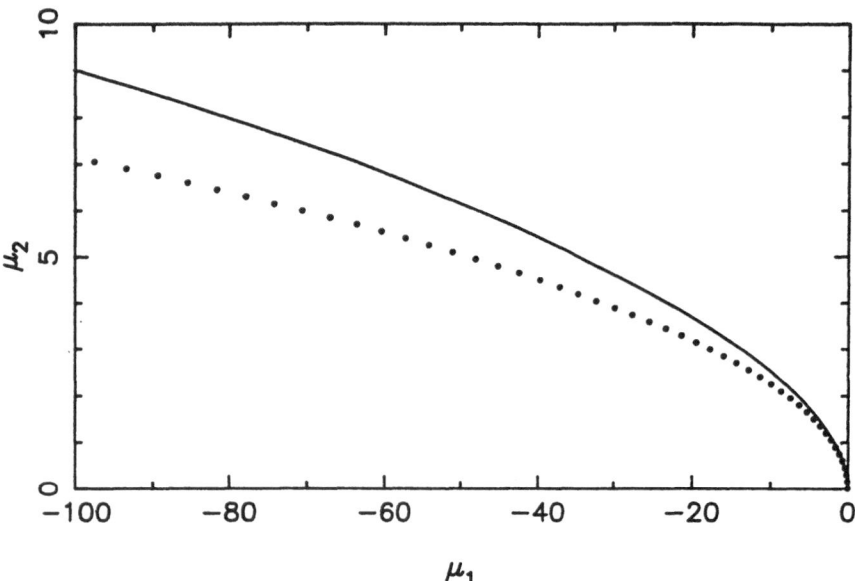

Figure 2. Curve of homoclinic orbits for the system (3.1). The dotted line represents the local approximation.

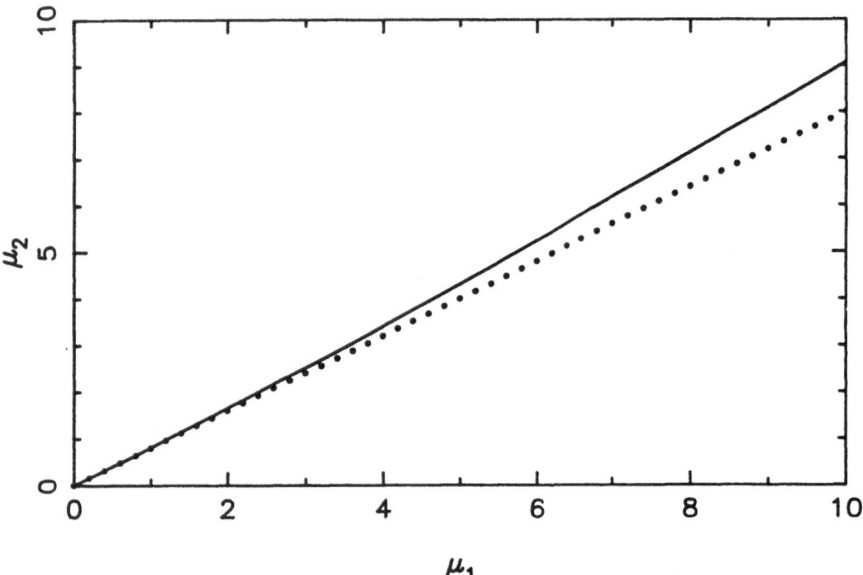

Figure 3. Curve of homoclinic connections for the system (3.2). The dotted line represents the local approximation.

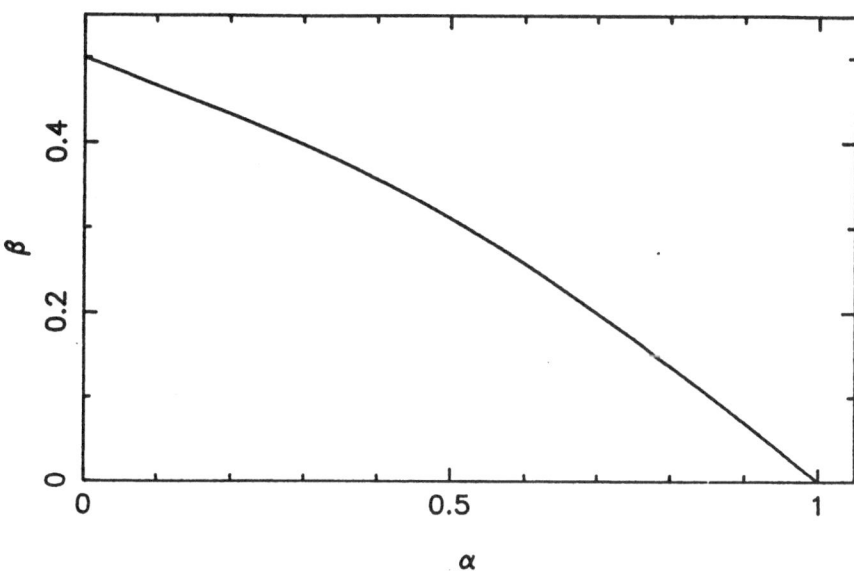

Figure 4. Curve of homoclinic orbits of the system (3.3).

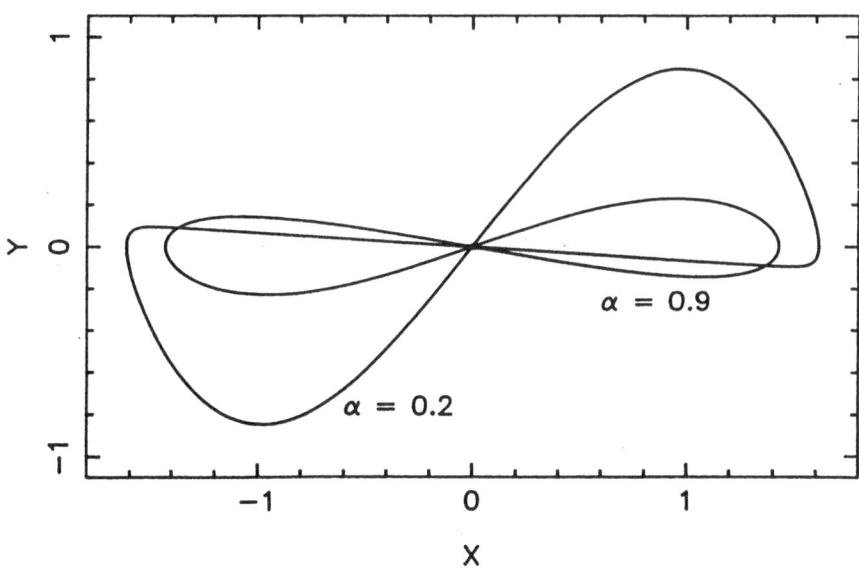

Figure 5. Homoclinic orbits of the system (3.3) for two values of the parameter $\alpha$.

in

$$u_t = u_{zz} + u(u - \mu)(1 - u) , \qquad z \in \mathbb{R}, \ t \geq 0 \qquad (3.4)$$

These waves are found as heteroclinic orbits connecting the saddle points $(0,0)$ and $(1,0)$ of the system

$$\dot{x} = y \qquad \dot{y} = -\lambda y + x(x - \mu)(x - 1) , \qquad \lambda, \ \mu \in \mathbb{R} \qquad (3.5)$$

This system has three equilibria on the x–axis: $(0,0)$, $(\mu,0)$ and $(1,0)$. Depending on the $\mu$–value, the relative position between these equilibria changes. In the three situations possible, both the external points are saddle points and the interior point, that is, $(0,0)$ if $\mu < 0$, $(\mu,0)$ if $0 < \mu < 1$ and $(1,0)$ if $\mu > 1$, is stable. A transcritical bifurcation arises when $\mu = 0$ or $\mu = 1$. The six curves obtained (only one of them corresponds to traveling waves) in the parameter plane (see fig. 6), with slopes as giving by Melnikov's method, represent the possible heteroclinic connections the system has. When two curves intersect, there exists a homoclinic loop. This situation appears three times ( $\mu = -1$, $1/2$, $2$ and $\lambda = 0$ ).

Let us consider, as the first tridimensional example, the system

$$\dot{x} = y \qquad \dot{y} = z \qquad \dot{z} = -z - by + cx - x^2 , \qquad b, \ c > 0 \qquad (3.6)$$

investigated in [1], [8]. It has two stationary points: $(0,0,0)$ and $(c,0,0)$. The origin is always a saddle point with two complex eigenvalues (saddle-focus) except in a small, cusplike region near $(b = c = 0)$ where the three eigenvalues are real. The other equilibrium is stable for $0 < c < b$ but loses stability when $c = b$ in a supercritical Hopf bifurcation. The corresponding homoclinic curve in the $cb$ plane is drawn in figure 7.

The last system we will study is the classic Lorenz equation [14] given by:

$$\dot{x} = \sigma(y - x) \qquad \dot{y} = \rho x - y - xz \qquad \dot{z} = -\beta z + xy , \qquad \sigma, \ \rho, \ \beta > 0 \qquad (3.7)$$

The origin is a stationary point for all parameter values; it is stable if $0 < \rho < 1$. At $\rho = 1$ there is a pitchfork bifurcation and, then, for $\rho > 1$ there are two other equilibria, $C_\pm = (\pm\sqrt{\beta(\rho - 1)}, \pm\sqrt{\beta(\rho - 1)}, \rho - 1)$.

First we find the homoclinic orbit, connecting the stable and unstable manifolds of the origin, that appears when $\rho \approx 13.926$, $\sigma = 10$ and $\beta = 8/3$. In this parameter region the three eigenvalues of the origin are real (two negative). We have continued the homoclinic curve fixing $\sigma = 10$ to obtain, in the $\beta\rho$ plane, the curve drawn in figure 8. We have also drawn homoclinic orbits for several values of $\beta$, projecting them on the $XZ$ plane (see fig. 9).

Second we have investigated, for small enough $\beta$ values, the symmetric heteroclinic orbits between the two fixed points $C_\pm$. As it is argued in [14] (see also [8]) when $\sigma = 10$, $\beta = 0.25$ a heteroclinic connection appears if $\rho \approx 487.16$. At these parameter values the linearized flow near the two equilibria $C_\pm$ has two complex eigenvalues (with positive real part) and one real (negative): these stationary

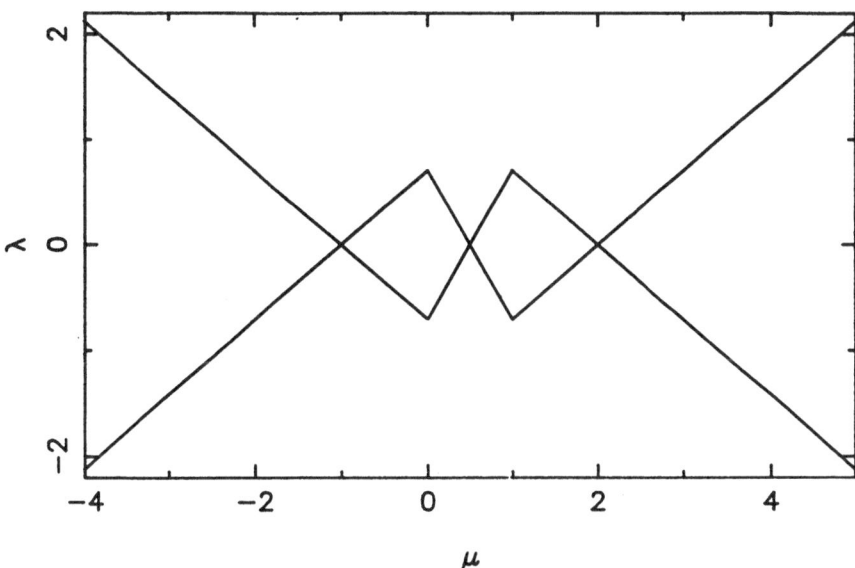

Figure 6. Curves of heteroclinic connections for the system (3.5).

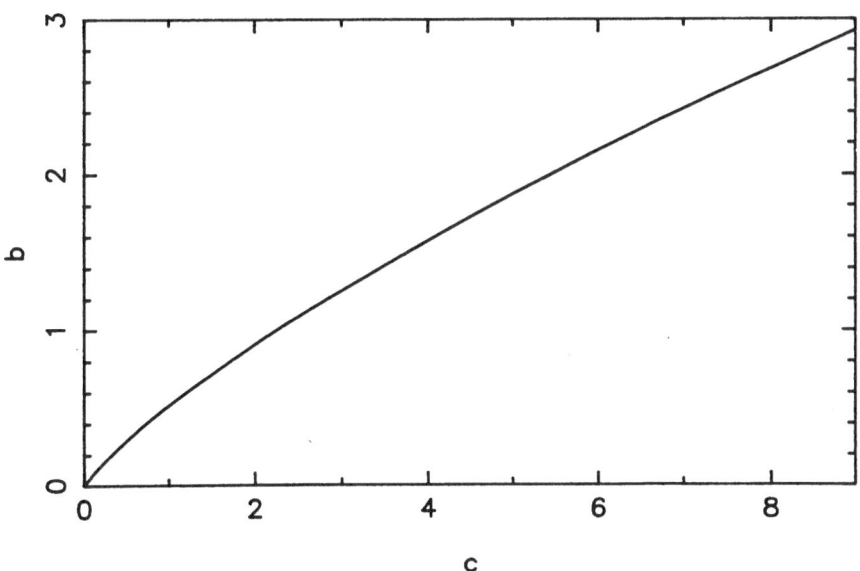

Figure 7. Curve of homoclinic orbits for the Arnéodo equation (3.6).

208

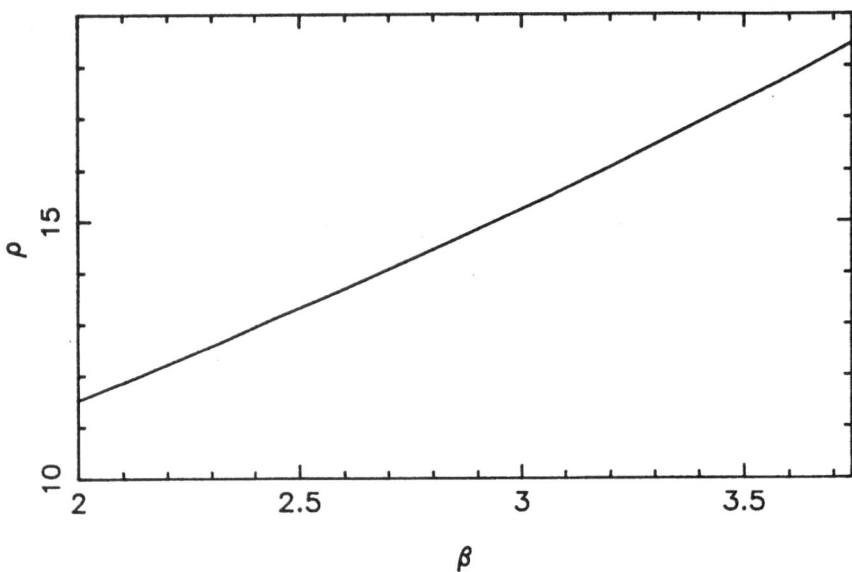

Figure 8. Curve of homoclinic connections for
the Lorenz equation (3.7) when $\sigma = 10$.

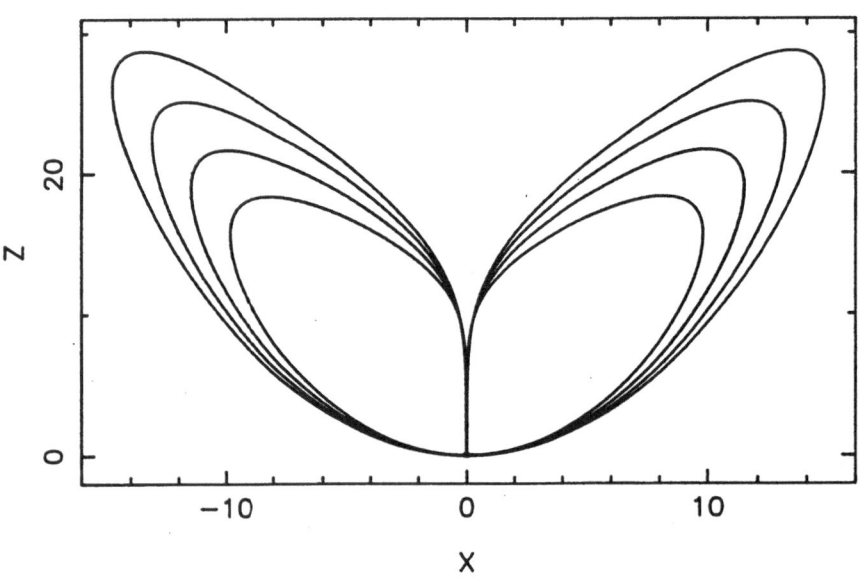

Figure 9. Homoclinic orbits of the Lorenz equation (3.7) for several values of $\beta$: 2.1, 8/3, 3.2, 3.7 and $\sigma = 10$.

points are of saddle–focus type. We have found that point in the parameter space and continued the heteroclinic curve fixing $\sigma = 10$ to obtain, in the $\beta\rho$ plane, the curve drawn in figure 10.

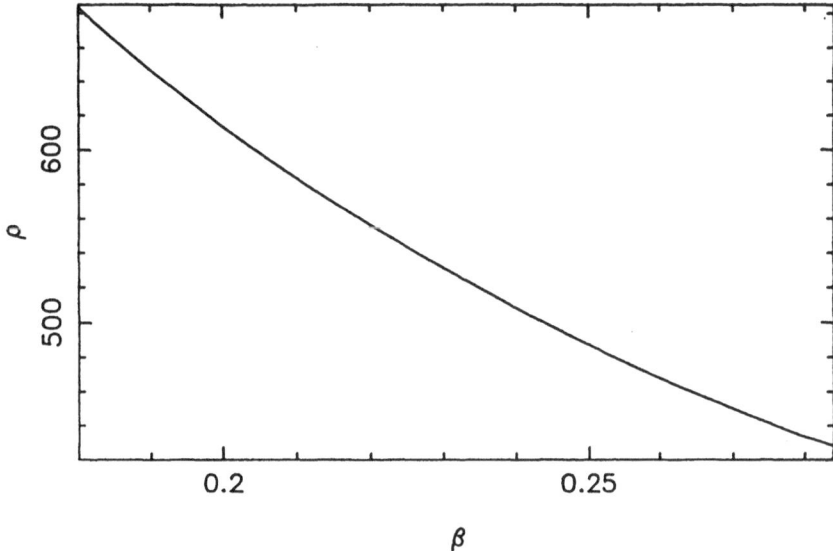

Figure 10. Curve of heteroclinic orbits for the
Lorenz equation (3.7) when $\sigma = 10$.

## Acknowledgements

Authors wish to thank to the Consejería de Educación y Ciencia de la Junta de Andalucía for partial financial support.

# REFERENCES

1. Arnéodo, A., Coullet, P., Spiegel, E., and Tresser, C. (1985) 'Asymptotic Chaos', Physica 14D, 327–347.

2. Beyn, W.-J. (1990) 'The Numerical Computation of Global Bifurcations: Homoclinic and Heteroclinic Orbits', in D. Roose (ed.), Continuation and Bifurcations: Numerical Techniques and Applications, Kluwer Academic Publishers, Dordrecht, (see this volume).

3. Coppel, W.A. (1978) Dichotomies in Stability Theory, Lecture Notes in Mathematics 629, Springer–Verlag, New York.

4. Fife, P.C. (1979) Mathematical Aspects of Reacting and Diffusing Systems, Lecture Notes in Biomathematics 28, Springer–Verlag, New York.

5. Freire, E., Gamero, E., Ponce, E., and Franquelo, L.G. (1989) 'An Algorithm for Symbolic Computation of Center Manifolds', in P. Gianni (ed.), Symbolic and Algebraic Computation, Lecture Notes in Computer Science, vol. 358, Springer–Verlag, New York, pp. 219–230.

6. Freire, E., Gamero, E., and Ponce, E. (1990) 'Symbolic Computation and Bifurcation Methods' in D. Roose (ed.), Continuation and Bifurcations: Numerical Techniques and Applications, Kluwer Academic Publishers, Dordrecht, (see this volume).

7. Gaspard, P., Kapral, R., and Nicolis, G. (1984) 'Bifurcation Phenomena near Homoclinic Systems: a Two Parameter Analysis', J. Stat. Phys., 35, 697–727.

8. Glendinning, P., and Sparrow, C. (1984) 'Local and Global Behavior near Homoclinic Orbits', J. Stat. Phys., 35, 645–696.

9. Gruendler, J. (1985) 'The existence of Homoclinic Orbits and the Method of Melnikov for Systems in $\mathbb{R}^n$', SIAM J. Math. Anal., 16, 907–931.

10. Guckenheimer, J., and Holmes, P.J. (1983) Nonlinear Oscillations, Dynamical Systems, and Bifurcations of Vector Fields, Springer–Verlag, New York.

11. Kuznetsov, Y.A. (1990) 'Computation of Invariant Manifold Bifurcations', in D. Roose (ed.), Continuation and Bifurcations: Numerical Techniques and Applications, Kluwer Academic Publishers, Dordrecht, (see this volume).

12. Palmer, K.J. (1984) 'Exponential Dichotomies and Transversal Homoclinic Points', J. Diff. Eqns., 55, pp. 225–256.

13. Rheinboldt, W.C., and Burkardt, J.V. (1983) 'A Locally Parametrized Continuation Process', ACM Trans. Math. Software, 9, pp. 215–235.

14. Sparrow, C. (1982) The Lorenz Equations: Bifurcations, Chaos, and Strange Attractors, Applied Mathematical Sciences 41, Springer–Verlag, New York.

15. Wiggins, S. (1988) Global Bifurcations and Chaos. Analytical Methods, Applied Mathematical Sciences 73, Springer–Verlag, New York.

# THE GLOBAL ATTRACTOR UNDER DISCRETISATION

ANDREW STUART
*School of Mathematical Sciences,*
*University of Bath,*
*Bath BA2 7AY, UK.*

ABSTRACT. The effect of temporal discretisation on dissipative differential equations is analysed. We discuss the effect of discretisation on the global attractor and survey some recent results in the area. The advantage of concentrating on $\omega$ and $\alpha$ limit sets (which are contained in the global attractor) is described. An analysis of spurious bifurcations in the $\omega$ and $\alpha$ limit sets is presented for linear multistep methods, using the time-step $\Delta t$ as the bifurcation parameter. The results arising from application of local bifurcation theory are shown to hold globally and a necessary and sufficient condition is derived for the non-existence of a particular class of spurious solutions, *for all* $\Delta t > 0$. The class of linear multistep methods satisfying this condition is fairly restricted since the underlying theory is very general and takes no account of any inherent structure in the underlying differential equations. Hence a method complementary to the bifurcation analysis is described, the aim being to construct methods for which spurious solutions do not exist *for* $\Delta t$ *sufficiently small*; for infinite dimensional dynamical systems the method relies on examining steady boundary value problems (which govern the existence of spurious solutions) in the singular limit correpsonding to $\Delta t \to 0_+$. The analysis we describe is helpful in the design of schemes for long-time simulations.

## 1 Introduction

The real world presents a variety of dynamical phenomena of bewildering complexity. These phenomena can often be modelled by means of differential equations. However, differential equations can rarely be solved in closed form and so it is often necessary to replace them by finite dimensional maps. These maps can exhibit a wide range of dynamical behaviour but it is not always clear which, if any, of the numerical observations are related to the real world. A necessary step in ascertaining the relationship of numerically generated dynamics to the real world is to study the dynamical properties of differential equations and their discretisations in conjunction. This is the approach taken here. The solutions of nonlinear maps are usually far easier to generate than the solutions of nonlinear differential equations and the price we pay is that the behaviour of the maps is often far more complicated than that of the underlying differential equations. Thus it is important to design schemes in which the effect of numerical artefacts on the dynamics is minimised.

*D. Roose et al. (eds.), Continuation and Bifurcations: Numerical Techniques and Applications, 211–225.*
© 1990 *Kluwer Academic Publishers.*

It is well known that a numerical method for an initial value problem which is convergent at a fixed time does not necessarily yield the same asymptotic behaviour as the underlying initial value problem, for fixed values of the time-step. For linear ordinary differential equations whose solutions decay with time it is necessary to operate the numerical method in the region of absolute stability of the scheme in order to obtain the correct asymptotic behaviour. This condition is stronger than that of zero stability, which is required for convergence. The issue of absolute stability and the construction of A-stable methods has been central to the development of numerical methods for linear initial value problems.

It is only relatively recently that analogous problems have been studied for the discretisation of nonlinear evolution equations, particularly those involving partial differential operators. Early work of interest includes the paper of Burrage and Butcher [2] in which AN and BN stability are defined for contractive nonlinear ODEs. The essential difficulty with the nonlinear problem is the dependence on initial conditions. The current growth of interest in the dynamics of numerical methods for nonlinear problems has been fuelled by the input of many ideas from dynamical systems. Here we discuss some of these recent results in a unified dynamical systems framework. The advantage of a dynamical systems approach to the numerical analysis of initial value problems is that it forces consideration of the flow generated by the numerical method. This is in contrast to classical numerical analysis which focusses on convergence of individual trajectories from a fixed initial condition.

In this paper we concentrate on dissipative differential equations: for simplicity we will take this to mean problems for which all trajectories converge into an absorbing set in the phase space after a finite time. (This definition is suitable in finite dimensions but sometimes inadequate in infinite dimensions [17].) In section 2 we review some definitions from continuous dynamical systems. Section 3 contains analogous definitions for discrete dynamical systems and a discussion of the relevance of these concepts to numerical analysis. It is shown, by means of an example, that the destruction of a global attractor under discretisation can occur when the unstable manifold of a spurious solution introduced by discretisation is connected to infinity. In section 4 we analyse spurious bifurcations in the $\omega$ and $\alpha$ limit sets for linear multistep methods. The limit sets are contained in the global attractor, if it exists. The analysis yields conditions necessary and sufficient for the non-existence of a particular class of spurious solutions, for all values of the discretisation parameter $\Delta t$. A complementary analysis, which applies only for $\Delta t$ sufficiently small, is described and applied to examples from discretisation of reaction-diffusion equations.

## 2 Continuous Evolution Semigroups

Consider an ODE in a Banach space

$$u_t = G(u), \tag{1}$$

together with an initial condition on $u$ at $t = 0$. Here $G(u) : B \to B'$ for two Banach spaces $B$ and $B'$ where, typically, $B \subset B'$. We now define the evolution semigroup $S(t)$ which maps the solution at a given time to a solution $t$ units of time later. This and subsequent definitions in sections 2 and 3 can be found in Temam [17].

**Definition 2.1** *A strongly continuous evolution semigroup is a mapping $S(t) : B \to B$ for some Banach space $B$ with $S(t)$ satisfying*

$$S(t + s) = S(t) \circ S(s) \; \forall t, s \geq 0, \tag{2}$$

$$S(0) = I, \tag{3}$$

*and*

$$S(t)u \; continuous \; in \; t \in [0, \infty) \; for \; each \; u \in B. \tag{4}$$

*The solution $u(t)$ of (1) then satisfies*

$$u(t) = S(t)u(0), \tag{5}$$

*where $u(0)$ is the initial condition. The action of $S(t)$ on a set $D \subseteq B$ is defined by*

$$S(t)D = \bigcup_{u(0) \in D} S(t)u(0).$$

The possible asymptotic states of $u(t)$ are captured in the $\omega$ and $\alpha$ limit sets which can be defined in terms of the evolution semigroup $S(t)$ as follows. Note that the $\alpha$ limit set of a point is not defined in general since infinite dimensional dissipative dynamical systems are usually not defined backwards in time – consider the heat equation for example. However, the $\alpha$ limit set may exist for specific choices of $u(0)$.

**Definition 2.2** *The $\omega$ limit set and the $\alpha$ limit set (when it exists) of a point $u(0)$ are defined respectively by*

$$\omega(u(0)) = \bigcap_{s \geq 0} \overline{\bigcup_{t \geq s} S(t)u(0)} \tag{6}$$

*and*

$$\alpha(u(0)) = \bigcap_{s \leq 0} \overline{\bigcup_{t \leq s} S(-t)^{-1}u(0)}. \tag{7}$$

*The $\omega$ and $\alpha$ limit sets of a set $D \subseteq B$ are defined analogously with $D$ replacing $u(0)$.*

**Discussion** Consider equation (6): we take the union of segments of trajectories starting at time $t \geq s$; then, taking the intersection over all $s \geq 0$, we eliminate the transient behaviour and we are left with information about the asymptotics of the evolution semigroup. Typical members of the $\omega$ and $\alpha$ limit sets include *steady solutions, periodic solutions, quasi-periodic solutions* and *strange attractors*.

We can now define the global attractor. This, if it exists, is a compact attractor which attracts the bounded sets of $B$ uniformly and whose basin of attraction is the whole space $B$. The global attractor is essentially comprised of members of the $\omega$ and $\alpha$ limit sets of points, together with the trajectories which connect them. Detailed discussion of the global attractor can be found in [7,9,17].

**Definition 2.3** *An attractor for the semigroup $S(t)$ is a set $A \subset B$ satisfying the following properties:*

*(i) A is a positively and negatively invariant set for the semigroup. (A positively and negatively invariant set satisfies $S(t)A = A, \forall t \geq 0$.)*

*(ii) A possesses an open neighbourhood U such that $\text{dist}(S(t)u(0), A) \to 0$ as $t \to \infty$, for all $u(0) \in U$. The distance from a point to a set is found by taking the infimum over the distances to all points in the set.*

*The largest open set satisfying (ii) is known as the basin of attraction of A.*

**Definition 2.4** *The global attractor is a compact attractor A which satisfies*

$$d(S(t)U, A) \to 0 \ as \ t \to \infty$$

*uniformly for any bounded set $U \subset B$. Here*

$$d(A_1, A_2) = \sup_{x \in A_1} \inf_{y \in A_2} d(x, y).$$

*The basin of attraction for A is the whole of B.*

**Examples 1** We study a very simple example which will become interesting under discretisation. Consider the ODE

$$u_t = -u^3, \tag{8}$$

with $u(0) \in \Re$. The global attractor for this differential equation is the singleton $\{0\}$ for which the conditions of Definition 2.4 are easily checked. The dynamics of (8) can be summarised as follows:

**Figure 1. The dynamics of (8)**

A slightly more interesting example is the ODE

$$u_t = u - u^3, \tag{9}$$

with $u(0) \in \Re$. Here the global attractor consists of the three equilibria $0, 1$ and $-1$ together with the heteroclinic orbits connecting 0 to $-1$ and to 1. The dynamics of (9) can be summarised as follows:

**Figure 2. The dynamics of (9)**

The first step towards proving the existence of a global attractor is the contruction of an *absorbing set* which all trajectories starting in a bounded set enter in a finite time. For a precise definition see [17]. The existence of an absorbing set is a *necessary* condition for the existence of a global attractor. Theorem 1.1 in Chapter I of [17] describes additional conditions required to ensure that the existence of an absorbing set is *sufficient* for the existence of a global attractor. Clearly it is important to determine what happens to an absorbing set under discretisation. We show in the next section that spurious members of the $\omega$ and $\alpha$ limit sets can destroy the absorbing set property. Consequently we study the existence of spurious solutions in section 5.

## 3 Discrete Evolutions Semigroups

Consider now a nonlinear map in a Banach space

$$U_{n+1} = \Phi(U_n), \tag{10}$$

together with an initial condition on $U_0$. Here $\Phi : B \to B$ for a Banach space $B$. We are particularly interested in the case where (10) forms an approximation to (1); we refrain from making a specific identification between $U_n$ and $u(t)$ because this will depend on the nature of the discretisation (whether or not it is a one step method, whether or not the elements of an infinite dimensional Banach space are approximated finite dimensionally *etc.*) Definitions analogous to 2.1,2.2,2.3 and 2.4 can be made as in [17]. Throgout this section $n$ and $m$ denote integers.

**Definition 3.1** *A discrete evolution semigroup is a mapping $S_n : B \to B$ for some Banach space $B$ with $S_n$ satisfying*

$$S_{n+m} = S_n \circ S_m \ \forall \ integer \ n, m \geq 0 \tag{11}$$

*and*

$$S_0 = I. \tag{12}$$

*The solution $U_n$ of (10) then satisfies*

$$U_n = S_n U_0. \tag{13}$$

*The action of $S_n$ on a set $D \subseteq B$ is defined by*

$$S_n D = \bigcup_{U_0 \in D} S_n U_0. \tag{14}$$

**Definition 3.2** *The $\omega$ limit set and the $\alpha$ limit set (when it exists) of a point $U_0$ are defined repsectively by*

$$\omega(U_0) = \bigcap_{m \geq 0} \overline{\bigcup_{n \geq m} S_n U_0} \tag{15}$$

*and*

$$\alpha(U_0) = \bigcap_{m \leq 0} \overline{\bigcup_{n \leq m} S_{-n}^{-1} U_0} \tag{16}$$

*The $\omega$ and $\alpha$ limit sets of a set $D \subseteq B$ are defined analogously with $D$ replacing $U_0$.*

**Definition 3.3** *An attractor for the semigroup $S_n$ is a set $A \subset B$ satisfying the following properties:*
*(i) $A$ is a positively and negatively invariant set for the semigroup. (A positively and negatively invariant set satisfies $S_n A = A \ \forall n \geq 0$.)*
*(ii) $A$ possesses an open neighbourhood $U$ such that $dist(S_n U_0, A) \rightarrow 0$ as $t \rightarrow \infty$, for all $U_0 \in U$. The distance from a point to a set is found by taking the infimum over the distances to all points in the set.*
*The largest open set satisfying (ii) is known as the basin of attraction of $A$.*

**Definition 3.4** *The global attractor is a compact attractor $A$ which satisfies*

$$d(S(t)U, A) \rightarrow 0 \ as \ t \rightarrow \infty$$

*uniformly for any bounded set $U \subset B$. Here*

$$d(A_1, A_2) = \sup_{x \in A_1} \inf_{y \in A_2} d(x, y).$$

*The basin of attraction for $A$ is the whole of $B$.*

Let us assume now that the discrete evolution semigroup $S_n$ forms an approximation to the continuous evolution semigroup $S(t)$ and that $\Delta t$ is the discretisation parameter. In the case where the underlying problem is infinite dimensional we shall suppress explicit reference to spatial discretisation, assume that a mesh refinement path has been chosen and assume that a suitable prolongation operator has been chosen. There are three fundamental questions which confront the numerical analyst:

(i) Are the $\omega$ and $\alpha$ limit sets for $S(t)$ and $S_n$ the same, or "close", in particular as $\Delta t \rightarrow 0$?

(ii) If $S(t)$ has a (global) attractor $A$ does $S_n$ have a (global) attractor $A_{\Delta t}$?

(iii) If the answer to (ii) is "yes", then does $A_{\Delta t} \rightarrow A$ as $\Delta t \rightarrow 0$?

The first question has been studied by a number of workers: in [6] conditions are derived which ensure that no spurious steady solutions are introduced by Runge-Kutta discretisation. In [10] the existence of spurious steady solutions is examined for Runge-Kutta, linear mulitstep and predictor-corrector methods. Spurious periodic solutions are often introduced by discretisation; in particular, spurious period 2 solutions in $n$ for the discrete semigroup are important for discretisations of evolution equations whose linear variational equations have real eigenvalues. Examples of this are given in [16], the background is surveyed in [14] and a complete theory described in [15,16]. Spurious invariant curves are also important and an instructive example of this is given in [1]. A unified approach to the existence of spurious members of the $\omega$ and $\alpha$ limit sets, using bifurcation theory, is contained in [11]. Some recent work by Elliott [5] describes a class of time-discretisation methods which preserve the Liapunov functional structure of certain evolution equations; this powerful approach prevents the existence of almost all spurious solutions, except spurious steady solutions introduced by *spatial* discretisation.

The second and third questions are more difficult and less work has been done on them. Kloeden and Lorenz examine the approximation of attracting sets in ODEs by one-step time-discretisations [12] and by multistep methods [13]. Hale and co-workers [8] have examined similar problems by different means. For a survey of further results see [9, p170.]

The answer to the first question is fundamental to the second and third questions since the $\omega$ and $\alpha$ limit sets of points are necessarily contained in the global attractor. Often a global attractor is destroyed by discretisation because there are orbits connecting a spurious member of the limit sets to infinity and hence the absorbing set property no longer holds; an example of this is given below.

In this paper we concentrate on question (i). Question (i) is related to both (ii) and (iii) and sheds light on those problems (see the example below). Furthermore, question (i) makes sense in those problems for which the dynamical system does not possess a global attractor, such as Hamiltonian systems and PDEs whose solutions blow-up in finite time.

**Example 2** Consider the ODE (8) under discretisation by the Euler method. We obtain

$$U_{n+1} = U_n - \Delta t U_n^3. \tag{17}$$

Note that the steady solution of the differential equation, 0, is preserved under discretisation. However a period 2 solution is introduced:

$$U_n = \sqrt{\frac{2}{\Delta t}}(-1)^n \tag{18}$$

is a solution of (17).

The spurious solution plays a very important role since it divides the phase space into regions in which the correct asymptotic behaviour is observed ($U_n \to 0$ as $n \to \infty$) and in which the scheme blows-up and solutions diverge to infinity. This is simple to see. Let $U_c = \sqrt{\frac{2}{\Delta t}}$. If $|U_0| > U_c$ then $1 - \Delta t U_0^2 < -1$. Hence, by (17) we have $|U_1| > |U_0|$. By induction, noting that there are no fixed points $> U_c$ of the map: $|U_n| \to |U_{n+1}|$, we deduce that $U_n \to \infty$ as $n \to \infty$. Similarly, if $|U_0| < U_c$ then $1 > 1 - \Delta t U_0^2 > -1$ and so $|U_1| < |U_0|$. By induction we deduce that $U_n \to 0$ as $n \to \infty$, since 0 is the only fixed point $< U_c$ of the map: $|U_n| \to |U_{n+1}|$.

We have constructed orbits connecting the spurious solution to 0 and to $\infty$. Thus the absorbing set property, which is necessary for the existence of a global attractor, has been destroyed by the unstable spurious solution. This spurious solution exists for any finite $\Delta t > 0$. Hence a global attractor does not exist for the discretisation. The dynamics of the map (17) are summarised in the following Figure which should be compared with Figure 1 for the underlying ODE:

218

**Figure 3. The dynamics of (17)**

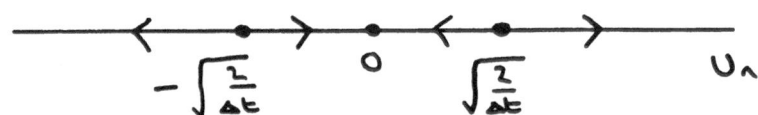

It is worth pointing out that, although the period 2 solution is obviously spurious and is also unstable it is still fundamental to an understanding of the equation (8) under the discretisation (17). Note also that the spurious solution approaches infinity in norm as $\Delta t \to 0$; this is typical of spurious solutions and is proved under quite general conditions in [16]. □

Further simple examples of spurious solutions can be found in section 2 of [16]. In higher dimensions the role of unstable spurious solutions is very similar to that demonstrated in the example above. A precise characterisation of the destination of all initial data is generally impossible but it is often the case that unstable spurious solutions have unstable manifolds which connect with infinity, thereby destroying the absorbing set property. See [16, Theorem 5.2] for an example of this in high dimensions. Hence, for the numerical solution of evolution equations where the long-time dynamics are of interest and where large classes of initial conditions are considered, it is very important to design schemes which minimise the effect of spurious members of the limit sets.

There are two approaches to this design criterion which we will consider here. Both rely conceptually on the idea of treating the approximating dynamical system as a bifurcation problem, with the discretisation parameter $\Delta t$ playing the role of bifurcation parameter. The first approach is to design schemes for which there are no spurious members (of a particular type) of the limit sets for *all* $\Delta t > 0$. This correponds to proving a global non-existence result for a particular class of branches of solutions. Unfortunately, this sometimes leads to schemes which are impractical for other reasons – solution of the nonlinear algebraic equations is prohibitively expensive, a maximum principle is difficult to enforce *etc.* A second approach is to establish that branches of a particular class of spurious solutions cannot extend back to *arbitrarily small positive* $\Delta t$. We consider both these approaches in the following section, where linear multistep methods are examined in detail.

## 4  Spurious Bifurcations In The $\omega$ and $\alpha$ Limit Sets

In this section we consider linear multistep methods for the solution of (1). These can be written in the general form

$$\sum_{k=0}^{M} \alpha_k U_{n+k} = \Delta t \sum_{k=0}^{M} \beta_k G(U_{n+k}). \tag{19}$$

Here $U_k$ approximates $u(k\Delta t)$. (Note that, to formulate this method as a one-step map in the form (10), it is necessary establish the solvability of the nonlinear equation for $U_{n+M}$

(if $\beta_M \neq 0$) and to consider a vector containing $M$ steps as a single unknown; see [11].) In this paper we consider only the effect of time-discretisation, although (1) may, of course, be a system of ODEs arising from the spatial discretisation of a PDE. We assume that (19) forms a consistent approximation of (1). The following polynomials will be useful.

**Definition 4.1** *We define*

$$\rho(z) = \sum_{k=0}^{M} \alpha_k z^k \ and \ \sigma(z) = \sum_{k=0}^{M} \beta_k z^k. \tag{20}$$

Definition 4.2 below is a generalisation of a definition contained in [6]. Roughly, *regular of degree 1* means no spurious steady solutions in the limit sets and *regular of degree 2* means no period 2 solutions in the limit sets. Note that period 2 solutions are always spurious and their importance in determining the dynamics of discretisations has been illustrated by means of example. See [16] for further examples. One could consider spurious solutions of higher periodicity but period one (steady) and period two solutions are particularly important since they bifurcate from steady solutions of the map as $\Delta t$ varies and are observed generically in systems with hyperbolic equilibria which are of saddle or of nodal type. Spurious invariant curves are also of interest since they are observed in systems with hyperbolic equilibria of spiral type – see [11].

**Definition 4.2**

*The numerical method (19) is regular of degree 1 if every fixed point $\hat{U} \in B$ of the map (19) satisfies $G(\hat{U}) = 0$, for all $\Delta t > 0$ and for all equations (1).*

*The numerical method (19) is regular of degree 2 if (19) does not admit period 2 solutions in n for all $\Delta t > 0$ and all equations (1).*

Theorem 4.3 characterises the regularity of linear multistep methods.

**Theorem 4.3**

(i) *The numerical method (19) is regular of degree 1.*

(ii) *If $\rho(-1) \neq 0$ the numerical method (19) is regular of degree 2 if and only if $\sigma(-1) = 0$.*

(iii) *If $\rho(-1) = 0$ the numerical method (19) is not regular of degree 2.*

**Proof** Part (i) is proved in [10]. Parts (ii) and (iii) are proved in [11]. For motivation we shall sketch the proof of (ii) for the $\theta$ method only; this simplifies the technicalities considerably without losing the central ideas of the proof. The $\theta$ method is

$$U_{n+1} - U_n = \Delta t[(1 - \theta)G(U_n) + \theta G(U_{n+1})]. \tag{21}$$

Here $\rho(z) = z - 1$ and $\sigma(z) = (1 - \theta) + \theta z$. Note that $\rho(-1) \neq 0$ and that $\sigma(-1) = 1 - 2\theta$. Let $U_{2n} = U$ and $U_{2n+1} = V$. Then period 2 solutions are pairs $U, V$ both in $B$ with $U \neq V$ satisfying

$$V - U = \Delta t[(1 - \theta)G(U) + \theta G(V)] \tag{22}$$

and

$$U - V = \Delta t[(1 - \theta)G(V) + \theta G(U)]. \tag{23}$$

Note that the steady solutions of the differential equation (1) are solutions of (22,23) with $U = V$.

To prove the *only if* part of (ii) we show that period 2 solutions bifurcate from steady solutions if $\sigma(-1) \neq 0$. Assume that (1) has a steady solution $\hat{U}$. Without loss of generality we let $\hat{U} = 0$ so that $G(0) = 0$. Note that $U = V = 0$ satisfies (22,23). Let $dG(0)$, the Frechet derivative of $G$ at 0, be non-singular and have a real, non-zero, simple eigenvalue $\eta$. Then a little calculation shows that the Frechet derivative of the system (22,23) is singular at

$$\Delta t = \frac{2}{(2\theta - 1)\eta} \quad and \quad U = V = 0. \tag{24}$$

Futhermore, this eigenvalue $\Delta t$ is simple; thus we deduce from Theorem 5.3 in Chapter 5 of [4] that period 2 solutions of (19) bifurcate from the trivial solution for $\theta \neq \frac{1}{2}$ (*i.e.* $\sigma(-1) \neq 0$.)

To prove the *if* part of (ii) subtract (23) from (22). This gives $V = U$ if $\theta = \frac{1}{2}$ (*i.e.* $\sigma(-1) = 0$) and so period 2 solutions do not exist. □

The result of Theorem 4.3 would seem to suggest that the optimal choice of linear multistep method is one for which $\sigma(-1) = 0$; in particular, for the $\theta$ method this requires $\theta = \frac{1}{2}$. (We restrict ourselves to discussion of the $\theta$ method henceforth.) However, often considerations other than spurious members of the limit sets come into play. For parabolic problems the maximum principle is of great importance and the choice $\theta = 1$ has many advantages from this point of view. For semilinear parabolic equations, the convergence of local attractors is proved in [8; Theorem 4.1] under the condition $\theta \in (\frac{1}{2}, 1]$ and the extra dissipativity afforded by this choice underlies the proof.

Thus the choice $\theta = \frac{1}{2}$ is often *not* made and it is important to examine what happens to branches of spurious solutions in this case. In general we know that branches of spurious solutions will exist if $\theta \neq \frac{1}{2}$ since they bifurcate from genuiune equilibria at the the critical value of $\Delta t$ given by (24). Thus it is important to design schemes for which the branches of spurious solutions cannot extend back to $\Delta t$ arbitrarily small. We indicate how the analysis of steady boundary value problems can shed light on the design of schemes appropriate to a particular equation. This is illustrated by means of two examples from the discretisation of reaction-diffusion equations.

**Example 3** Consider the equation

$$u_t = u_{xx} - u^p, \tag{25}$$

where $p$ is odd. For simplicity consider the Dirichlet boundary conditions

$$u(0, t) = u(1, t) = 0. \tag{26}$$

The solution of this problem is defined for all time and the global attractor is the trivial solution 0. Let us assume that we apply the $\theta$ method to (25). Thus period 2 solutions satisfy (22,23) where $G(u) = u_{xx} - u^p$. Since $p$ is odd we can examine the existence of period 2 solutions satisfying the $Z_2$ symmetry $V = -U$; equation (22) yields

$$\Delta t(1 - 2\theta)[U_{xx} - U^p] + 2U = 0, \tag{27}$$

with boundary conditions

$$U(0) = U(1) = 0. \tag{28}$$

We are interested in choosing schemes for which there are no solutions of (27,28) as $\Delta t \to 0_+$. Multiplying by $U$ and integrating by parts we obtain

$$- \Delta t(1 - 2\theta)[\int_0^1 U_x^2 + U^{p+1} dx] + 2 \int_0^1 U^2 dx = 0. \tag{29}$$

If $\theta > \frac{1}{2}$ the left-hand side is positive for $\Delta t$ positive and hence no spurious solutions exist for $\Delta t > 0$. On the other hand, if $\theta < \frac{1}{2}$ it may be shown that solutions of (27,28) can exist for $\Delta t$ arbitrarily small and positive [3]. Hence $\theta > \frac{1}{2}$ is a superior choice to $\theta < \frac{1}{2}$. Note that this result is in accordance with Theorem 4.1 of [8], alluded to above; it shows that the convergence of attractors is intimately related to the non-existence of spurious solutions. □

In practice equation (25) will be discretised in space as well as in time. However, the semi-discrete argument given sheds light on the appropriate choice of fully discrete scheme: we now show that any solution sequence satisfying the backward Euler scheme ($\theta = 1$) coupled with the usual centred differences in space must converge to zero as $n \to \infty$. *That is, the global attractor is preserved under discretisation.* This is strongly related to the fact that the backward Euler scheme does not possses period 2 solutions in $n$.

**Example 3 Continued** Let $u_j^n$ denote our approximation to $u(j\Delta x, n\Delta t)$. Set $J\Delta x = 1$ and $r = \frac{\Delta t}{\Delta x^2}$. The backward Euler scheme coupled with centred differences in space yields, for $j = 1, \ldots, J - 1$

$$u_j^{n+1} = u_j^n + r\delta^2 u_j^{n+1} - \Delta t(u_j^{n+1})^p. \tag{30}$$

The boundary conditions are

$$u_0^n = u_J^n = 0. \tag{31}$$

Let

$$u_{max}^n = \max_{0 \le j \le J} u_j^n \tag{32}$$

and

$$u_{min}^n = \min_{0 \le j \le J} u_j^n. \tag{33}$$

Re-arranging equation (30) gives u, for $1 \le j \le J - 1$,

$$u_j^{n+1}[1 + 2r + \Delta t(u_j^{n+1})^{p-1}] \le u_j^n + ru_{j+1}^{n+1} + ru_{j-1}^{n+1} \tag{34}$$

$$\le u_{max}^n + 2ru_{max}^{n+1}. \tag{35}$$

If $u_{max}^{n+1}$ is attained for $1 \leq j \leq J - 1$ we find that

$$[1 + \Delta t(u_{max}^{n+1})^{p-1}]u_{max}^{n+1} \leq u_{max}^n. \tag{36}$$

If $u_{max}^{n+1}$ is attained for $j = 0, J$ then clearly $u_{max}^{n+1} = 0$. Similarly, by considering $-u_j^n$ we can show that, if $u_{min}^{n+1}$ is attained for $1 \leq j \leq J - 1$,

$$[1 + \Delta t(u_{min}^{n+1})^{p-1}]u_{min}^{n+1} \geq u_{min}^n. \tag{37}$$

If $u_{min}^{n+1}$ is attained for $j = 0, J$ then clearly $u_{min}^{n+1} = 0$. Combining (36,37) and the fact that the maximum (resp. minimum) is non-negative (resp. non-positive) we deduce that $u_j^n \to 0$ as $n \to \infty$. (We use the fact that $p$ is odd.) Thus the global attractor has been preserved under discretisation by the backward Euler scheme. Such a result *cannot be proved for the forward Euler scheme* ($\theta = 0$) without introducing a restriction on $\Delta t$ in terms of the initial data – see Theorem 5.2 in [16]. Such a restriction is necessary to avoid the effect of the spurious period 2 solutions shown to exist for $\theta = 0$. $\square$

Example 3 is particularly simple since we could show non-existence of spurious solutions for *all* $\Delta t > 0$ when $\theta > \frac{1}{2}$. In general this is not possible since for $\theta > \frac{1}{2}$ spurious periodic solutions always bifurcate from linearly unstable equilibria at positive values of $\Delta t$. (This follows from equation (24) with $\eta > 0$ which implies that the equilibrium is unstable; note that in Example 3 the only equilibrium is 0 and it is stable so that bifurcation does not occur). However, it is always be possible to write down a boundary value problem governing the existence of spurious period 2 solutions. Analysing the existence of solutions for this problem in the limit $\Delta t \to 0_+$ can yield practical guidelines for the choice of scheme. We illustrate this by means of a more involved example.

**Example 4** Consider the equation

$$u_t = u_{xx} - f(u), \tag{38}$$

where

$$f(u) = \sum_{j=0}^{p} a_j u^{2j+1}. \tag{39}$$

The boundary conditions are

$$u(0,t) = u(1,t) = 0. \tag{40}$$

Here $a_p > 0$; under this assumption, it is shown in Chapter III of [17] that the problem (38-40) possesses a global attractor. We now show that the backward Euler method is an appropriate discretisation of this equation since branches of spurious period 2 solutions cannot extend back to $\Delta t$ arbitrarily small. The argument is easily extended to cope with all $\theta \in (\frac{1}{2}, 1]$.

The backward Euler discretisation of (38) gives

$$U^{n+1} - U^n = \Delta t[U_{xx}^{n+1} - f(U^{n+1})], \tag{41}$$

with boundary conditions

$$U^n(0) = U^n(1) = 0. \tag{42}$$

Seeking period 2 solutions with the $Z_2$ symmetry $U^{2n+1} = -U^{2n} = U$ we obtain

$$- U_{xx} + f(U) + \frac{2}{\Delta t} U = 0, \tag{43}$$

with boundary conditions

$$U(0) = U(1) = 0. \tag{44}$$

It is our aim to show that (43,44) has no non-trivial solutions for $\Delta t$ sufficiently small and positive.

Multiplying (43) by $U$ and integrating by parts we obtain

$$\int_0^1 [U_x^2 + U f(U) + \frac{2}{\Delta t} U^2] dx = 0. \tag{45}$$

The Young inequality gives us

$$\frac{1}{2} a_p U^{2p+2} - c \leq U f(U).$$

Using this in (45) we obtain

$$\int_0^1 U^2 dx \leq \frac{c \Delta t}{2} \quad \text{and} \quad \int_0^1 U^{2p+2} dx \leq \frac{2c}{a_p}. \tag{46}$$

Applying the Hölder inequality we obtain a further bound

$$\int_0^1 U^2 dx \leq (2c/a_p)^{\frac{1}{p+1}}. \tag{47}$$

Whilst (46) and (47) are useful in establishing where spurious solutions can be found in function and parameter space, they are not sufficient to establish non-existence for $\Delta t$ sufficiently small. This we now do.

Equation (45) together with (39) gives

$$0 \geq \int_0^1 U^2 [\frac{2}{\Delta t} + \sum_{j=0}^p a_j U^{2j}] dx \tag{48}$$

Because $a_p > 0$ it follows that the sum is bounded from below independently of $\Delta t$, since the $a_j's$ do not involve $\Delta t$. Hence, by choosing $\Delta t$ sufficiently small, the integrand can be made positive and we deduce that non-trivial solutions of (43,44) cannot exist.

## 5  Conclusions

We have examined dissipative evolution problems and their discretisations in a unified fashion. It has been shown that the property of possessing a global attractor can be destroyed by the introduction of spurious memebers of the $\omega$ and $\alpha$ limit sets when discretisation is performed. This destruction of the global attractor occurs when the unstable manifold of the spurious solution is connected to infinity. The existence of spurious solutions in linear multistep methods has been examined by two methods both of which treat the numerical method as a dynamical system parameterised by the time-step $\Delta t$. The first corresponds to a global theory on the existence of branches of spurious solutions for *all values of* $\Delta t$. It is based around the examination of bifurcation of spurious solutions from genuiune members of the limit sets. The second method corresponds to an analysis of the existence of spurious solutions in the singular limit $\Delta t \to 0_+$ : it relies on examining boundary value problems and proving the non-existence of solutions *for $\Delta t$ sufficiently small.*

# 6  Acknowledgements

I am grateful to Endre Süli and John Toland for helpful discussions concerning the material in this article.

# 7  References

[1] **F.Brezzi, S.Ushiki and H.Fujii** *Real and ghost bifurcation dynamics in difference schemes for ordinary differential equations.* Appears in Numerical Methods for Bifurcation Problems, Eds: T.Kupper, H.D.Mittleman and H.Weber. Birkhauser-Verlag, Boston, 1984.

[2] **K.Burrage and J.C.Butcher** *Stability criteria for implicit Runge-Kutta methods.* SIAM J. Num. Anal. **16**(1979), 46-57.

[3] **N.Chafee and E.Infante** *A bifurcation problem for a nonlinear parabolic equation.* J. Applic. Anal., 4(1974), 17-37.

[4] **S.N.Chow and J.K.Hale** *Methods of Bifurcation Theory.* Springer Verlag, New York, 1982.

[5] **C.M.Elliott** In preparation.

[6] **E.Hairer, A.Iserles and J.M.Sanz-Serna** *Equilibria of Runge-Kutta methods.* University of Cambridge, DAMTP Report 1989/NA4.

[7] **J.K.Hale, L.T.Magalhaes and W.M.Oliva** *An Introduction to Infinite Dimensional Dynamical Sytems - Geometric Theory.* Springer Lecture Notes in Applied Mathematics, 47. 1984.

[8] **J.K.Hale, X.-B.Lin and G.Raugel** *Upper semicontinuity of attractors for approximations of semigroups and partial differential equations.* Math. Comp. **181**(1988), 89-123.

[9] **J.K.Hale** *Asymptotic Behaviour of Dissipative Dynamical Systems.* AMS Mathematical Surveys and Monographs # 25, 1988.

[10] **A.Iserles** *Stability and dynamics of numerical methods for nonlinear ordinary differential equations.* IMA J. Num. Anal. **10**(1990), 1-30.

[11] **A.Iserles, A.T.Peplow and A.M.Stuart** *A unified approach to spurious solutions introduced by time discretisation.* In preparation.

[12] **P.E.Kloeden and J.Lorenz** *Stable attracting sets in dynamical systems and in their one-step discretisations* SIAM J. Num. Anal. **23**(1986), 986-995.

[13] **P.E.Kloeden and J.Lorenz** *A note on multistep methods and attracting sets of dynamical systems.* To appear in Numer. Math.

[14] **A.M.Stuart** *Nonlinear instability in dissipative finite difference schemes.* SIAM Review **31**(1989), 191-220.

[15] **A.M.Stuart** *Linear instability implies spurious periodic solutions.* IMA J. Num. Anal. **9**(1989), 465-486.

[16] **A.M.Stuart and A.T.Peplow** *The dynamics of the theta method.* Submitted to SIAM J. Sci. Stat. Comp.

[17] **R.Temam** *Infinite Dimensional Dynamical Systems in Mathematics and Physics.* Applied Mathematical Sciences # 68, Springer-Verlag, 1988.

# THE NUMERICAL DETECTION OF HOPF BIFURCATION POINTS

G. MOORE,
*Department of Mathematics,*
*Imperial College,*
*Queens's Gate,*
*LONDON SW7 2BZ.*

T. J. GARRATT, A. SPENCE,
*School of Mathematical Sciences,*
*University of Bath,*
*Claverton Down,*
*Bath. AVON. BA2 7AY.*

ABSTRACT.During the computation of a one-parameter set of steady solutions to a time dependent problem, it is often of great interest to know when periodic behaviour can occur. This Hopf bifurcation is characterised by the linearisation about a steady state solution possessing a purely imaginary pair of eigenvalues. The cheap and reliable detection of such bifurcation is not however straightforward. The present paper examines the problem and suggests various strategies for solving it.

## 1 Introduction

Consider the parameter-dependent system

$$\frac{d\mathbf{u}}{dt} = \mathbf{f}(\mathbf{u}, s) \qquad \mathbf{f} : \Re^n \times \Re \to \Re^n, \tag{1}$$

where $\mathbf{f}$ may have been obtained by means of a spatial discretisation of a differential operator. The simplest solutions of (1) are the steady state solutions obtained from

$$\mathbf{f}(\mathbf{u}, s) = \mathbf{0}. \tag{2}$$

If the Jacobian matrix

$$A(s) := \mathbf{f}_{\mathbf{u}}(\mathbf{u}, s), \tag{3}$$

is non-singular then the Implicit Function Theorem guarantees that (locally) the solution set of (2) consists of a unique solution curve parametrisable by s, i.e. we may write the solution set as $(\mathbf{u}^*(s), s)$. This curve would usually be computed by a variant of Newton's method involving the factorisation of $A(s)$ or its approximation. Any change in sign of $\det(A(s))$ is thus observed while computing the solution curve and this information is often very useful for detecting the bifurcation of steady solutions.

The next simplest solutions of (1) are those which are periodic in time. If we wish to detect when periodic solutions of (1) branch away from a solution curve of (2), i.e. Hopf bifurcation, then this is revealed (ignoring some additional conditions) by a pair of complex conjugate eigenvalues of $A(s)$ crossing the imaginary axis. A succession of LU factorisations of $A(s)$, however, does not generally give this information, i.e. a change in the relative number of eigenvalues of $A(s)$ with positive or negative real part. The purpose of this paper is thus to consider the numerical linear algebra problem of efficiently detecting the above eigenvalue behaviour in a one-parameter family of $n \times n$ matrices.

In the standard continuation/bifurcation code [2], Hopf bifurcation is detected by computing all the eigenvalues of $A(s)$, by means of the unsymmetric QR algorithm, at different

*D. Roose et al. (eds.), Continuation and Bifurcations: Numerical Techniques and Applications, 227–246.*

values of $s$. Although acceptable for very small problems, this strategy has a number of defects.

(i) A different problem is solved, i.e. all the eigenvalues of $A(s)$ are obtained rather than just a count of those eigenvalues with positive real part. The QR algorithm has a wonderful reputation as a solver of the former problem but that does not mean it is the best for the latter. (In this respect it is interesting that the latter problem may be solved by a finite procedure, as is shown in section 4, while it is well-known that, in general for $n \geq 5$, the computation of a single eigenvalue is an infinite procedure.) Also it should be remembered that our problem only really requires a low level of accuracy (say 1 significant figure!).

(ii) The QR algorithm does not make use of the fact that we are in a continuation framework and it can be assumed that information will be available from a previous value of $s$. The most obvious attempt to remedy this is to use previously computed eigenvalues as shifts, but the QR method is so good that little or nothing is gained by this change. In general it can be stated that algorithms which rely on the reduction of a matrix to more compact form have inherent difficulty in using previous information, in contrast to iterative methods which we shall consider in section 4.

(iii) The QR algorithm does not retain any band structure that $A(s)$ may possess (apart from Hessenberg of course). This is destroyed both by the reduction to Hessenberg form or in the direct application of QR to $A(s)$. It would seem that the only way to avoid this loss of sparsity is to use the LR method without interchanges [21, 29]. This is usually severely frowned upon since the method can be numerically unstable and may breakdown. If, however, only very modest accuracy is required and the occasional breakdown is tolerable in a continuation framework then it should be considered since previously computed eigenvalues may be used as shifts.

We shall investigate approaches which try to avoid some of these defects. In this context it is useful to keep in mind the three possible classes of matrix to which $A$ may belong. (The $s$ dependence is omitted from now on).

(a) Small dense matrices for which standard factorisations are acceptable.

(b) Larger banded matrices for which the band structure must be preserved.

(c) Huge sparse matrices for which one can only form the product $Ax$.

We shall be chiefly interested in (a) and (b), although a number of observations will apply to (c) as well.

## 2   Classical methods based on the characteristic polynomial

The classical methods for counting the number of eigenvalues of a matrix which have positive real part, eg. the Routh-Hurwitz criterion [5, 9, 6, 7], rely on working with the characteristic polynomial. These methods were developed to deal with very small matrices by hand computation. In this section we investigate whether it is possible to apply these techniques to larger problems. Of course it is well known that it is dangerous to determine the coefficients of the characteristic polynomial itself but that the evaluation of

$$p_n(\lambda) := \det(A - \lambda I) \qquad (4)$$

for particular values of $\lambda$ can be carried out perfectly safely. The main consideration is then how efficiently this evaluation may be performed. A general matrix $A$ can be reduced to Hessenberg form by either orthogonal or stabilised similarity transformations and then the standard recurrence relation for the determinant of a Hessenberg matrix may be used to obtain $p_n(\lambda)$. This approach also enables us to evaluate derivatives of $p_n(\lambda)$ if required and also to avoid the use of complex arithmetic for complex $\lambda$ by working with quadratic factors. Additionally it is possible, by accepting a certain loss in accuracy which we do not mind [29], to further reduce the Hessenberg matrix to tridiagonal form. The standard recurrence relation for the determinant of a tridiagonal matrix may then be used. For future reference we state this recurrence here:

$$
\begin{aligned}
p_0(\lambda) &= 1, \\
p_1(\lambda) &= \alpha_1 - \lambda, \\
p_j(\lambda) &= (\alpha_j - \lambda)p_{j-1}(\lambda) - \beta_j p_{j-2}(\lambda), \qquad j = 2, \dots, n,
\end{aligned}
\qquad (5)
$$

with $\alpha_j := a_{jj}, \beta_j := a_{j,j+1}a_{j+1,j}$ and $p_j(\lambda)$ being the determinant of the leading principle $j \times j$ minor of $A - \lambda I$. Note that the reduction to Hessenberg or tridiagonal form only needs to be done once and then the recurrence relation is used to evaluate $p_n(\lambda)$ at the required points. On the other hand, if $A$ is a banded matrix it cannot in general be stably reduced to Hessenberg form without destroying the band and neither is there an efficient generalisation of the recurrence relation. Consequently the accepted method for the evaluation of $p_n(\lambda)$ in the banded case is to use LU factorisation of $A - \lambda I$ with partial pivoting. This approach however has the drawbacks that :-

(a) each value of $\lambda$ requires a new factorisation,

(b) complex $\lambda$ require the use of complex arithmetic,

(c) derivatives of $p_n(\lambda)$ are not available.

Once an efficient method for evaluating $p_n(\lambda)$ is available, it is possible to use iterative methods to solve $p_n(\lambda) = 0$. This idea will be considered further in section 4, here we restrict attention to the classical methods for counting the zeros of $p_n(\lambda)$ with positive real part.

The fact that $p_n(\lambda)$ is an analytic function of $\lambda$ enables us to use complex function theory to obtain a formula for the number of eigenvalues of $A$ in the right half-plane, $q$ say. Thus if C is the semi-circle

$$C := \left\{ it, -R \le t \le R \right\} \cup \left\{ Re^{i\theta}, -\frac{\pi}{2} \le \theta \le \frac{\pi}{2} \right\}$$

described in an anti-clockwise direction, then, so long as no eigenvalues of $A$ are purely imaginary and $R$ is greater than the spectral radius of $A$,

$$q = \frac{1}{2\pi i} \oint_C \frac{p_n'(z)}{p_n(z)} dz. \qquad (6)$$

Alternatively we can also say that

$$q = \frac{1}{2\pi} \{\text{change in } \arg(p_n(z)) \text{ around } C\} \tag{7}$$

or

$$n - 2q = \frac{1}{\pi} \{\text{change in } \arg(p_n(i\mu)) \text{ from } \mu = -\infty \text{ to } \mu = +\infty\}. \tag{8}$$

Although (7) is a very nice theoretical result, it does not seem to provide a practical algorithm. Not only would a large number of evaluations of $p_n(z)$ be required but also $\arg(p_n(z))$ often changes rapidly, which indicates that this is not a suitable formula for numerical computation.

The most well-known classical method for computing $q$, which leads to the Routh-Hurwitz criterion for example, is derived from (8). First we write

$$i^n p_n(i\mu) := p_n^R(\mu) + i p_n^I(\mu),$$

where $\mu$ is a real variable and $p_n^R$ and $p_n^I$ are real polynomials, in order to ensure that $p_n^R$ is exactly of degree $n$ and $p_n^I$ has lesser degree. Second we use the fact that

$$\tan[\arg(i^n p_n(i\mu))] = \frac{p_n^I(\mu)}{p_n^R(\mu)} \tag{9}$$

to deduce that $p_n^I(\mu)/p_n^R(\mu)$ jumps from $+\infty$ to $-\infty$ (at $\arg(i^n p_n(i\mu)) = (2m \pm \frac{1}{2})\pi$) when $\arg(i^n p_n(i\mu))$ is increasing and correspondingly jumps from $-\infty$ to $+\infty$ when $\arg(i^n p_n(i\mu))$ is decreasing. Consequently we may easily deduce from (8) that

$$n - 2q = \left[\text{Number of jumps of } p_n^I(\mu)/p_n^R(\mu) \text{ from } +\infty \text{ to } -\infty\right] -$$
$$\left[\text{Number of jumps of } p_n^I(\mu)/p_n^R(\mu) \text{ from } -\infty \text{ to } +\infty\right] \tag{10}$$
$$\text{as } \mu \text{ moves from } -\infty \text{ to } +\infty.$$

This last expression can be calculated by obtaining a Sturm sequence of polynomials of decreasing degree via the Euclidean algorithm applied to $p_n^R(\mu)$ and $p_n^I(\mu)$, i.e.

$$\begin{aligned} g_1(\mu) &= p_n^R(\mu), \\ g_2(\mu) &= p_n^I(\mu), \\ g_{i-1}(\mu) &= d_i(\mu)g_i(\mu) - g_{i+1}(\mu), \qquad i = 2, \ldots, m-1, \\ g_{m-1}(\mu) &= d_m(\mu)g_m(\mu), \end{aligned}$$

where $g_{i+1}(\mu)$ is the remainder obtained after dividing $g_{i-1}(\mu)$ by $g_i(\mu)$ and $g_m(\mu)$ is the greatest common divisor of $p_n^R(\mu)$ and $p_n^I(\mu)$. If $g_m(\mu)$ is of constant sign for $\mu \in (-\infty, \infty)$ then it follows that

$$g_i(\mu^*) = 0 \Rightarrow g_{i-1}(\mu^*)g_{i+1}(\mu^*) < 0 \qquad i = 2, \ldots, m-1.$$

Consequently $V(\mu)$, the number of sign changes in the sequence

$$\{g_1(\mu), g_2(\mu), \ldots, g_m(\mu)\},$$

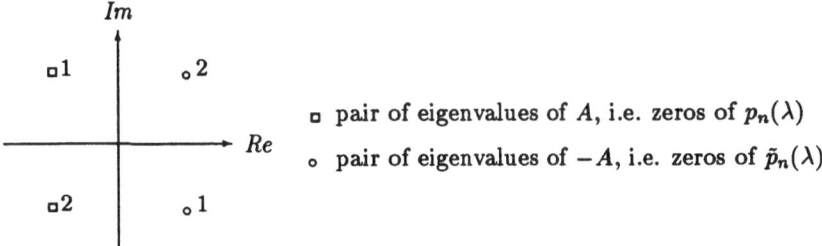

Figure 1: Reflection of complex conjugate eigenvalues

can only vary at a zero of $p_n^R(\mu)$ and then it will either increase by one ( if $p_n^I(\mu)/p_n^R(\mu)$ changes from $+\infty$ to $-\infty$ ) or decrease by one ( if $p_n^I(\mu)/p_n^R(\mu)$ changes from $-\infty$ to $+\infty$). Hence from (10) we must have

$$n - 2q = V(\infty) - V(-\infty). \tag{11}$$

On the other hand, even if $g_m(\mu)$ is not of constant sign it must still be a factor of all the $g_i(\mu)$'s. Hence neither $V(\mu)$ nor $p_n^I(\mu)/p_n^R(\mu)$ is affected by this factor and so (11) can again be deduced from (10). Classically (11) has been used by explicitly calculating the sequence $\{g_i(\mu)\}$ and looking at the signs of the coefficients of the leading terms. This is sufficient to determine $V(\pm\infty)$ but is unacceptable as a numerical algorithm. As a first step towards a practical procedure, a recurrence relation for $\{g_i(\mu)\}$ would have to be developed. Even in the tridiagonal case however (with $n = 0(\bmod 4)$ for simplicity); when $p_n^R(\mu)$ and $p_n^I(\mu)$ may be defined using (5), i.e.

$$p_j^R(\mu) = \alpha_j p_{j-1}^R(\mu) + \mu p_{j-1}^I(\mu) - \beta_j p_{j-2}^R(\mu), \quad j = 2, \ldots, n,$$
$$p_0^R(\mu) = 1 \quad p_1^R(\mu) = \alpha_1;$$
$$p_j^I(\mu) = \alpha_j p_{j-1}^I(\mu) - \mu p_{j-1}^R(\mu) - \beta_j p_{j-2}^I(\mu), \quad j = 2, \ldots, n,$$
$$p_0^I(\mu) = 0 \quad p_1^I(\mu) = -\mu;$$

it is not clear how to manage this. Thus regretfully we must conclude that Routh-Hurwitz type methods do not at present have an acceptable numerical implementation.

A different classical method for computing $q$ is based on an algorithm of Schur [9, 6, 7]. If we define

$$\tilde{p}_n(\lambda) := \det(A + \lambda I), \tag{12}$$

then the zeros of $\tilde{p}_n(\lambda)$ are those of $p_n$, c.f. (4), rotated through $180°$ about the origin in the complex plane (see Figure 1). Since these zeros occur in complex conjugate pairs it is as if they have been reflected across the imaginary axis. This symmetry between the zeros of $p_n$ and $\tilde{p}_n$ means that $\tilde{p}_n$ must have $q$ zeros with negative real part and that

$$|p_n(i\mu)| = |\tilde{p}_n(i\mu)| \tag{13}$$

for all real $\mu$. If however this last result is true on the real axis, i.e.

$$|p_n(\mu)| = |\tilde{p}_n(\mu)| \tag{14}$$

for all real $\mu$, then $p_n$ must either consist of only even powers or only odd powers. In each case we may deduce that $p_n$ has exactly as many zeros with positive real part as zeros with negative real part. On the other hand, if (14) does not hold there will be an infinite number of non-zero $\theta \in \Re$ such that $|p_n(\theta)| \neq |\tilde{p}_n(\theta)|$. Fixing on one particular $\theta$, we may write

$$\tilde{p}_n(\theta)p_n(\lambda) - p_n(\theta)\tilde{p}_n(\lambda) = (\lambda - \theta)q_{n-1}(\lambda), \tag{15}$$

because the l.h.s. of (15) is an $n^{th}$ degree polynomial in $\lambda$ which must have a root $\theta$. Now there are two possibilities.

(i) If $|\tilde{p}_n(\theta)| > |p_n(\theta)|$ then, by following the proof of Rouché's theorem [1], we may deduce that

$$p_n(\lambda) - \frac{p_n(\theta)}{\tilde{p}_n(\theta)}\tilde{p}_n(\lambda)$$

has the same number of roots with positive/negative real part as $p_n(\lambda)$. Consequently if $\theta < 0$ then $q_{n-1}(\lambda)$ has the same number of zeros with positive real part as $p_n(\lambda)$, while if $\theta > 0$ it has one less.

(ii) If $|\tilde{p}_n(\theta)| < |p_n(\theta)|$ then in similar fashion we may deduce that

$$\tilde{p}_n(\lambda) - \frac{\tilde{p}_n(\theta)}{p_n(\theta)}p_n(\lambda)$$

has the same number of roots with positive/negative real part as $\tilde{p}_n(\lambda)$. Hence if $\theta > 0$ then $q_{n-1}(\lambda)$ has the same number of zeros with negative real part as $\tilde{p}_n(\lambda)$, while if $\theta < 0$ it has one less.

We can now repeat the whole idea with the lower degree polynomial $q_{n-1}(\lambda)$ and thus eventually obtain the required information about the zeros of $p_n(\lambda)$. To sum up, it seems that there is greater hope that this classical method (compared with the previous one) may lead to a practical algorithm. In particular, if $p_n(\lambda)$ is given by a recurrence relation then we may also obtain a recurrence relation for $q_{n-1}(\lambda)$. This is illustrated by a possible procedure below for tridiagonal $A$, c.f. (5), where $\{s^{(k)}(\lambda)\}$ is our sequence of polynomials of degree $n - k$, i.e. $s^{(1)}(\lambda) \equiv q_{n-1}(\lambda)$ above.

1   $s^{(k)}(\lambda) = p^{(k)}(\lambda) + \tilde{p}^{(k)}(\lambda)$

where $p^{(k)}(\lambda) = p^{(k)}_{n-k}(\lambda)$ with

$$p^{(k)}_j(\lambda) = (\alpha_{k+j} - \lambda)p^{(k)}_{j-1}(\lambda) - \beta_{k+j}p^{(k)}_{j-2}(\lambda) + A^{(k)}_j, \quad 1 \leq j \leq n - k,$$

$$p^{(k)}_0(\lambda) = \gamma^{(k)}, \qquad p^{(k)}_{-1}(\lambda) = 0,$$

and $\tilde{p}^{(k)}(\lambda) = \tilde{p}^{(k)}_{n-k}(\lambda)$ with

$$\tilde{p}^{(k)}_j(\lambda) = (\alpha_{k+j} + \lambda)\tilde{p}^{(k)}_{j-1}(\lambda) - \beta_{k+j}\tilde{p}^{(k)}_{j-2}(\lambda) + B^{(k)}_j, \quad 1 \leq j \leq n - k,$$

$$\tilde{p}^{(k)}_0(\lambda) = \delta^{(k)}, \qquad \tilde{p}^{(k)}_{-1}(\lambda) = 0.$$

Initially (for $k = 0$) we would have $\gamma^{(0)} = 1$, $\delta^{(0)} = 0$ and $A^{(0)}_j = B^{(0)}_j = 0, 1 \leq j \leq n$, so that $\tilde{p}^{(0)}(\lambda)$ is identically zero.

2 Choose real $\theta^{(k)} \neq 0$ such that $|s^{(k)}(\theta^{(k)})| \neq |s^{(k)}(-\theta^{(k)})|$ and set

$$s^{(k+1)}(\lambda) = \frac{s^{(k)}(-\theta^{(k)})s^{(k)}(\lambda) - s^{(k)}(\theta^{(k)})s^{(k)}(-\lambda)}{\lambda - \theta^{(k)}}.$$

Since $s^{(k+1)}(\lambda) = p^{(k+1)}(\lambda) + \tilde{p}^{(k+1)}(\lambda)$, we wish to define $\gamma^{(k+1)}, \delta^{(k+1)}$ and $A_j^{(k+1)}, B_j^{(k+1)}$ $1 \leq j \leq n - k - 1$, in terms of the corresponding data at the $k^{th}$ step.

3   (i) $p_j^{(k)}(\lambda) = (\lambda - \theta^{(k)})t_{j-1}^{(k)}(\lambda) + C_j^{(k)}, \quad 0 \leq j \leq n - k,$
where

$$t_j^{(k)}(\lambda) = (\alpha_{k+1+j} - \lambda)t_{j-1}^{(k)}(\lambda) - \beta_{k+1+j}t_{j-2}^{(k)}(\lambda) - C_j^{(k)}, \quad 1 \leq j \leq n - k - 1,$$
$$t_0^{(k)}(\lambda) = -\gamma^{(k)}, \qquad t_{-1}^{(k)}(\lambda) = 0,$$

and

$$C_j^{(k)} = (\alpha_{k+j} - \theta^{(k)})C_{j-1}^{(k)} - \beta_{k+j}C_{j-2}^{(k)} - A_j^{(k)}, \quad 1 \leq j \leq n - k,$$
$$C_0^{(k)} = \gamma^{(k)} \qquad C_{-1}^{(k)} = 0.$$

(ii) $p_j^{(k)}(-\lambda) = (\lambda - \theta^{(k)})u_{j-1}^{(k)}(\lambda) + D_j^{(k)}, \quad 0 \leq j \leq n - k,$
where

$$u_j^{(k)}(\lambda) = (\alpha_{k+1+j} + \lambda)u_{j-1}^{(k)}(\lambda) - \beta_{k+1+j}u_{j-2}^{(k)}(\lambda) + D_j^{(k)}, \quad 1 \leq j \leq n - k - 1,$$
$$u_0^{(k)}(\lambda) = \gamma^{(k)} \qquad u_{-1}^{(k)}(\lambda) = 0,$$

and

$$D_j^{(k)} = (\alpha_{k+j} + \theta^{(k)})D_{j-1}^{(k)} - \beta_{k+j}D_{j-2}^{(k)} + A_j^{(k)}, \quad 1 \leq j \leq n - k,$$
$$D_0^{(k)} = \gamma^{(k)} \qquad D_{-1}^{(k)} = 0.$$

(iii) $\tilde{p}_j^{(k)}(\lambda) = (\lambda - \theta^{(k)})v_{j-1}^{(k)}(\lambda) + E_j^{(k)}, \quad 0 \leq j \leq n - k,$
where

$$v_j^{(k)}(\lambda) = (\alpha_{k+1+j} + \lambda)v_{j-1}^{(k)}(\lambda) - \beta_{k+1+j}v_{j-2}^{(k)}(\lambda) + E_j^{(k)}, \quad 1 \leq j \leq n - k - 1,$$
$$v_0^{(k)}(\lambda) = \delta^{(k)} \qquad v_{-1}^{(k)}(\lambda) = 0,$$

and

$$E_j^{(k)} = (\alpha_{k+j} + \theta^{(k)})E_{j-1}^{(k)} - \beta_{k+j}E_{j-2}^{(k)} + B_j^{(k)}, \quad 1 \leq j \leq n - k,$$
$$E_0^{(k)} = \delta^{(k)} \qquad E_{-1}^{(k)} = 0.$$

(iv) $\tilde{p}_j^{(k)}(-\lambda) = (\lambda - \theta^{(k)})w_{j-1}^{(k)}(\lambda) + F_j^{(k)}, \quad 0 \leq j \leq n - k,$
where

$$w_j^{(k)}(\lambda) = (\alpha_{k+1+j} - \lambda)w_{j-1}^{(k)}(\lambda) - \beta_{k+1+j}w_{j-2}^{(k)}(\lambda) - F_j^{(k)}, \quad 1 \leq j \leq n - k - 1,$$
$$w_0^{(k)}(\lambda) = -\delta^{(k)} \qquad w_{-1}^{(k)}(\lambda) = 0,$$

and

$$F_j^{(k)} = (\alpha_{k+j} - \theta^{(k)})F_{j-1}^{(k)} - \beta_{k+j}F_{j-2}^{(k)} - B_j^{(k)}, \quad 1 \le j \le n - k,$$
$$F_0^{(k)} = \delta^{(k)} \quad F_{-1}^{(k)} = 0.$$

4 Hence

$$
\begin{aligned}
\gamma^{(k+1)} &= s^{(k)}(\theta^{(k)})\delta^{(k)} - s^{(k)}(-\theta^{(k)})\gamma^{(k)}, \\
A_j^{(k+1)} &= s^{(k)}(\theta^{(k)})F_j^{(k)} - s^{(k)}(-\theta^{(k)})C_j^{(k)}, \quad 1 \le j \le n - k - 1, \\
\delta^{(k+1)} &= s^{(k)}(-\theta^{(k)})\gamma^{(k)} - s^{(k)}(\theta^{(k)})\delta^{(k)}, \\
B_j^{(k+1)} &= s^{(k)}(-\theta^{(k)})D_j^{(k)} - s^{(k)}(\theta^{(k)})E_j^{(k)}, \quad 1 \le j \le n - k - 1.
\end{aligned}
$$

There still remain, however, a number of obstacles to be overcome before this approach can be recommended for certain problems. In particular,

(a) how can we decide when $|s^{(k)}(\mu)| = |s^{(k)}(-\mu)|$ to an acceptable accuracy for all real $\mu$,

(b) presumably $\theta^{(k)}$ should be chosen so that 'element growth' is controlled during the computation.

Nevertheless this method seems worthy of further investigation.

## 3 Generalised spectrum-slicing

The problem we are trying to solve, i.e. the satisfactory numerical determination of the number of eigenvalues with positive real part belonging to a given matrix $A$, has a very satisfactory solution when $A$ is symmetric. In this case the eigenvalue are of course real and it is well-known that the *inertia* of $A$, the triple $(p, q, r)$ of numbers of negative, positive and zero eigenvalues respectively, is invariant under any *congruence* transformation, i.e.

$$A \to SAS^T$$

where $S$ is non-singular [16]. Hence if $A$ has a factorisation

$$A = LDL^T, \tag{16}$$

where $L$ is unit lower triangular and $D$ is diagonal, we may determine the inertia of $A$ from the signs of $D$'s diagonal elements. This technique has been strongly recommended in [16] under the name *spectrum-slicing*. Although the factorisation (16) may not exist or may be numerically unreliable, this can be detected during the computation. Thereupon we can make use of the corresponding factorisation of $A - \mu I$ for suitably small $|\mu|$, because this will only be unsatisfactory (for our requirements) in very small intervals about those values of $\mu$ which make the leading principle minors of $A - \mu I$ singular. Two strong advantages of this method are it's efficiency, being roughly equivalent to an LU factorisation, and the fact that banded $A$ leads to correspondingly banded L.

A generalisation of the above result to give the number of eigenvalues with negative, positive or zero part of a general matrix $A$ would be very acceptable, but unfortunately such a theorem is not known. It is, however, possible to make two slightly more general statements by considering the symmetric part of $A$, i.e. $\frac{1}{2}(A + A^T)$. A simple application of Liapunov functions [11] then shows that:-

"$\frac{1}{2}(A + A^T)$ has $n\{\pm\}$ eigenvalues $\Rightarrow$

$A$ has $n$ eigenvalues with $\{\pm\}$ real part."

The converse, however, is not true and neither is there a generalisation when the eigenvalues of $\frac{1}{2}(A + A^T)$ are of different sign. Nevertheless, under the additional assumption that $A$ is *normal*, the required result

$$\text{``}\tfrac{1}{2}(A + A^T) \ has \ \begin{Bmatrix} p - \\ q + \\ r \ 0 \end{Bmatrix} \ eigenvalues \Rightarrow$$

$$A \ has \ \begin{Bmatrix} p \\ q \\ r \end{Bmatrix} \ eigenvalues \ with \ \begin{Bmatrix} - \\ + \\ 0 \end{Bmatrix} \ real \ part \text{ ''}$$

follows trivially from the fact that the common unitary set of eigenvectors for $A$ and $A^T$ possess corresponding eigenvalues $\lambda$ and $\overline{\lambda}$ respectively, i.e. the corresponding eigenvalue of $\frac{1}{2}(A + A^T)$ is the real part of $\lambda$. Hence for a normal $A$ we can just apply spectrum-slicing to $\frac{1}{2}(A + A^T)$.

In pursuit of further generalisation, the natural step now is to reduce an arbitrary matrix to normal form by similarity transformations so that the above procedure may be applied. That this can be achieved relies on the result that

$$\|\hat{S}A\hat{S}^{-1}\|_E = \min\left\{\|SAS^{-1}\|_E : S \text{ non-singular}\right\} \tag{17}$$

implies that $\hat{S}A\hat{S}^{-1}$ is a normal matrix [6]. (This minimum only exists for non-defective $A$, otherwise we have to use inf $\{\|SAS^{-1}\|_E : S$ non-singular$\}$ because $S$ becomes singular in the limit.) This approach is the basis of several related ideas for extending the Jacobi method to non-normal matrices [22, 20, 28, 3], although then it is combined with an orthogonal (or unitary) reduction to diagonal or quasi-diagonal form. Here, however, we are only concerned with (17) and we note that the Euler-Lagrange equation of this problem is

$$A_s A_s^T - A_s^T A_s = 0, \qquad\qquad A_s \equiv SAS^{-1},$$

i.e. one of the alternative characterisations of a normal matrix signifies a zero gradient.

To solve (17) we can attempt to apply one of the standard optimisation procedures and generate a sequence $\{A_k\}$ such that

$$A_{k+1} = S_k A_k S_k^{-1}, \qquad k \geq 0, \qquad A_0 = A, \tag{18}$$

with $\|A_{k+1}\|_E < \|A_k\|_E$. To date, however, only the very simplest idea of minimising in successive 'co-ordinate directions' has been suggested [27]. This algorithm is derived by first noting that we may replace (17) by

$$\|\hat{L}A\hat{L}^{-1}\|_E = \min\left\{\|LAL^{-1}\|_E : L \text{ lower triangular and non-singular}\right\}, \tag{19}$$

because $S$ may be decomposed into $QL$ with $Q$ orthogonal and orthogonal transformations do not affect $\|\cdot\|_E$. Hence (18) becomes

$$A_{k+1} = L_k A_k L_k^{-1}, \quad k \geq 0, \quad A_0 = A. \tag{20}$$

The matrices $\{L_k\}$ are chosen from the basic set

$$L(i,j;\beta) = I + \beta e_i e_j^T, \quad n \geq i \geq j \geq 1,$$

with $(i,j)$ fixed for each $k$ and run through in some cyclic fashion as $k$ varies. (Alternatively $(i,j)$ could be determined through trying to maximise the reduction in $\|\cdot\|_E$ achieved). At each $k$-step we then choose $\beta$ to minimise

$$\|L(i,j;\beta)A_k L(i,j;\beta)^{-1}\|_E$$

and then set $L_k = L(i,j;\beta)$. A convergence proof is given in [27]. This algorithm has the two advantages that:-

(a) lower Hessenburg structure is preserved by (20) and so $A$ can initially be reduced to lower Hessenberg form;

(b) it should often be possible to use the minimum $\hat{L}$ from the previous continuation step as a good 'initial guess' (although this has yet to be practically tested).

The two main disadvantages are that:-

(a) the convergence is no better than linear;

(b) any band structure $A$ possesses is not preserved.

In conclusion we may say that to compete with the QR method an algorithm for (17) is needed which

(i) maintains any band-structure of $A$, or

(ii) converges more rapidly than linear.

Thus further work on the approach considered in this section should follow one of these lines.

## 4  Iterative methods

In this section we wish to consider iterative methods for computing eigenvalues of $A$ or for deducing that these eigenvalues must lie in a certain region of the complex plane. Obviously it is this class of methods which have most hope of using our continuation framework efficiently because not only will good starting approximations be available but also our choice of termination criteria can reflect the fact that only modest accuracy is necessary. We shall be concerned with vector iterations but first mention that the eigenvalues of $A$ may be continued by iterating with the characteristic polynomial. This approach does not depend on the condition of the eigenvector system but in general it is necessary to follow all the eigenvalues because we do not know which will cross the imaginary axis. Working with all $n$ eigenvalues is acceptable for dense matrices because $A$ may first be reduced to tridiagonal form by similarity transformations, then the function and derivative evaluations required by (say) Laguerre's method [17] will be very cheap. For band matrices this idea is relatively much more expensive.

## 4.1 STABLE STEADY SOLUTIONS

In this subsection we shall assume that all the eigenvalues of $A$ have negative real part and we wish to detect when a first crossing of the imaginary axis occurs. This can be achieved by considering the size of the spectral radius $\rho(B)$ of the transformed matrix

$$B := (I - \alpha A)^{-1}(I + \alpha A), \tag{21}$$

where $\alpha$ is a positive constant. The complex mapping

$$w = \frac{1 + \alpha z}{1 - \alpha z},$$

transforms circles with {centre,radius} according to

| | z-plane | w-plane |
|---|---|---|
| $0 \leq R < 1$ | $\{(R^2 + 1)/\alpha(R^2 - 1), 2R/\alpha(1 - R^2)\}$ | $0, R$ |
| | Imaginary axis | $\{0, 1\}$ |
| $R > 1$ | $\{(R^2 + 1)/\alpha(R^2 - 1), 2R/\alpha(R^2 - 1)\}$ | $\{0, R\}$ |

and so the left half of the $z$-plane is mapped onto the interior of the unit disc in the $w$-plane and similarly the right half of the $z$-plane onto the exterior of this disc. Consequently $\rho(B) < 1$ iff all the eigenvalues of $A$ have negative real part and hence possible Hopf bifurcation points may be detected by noting when $\rho(B)$ exceeds unity. Even if $B$ were symmetric and positive definite however it is not a trivial task to achieve this efficiently [18, 15]. The simplest approach is to apply the power method to $B$ as recommended in [4], i.e.

(i) Solve $(I - \alpha A)\mathbf{x}_k = (I + \alpha A)\mathbf{x}_{k-1}$ for $\mathbf{x}_k$.

(ii) Normalise $\mathbf{x}_k$. (22)

This only requires a single LU factorisation of $I - \alpha A$ and so is acceptable for banded $A$. The constant $\alpha$ is needed to remedy a poor scaling of $A$ which would cluster many eigenvalues of $B$ with modulus $\approx 1$. A reasonable choice is $\alpha \approx 1/\rho(A)$ and this can be maintained during the continuation process.

If the above factorisation of $I - \alpha A$ is impractical then we could just apply the power method to

$$\tilde{B} := I + \alpha A, \tag{23}$$

for small $\alpha$. The idea here is that the eigenvalues of $A$ within the ball centered at $-\alpha^{-1}$ and of radius $\alpha^{-1}$ will transform to eigenvalues of $\tilde{B}$ with modulus less than 1. Hence $\rho(\tilde{B}) < 1$ implies that all of $A$'s eigenvalues are in the left half-plane while the smaller $\alpha$ the less likely it is that an eigenvalue of $A$ with negative real part transforms to an eigenvalue of $\tilde{B}$ with modulus greater than 1. For small $\alpha$, however, all the eigenvalues of $\tilde{B}$ are close to unity. It is noteworthy that the power method applied to both (21) and (23) may be regarded as a discretisation of the autonomous system

$$\dot{\mathbf{x}} = A\mathbf{x}, \tag{24}$$

the former being the trapezoidal rule with step-length $h = 2\alpha$ and the latter being the simple explicit Euler method with $h = \alpha$. Thus the eigenvalue transformations above correspond to well-known stability results. Of course other discretisations are also possible but in general higher-order methods will be unnecessarily expensive since our aim is not to solve (24) accurately. One idea is to approximate $(1 + z)/(1 - z)$ by a polynomial, e.g. $(1 + z)(1 + z + z^2 + ...)$.

The main problem with the simple power method is the rate of convergence which is proportional to $|\lambda_2/\lambda_1|$, where $\lambda_1$ is the dominant eigenvalue and $\lambda_2$ the second largest eigenvalue in modulus. If this ratio is close to 1 then convergence will be slow. Another drawback to the power method is that we should not use a starting value $\mathbf{x}_0$ obtained from the previous continuation step because this can cause problems when the dominant eigenvalue changes. An improvement over the power method is to work with $r$ vectors simultaneously in an $n \times r$ matrix $X_k$ with orthogonal columns, rather than with a single vector $\mathbf{x}_k$, and allow $r$ to change at each continuation step. This leads to the approximation of the dominant invariant subspace of $B$ by the accelerated orthogonal iteration method [26]:

| | | |
|---|---|---|
| (i.) | $Z_k = BX_{k-1}$ | Iteration step. |
| (ii.) | $Q_k R_k = Z_k$ | QR factorisation of $Z_k$. |
| (iii.) | $C_k = Q_k^T B Q_k$ | Restriction of $B$ to span$\{Q_k\}$. |
| (iv.) | $T_k = V_k^T C_k V_k$ | Orthogonal reduction of $C_k$ to quasi-triangular form. |
| (v.) | $X_k = Q_k V_k$. | |

The $r$-dimensional subspaces spanned by the columns of $X_k$ will converge to the dominant $r$-dimensional invariant subspace of $B$ provided that $|\lambda_r| > |\lambda_{r+1}|$. It is more useful to us that, provided $|\lambda_1| > |\lambda_{r+1}|$, $\rho(T_k)$ will converge to $\rho(B)$ at a rate proportional to $|\lambda_{r+1}/\lambda_1|$. (The aim of course is that this should be significantly smaller than $|\lambda_2/\lambda_1|$.) At each continuation step the starting matrix $X_0$ should retain or reject columns from the previous iteration according to the relative size of the eigenvalues of $T_k$. $X_0$ should also contain a random additional column in order to detect easily an eigenvalue of $B$ increasing in modulus. Thus $r$ may increase or decrease from step to step.

The use of the power method or orthogonal iteration leads naturally to a discussion of termination criteria. If we have only a $p$-dimensional approximate right invariant subspace $\mathcal{N}$ (where $1 \leq p \leq r$ is the orthogonal iteration method), spanned by the orthonormal columns of an $n \times p$ matrix $X$, then our measure of $\mathcal{N}$'s 'invariance' must be

$$R := AX - X(X^T A X). \tag{25}$$

In other words the columns of $R$ span the orthogonal projection of $A\mathcal{N}$ onto $\mathcal{N}^\perp$ and

$$\|R\|_2 = \max_{u \in \mathcal{N}, \, v \in \mathcal{N}^\perp} \left\{ \frac{|v^T A u|}{\|v\|_2 \|u\|_2} \right\}.$$

Correspondingly the eigenvalues of the $p \times p$ matrix $X^T A X$, the Rayleigh quotient or matrix representation of the compression of $A$ to $\mathcal{N}$, will give approximations to some of the eigenvalues of $A$ and we would expect these approximations to improve as $R$ decreases in size. Directly from (25) one can deduce the backwards error analysis result that the matrix $A - E$, where $E := RX^T$ with $\|E\|_2 = \|R\|_2$ and $\|E\|_F = \|R\|_F$, has $\mathcal{N}$ as an invariant subspace and associated eigenvalues given by the spectrum of $X^T A X$, and it can easily be shown that no smaller perturbation of $A$ in either the spectral or Frobenius norms can have this property. In contrast, however, to the symmetric matrix case this perturbation result cannot be followed be easily computable eigenvalue error bounds. Those which are available rely on knowing the distance to the rest of the spectrum and in general the required quantities will not be computable. Hence the iteration can only be terminated when the size of the perturbation $E$, i.e. the size of the residual $R$, satisfies some tolerance.

The above considerations allow us to appreciate the advantages of computing an approximation to the dominant left invariant subspace as well. This can be achieved by using the orthogonal iteration method with $A^T$ or by using inverse iteration with $A^T - \mu I$ where $\mu$ is an eigenvalue of $X^T A X$. The advantages of doing this extra work are that eigenvalue approximations which converge quadratically are now available and also that more useful error criteria can be used. Thus suppose $\mathcal{M}$ is a $p$-dimensional approximate left invariant subspace of $A$, which is spanned by the orthonormal columns of an $n \times p$ matrix $Y$. How close $Y^T X$ is to singularity, i.e. the smallest singular value, can be used to test whether $\mathcal{M}$ and $\mathcal{N}$ approximate corresponding invariant subspaces and whether $p$ should be increased [10]. We should naturally seek approximate eigenvalues of $A$ by considering its compression to a mapping from $\mathcal{N}$ to $\mathcal{N}$ with $\mathcal{N}$ and $\mathcal{M}^\perp$ regarded as complementary subspaces, i.e. by solving the generalised eigenvalue problem

$$Y^T A X - \mu Y^T X = 0. \tag{26}$$

This is because if $X_E$ and $Y_E$ were exact right and left invariant subspaces of $A$ and

$$X = X_E + U,$$
$$Y = Y_E + V,$$

with $U, V$ small and $Y_E^T U = V^T X_E = 0$ then (26) becomes

$$(Y_E^T A X_E + V A U) - \mu(Y_E^T X_E + V^T U) = 0,$$

i.e. a quadratic perturbation of the exact eigenvalue formula $Y_E^T A X_E = \lambda_E Y_E^T X_E$. Hence this time our measure of the invariance of $\mathcal{M}$ and $\mathcal{N}$ should be the two residuals

$$\begin{aligned} R &:= AX - X\{(Y^T X)^{-1} Y^T A X\}, \\ S^T &:= Y^T A - \{Y^T A X (Y^T X)^{-1}\} Y^T, \end{aligned} \tag{27}$$

with the columns of $R$ now spanning the oblique projection of $A\mathcal{N}$ onto $\mathcal{N}$ and the columns of $S$ spanning the oblique projection of $A^T\mathcal{M}$ onto $\mathcal{M}$. This leads again to the backward error analysis result that the matrix $A - E$, where now

$$E := RX^T + YS^T$$

with $\|E\|_2 = \|R\|_2 + \|S\|_2$ and $\|E\|_F^2 = \|R\|_F^2 + \|S\|_F^2$, has $\mathcal{M}$ and $\mathcal{N}$ as corresponding left and right invariant subspaces with associated eigenvalues given by the solutions of (26) and no smaller perturbation of $A$ in either the spectral or Frobenius norms has these properties [10]. Again however we cannot follow this perturbation result with easily computable error bounds. Nevertheless, with approximations to both left and right invariant subspaces available, we now have the first order result that the computed eigenvalues (those of $A - E$) differ from the exact eigenvalues (those of $A \equiv (A - E) + E$) by the eigenvalues of $(Y^TX)^{-1}Y^TEX$. This is inapplicable directly because our computed eigenvalues are obtained by a generalised Rayleigh quotient procedure which gives second order accuracy and thus $Y^TEX$ is zero by definition of $E$. It does indicate, however, that our error depends on powers of the sizes of $E$ and $(Y^TX)^{-1}$ and so

$$(\|R\|_2 + \|S\|_2)^2/\sigma_{min}^2(Y^TX) \text{ or } (\|R\|_F^2 + \|S\|_F^2))/\sigma_{min}^2(Y^TX),$$

is a suitable stopping criterion.

We finally wish to consider Krylov subspace methods for solving our problem. It has been shown that these are often superior to subspace iteration [14]. We first look at the Arnoldi method which works with approximate right invariant subspaces. For a given unit starting vector $\mathbf{q}_1$, the Krylov subspace of dimension $k$ for $A$ is defined to be the subspace spanned by the vectors $\{\mathbf{q}_1, A\mathbf{q}_1, \ldots, A_{k-1}\mathbf{q}_1\}$. (Any linear dependence of these vectors immediately gives us an exact invariant subspace of $A$). An orthonormal basis $\{\mathbf{q}_1, \mathbf{q}_2, \ldots, \mathbf{q}_k\}$ for this subspace may be obtained by Gram-Schmidt orthogonalisation, i.e.

$$h_{j,j-1}\mathbf{q}_j := A\mathbf{q}_{j-1} - \sum_{i=1}^{j-1} h_{i,j-1}\mathbf{q}_i, \quad j = 2, \ldots, k+1, \tag{28}$$

$$h_{i,j-1} = \mathbf{q}_i^T A\mathbf{q}_{j-1},$$

$$h_{j,j-1} = 1/\|A\mathbf{q}_{j-1} - \sum_{i=1}^{j-1} h_{i,j-1}\mathbf{q}_i\|_2.$$

(We refer to [23] for considerations of the accuracy of the orthogonalisation.) If $Q$ is the $n \times k$ matrix with orthonormal columns $\mathbf{q}_1, \ldots, \mathbf{q}_k$ then $Q^TAQ$, the matrix representation with respect to the basis of the compression of $A$ of the Krylov subspace, is the $k \times k$ upper Hessenberg matrix $H$ whose elements are given by (28). The eigenvalues of $H$, which can be computed by the QR method, may be regarded as approximations to the eigenvalues of $A$ and they will be exact if $h_{k+1,k} = 0$, i.e. the Krylov subspace is an invariant subspace of $A$. In general, however, we will have instead that

$$AQ = QH + h_{k+1,k}\mathbf{q}_{k+1}\mathbf{e}_k^T, \tag{29}$$

where $e_k$ is the $k^{th}$ unit vector of order $k$. If the eigenvalues of $A$ are $\lambda_1, \ldots \lambda_n$ then the accuracy of the approximation of $\lambda_i$ by the spectrum of $H$ depends on

$$\min_{p \in \mathcal{P}_{k-1} \, p(\lambda_i)=1} \left\{ \max_{\substack{j=1,\ldots,n \\ j \neq i}} |p(\lambda_j)| \right\}. \tag{30}$$

[23]. Hence those eigenvalues approximated well for small $k$ depends on the overall spread of $\lambda_1, \ldots, \lambda_n$ and need not include the dominant eigenvalues. Thus it is unclear whether one should apply the Arnoldi method to $B$ in (21) (if this is possible) or just to $A$ and then look for the eigenvalue of $H$ with largest real part. (Note that $\bar{B}$ in (23) will generate the same Krylov subspace as $A$ itself). We can also attempt to force the Arnoldi method to produce the required eigenvalues early by utilising our continuation framework. The eigenvalue approximations generated at the previous continuation step can be used in several ways to make (30) smaller for wanted eigenvalues $\lambda_i$. The following two approaches have been recommended in the literature.

(i) Construct an ellipse which contains all the eigenvalues of $A$ at the previous continuation step apart from the one (or the few) with largest real part. Let $p_s^{(T)}$ be the scaled Tchebychev polynomial of degree $s$ with smallest $\infty$-norm on this ellipse [12, 13, 8]. At the present step, apply the Arnoldi method to $p_s^{(T)}(A)$ rather than $A$ itself, making use of the three-term recurrence relation for Tchebychev polynomials to calculate $p_s^{(T)}(A)\mathbf{u}$ efficiently.

(ii) Construct a polygonal approximation to the convex hull of the eigenvalues of $A$ at the previous continuation step apart from the one (or the few) with largest real part. Let $p_s^{(LS)}$ be the scaled polynomial of degree $s$ with smallest 2-norm on this polygon [24]. At the present step, apply the Arnoldi method to $p_s^{(LS)}(A)$ rather than $A$ itself, making use of the three-term recurrence relation for orthogonal polynomials to calculate $p_s^{(LS)}(A)\mathbf{u}$ efficiently.

In either case of course there are a number of important practical decisions to be made and these are considered in [25].

In practice residuals must be used to test the accuracy of approximations derived from the Arnoldi process. If $V$ is a $k \times p$ matrix whose orthonormal columns span an invariant subspace of $H$, i.e.

$$HV = VT$$

where $T$ is a $p \times p$ quasi-triangular matrix, then the columns of the $n \times p$ matrix $X := QV$ span an approximate invariant subspace of $A$ with (29) giving the fundamental residual equation

$$AX - XT = h_{k+1,k}\mathbf{q}_{k+1}e_k^T V. \tag{31}$$

Hence we have the residual norm

$$\|R\|_2 = |h_{k+1,k}| \|\mathbf{z}\|_2,$$

where $\mathbf{z} \in \Re^p$ with components $V_{kj}, j = 1, \ldots, p$, which can be used as a termination test. A significant practical drawback of the Arnoldi method is that storage is required for all the vectors $\mathbf{q}_1, \ldots, \mathbf{q}_k$. In [23] several methods are proposed for avoiding this overhead.

(a) The Arnoldi process is terminated after at most $m$ steps. If the current approximations are not good enough then a new process is started with the new initial vector $\mathbf{q}_1$ being a combination of the current most dominant eigenvalues or eigenvalue with largest real part.

(b) Each vector $\mathbf{q}_j$ is only orthogonalised against the previous $r$ vectors $\mathbf{q}_{j-1}, \ldots, \mathbf{q}_{j-r}$ instead of $\mathbf{q}_1, \ldots, \mathbf{q}_{j-1}$. Hence only the previous $r$ vectors need to be retained at any step in this 'incomplete orthogonalisation' method.

The alternative Krylov subspace method is the unsymmetric Lanczos algorithm based on approximate left and right invariant subspaces. Thus we have two unit starting vectors, $\mathbf{p}_1$ and $\mathbf{q}_1$, which generate two $k$ dimensional Krylov subspaces spanned by $\{\mathbf{p}_1, A^T\mathbf{p}_1, \ldots (A^T)^{k-1}\mathbf{p}_1\}$ and $\{\mathbf{q}_1, A\mathbf{q}_1, \ldots, A_{k-1}\mathbf{q}_1\}$ respectively. (Again any linear dependence immediately indicates an exact invariant subspace.) A pair of *bi-orthonormal* bases $\{\mathbf{p}_1, \ldots, \mathbf{p}_k\}$ and $\{\mathbf{q}_1, \ldots, \mathbf{q}_k\}$ for these two subspaces is then constructed using the formulae

$$\gamma_{j,j-1}\mathbf{p}_j := A^T\mathbf{p}_{j-1} - \sum_{i=1}^{j-1}\gamma_{i,j-1}\mathbf{p}_i, \qquad \beta_{j,j-1}\mathbf{q}_j := A\mathbf{q}_{j-1} - \sum_{i=1}^{j-1}\beta_{i,j-1}\mathbf{q}_i,$$
$$j = 2, \ldots, k+1,$$
$$\gamma_{i,j-1} = \mathbf{q}_i^T A^T \mathbf{p}_{j-1} / \mathbf{q}_i^T \mathbf{p}_i, \qquad \beta_{i,j-1} = \mathbf{p}_i^T A \mathbf{q}_{j-1} / \mathbf{p}_i^T \mathbf{q}_i, \qquad (32)$$
$$i = 1, \ldots j-1,$$
$$\gamma_{j,j-1} = \|A^T\mathbf{p}_{j-1} - \sum_{i=1}^{j-1}\gamma_{i,j-1}\mathbf{p}_i\|_2^{-1}, \qquad \beta_{j,j-1} = \|A\mathbf{q}_{j-1} - \sum_{i=1}^{j-1}\beta_{i,j-1}\mathbf{q}_i\|_2^{-1},$$

(It is here that the algorithm may breakdown or suffer from numerical instability because of $\mathbf{p}_i^T\mathbf{q}_i \approx 0$. A remedy for this difficulty is proposed in [19].) Using the biorthogonality properties produces the significant simplification

$$\gamma_2\mathbf{p}_2 := (\mathbf{q}_1^T\mathbf{p}_1)A^T\mathbf{p}_1 - \alpha_1\mathbf{p}_1$$
$$\beta_2\mathbf{q}_2 := (\mathbf{p}_1^T\mathbf{q}_1)A\mathbf{q}_1 - \alpha_1\mathbf{q}_1,$$
$$\gamma_j\mathbf{p}_j := (\mathbf{q}_{j-1}^T\mathbf{p}_{j-1})A^T\mathbf{p}_{j-1} - \alpha_{j-1}\mathbf{p}_{j-1} - \beta_{j-1}\mathbf{p}_{j-2},$$
$$\beta_j\mathbf{q}_j := (\mathbf{p}_{j-1}^T\mathbf{q}_{j-1})A\mathbf{q}_{j-1} - \alpha_{j-1}\mathbf{q}_{j-1} - \gamma_{j-1}\mathbf{q}_{j-2},$$
$$j = 3, \ldots, k+1,$$
$$\alpha_j = \mathbf{p}_j^T A\mathbf{q}_j, \qquad j = 1, \ldots, k$$
$$\gamma_2 = \|(\mathbf{q}_1^T\mathbf{p}_1)A^T\mathbf{p}_1 - \alpha_1\mathbf{p}_1\|_2^{-1}, \qquad \beta_2 = \|(\mathbf{p}_1^T\mathbf{q}_1)A\mathbf{q}_1 - \alpha_1\mathbf{q}_1\|_2^{-1}, \qquad (33)$$
$$\gamma_j = \|(\mathbf{q}_{j-1}^T\mathbf{p}_{j-1})A^T\mathbf{p}_{j-1} - \alpha_{j-1}\mathbf{p}_{j-1} - \gamma_{j-2}\mathbf{p}_{j-2}\|_2^{-1},$$
$$\beta_j = \|(\mathbf{p}_{j-1}^T\mathbf{q}_{j-1})A\mathbf{q}_{j-1} - \alpha_{j-1}\mathbf{q}_{j-1} - \beta_{j-2}\mathbf{q}_{j-2}\|_2^{-1},$$
$$j = 3, \ldots, k+1.$$

Hence if $P$ and $Q$ are $n \times k$ matrices with columns $\mathbf{p}_1, \ldots, \mathbf{p}_k$ and $\mathbf{q}_1, \ldots, \mathbf{q}_n$ respectively then the compression of $A$ to the subspace spanned by the $\mathbf{q}_j$'s is represented by the $k \times k$ tridiagonal matrix $(P^TQ)^{-1}J$ with respect to these bases. Here $J$ is the tridiagonal matrix with diagonal elements $\alpha_1, \ldots, \alpha_n$ and off-diagonal elements $\beta_2, \ldots, \beta_k$ and $\gamma_2, \ldots, \gamma_k$. The solutions of the generalised eigenvalue problem

$$det(J - \mu P^TQ) = 0, \qquad (34)$$

which can be computed by Laguerre's method, may be regarded as approximations to the eigenvalues of $A$ and they will be exact if $\beta_{k+1}\gamma_{k+1} = 0$, i.e. if either Krylov subspace is a (left, respectively right) invariant subspace of $A$. In general we will have instead that

$$
\begin{aligned}
AQ &= Q(P^TQ)^{-1}J + \beta_{k+1}\mathbf{q}_{k+1}\mathbf{e}_k^T, \\
P^TA &= J(P^TQ)^{-1}P^T + \gamma_{k+1}\mathbf{e}_k\mathbf{p}_{k+1}^T.
\end{aligned}
\tag{35}
$$

The accuracy of the approximations now depends on the square of (30) but we have the same choices as for Arnoldi's method in trying to make the Lanczos process produce good approximations to the required eigenvalues quickly. (Note however that there is no storage problem now because only the previous two $\mathbf{p}_j$'s and $\mathbf{q}_j$'s need be retained.) In practice residuals must be used to test the algorithm for termination. If $U$ and $V$ are $k \times p$ matrices whose columns span corresponding left and right invariant subspaces for the generalised problem of (34), i.e.

$$
JV = (P^TQ)VD \qquad U^TJ = DU^T(P^TQ)
$$

where $D$ is a quasi-diagonal $p \times p$ matrix, then the columns of $Y := PU$ and $X := QV$ span approximate corresponding invariant subspaces of $A$. The eigenvalues of $D$ are the double-accuracy generalised Rayleigh quotient approximations given by (26) and, through (35), the residuals of (27) are

$$
\begin{aligned}
\text{(i)} \quad & R := \beta_{k+1}\mathbf{q}_{k+1}\mathbf{e}_k^T V, \\
\text{(ii)} \quad & S^T := \gamma_{k+1}U^T\mathbf{e}_k\mathbf{p}_{k+1}^T,
\end{aligned}
\tag{36}
$$

Hence we may apply the previous termination criteria of using

$$
\begin{aligned}
\text{(i)} \quad & \|R\|_2 := |\beta_{k+1}|\|\mathbf{z}\|_2 \\
\text{(ii)} \quad & \|S\|_2 := |\gamma_{k+1}|\|\mathbf{w}\|_2
\end{aligned}
$$

where $\mathbf{z}, \mathbf{w} \in \Re^p$ with components $V_{kj}, j = 1, \ldots, p$ and $U_{kj}, j = 1, \ldots, p$ respectively. In [10] a sophisticated three stage test is recommended.

(a) Increase $p$ to consider a cluster of eigenvalues of $J$ so that they are not too ill-conditioned, i.e. $U^TV$ is not too close to singularity.

(b) Test that the proposed eigenvalues of $J$ have settled down in the Lanczos process, i.e. continuing the process will only change these eigenvalues slightly.

(c) Use (36) to test the proposed eigenvalues as eigenvalues of $A$.

## 4.2 UNSTABLE STEADY SOLUTIONS

In this subsection we shall assume that all the eigenvalues of $A$ *apart from a relatively small number* have negative real part. Apart from the stable case, this situation is the most common in practice and we shall examine how the methods of the previous subsection can be adapted.

If we can work with the transformed matrix $B$ of (21) then the accelerated orthogonal iteration method may still be used. Now, however, $r$ must be increased so that all the eigenvalues with positive real part are computed and a good starting approximation to the required invariant subspace is available from the previous continuation step. Alternatively, it may be more efficient to calculate the eigenvalues with positive real part rapidly and accurately by inverse iteration and then to use the method for stable steady solutions on an implicitly deflated $A$.

If we apply Arnoldi's method to this problem there are several possibilities for using the eigenvalue approximations from the previous continuation step to enhance the convergence towards the required eigenvalues at the present step.

(i) We can construct an ellipse or polygon containing all the eigenvalues of $A$ at the previous continuation step, apart from those with positive real part and the one (or the few) with smallest negative real part. Then Arnoldi's method can be applied to $p_s^{(T)}(A)$ or $p_s^{(LS)}(A)$ again. Additionally the starting vector for the Arnoldi iteration can be taken as a linear combination of the eigenvectors corresponding to previous eigenvalues with positive real part plus a random orthogonal vector.

(ii) The eigenvalues with positive real part can be computed one at a time by several runs of Arnoldi's method with the pre-conditioning ellipse/polygon changing for each run. Initially we would use an ellipse or polygon containing all the eigenvalues from the previous continuation step apart from that with the most positive real part (or a complex·conjugate pair) and the starting vector would be the corresponding eigenvector. The second run of Arnoldi's method would then apply the same procedure to an implicitly deflated $A$, the latter's eigenvalue with most positive real part being the eigenvalue of $A$ with second most positive real part; and so on. Computing the eigenvalues one-by-one produces more rapidly convergent Arnoldi iterations but this must be balanced against the several runs required and the higher accuracy necessary for the deflation to succeed.

(iii) A third possibility is to apply Arnoldi's method in block form and produce a block-Hessenberg matrix; the size of the block being one plus the number of eigenvalues with positive real part at the previous continuation step. The ellipse or polygon would be constructed as in (i) and the block of starting vectors would span the invariant subspace corresponding to the previous eigenvalues with positive real part. Once again the assessment of this approach must depend on comparing the more rapidly convergent iteration with the extra work involved in computing the eigenvalues of block-Hessenberg matrices.

Finally, of course, these ideas may also be applied to the Lanczos method since we are only working with the Krylov subspaces of various polynomials of $A$.

# References

[1] L. V. Ahlfors. *Complex Analysis. McGraw Hill, New York*, 1966.

[2] E. J. Doedel and J. P. Kernevez. *AUTO: software for continuation and bifurcation problems in ordinary differential equations. Appl. Math. Tech. Rep., Cal. Tech.*, 1986.

[3] P. J. Eberlein. *A Jacobi-like method for the automatic computation of eigenvalues and eigenvectors of an abitrary matrix. J. Soc. Ind. Appl. Math.*, 10, pages 74–88, 1962.

[4] J.N. Franklin. *Matrix Theory. Prentice-Hall, New Jersey*, 1968.

[5] F. R. Gantmacher. *Matrix Theory Vol. II. Chelsea Publishing Co., New York*, 1959.

[6] P. Henrici. *Bounds for iterates, inverses, spectral variation and fields of values of non-normal matrices. Numer. Math.*, 4, pages 24–40, 1962.

[7] P. Henrici. *Applied and Computational Complex Analysis Vol. 1. Wiley-Interscience, New York*, 1974.

[8] D. Ho, F. Chatelin, and M. Bennani. *Arnoldi-Tchebyshev procedure for large scale nonsymmetric matrices. Math. Mod. Num. Anal.*, 24, pages 53–65, 1990.

[9] A. S. Householder. *The Numerical Treatment of a Single Nonlinear Equation. McGraw Hill, New York*, 1970.

[10] W. Kahan, B.N. Parlett, and E. Jiang. *Residual bounds on approximate eigensystems of non-normal matrices. SIAM J. Numer. Anal.*, 19, pages 470–484, 1982.

[11] P. Lancaster. *Theory of Matrices. Academic, New York*, 1969.

[12] T.A. Manteuffel. *The Tchebyshev iteration for nonsymmetric linear systems. Numer. Math.*, 28, pages 307–327, 1977.

[13] T.A. Manteuffel. *Adaptive procedure for estimation of parameter for the non-symmetric Tschebyshev iteration. Numer. Math.*, 31, pages 183–208, 1978.

[14] B. Nour-Omid, B.N. Parlett, and R. Taylor. *Lanczoz versus subspace iteration for the solution of eigenvalue problems. Int. J. Numer. Meth. Eng.*, pages 859–871, 1983.

[15] D. O'Leary, G.W. Stewart, and J.S. Vandergraft. *Estimating the largest eigenvalue of a positive definite matrix. Math. Comp.*, 33, pages 1289–1292, 1979.

[16] B. N. Parlett. *The Symmetric Eigenvalue Problem. Prentice Hall, New Jersey*, 1980.

[17] B.N. Parlett. *Laguerre's method applied to the matrix eigenvalue problem. Math. Comp.*, 18, pages 464–487, 1964.

[18] B.N. Parlett, H. Simon, and L.M. Stringer. *On estimating the largest eigenvalue with the Lanczos algorithm. Math. Comp.*, 33, pages 153–165, 1982.

[19] B.N. Parlett, D.R. Taylor, and Z.A. Liu. *A look-ahead Lanzcos algorithm for unsymmetric matrices. Math. Comp.*, 44, pages 105–124, 1985.

[20] A. Ruhe. *On the quadratic convergence of a generalisation of the Jacobi method to arbitrary matrices. BIT*, 8, pages 210–231, 1968.

[21] H. Rutishauser. *Solution of eigenvalue problems with the LR transformation. Nat. Bur. Stand. Appl. Math., Ser. 49*, pages 47–81, 1958.

[22] H. Rutishauser. *Une methóde pour le calcul des valeurs propres des matrices non-symétriques. Comptes Rendus, 259*, page 2758, 1964.

[23] Y. Saad. *Variations of Arnoldi's method for computing eigenelements of large unsymmetric matrices. Lin. Alg. & Appl., 34*, pages 269–295, 1980.

[24] Y. Saad. *Least squares polynomials in the complex plane with applications to solving sparse non-symmetric matrix problems. Research Report YALEU/DCS/RR-276*, 1983.

[25] Y. Saad. *Chebyshev acceleration techniques for solving nonsymmetric eigenvalue problems. Math. Comp., 42,*, pages 567–588, 1984.

[26] G.W. Stewart. *Simultaneous iteration for computing invariant subspaces of non-hermitian matrices. Numer. Math., 25*, pages 123–136, 1976.

[27] F. Tibor. *Normal equivalent to an arbitrary diagonalisable matrix. Lin. Alg. & Appl., 51*, pages 153–162, 1983.

[28] K. Veselic. *On a class of Jacobi-like procedures for diagonalising arbitrary real matrices. Numer. Math., 33*, pages 157–172, 1979.

[29] J. H. Wilkinson. *The Algebraic Eigenvalue Problem. Oxford Univ. Press, Oxford*, 1965.

# A NEWTON-LIKE METHOD FOR SIMPLE BIFURCATION PROBLEMS WITH APPLICATION TO LARGE SPARSE SYSTEMS

Z. MEI
*Department of Mathematics*
*University of Marburg*
*D-3550 Marburg/Lahn*
*F. R. Germany*

ABSTRACT. A special extended system is introduced for simple bifurcation problems. The block structure of this system allows us to approximate its nonsingular solutions with a simplified Newton-like method which is related to the modifications of Newton's method for the singular problems. Rank-1 corrections are discussed for large sparse problems to reduce the computational cost.

## 1. Introduction

Let $G : \mathbf{R}^n \times \mathbf{R} \to \mathbf{R}^n$ be a smooth mapping. We consider the parameter-dependent equation

$$G(u, \lambda) = 0 \tag{1.1}$$

and a numerical approximation of its simple bifurcation points. The various methods for this problem in the literature are divided mainly into two classes. The one comes from direct modifications of Newton's method for (1.1), see e.g. Decker/Kelley [6], Griewank [7], Hoy [8], Hoy/Schwetlick [9], Jepson/Decker [10] and Keller [11], etc. The other consists of regularizations of (1.1) by various extended systems and approximations of their nonsingular solutions which include the bifurcation point of (1.1) as well as additional knowledge about the singularity of (1.1). The structure of the extended systems are often utilized to reduce the computational cost, see e.g. Allgower/Böhmer [1], Beyn [3], Spence/Jepson [18], Moore [12], Moore/Spence [13], Weber/Werner [19], Werner [20] and Werner/Spence [21], etc. Obviously, these are two different, but closely related ways in approximating bifurcation points. In this paper we want to consider a special extended system and a Newton-like method for (1.1) which describe a relationship between the above two approaches. For large sparse nonlinear problems we propose rank-1 corrections in the Newton-like method to make use of the sparsity of the original problem.

For simplicity, we assume that the mapping $G$ is $C^3$- continuous and $(u_0, \lambda_0) \in \mathbf{R}^n \times \mathbf{R}$ satisfies

H1) $$G_0 := G(u_0, \lambda_0) = 0; \tag{1.2}$$

247

*D. Roose et al. (eds.), Continuation and Bifurcations: Numerical Techniques and Applications, 247–259.*
© 1990 *Kluwer Academic Publishers.*

**H2)** Zero is a simple eigenvalue of $D_u G_0$ with algebraic multiplicity 1 and

$$a) \quad \dim(N(D_u G_0)) = 1, \qquad b) \quad D_\lambda G_0 \in R(D_u G_0). \tag{1.3}$$

Under the conditions **H1-H2)**, there are elements $\phi$, $\phi^* \in \mathbf{R}^n$, such that

$$\begin{cases} N(D_u G_0) = span[\phi], & N(D_u G_0^T) = span[\phi^*] \\ (\phi, \phi) = 1, & (\phi, \phi^*) = 1 \end{cases} \tag{1.4}$$

and

$$\begin{cases} R(D_u G_0) = \{u \in \mathbf{R}^n \mid (\phi^*, u) = 0\}, \\ R(D_u G_0^T) = \{u \in \mathbf{R}^n \mid (\phi, u) = 0\}, \end{cases} \tag{1.5}$$

where $(\cdot, \cdot)$ represents the inner product in $\mathbf{R}^n$. Moreover, we have the direct decompositions (cf. Rabier [15])

$$\mathbf{R}^n = N(D_u G_0) \oplus R(D_u G_0) = N(D_u G_0^T) \oplus R(D_u G_0^T). \tag{1.6}$$

The statement (1.3) and (1.6) show that there is a unique $v_0 \in R(D_u G_0)$ satisfying

$$D_u G_0 v_0 + D_\lambda G_0 = 0. \tag{1.7}$$

Let

$$\alpha^* = (\phi^*, D_{uu} G_0 \phi \phi), \quad \beta^* = (\phi^*, D_u D G_0(v_0, 1)\phi), \quad \gamma^* = (\phi^*, D^2 G_0(v_0, 1)^2).$$

As it is well known, if the conditions **H1-H2)** and

**H3)** $$\mathcal{D}^* := \beta^{*2} - \alpha^* \cdot \gamma^* > 0 \tag{1.8}$$

hold, $(u_0, \lambda_0)$ is called a simple bifurcation point of (1.1), see e.g. Brezzi/Rappaz/Raviart [4], Moore [12] and Weber/Werner [19].

**Lemma 1.1:** The condition **H3)** is independent of norms of the elements $\phi, \phi^*$ and the constant $c \in \mathbf{R}$ in the solution $v_0 + c\phi$ of (1.7).

**Proof:** For all $a, b, c \in \mathbf{R}$, $a \cdot b \neq 0$, the elements $a\phi$, $b\phi^*$ provide new bases for the spaces $N(D_u G_0)$, $N(D_u G_0^T)$ respectively and $v_0 + c\phi$ is still a solution of (1.7). In this case, we see

$$\begin{aligned} \mathcal{D} := \\ &= (b\phi^*, D_u G_0(v_0 + c\phi, 1)a\phi)^2 - (b\phi^*, D_{uu} G_0(a\phi)^2) \cdot (b\phi^*, D^2 G_0(v_0 + c\phi, 1)^2) \\ &= a^2 b^2 [(\beta^* + c\alpha^*)^2 - \alpha^*(\gamma^* + 2c\beta^* + c^2\alpha^*)] \\ &= a^2 b^2 \mathcal{D}^*. \end{aligned}$$

This proves our conclusion. ∎

In the sequel, we will assume H1-H3) and find the solution $v_0$ of (1.7) satisfying

$$(\phi, v_0) = 0. \tag{1.9}$$

## 2. A Special Extended System

To approximate the bifurcation point $(u_0, \lambda_0)$ and the null vectors $\phi, \phi^*, v_0$ which are required usually in the path following and branch switching, we will set up a special extended system for (1.1) in the neighborhood of $(u_0, \lambda_0)$. For the other extended systems one can see, e.g. Allgower/Böhmer [1], Moore [12], Weber/Werner [19] and Werner/Spence [21]. To this end, we define

$$X := \mathbf{R}^{n+1} \times (\mathbf{R}^n)^3, \quad x := (u, \lambda, u_1, u_2, u_3) \in X \tag{2.1}$$

with the product norm

$$\|x\|_X := (\|u\|^2 + |\lambda|^2 + \|u_1\|^2 + \|u_2\|^2 + \|u_3\|^2)^{1/2}, \tag{2.2}$$

where $\|\cdot\|$ is a norm in $\mathbf{R}^n$. We introduce a system in $X$

$$H(x) := \begin{pmatrix} G(u, \lambda) + (u_3, D_u G \cdot u_1) u_1 \\ (u_3, D_u G \cdot u_2 + D_\lambda G) \\ D_u G \cdot u_1 + [(u_1, u_1) - 1] u_1/2 \\ D_u G \cdot u_2 + D_\lambda G + (u_1, u_2) u_1 \\ D_u G^T \cdot u_3 + [(u_1, u_3) - 1] u_1 \end{pmatrix} = 0. \tag{2.3}$$

It is evident that the mapping $H$ is well defined and $C^2$-continuous in $X$. Let

$$x^* := (u_0, \lambda_0, \phi, v_0, \phi^*). \tag{2.4}$$

**Theorem 2.1:** $x^*$ is a nonsingular solution of (2.3).

**Proof:** It follows from the definition of $H$ and the statements (1.4-9) that

$$H(x^*) = 0. \tag{2.5}$$

On the other hand, a simple calculation yields

$$DH(x^*) = \begin{pmatrix} A(x^*) & 0 & 0 & 0 \\ DD_u G_0 \phi & D_u G_0 + \phi \phi^T & 0 & 0 \\ D^2 G_0(v_0, 1) & \phi v_0^T & D_u G_0 + \phi \phi^T & 0 \\ DD_u G_0^T \phi^* & \phi \phi^{*T} & 0 & D_u G_0^T + \phi \phi^T \end{pmatrix}, \tag{2.6}$$

where

$$A(x) := \begin{pmatrix} D_u G + (u_3, D_{uu} G u_1) u_1 & D_\lambda G + (u_3, D_{u\lambda} G u_1) u_1 \\ (u_3, D_u DG(u_2, 1)) & (u_3, D_\lambda DG(u_2, 1)) \end{pmatrix}. \tag{2.7}$$

We claim that the matrices $A(x^*) \in \mathbf{R}^{(n+1)\times(n+1)}$ and $D_uG_0 + \phi\phi^T \in \mathbf{R}^{n\times n}$ are nonsingular. To prove this assertion we consider the homogeneous the equation

$$(D_uG_0 + \phi\phi^T)u = 0. \tag{2.8}$$

Firstly, we derive from (1.4) and (2.8)

$$(\phi^*, (D_uG_0 + \phi\phi^T)u) = \phi^T u = 0. \tag{2.9}$$

Hence, taking it back into (2.8), one obtains

$$D_uG_0u = 0.$$

The statement (1.4) shows that $u = c\phi$. Together with (2.9), we get $c = 0$. That means, (2.8) has only the trivial solution $u = 0$ and the matrix $D_uG_0 + \phi\phi^T$ is nonsingular. Similarly, we prove the nonsingularity of $A(x^*)$ by considering the equation

$$A(x^*) \cdot (u, \lambda)^T = 0$$

for $(u, \lambda)$ in $\mathbf{R}^n \times \mathbf{R}$. More precisely,

$$D_uG_0u + D_\lambda G_0\lambda + (\phi^*, D_uDG_0(u,\lambda)\phi)\phi = 0, \tag{2.10a}$$
$$(\phi^*, D^2G_0(v_0,1)(u,\lambda)) = 0. \tag{2.10b}$$

Since $D_\lambda G_0 = -D_uG_0v_0 \in R(D_uG_0)$, the decompositions of $\mathbf{R}^n$ in (1.6) and the equation (2.10a) imply

$$D_uG_0u + D_\lambda G_0\lambda = D_uG_0(u - \lambda v_0) = 0, \tag{2.11a}$$
$$(\phi^*, D_uDG_0(u,\lambda)\phi) = 0. \tag{2.11b}$$

Then, it follows from (2.11a), (1.4) and (1.5) $u = \mu\phi + \lambda v_0$, $\mu, \lambda \in \mathbf{R}$. Substituting it into (2.10b) and (2.11b), we derive a system for $\mu, \lambda$ in $\mathbf{R}^2$

$$\begin{cases} \beta^*\mu + \gamma^*\lambda = 0, \\ \alpha^*\mu + \beta^*\lambda = 0. \end{cases}$$

According to the condition **H3)**, this system has only the trivial solution $\mu = \lambda = 0$. Consequently, $(u, \lambda) = (0,0)$ is the unique solution of (2.10) and the matrices $A(x^*)$, $DH(x^*)$ are nonsingular. ∎

Since $DH(x^*)$ in (2.6) is a block lower triangular matrix, its inverse has the same structure and can be expressed explicitly by the inverses of $A(x^*)$ and $D_uG_0 + \phi\phi^T$. We will utilize this property later in approximating the solution $x^*$ of (2.3). At the end of this section, we mention that the extended system

$$\tilde{H}(x) := \begin{pmatrix} G(u,\lambda) + (u_3, DG \cdot (u_2,1))u_1 \\ (u_3, D_uG \cdot u_1) \\ D_uG \cdot u_1 + [(u_1,u_1) - 1]u_1/2 \\ D_uG \cdot u_2 + D_\lambda G + (u_1,u_2)u_1 \\ D_uG^T \cdot u_3 + [(u_1,u_3) - 1]u_1 \end{pmatrix} = 0. \tag{2.12}$$

has exactly the same properties as (2.3).

**Remark 2.1:** If the matrix $D_u G_0$ is symmetric, then $\phi = \phi^*$ and the extended system (2.3) can be reduced to

$$
H_r(x_r) := \begin{pmatrix} G(u,\lambda) + (u_1, D_u G \cdot u_1)u_1 \\ (u_1, D_u G \cdot u_2 + D_\lambda G) \\ D_u G \cdot u_1 + [(u_1, u_1) - 1]u_1/2 \\ D_u G \cdot u_2 + D_\lambda G + (u_1, u_2)u_1 \end{pmatrix} = 0, \tag{2.13}
$$

where $x_r := (u, \lambda, u_1, u_2) \in X_r := \mathbf{R}^n \times \mathbf{R} \times (\mathbf{R}^n)^2$.

## 3. Quadratically Convergent Newton-like Methods

Since $x^*$ is a nonsingular solution of (2.3), it can be approximated by various numerical methods, see e.g. Moore [12], Spence/Jepson [18] and Weber/Werner [19], etc. We consider here a Newton-like method using the triangular structure of $DH(x^*)$.

For $\delta > 0$, we define

$$
B(x^*, \delta) := \{x \in X, \quad \|x - x^*\| \le \delta\}, \tag{3.1}
$$
$$
\Delta x^k := (\Delta u^k, \Delta \lambda^k, \Delta u_1^k, \Delta u_2^k, \Delta u_3^k) = x^{k+1} - x^k, \quad k \in \mathbf{N}. \tag{3.2}
$$

Let $M_1 : X \to \mathbf{R}^{(n+1)\times(n+1)}$, $M_2 : X \to \mathbf{R}^{n\times n}$ be two $C^2$-continuous matrix mappings and satisfy

$$
M_1(x^*) = A(x^*)^{-1}, \quad M_2(x^*) = (D_u G_0 + \phi\phi^T)^{-1}, \tag{3.3}
$$

where $x^* \in X$ is defined by (2.4). Then, we make up a matrix mapping $M : X \to L(X,X) := \mathbf{R}^{(4n+1)\times(4n+1)}$

$$
M(x) := \tag{3.4}
$$
$$
\begin{pmatrix} M_1 & 0 & 0 & 0 \\ -M_2 D D_u G u_1 M_1 & M_2 & 0 & 0 \\ M_2[-D^2 G(u_2,1) + u_1 u_2^T M_2 D D_u G u_1]M_1 & -M_2 u_1 u_2^T M_2 & M_2 & 0 \\ M_2^T[-(D_u D G M_1)^T u_3 + u_1 u_3^T M_2 D_u D G u_1 M_1] & -M_2^T u_1 u_1^T M_2 & 0 & M_2^T \end{pmatrix}.
$$

The continuity of the mappings $M_1, M_2$ and $G$ show that the mapping $M$ is well defined and $C^1$-continuous in $X$. Furthermore, it is easy to see

**Lemma 3.1:** If the condition (3.3) holds, then

$$
M(x^*) = DH(x^*)^{-1}. \tag{3.5}
$$

Using $M$, we approximate the nonsingular solution $x^*$ of (2.3) with the following Newton-like method

$$
\begin{cases} FOR \quad x^0 \in B(x^*, \delta), \ k = 0, 1, \cdots, \quad DO \\ \qquad\qquad x^{k+1} = x^k - M(x^k) \cdot H(x^k). \end{cases} \tag{3.6}
$$

**Theorem 3.1:** If the conditions **H1-H3)** and (3.3) are satisfied, the iteration (3.6) converges locally quadratically to $x^*$.

**Proof:** We define a smooth mapping $F : X \to X$

$$F(x) := x - M(x) \cdot H(x). \tag{3.7}$$

The statement (2.5) shows

$$F(x^*) = x^*.$$

On the other hand, the statements (2.5), (3.5) yield

$$DF(x^*) = 0.$$

Therefore, the fixed point iteration (3.6) for $F$ is locally quadratically convergent (cf. Ortega/Rheinboldt [14]). ■

In practice, the iteration (3.6) is carried out blockwise as follows.

**Algorithm 3.1:** Let $x^0 \in B(x^*, \delta)$ be a starting value. $FOR \quad k = 0, 1, \ldots, \quad DO$

**Step 1)** Compute

$$(\Delta u^k, \Delta \lambda^k) = -M_1(x^k) \cdot (G^k + (u_3^k, D_u G^k u_1^k) u_1^k, (u_3^k, DG^k(u_2^k, 1)));$$

**Step 2)** Using $(\Delta u^k, \Delta \lambda^k)$, calculate

$$\Delta u_1^k = -M_2(x^k) \cdot [D_u G^k u_1^k + ((u_1^k, u_1^k) - 1) u_1^k / 2 + D_u DG^k (\Delta u^k, \Delta \lambda^k) u_1^k];$$

**Step 3)** Using $(\Delta u^k, \Delta \lambda^k)$, $\Delta u_1^k$, compute

$$\Delta u_2^k = -M_2(x^k) \cdot [D_u G^k(u_2^k, 1) + (u_1^{k+1}, u_2^k) u_1^k + D_u DG^k (\Delta u^k, \Delta \lambda^k) u_2^k];$$

**Step 4)** Using$(\Delta u^k, \Delta \lambda^k)$, $\Delta u_1^k$, compute

$$\Delta u_3^k = -M_2(x^k)^T \cdot [D_u G^{kT} u_3^k + ((u_1^{k+1}, u_3^k) - 1) u_1^k + (D_u DG^k (\Delta u^k, \Delta \lambda^k))^T u_3^k];$$

**Step5)** Let $x^{k+1} = x^k + \Delta x^k$, go back to 1) untill the given criteria are satisfied.

In the above algorithm, the computations of the terms $(\tilde{u}, \tilde{\lambda})^T = M_1(x)(v, \beta)^T$ and $\tilde{u}_1 = M_2(x) v_1$ are done by solving the system

$$M_1(x)^{-1} (\tilde{u}, \tilde{\lambda})^T = (v, \beta)^T \tag{3.8}$$

and correspondingly

$$M_2(x)^{-1} \tilde{u}_1 = v_1. \tag{3.9}$$

That means, at each k-th step in (3.6) we need to solve (3.8) once and (3.9) three times. Based on the properties of the original problem (1.1), one can choose suitable matrix mappings $M_1(x)$ and $M_2(x)$ to make (3.6) more effectively. For example, by choosing

$$M_1(x) = A(x)^{-1}, \qquad M_2(x) = (D_u G(u, \lambda) + u_1 u_1^T)^{-1}$$

in the neighborhood of $x^*$, we derive a simplificaton of Algorithm 3.1.

**Algorithm 3.2:** Let $x^0 \in B(x^*, \delta)$ be a starting value. $FOR \quad k = 0, 1, \cdots, \quad DO$

$$(u^{k+1}, \lambda^{k+1})^T = (u^k, \lambda^k)^T - M_1(x^k) \cdot (G^k + (u_3^k, D_u G^k u_1^k)u_1^k, (u_3^k, DG^k(u_2^k, 1))^T;$$
$$u_1^{k+1} = M_2(x^k) \cdot [(1 + (u_1^k, u_1^k))u_1^k/2 - D_u DG^k(\Delta u^k, \Delta \lambda^k)u_1^k];$$
$$u_2^{k+1} = -M_2(x^k) \cdot [D_\lambda G^k + (\Delta u_1^k, u_2^k))u_1^k + D_u DG^k(\Delta u^k, \Delta \lambda^k)u_2^k];$$
$$u_3^{k+1} = -M_2(x^k)^T \cdot [(-1 + (\Delta u_1^k, u_3^k))u_1^k + (D_u DG^k(\Delta u^k, \Delta \lambda^k))^T u_3^k].$$

**Remark 3.1:** If $M_1(x) = A(x)^{-1}$ holds, the system (3.8) can be solved by a block-elimination method as in Chan [5].

**Theorem 3.2:** If $\alpha^* \neq 0$, the matrix $D_u G_0 + (\phi^*, D_{uu} G_0 \phi \cdot) \phi$ is nonsingular.

**Proof:** We show the homogeneous equation

$$(D_u G_0 + (\phi^*, D_{uu} G_0 \phi \cdot) \phi) u = 0$$

has only the trivial solution $u = 0$. Taking an inner product with $\phi^*$ at its both sides, we derive

$$(\phi^*, D_{uu} G_0 \phi u) = 0. \tag{3.10}$$

Hence $D_u G_0 u = 0$. That means $u = c\phi$, $c \in \mathbf{R}$. The statement (3.10) and $\alpha^* \neq 0$ imply that $c = 0$, i.e. $u = 0$. ∎

**Remark 3.2:** Similarly, if $\beta^* \neq 0$, the matrix $D_u G_0 + (\phi^*, D_u DG_0(v_0, 1) \cdot) \phi$ is nonsingular. This is used in the analysis of the extended system (2.12).

The matrix $A(x^*)$ is nonsingular if and only if the condition **H3)** holds and the matrix $D_u G_0 + (\phi^*, D_{uu} G_0 \phi \cdot) \phi$ is nonsingular, see e.g. Keller [11]. To describe the inverse of $A(x^*)$ explicitly, we define the functionals

$$\alpha(x) := u_3^T D_{uu} G u_1 u_2, \quad \beta(x) := u_3^T D_u DG(u_2, 1) u_1, \tag{3.11a}$$
$$\gamma(x) := u_3^T D^2 G(u_2, 1)^2, \quad \mathcal{D}(x) := \beta(x)^2 - \alpha(x) \cdot \gamma(x) \tag{3.11b}$$

and the matrix functions

$$B(x) := [I - \alpha(x)^{-1} u_1 (u_3^T D_{uu} G u_1 - u_1^T)](D_u G + u_1 u_1^T)^{-1} \in \mathbf{R}^{n \times n},$$

$$M_1(x) := \tag{3.12}$$

$$\begin{pmatrix} [I - \mathcal{D}^{-1}(\beta(x) u_1 - \alpha(x) u_2) u_3^T D_u DG(u_2, 1)] B(x) & \mathcal{D}^{-1}(\beta(x) u_1 - \alpha(x) u_2) \\ \alpha(x) \mathcal{D}^{-1} u_3^T D_u DG(u_2, 1) B(x), & -\alpha(x) \mathcal{D}^{-1} \end{pmatrix}$$

with $M_1(x) \in \mathbf{R}^{(n+1) \times (n+1)}$ for $x \in X$. Obviously, these functions are well defined and $C^1$-continuous. Furthermore,

$$\alpha(x^*) = \alpha^*, \quad \beta(x^*) = \beta^*, \quad \gamma(x^*) = \gamma^*, \quad \mathcal{D}(x^*) = \mathcal{D}^*.$$

**Lemma 3.2:** For $x^*$ in (2.4), if $\alpha^* \neq 0$, $\mathcal{D}^* \neq 0$, then

$$B(x^*) = (D_u G_0 + \phi\phi^{*T} D_{uu} G_0 \phi)^{-1} \tag{3.13}$$

and $M_1(x^*)$ satisfies the condition (3.3).

**Proof:** Using the statements (1.4-6), one can verify directly that

$$(D_u G_0 + \phi\phi^{*T} D_{uu} G_0 \phi) \cdot B(x^*) =$$
$$= [D_u G_0 - \alpha^{*-1} \phi\phi^{*T} D_{uu} G_0 \phi\phi(\phi^{*T} D_{uu} G_0 \phi - \phi^T) + \phi\phi^{*T} D_{uu} G_0 \phi](D_u G_0 + \phi\phi^T)^{-1}$$
$$= I.$$

Similarly, we have

$$A(x^*) \cdot M_1(x^*) = I_{(n+1)\times(n+1)}. \qquad \blacksquare$$

For $M_1(x)$ in (3.12) and $(v,b)^T$, $(w,c)^T \in \mathbf{R}^{n+1}$, the matrix-vector multiplication

$$(v,b)^T = M_1(x)(w,c)^T \tag{3.14}$$

in Algorithm 3.2 is computed in three steps. Firstly, we solve the system

$$(D_u G + u_1 u_1^T)\tilde{v} = w \tag{3.15}$$

for $\tilde{v} \in \mathbf{R}^n$. Then, we calculate $\alpha(x)$, $\beta(x)$, $\gamma(x)$ and $u_3^T D_u DG(u_2, 1)\tilde{v}$, $u_3^T D_{uu} G u_1 \tilde{v}, \cdots$ which can be approximated by difference quotients as in Griewank [7]. Finally, $(v,b)^T$ follows directly from the definition of $M_1(x)$ in (3.12). Since the matrix $D_u G + u_1 u_1^T$ is also used in the other steps, the main computational work of Algorithm 3.2 is to solve the system (3.15) four times with the corresponding right hand sides.

## 4. Rank-1 Corrections

In many practical problems, the derivative $D_u G$ has some special structure, e.g. bandedness, sparsity etc. To utilize these properities in the quadratically convergent Newton-like methods in Section 3, we consider rank-1 corrections of $D_u G$ in this section. Rank-1 corrections were first used to approximate simple bifurcation points of (1.1) with stimulating results in Moore [12], Moore/Spence [13] and in Rheinboldt [16], Rheinboldt/Burkadt [17] by dropping or replacing one suitable column (row) of $D_u G$ with a proper vector in $\mathbf{R}^n$. We discuss here some prior rank-1 corrections of $D_u G$. In this way, the fast solvers for sparse, banded, $\cdots$ matrix in the available software packages, eg. Ellpack, Linpack, Nag etc. can be employed in Algorithm 3.1 to reduce the computational cost, especially if (1.1) arises from a discretization of a differential equation.

Denote

$$\phi = (\phi_1, \phi_2, \cdots, \phi_n)^T, \qquad \phi^* = (\phi_1^*, \phi_2^*, \cdots, \phi_n^*)^T \in \mathbf{R}^n.$$

**Lemma 4.1:** If the conditions **H1-H2)** are satisfied, then the index set

$$ID := \{i \in \mathbf{N} \quad | \quad \phi_i \cdot \phi_i^* \neq 0\} \tag{4.1}$$

is nonempty. Furthermore, if $0 \neq \omega \in \mathbf{R}, j \in ID$, the matrix $D_u G_0 + \omega e_j \cdot e_j^T$ is nonsingular.

**Proof:** The fact $ID \neq \emptyset$ follows directly from the statement (1.4). Concerning with the nonsingularity of $D_u G_0 + \omega e_j \cdot e_j^T$, we consider the equation

$$(D_u G_0 + \omega e_j \cdot e_j^T)u = 0 \tag{4.2}$$

for $u$ in $\mathbf{R}^n$. Taking an inner product with $\phi^*$, one gets $\omega \phi_j^*(e_j^T \cdot u) = 0$. Since $j \in ID$ and $\omega \neq 0$, we have $\omega \cdot \phi_j^* \neq 0$. Hence

$$e_j^T \cdot u = 0. \tag{4.3}$$

Substituting it into (4.2), we obtain $u = c\phi$, $c \in \mathbf{R}$. On the other hand, the equation (4.3) yields $c\phi_j = 0$. Again, the statement $j \in ID$ implies that $c = 0$, i.e., $u = 0$ is the unique solution of (4.2). This proves our conclusion. ∎

The above rank-1 corrections of $D_u G$ occur only on its diagonal and $D_u G_0 + \omega e_j \cdot e_j^T$ has exactly the same structure as $D_u G$, e.g. symmetry, bandedness, sparsity, etc. According to the $C^3$-continuity of $G$ and Lemma 4.1, there is a constant $\delta_0 > 0$, such that $D_u G_0 + \omega e_j \cdot e_j^T$ is bounded invertible in $B(x^*, \delta_0)$ (cf. (3.1)).

**Remark 4.1:** If $D_u G_0$ is semi-positive definite and $\omega > 0$, then $D_u G_0 + \omega e_j \cdot e_j^T$ is too.

**Lemma 4.2:** Under the conditions **H1-H2**) we have for all $0 \neq \omega \in \mathbf{R}$ and $j \in ID$

$$e_j^T \cdot (D_u G_0 + \omega e_j \cdot e_j^T)^{-1} \cdot (I - \phi \cdot \phi^{*T}) = 0, \tag{4.4a}$$

$$e_j^T \cdot (D_u G_0^T + \omega e_j \cdot e_j^T)^{-1} \cdot (I - \phi^* \cdot \phi^T) = 0. \tag{4.4b}$$

**Proof:** Since $D_u G_0 + \omega e_j \cdot e_j^T$ is invertible for all $0 \neq \omega \in \mathbf{R}$ and $j \in ID$, for $v \in \mathbf{R}^n$ arbitrary we define

$$u := (D_u G_0 + \omega e_j \cdot e_j^T)^{-1} \cdot (I - \phi \cdot \phi^{*T})v.$$

Then,

$$(D_u G_0 + \omega e_j \cdot e_j^T)u = (I - \phi \cdot \phi^{*T})v.$$

Taking an inner product at its both sides with $\phi^*$, one gets $\omega \phi_j^*(e_j^T \cdot u) = 0$. Due to $\omega \neq 0$ and $\phi_j^* \neq 0$, we have $e_j^T \cdot u = 0$. In other words,

$$e_j^T \cdot (D_u G_0 + \omega e_j \cdot e_j^T)^{-1} \cdot (I - \phi \cdot \phi^{*T}) \cdot v = 0, \quad \forall v \in \mathbf{R}^n.$$

This proves (4.4a). The statement (4.4b) can be proven in the same manner. ∎

Usually, the parameter $\omega \in \mathbf{R}$ is chosen to make the condition number of $D_u G_0 + \omega e_j \cdot e_j^T$ as small as possible. If $D_u G$ is symmetric and positive definite, then $\omega > 0$ is expected. Now, we introduce three matrix functions in $X$

$$M_0(x) := (I - u_1 \cdot u_1^T)(D_u G + \omega e_j \cdot e_j^T)^{-1}(I - u_1 \cdot u_3^T),$$

$$M_1(x) := \tag{4.5}$$

$$\mathcal{D}^{-1} \cdot \begin{pmatrix} l_1(x)u_3^T(I + D_{uu}Gu_1 \cdot M_0) + [\mathcal{D}I - l_2(x)u_3^T D_u DG(u_2, 1)] \cdot M_0 & l_2(x) \\ \beta u_3^T + u_3^T[\alpha D_u DG(u_2, 1) - \beta D_{uu}Gu_1] \cdot M_0 & \alpha \end{pmatrix}$$

and

$$M_2(x) := M_0(x) + u_1 \cdot u_3^T, \tag{4.6}$$

where $l_1(x) := \beta u_2 - \gamma u_1$, $l_2(x) := \beta u_1 - \alpha u_2$ and $M_0(x), M_2(x) \in \mathbf{R}^{n \times n}$, $M_1(x) \in \mathbf{R}^{(n+1) \times (n+1)}$. The functions $\alpha(x), \beta(x), \gamma(x)$ are defined by (3.11) for $x \in X$.

Since $\mathcal{D}(x^*) = \mathcal{D}^* \neq 0$ and $G$ is $C^3$-continuous, the mappings $M_i, i = 0, 1, 2$ are all well defined and $C^1$-continuous in the neighborhood of $x^*$. Furthermore,

**Theorem 4.1:** Let the conditions **H1-H3**) hold. Then the mappings $M_1, M_2$ in (4.5-6) satisfy the condition (3.3).

**Proof:** Let us do directly the multiplication

$$(D_u G_0 + \phi \cdot \phi^T) \cdot M_2(x^*) =$$
$$= (D_u G_0 + \phi \cdot \phi^T) \cdot (D_u G_0 + \omega e_j \cdot e_j^T)^{-1} \cdot (I - \phi \cdot \phi^{*T}) + \phi \cdot \phi^{*T}$$
$$= I - \phi \cdot \phi^{*T} - \omega e_j \cdot e_j^T \cdot (D_u G_0 + \omega e_j \cdot e_j^T)^{-1} \cdot (I - \phi \cdot \phi^{*T}) + \phi \cdot \phi^{*T}$$
$$= I.$$

Similarly, after elementary and tedious calculations one gets

$$A(x^*) \cdot M_1(x^*) = I_{(n+1) \times (n+1)}. \qquad \blacksquare$$

Based on the mappings $M_1(x), M_2(x)$ in (4.5-6), we can utilize the special sturcture (sparsity, bandedness, $\cdots$ ) of (1.1) to calculate the terms $M_1(x)(w,b)^T$, $M_2(x)w$ in Algorithm 3.1. As an example, we consider here computations of the matrix vector multiplications of Algorithm 3.1 in a general form. For $(w,b) \in \mathbf{R}^n \times \mathbf{R}$, $x \in X$, let

$$(v, a)^T := M_1(x) \cdot (w, b)^T.$$

Firstly, we solve the equation

$$(D_u G + \omega e_j \cdot e_j^T)\tilde{v} = (I - u_1 \cdot u_1^T)w \tag{4.7}$$

for $\tilde{v}$. Then from $\hat{v} := (I - u_1 \cdot u_1^T)\tilde{v} = M_0(x) \cdot w$, one gets

$$\begin{cases} v = \mathcal{D}^{-1}[l_1(x)u_3^T(w - D_{uu}Gu_1\hat{v}) + \mathcal{D}\hat{v} - l_2(x)(u_3^T D_u DG(u_2, 1)\hat{v} - b)] \\ a = \alpha(u_3^T D_u DG(u_2, 1)\hat{v} + b) + \beta u_3^T(w - D_{uu}Gu_1\hat{v}). \end{cases}$$

Similarly, the product $v = M_2(x) \cdot w$ follows from a solution of the system (4.9) with the corresponding right hand side and two inner products in $\mathbf{R}^n$

$$v = (I - u_1 \cdot u_1^T)\tilde{v} + u_1 \cdot u_3^T w.$$

The terms $\alpha$, $\beta$, $\gamma$ and $u_3^T w$, $u_3^T D_{uu}Gu_1\hat{v}, \cdots$ can be computed directly or approximated by difference quotients as in Griewank [7]. The main computational work in Algorithm 3.1 are then reduced to four times solutions of (4.7). If the original problem (1.1) has sparse or banded sturcture, fast solvers can be employed to accomplish this process (cf. Chan [5]).

**Remark 4.2:** Since the index set $ID$ is usually prior unknown, it is approximated and updated during the iteration by

$$ID_k := \{i \in \mathbb{N} \mid u_{1i}^k \cdot u_{3i}^k \neq 0\}, \quad k = 0, 1, \cdots.$$

At the same time, the constant $\omega$ will also be updated to make the condition number of $D_u G^k + \omega e_j \cdot e_j^T$ as small as possible. The initial values for Algorithms 3.1, 3.2 are usually obtained by continuation methods during the path following.

## 5. A Simple Numerical Example

We consider a simplified buckling problem in Allgower/Chien [2].

$$\begin{cases} \Delta u + \lambda sin(u) = 0 & in \ \Omega := [0,1] \times [0,1] \\ u = 0 & on \ \partial\Omega. \end{cases} \tag{5.1}$$

It is well known that $(u_0, \lambda_0) := (0, 2\pi^2)$ is a simple bifurcation point of (5.1) and

$$\phi = \phi^* = 2sin\pi x sin\pi y, \quad v_0 = 0.$$

Discretizing (5.1) with five-point difference scheme, we get

$$G(u, \lambda) := Au + \lambda f(u) = 0, \tag{5.2}$$

where $h = 1/(N+1)$ and

$$u := (u_{11}, \cdots, u_{N1}, \cdots, u_{NN})^T,$$
$$f(u) := h^2(sinu_{11}, \cdots, sinu_{NN})^T,$$

and

$$A := \begin{pmatrix} D & I & & & \\ I & D & I & & \\ & \ddots & \ddots & \ddots & \\ & & \ddots & \ddots & I \\ & & & I & D \end{pmatrix}, \quad D := \begin{pmatrix} -4 & 1 & & & \\ 1 & -4 & 1 & & \\ & \ddots & \ddots & \ddots & \\ & & \ddots & \ddots & 1 \\ & & & 1 & -4 \end{pmatrix}.$$

Since the matrix $D_u G(u, \lambda)$ is symmetric, we use an extended system in the form (2.13). More precisely, we consider the system

$$H(x) := \begin{pmatrix} Au + \lambda f(u) + u_1 \cdot u_1^T(A + \lambda f'(u))u_1 \\ u_1^T(Au_2 + \lambda Df(u)u_2 + f(u)) \\ Au_1 + \lambda f'(u)u_1 + (u_1^T u_1 - 1)u_1/2 \\ Au_2 + \lambda f'(u)u_2 + f(u) + u_1^T u_2 \end{pmatrix} = 0.$$

For this example the functions $\alpha(x)$, $\beta(x)$, $\gamma(x)$, $u_1^T f''(u)vw$, $\cdots$ can be computed explicitly and $ID = \{1, 2, \cdots, N^2\}$. Taking a starting value as discrete values of the functions

$$u^0 = x(1-y), \quad u_1^0 = 2sin\pi x sin\pi y + 0.5x,$$
$$u_2^0 = 0.5x, \quad \lambda^0 = 18,$$

and $N = 20$, $j = 105$, we approximate $x^* := (0, 2\pi^2, \phi, 0)$ with Algorithm 3.1 and get convergent results, see Table 5.1.

TABLE 5.1

| $k$ | $\lambda^k$ | $\|u^k - 0\|$ | $\|u_1^k - \phi\|$ | $\|u_2^k - 0\|$ |
|---|---|---|---|---|
| 1 | 14.557284 | 0.062868 | 0.070813 | 0.027669 |
| 2 | 17.270507 | 0.024838 | 0.029792 | 0.003089 |
| 3 | 18.691284 | 0.008693 | 0.005503 | 0.000125 |
| 4 | 19.494918 | 0.000917 | 0.000320 | 0.000009 |
| 5 | 19.605178 | 0.000006 | 0.000004 | 0.000001 |

**Acknowledgement:** I am grateful to Priv. Doz. Dr. B. Schmitt for several discussions on rank-1 corrections and Friedrich Naumann Foundation, F. R. Germany for the financial support.

## References

[1] Allgower, E. L. and Böhmer, K. (1988) 'Resolving singular nonlinear equations', Rocky Mountain J. of Math. 18, 225-268

[2] Allgower, E. L. and Chien, C.-S. (1986) 'Continuation and local perturbation for multiple bifurcations', SIAM J. Sci. Stat. Comput. 7, 1265-1281

[3] Beyn, W.-J. (1984) 'Defining equations for singular solutions and numerical applications', in T. Küpper, H. D. Mittelmann, H. Weber (eds.), Numerical Methods for Bifurcation Problems, ISNM 70, Birkhäuser, Basel, pp. 57-67

[4] Brezzi, F. Rappaz, J. and Raviart, P.-A. (1981) 'Finite dimensional approximation of nonlinear problems, Part 3: Simple bifurcation points', Numer. Math. 38, 1-30

[5] Chan, T. F. (1984) 'Deflation techniques and block elimination algorithms for solving bordered singular systems', SIAM J. Sci. Stat. Comput. 5, 121-134

[6] Decker, D. W. and Kelley, C. T. (1985) 'Expanded convergence domains for Newton's method at nearly singular roots', SIAM J. Sci. Stat. Comput. 6, 951-966

[7] Griewank, A. (1985) 'On solving nonlinear equations with simple singularities or nearly singular solutions', SIAM Review 27, 537-563

[8] Hoy, A. (1989) ' An efficiently implementable Gauss-Newton-like method for solving singular nonlinear equations', Computing 41, 107-122

[9] Hoy, A. and Schwetlick, H. (1989) Some superlinearly convergent methods for solving singular nonlinear equations', in E. L. Allgower and K.Georg (eds.), Computational Solutions of Nonlinear Systems of Equations, AMS, Providence

[10] Jepson, A. D. and Decker, A. W. (1986) 'Convergence cones near bifurcation', SIAM J. Numer. Anal. 23, 959-975

[11] Keller, H. B. 'Numerical solution of bifurcation and nonlinear eigenvalue problems', in P. H. Rabinowitz (ed.), Application of Bifurcation Theory, Academic Press, New York pp. 359-384

[12] Moore, G. (1980) ' The numerical treatment of non-trivial bifurcation points', Numer. Funct. Anal. Optimz. 2, 441-472

[13] Moore, G. and Spence. A. (1980) 'The calculation of turning points of nonlinear equations', SIAM J. Numer. Anal. 17, 567-576

[14] Ortega, J. M. and Rheinboldt, W. C. (1970) Iterative Solution of Nonlinear Equations in Several Variables, Academic Press, New York

[15] Rabier, P. (1985) Topics in One-parameter Bifurcation Problems, Springer - Verlag, Heidelberg

[16] Rheinboldt, W. C. (1986) Numerical Analysis of Parametrized Nonlinear Equations, John Wiley & Sons, New York

[17] Rheinboldt, W. C. and Burkardt, J. V. (1983) 'A locally parametrized continuation process', ACM Trans. Software 9, 215-235

[18] Spence, A. and Jepson, A. (1982) 'The numerical computations of turning points of nonlinear equations', in C. Baker and G. Miller (eds.), Treatment of Integral Equations by Numerical Methods, Academic Press, New York, pp. 169-183

[19] Weber, H. and Werner W. (1981) 'On the accurate determination of non-isolated solutions of nonlinear equations', Computing 26, 315-326

[20] Werner, B. (1984) 'Regular systems for bifurcation points with underlying symmetries', in T. Küpper, H. D. Mittelmann, H. Weber (eds.), Numerical Methods for Bifurcation Problems, ISNM 70, Birkhäuser, Basel, pp. 562-574

[21] Werner, B. and Spence, A. (1984) 'The computation of symmetry-breaking bifurcation points', SIAM J. Numer. Anal. 21, 388-399

# ASPECTS OF CONTINUATION SOFTWARE

W.C. RHEINBOLDT
*Dept. of Mathematics*
*and Statistics*
*University of Pittsburgh*
*Pittsburgh, PA 15260*
*U.S.A.*

D. ROOSE
*Dept. Computer Science*
*K.U. Leuven*
*Celestijnenlaan 200A*
*B-3030 Heverlee*
*Belgium*

R. SEYDEL
*Applied Mathematics*
*University at Würzburg*
*Am Hubland*
*D-8700 Würzburg*
*Fed. Rep. Germany*

ABSTRACT. In recent years, many continuation algorithms have been implemented. In order to assess the various aspects of these codes, this paper presents a list of features and options that appear to be necessary or desirable for continuation codes. With this it is hoped to provide a framework for writing further such codes, and for judging their differences.

## 1. Introduction

During the past ten or fifteen years many methods have been proposed for the numerical solution of nonlinear problems. This includes, in particular, the solution of parameter dependent nonlinear equations by continuation techniques and the related methods for bifurcation and stability analysis; in alphabetical order, and without claiming completeness, we list here some relevant packages: ALCON [3], AUTO [4], BIFPACK [12], BISTAB [14], CONEX [10], CONSOL [6], HOMPACK [13], PATH [5], PITCON [9], PLTMG [2]. Some of these codes are in the nature of packages that deal with several aspects of the problem while others concentrate only on specific aspects.

At the present state of the art, it appears to be impossible to compare the above packages, because their aims differ so widely. What can and should be done is

(1) to discuss which options are necessary, recommended, or convenient for continuation codes,

(2) to collect and provide a set of test examples that may be the basis of computational tests,

(3) to compare the performance of codes for handling specific tasks.

The present paper proposes a basis for possible standards for (1); the tasks (2) or (3) are beyond the scope of this presentation. A standardization of options will be essential for any comparisons. We shall not give introductions to the various methods; for this we refer to the original literature. (A general introduction into the phenomena and methods is given in [7],[11]). Hence, this paper addresses the specialist rather than the novice.

The outline of the paper is as follows: First, assumptions and aims will be stated (Section 2). Then in Sections 3 through 5 we arrange options and features according to the requirements of the *problem* (Section 3), the requirements of the *user* (Section 4), and the requirements of the *algorithms* (Section 5). In this way, we arrive at a classification that appears natural. The sections are characterized

*D. Roose et al. (eds.), Continuation and Bifurcations: Numerical Techniques and Applications,* 261–268.
© 1990 *Kluwer Academic Publishers.*

by different levels of necessity. The requirements of the problem are *indispensable*. The requirements of the user have *to be judged individually* according to the amount of overhead that is tolerated in relation to the degree a specific option is essential. The requirements of the algorithm should be *as limited as possible*.

Clearly, it cannot be expected that there will be a general consensus on all topics; the paper is principally intended to provide a basis for further discussion.

## 2. Assumptions and Aims

Let the problem be described by a system of equations

$$f(z) = 0 \ , \ z \in \mathcal{R}^{n+1} \ , \ f(z) \in \mathcal{R}^n \ . \tag{1}$$

Under natural conditions on $f$ the solution set defines a one-dimensional manifold in $\mathcal{R}^{n+1}$. Segments of this manifold may be parametrized by one of the components of $z$ (say $z_k$); that is, on such a segment the remaining $n$ components can be calculated in dependence of $z_k$. Frequently, in applications, there is at least one specific parameter $\lambda$, which is then denoted by a character different from the other components. In that case (1) has the form

$$f(y, \lambda) = 0 \ , \ y \in \mathcal{R}^n \ , \ \lambda \in \mathcal{R} \ , \ f(y, \lambda) \in \mathcal{R}^n \ . \tag{2}$$

Many practical problems contain not just one but several parameters in which case the solution set is a higher-dimensional manifold. Frequently, it is desirable to follow a one-dimensional path on such a manifold usually defined by specific augmentations of the given mapping $f$. This is covered by the above setting. But if the higher-dimensional manifold itself is to be approximated, other methods than continuation are needed, see [1], [8].

The application sometimes determines a subset $\mathcal{D} \subset \mathcal{R}^{n+1}$ that the solutions should not leave. Such windows may be needed to limit the computational cost or to avoid that the solution enters regions where it loses its physical meaning. In particular, such a *window* is frequently given for the parameter, $\lambda_a \leq \lambda \leq \lambda_b$ .

Two assumptions are basic for continuation methods applied to a problem (1):

C-1 CONDITION ON THE FUNCTION $f$

We assume that the function $f$ is at least continuously differentiable on some open set $S$ of $\mathcal{R}^{n+1}$. Let $reg(f, S)$ be the set of all points in $S$ where the derivative $Df(z)$ has full rank $n$. Then the solution set $\mathcal{M} = \{z \in reg(f, S); f(z) = 0\}$ is a one dimensional sub-manifold of $\mathcal{R}^{n+1}$.

C-2 ONE SOLUTION OF THE EQUATIONS IS AVAILABLE.

We assume to know $z^{(0)} = (y^{(0)}, \lambda^{(0)})$ on $\mathcal{M}$. Then the continuation method is expected to produce points approximating that connected component of $\mathcal{M}$ which contains $z^{(0)}$.

The requirements of the problem, of the user, and of the method strongly depend on the *aims* of the process. We list here several aims which are not mutually exclusive. More specifically, aims A-S1, A-S2, and A-S3 address the computation of specific solutions, whereas A-B1 and A-B2 concern the calculation of entire branches.

A-S1 : It is desired to obtain *one specific* solution, usually called the *target point*. This situation is typical, for instance, when a homotopy approach is used.

A-S2 : It is desired to obtain solutions for a specified *sequence* of values of a particular variable, as, for instance, the parameter. This aim can be considered a special case of A-S1, namely, as a request for a sequence of target points. We list it separately because it may involve different input and different step controls.

A-S3 : The goal of the computation is to determine those points where certain *properties* of the solutions change, as, for example, where stability is lost. In other words, the aim is here to reach *critical* solutions, such as bifurcation points or limit points.

A-B1 : A branch is of interest; that is, a connected subset of the manifold $\mathcal{M}$ is to be approximated computationally. This may involve the computation of a sequence of points along the branch or the approximation of the branch by one or several smooth *functions*. Evidently, if a sequence of points is computed their selection will be differently controlled than in A-S2.

A-B2 : *All* branches of $\mathcal{M}$ within a specified window are of interest. This places special emphasis on detecting bifurcation points and on branch switching.

It should be emphasized that each one of these computational aims must include some definition of the approximation errors of the desired results. These error definitions tend to be different in each case which, in turn, may influence the controls of the computational process. The output of the results should be accompanied by error estimates. For example, if the output of A-B1 is a smooth function approximating the branch, it will have to be verified that the approximation error meets a prescribed tolerance. Since such a verification tends to be difficult, the standard approach to A-B1 appears to be to produce a sequence of points where it is known that the residual of $f$ is small.

## 3. Definition of a Continuation Run

Between two consecutive interventions of the user the code performs a continuation *run* which produces output in accordance with the particular chosen aims. For the specification of this computational task the user has to provide a number of items that, in essence, define the chosen continuation *problem*. We categorize here the standard items needed in each such problem.

P-1 FUNCTIONS
A subroutine has to be set up for the evaluation of $f(z)$ for a given $z$. This does not imply that expressions for $f$ need to be known explicitly. In fact, frequently in applications $f(z)$ is the output of some other program, as, for example, some differential equations solver. The subroutine for $f$ should provide appropriate error returns which signal whether $f$ has been evaluated correctly or not.

P-2 STARTING SOLUTION
The point $z^{(0)}$ on $\mathcal{M}$ required by C-2 has to be provided. This may or may not be a restrictive assumption. Sometimes, if only an approximate solution is known, then a locally convergent iterative process can be applied to compute $z^{(0)}$. In other cases one of the homotopy approaches may be attempted.

P-3 ERROR TOLERANCE

Specific error measures and appropriate tolerances have to be provided that correspond to the particular chosen aim of the computation. For instance, this should most likely include termination tolerances for the corrector iteration as well as tolerances for the approximation error. In addition, the specification of the approximation error must include also an identification of the particular local coordinate system in which the error is to be calculated. The precise specification of the approximation error for the desired results of the different aims is by no means a simple problem which, in fact, still requires further research. Up to now, typically, only one generic error tolerance is prescribed for all phases of the process and all different aims, and the approximation error is not specified precisely.

P-4 INITIAL DIRECTION

The initial direction in which the branch is to be followed has to be generally considered an input. Deriving an initial direction from a comparison of the initial solution with a target solution is in many cases not sufficient.

P-5 AIM-DEPENDENT INPUT

The specification of the underlying aims basically amounts to setting up appropriate output and termination criteria. The code must incorporate suitable decision procedures to respond to the different possible events in whatever order they occur. For example, decisions have to be made when a target point is encountered, or the window for the variables has been left. Further details will be included in Section 4.

## 4. User Control of the Process

Typically, after each continuation run, the user judges the output and decides which aims are to be pursued in the next continuation run, or which other software might have to be applied. Here extended codes may be designed to take over some of these intermediate user-decisions. Even though the overall control remains with the *user* such codes are often called "black boxes" or "expert systems". There is a wide range of potential levels of control, depending on what portion of the routine work is to be delegated to a black box, and to what extent the user is required to take over decisions. The interactive work is often greatly facilitated when the control uses menu-driven dialogues.

In this section we list some of the options and features that are helpful to control the process. Most of them fall under the above category P-5. Not all of these options are essential, some of them may be dispensible but they are usually convenient. The number of implemented options severely affects the overhead of a continuation code. It should also be emphasized that most of the options are highly related with the approximation errors.

U-1 TERMINATION CRITERIA

U1.1 *Target point.* A target point is defined by an index $k$ and a target value $\eta$ such that the continuation terminates with $z_k = \eta$. This option enables to reach a specific solution. If the code is asked to reach a turning point then it is only necessary to give the index $k$ of the component, with respect to which the turning point is sought. If no target point is specified the continuation procedure requires some other termination criteria, see e.g. U1.3.

U1.2 *Window for the variables.* Intervals for some or all of the components of $z$ may be defined with the effect that the continuation has to stop when a solution outside this "window" has been calculated.

U1.3 *Number of continuation steps.* In interactive work, it is often reasonable to calculate branches in a piecewise fashion. This can be implemented by a halt after a suitable prescribed number of continuation steps. Together with a fixed step length, this is a convenient input for aim A-S2. In other cases, the number of continuation steps has the character of a *maximum* number, serving as an "emergency brake". Prescribing a maximum number of continuation steps is also important, for instance, if closed-loop branches are traced, or the response times in interactive work are to be limited.

## U-2 LEVELS OF STEP CONTROL

A *step* is related to the difference between two consecutively calculated solution points on $\mathcal{M}$. Two items characterize a step. The first specifies how the two solutions are to be compared by choosing a local coordinate system (local parametrization) of $\mathcal{M}$. Examples are the pseudo-arclength parametrization, the choice of a particular component $z_k$ , and the local coordinate system induced by some secant line. The second item is the step *length*; that is, the distance between the two points measured in the local coordinate system. Different levels of step control can be achieved by allowing both items to vary, keeping both fixed, or fixing one and leaving the other variable.

U2.1 *Variable steps.* This is the standard option where the local coordinate system is varied, and the step length is adjusted by some step-control algorithm. This option allows especially to overcome sharp changes of the curvature, and, for instance, to pass around turning points.

U2.2 *Fixed and equidistant steps.* This level of control keeps the coordinate system fixed, and maintains also an equidistant step length. This option matches aim A-S2. It is usually applied in a piecewise manner; that is, for a small number of continuation steps at a time.

U2.3 *Variable step length controlling only one fixed choice of component.* Here the step length is adjusted, but the parametrization is not changed. This is sometimes used as a brute-force method to approximate turning points or points where one of the components is stationary with respect to $\lambda$.

U2.4 *Step window.* Whenever variable steps are desired it is frequently necessary to restrict the step-length to some fixed window. The upper and lower limits of this window have different character.
*Minimum step length.* When the step control proposes a step length smaller than this minimum value, the continuation either stops, or continues to trace the branch using the minimal step length. When U2.3 is applied, only the former decision makes sense.
*Maximum step length.* This control directly affects the reliability of the process and is often used to reduce the sensitivity to undesirable branch jumping (see also A1-5 below).

## U-3 BIFURCATION CONTROL

The handling of bifurcation points and turning points may require decisions from among a wide range of possible actions.

U3.1 *Detection of critical points.* This option requires the code to detect the presence of turning points or bifurcation points without necessarily calculating these points more accurately. The detection of turning points is generally rather simple and involves little expense. The detection of bifurcation points requires the design of suitable *test functions* that vanish at these points.

U3.2  *Calculation of critical points.* This option requires the calculation of a critical point detected under U3.1 within the prescribed error tolerances. The calculation of turning points and, when test functions are used, also that of bifurcation points can be reduced to a one-dimensional root finding problem. Other methods for approximating bifurcation points or turning points require the solution of suitably augmented systems of equations.

U3.3  *Branch switching.* This option requires the computation of a "first" solution point on any branch emanating from a bifurcation point. This point then has to be added onto a stack for later use.

U3.4  *Branch directions.* This option requires the computation of the tangent directions of all branches emanating from a bifurcation point. It is of course closely related to U3.3; each option can be used as a basis for an algorithm for the other one.

U3.5  *Classification of critical points.* This option requires the code to classify any detected critical point. This is generally a simple task for simple turning points or simple bifurcation points, but becomes quickly very costly when higher-order singularities are involved.

U-4 OUTPUT

U4.1  *Data files.* Data files are required to store all calculated solution points, including calculated branch points. In addition, they have to record all connectivity information between these points and any information on the type of solution. Data files must contain enough information to permit the possible use of postprocessing codes, and to allow for the application of visualization procedures, in particular those providing different projections. When calculated also results of any stability analysis and of bifurcation test functions have to be included in the data files. The proper organization of such data files is far from being trivial and, in fact, requires still further research.

U4.2  *Print files.* Print files include selected information from the data files, as well as information about the performance behavior of the code, such as data about the number of function evaluations, the convergence behavior, etc. There should be some control over the amount of printed output during any specific run.

U4.3  *Graphical display.* Graphical display is important. For this separate graphics packages are usually applied; their input files have to be created from the data files. This frequently requires the use of interpolation or smoothing routines which in turn demand again some attention to the approximation errors.

## 5. Requirements of the Algorithm

Here we list limitations due to the specific nature of the of *algorithms* implemented in the code. In general, a continuation code should involve as few such restrictions as possible.

Al-1 GENERALITY OF EQUATIONS

The continuation code should permit the user to adjust data items in the function routine (see P-1) between continuation runs without the need for recompiling or relinking. In particular, if the given problem involves more than one parameter, it is desirable to introduce controls in the function routine which allow the active parameter to be switched between runs (either automatically or under user control). This represents a very helpful device in handling a multi-parameter setting.

Al-2 JACOBIAN

Many continuation algorithms require the availability of the first-order derivatives of $f$, or even of higher-order derivatives. Implementations of such derivatives may be relatively simple when explicit formulas for $f$ are known, or when $f$ is the discretization of some nonlinear operator for which derivatives and their discretization can be readily computed. But in other problems derivatives of $f$ may be very difficult to compute and for such situations the continuation code should incorporate routines for finite difference approximations of the needed derivatives.

Al-3 MODULAR STRUCTURE

As for most applications-software, a modular structure of the continuation code facilitates an exchange of parts of the numerical software. For instance, in many applications, the linear equations arising in the algorithm involve large sparse matrices for which sparse-solvers are desired. Of course, there tends to exist a general conflict between a finely-divided modular structure and the efficient handling of some of the options and features listed above, especially those of Section 4. For instance, test functions of the type indicated in U3.1 may depend critically on a specific implementation of the corrector process and hence cannot be departed readily from it. The design of an efficient modular structure for continuation packages still remains open for more in-depth studies.

Al-4 INITIAL STEP

Frequently, continuation algorithms require the length of the initial step as input. This may be a restrictive assumption which appears to be avoidable by the introduction of appropriate start-up procedures. Note, however, that when the initial step indeed is provided by the user then its sign offers a simple device for specifying the direction of the continuation (see P-4 above).

Al-5 CODE-SPECIFIC INTERNAL CONTROLS

Any code incorporates numerous internal control settings and can be fine-tuned for a specific problem by adjusting these settings. Generally, it is desirable that the performance of the code is, as much as possible, unaffected by small changes of these settings; in other words, the need for tuning the controls should be the exception rather than the rule. It is highly desirable that special attention be paid to reducing any high sensitivity to specific control settings. For example, as noted in U2.4 a maximum step length is often used to improve reliability. Usually such an upper bound simply restricts the absolute length of the step which tends to lead to a sensitivity to the specific choice of the bound. This sensitivity can often be reduced by adding a restriction to the relative length of the step when compared to the norm of the solution.

*Acknowledgements.* Part of this work was done during the second author's visit as a A.v.Humboldt fellow at the Universität Würzburg; this paper was finished during the third author's visit at the University of Pittsburgh. The authors are grateful to the Humboldt foundation, and to the University of Pittsburgh for supporting this research.

## References

[1] Allgower, E.L., and Schmidt, P.H. (1985) An algorithm for piecewise-linear approximation of an implicitly defined manifold. SIAM J. Numer. Anal. 22, 322-346.

[2] Bank, R.E. (1988) PLTMG User's Guide - Edition 5.0. University of California, La Jolla.

[3] Deuflhard, P., Fiedler, B., and Kunkel, P. (1987) Efficient numerical path-following beyond critical points. SIAM J.Numer.Anal. 24, 912-927.

[4] Doedel, E. (1986) AUTO: Software for continuation and bifurcation problems in ordinary differential equations. California Institute of Technology, Pasadena.

[5] Kaas-Peterson, C. (1987) PATH - User's Guide . University at Leeds, Centre for nonlinear studies.

[6] Morgan, A. (1987) Solving Polynomial Systems Using Continuation. Prentice Hall, Englewood.

[7] Rheinboldt, W.C. (1986) Numerical Analysis of Parametrized Nonlinear Equations. J.Wiley, New York.

[8] Rheinboldt, W.C. (1988) On the computation of multi-dimensional solution manifolds of parametrized equaitions. Numer.Math. 53, 165-181.

[9] Rheinboldt, W.C., Burkardt, J. (1983) A locally parametrized continuation process. ACM Transactions of Math. Software 9, 215-235.

[10] Rosendorf, P., Orsag, J., Schreiber, I., and Marek, M. (1989) Interactive System for Studies in Nonlinear Dynamics. Prague Institute of Chemical Technology, Prague.

[11] Seydel, R. (1988) From Equilibrium to Chaos. Practical Bifurcation and Stability Analysis. Elsevier, New York.

[12] Seydel, R. (1989) BIFPACK, a program package for continuation, bifurcation, and stability analysis. University at Würzburg.

[13] Watson, L.T., Billups, S.C., and Morgan, A.P. (1987) HOMPACK: A suite of codes for globally convergent homotopy algorithms. ACM Transactions on Math. Software 13 , 281-310.

[14] Wood, E.F., Kempf, J.A., and Mehra, R.K. (1984) BISTAB: A portable bifurcation and stability analysis package. Appl. Math. Comp. 15 (1984) 343-355.

# INTERACTIVE SYSTEM FOR STUDIES IN NONLINEAR DYNAMICS

P. ROSENDORF, J. ORSAG, I. SCHREIBER AND M. MAREK
*Dept. of Chemical Engineering*
*Prague Institute of Chemical Technology*
*Suchbatarova 5*
*166 28 Prague 6*
*Czechoslovakia*

## 1. Introduction

Numerical methods for the analysis of nonlinear dynamical systems including methods based on continuation techniques for obtaining curves of stationary or periodic solutions, bifurcation points and limit points in dependence on parameters have been developed and discussed in several available textbooks [1-4,6,7] and in the proceedings of conferences, cf. e.g. [5]. Original software for the analysis of the dependence of stationary solutions on a parameter [1,4] and of periodic solutions on a parameter[4,7], were also published. Productive application of such a software requires relatively deep knowledge both of specialized numerical methods and of the theory of nonlinear dynamical systems. Also the organization of computations and the evaluation of computed results can often be quite complicated. Here we describe our attempt to test the idea whether a unification of available numerical algorithms with the means and approaches of logical programming and knowledge engineering could help to increase productivity of the analysis of a nonlinear dynamical system.

An expert open modular system - CONEX (**CON**tinuation **EX**pert system) has been developed. First, a brief overview of the CONEX structure and functions will be given. Then an example of the application of the CONEX for two problems is given:

    a) the problem of a model in heterogeneous catalysis
    b) the problem of a forced impact oscillator.

*D. Roose et al. (eds.), Continuation and Bifurcations: Numerical Techniques and Applications, 269–282.*
© 1990 *Kluwer Academic Publishers.*

CONEX enables an interactive use of routines for continuation and bifurcation analysis of model equations, the construction of solution branches and their analysis. The CONEX provides users with:

1. an automatic generation of subroutines in FORTRAN 77 for model-dependent parts of the problem (computation of right-hand sides, Jacobian matrix, etc.) by means of symbolic manipulations,
2. interactive data input controlled by menus,
3. formation and continuous updating of a database of the studied problem, suggestions on further steps of analysis,
4. interactive analysis of the obtained results stored in the database including various ways of graphical representation of results.

The CONEX can automatically prepare the actual continuation program in a runable form, control its execution and incorporate results into the database. It works with a library of numerical subroutines; however, only those necessary for the given task are used during the generation of the actual program. The minimal hardware requirement is an IBM PC/XT compatible computer with a hard disc and 640 kB of RAM. A numerical coprocessor and an output graphical device are useful for an effective work. Numerical computations can be executed on a mainframe computer.

## 2. Structure of C O N E X

The CONEX can be divided into several relatively independent parts shown in the simplified scheme in Figure 1.

The different parts of CONEX, shown in this figure, are in principle of two different types - a control program and databases (either knowledge databases or ordinary databases).

The control program, which is built from a number of specialized parts, is responsible for the dispatching of processes, the formation of a user interface and an error handling. The dispatcher is the only permanently resident part of the CONEX. It is written in ASSEMBLER, calls all other modules and programs and allocates memory and other resources for them. Since its size is about 10 kB, there are low memory restrictions on the programs running under the CONEX. The dispatcher calls modules sequentially on the basis of a special description file which can be either generated by the CONEX automatically or can be created by the user in any ASCII editor. The content of the description file is in principle a very simple job control language which involves sequencing of jobs, conditional execution of jobs and conditional execution of the error recovery routine. It includes means for the redirection of inputs and outputs.

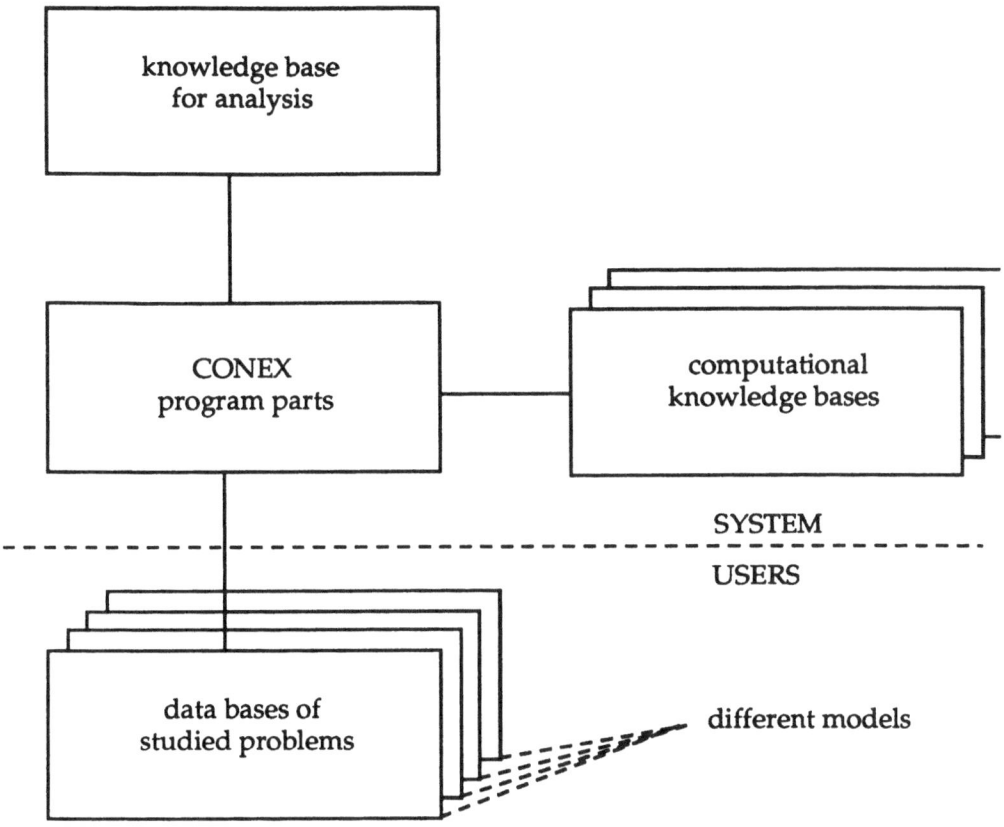

Figure 1. Overall structure of C O N E X

The user interface provides comfortable input of all the data defined by the user. Because of a large variety of the possible types of the data (menu choices, numbers, equations etc.), this part of the CONEX consists of a number of modules. The module MAINM which is written in PROLOG builds the description file for the dispatcher on the basis of a number of user choices from available menus. It also handles the creation and/or setting of the model; as it works on a special tree of offerings it can be very easily customized. From the user point of view, a model is a problem expressed as a set of equations and written down in a symbolic form (in notation similar to mathematical rules) as input to the program. The names of the variables are defined by the user. Symbolic terms can be introduced and there is no restriction on their nesting. After the input, the module performs checking of the validity of defined expressions, and their simplification if it is possible.

The input of the data for the numerical programs is provided by the DATED module. This module works on the basis of the description of the structure of the input data for the particular continuation program. This description is generated during the generation of the main program and of the necessary subprograms on the basis of the choice of the type of continuation and the type of the model.

Several types of output are available in the CONEX during the actual computations. Standard output files are mandatory. The user can choose either from a graphical or a text (numerical) representation of results on the console. Different levels of the text output are available. Graphical representation of results (2-D and 3-D graphics) is based on the GKS standard [3]. An interactive graphics program which is used for the graphical representation of results stored in the database of the studied problem is also based on the GKS standard.

## 2.1. DATABASES

Three different databases are included in the CONEX (cf. Fig.1) -the Computational Knowledge Base (CKB), the Problem DataBase (PDB), and the Problem Analysis Knowledge Base (PAKB).

2.1.1.*Computational knowledge base (CKB)*. The CKB consists, firstly, of libraries necessary to build a proper continuation program (e.g. the library of continuation routines, library of graphics support routines, etc.), and of a set of system routines (e.g. compiler, linker, editor, etc.) and, secondly, of the knowledge base containing information on the use of the above routines and libraries.

The main program for solution of the given problem is generated on the basis of the model equations and the chosen type of continuation. Also other options (e.g. inclusion of graphics etc.) affect the generation of the main program. At the same time other subroutines (default data initialization, evaluation of right hand sides, evaluation of Jacobian matrices etc.) are generated. Methods of symbolic differentiation and symbolic simplification of algebraic expressions are used, for example, to generate the module for the evaluation of the Jacobian matrix. In addition to the above FORTRAN 77 programs, also the description of the structure of the input and output data sets and necessary information for invoking the compiler and linker are generated. The description of the input data set structure is used by an interactive data editor. The description of the output data set is included into the database of the studied problem only after successful computation and analysis of the computed data.

There is a possibility to generate full source code of the particular continuation program based on the source code libraries. This feature enables creation of specialized stand- alone continuation programs. The continuation library in the present form includes software for the following tasks :

a) <u>stationary solutions of ordinary differential equations (ODE)</u>
- continuation of nonsingular stationary solutions
- continuation of limit points
- continuation of Hopf bifurcation points

b) <u>periodic solutions of difference equations</u>
- continuation of nonsingular periodic solutions
- continuation of limit points
- continuation of period doubling bifurcation points
- continuation of Hopf points

c) <u>periodic solutions of periodically perturbed ODE's</u>
- continuation of nonsingular periodic solutions
- continuation of limit points
- continuation of period doubling bifurcation points
- continuation of Hopf bifurcation points

d) <u>periodic solutions of autonomous ODE's</u>
- continuation of nonsingular periodic solutions
- continuation of limit points
- continuation of period doubling bifurcation points
- continuation of Hopf bifurcation points

e) <u>analysis of the computed continuation curves</u>
- computation of the slope of the continuation curve at the braching point of a stationary solution
- computation of the slope of the continuation curve at the branching point of a periodic solution
- computation of the slope of the continuation curve at the Hopf bifurcation point.

The CKB has been designed such that it permits easy modification for other applications.

2.1.2.*Problem database (PDB)*. The PDB is formed both during the communication between the CONEX and the user, and during the computations (automatically). It consists of three parts : the first one is formed by the output data sets, the second one contains names of the output data sets, description of their structure and information on the studied problem resulting from an automatic analysis of data sets and/or heuristics. The third part contains knowledge derived from computational experience, together with additional information derived from the structure of the model equations and from the analysis of the output data sets. The second one is organized in a SQL database into the following tables :

1. FILES   - contains description of output datasets (the name of the dataset, names of continuation parameters etc.)
           - contains initial values of all model variables  and continuation parameters
           - contains data required by the used  continuation method (accuracy, step length, etc.)
           - contains minimal values of model  variables  and continuation parameters in the dataset
           - contains maximal values of model variables and continuation parameters in the dataset

2. ALIAS   - contains user defined names of datasets

3. POINTS - contains special points (not only bifurcation  points, but also points on boundaries) (e.g.  pointer to the dataset where the point is  contained and some information about the type  of the special point)
           - contains coordinates of special points
           - contains accuracy of the computed special  point

All information in this database can be retrieved via special CONEX functions or directly by the SQL commands (interactive or batch). The PKB enables the user to make decisions on further steps of the model analysis (rebuilding of the model, etc.).

2.1.3.*Problem analysis knowledge base (PAKB)*. The PAKB consists of two parts. The first one contains algorithms and subroutines for an analysis of the output data, location of specific points on solution curves, e.g. limit points, branching points, etc.). The second part contains general rules of inference from the obtained databases. The PAKB also enables the user to study the data contained in the problem database (generation of tables, graphs, integration etc.). The programs analyzing the output data file can be divided into two parts. The first one, the preliminary analysis program, is run to determine whether the data files contain useful information and to prepare them for the run of the main analysis program; secondly, the main analysis program searches data files for special points (e.g. bifurcation points), and incorporates the information obtained into the database of the given problem. The necessary information about special points is derived mostly from the eigenvalues of the Jacobian matrices and from the changes of directions of the continuation parameters. The information resulting from the analysis is added to the problem database.

## 2.2. CONEX FROM THE USER'S POINT OF VIEW

A prospective user views CONEX as a compact system for studies of nonlinear dynamical systems of defined types. The tasks that can be performed by the CONEX are summarized on Fig.2. Almost all steps can be performed independently but only certain sequences of them are reasonable, a typical session in the course of the study of a chosen model occurs as follows:

1. The first step involves the formation of a model. The user has to write down the model in the form of a set of equations and has to choose continuation parameters. These equations are used after a modification as an input to the CONEX.
2. In the next step the user may obtain recommendations how to proceed further.
3. On the basis of recommendations the user makes a choice of the type of continuation.
4. After this choice the generation of all model dependent parts of the continuation program is performed.
5. Necessary input datasets are introduced by the user. In general, there are three types of data required: model parameters, parameters specific for the chosen continuation method and starting points. Help and recommended values can be obtained from the CONEX during the definition of all types of input data and starting points. The model parameters must be fully defined by the user.
6. In this step the CONEX makes an estimation of the required computer power and the user can take a decision on which computer the computations will be done.
7. Choice of a computer by the user.
8. If computations will be run on a PC, compiling and linking to form a runable continuation program is made. In other cases, programs, data and the job control file are uploaded to the host computer and the job control file is started on the remote computer.

276

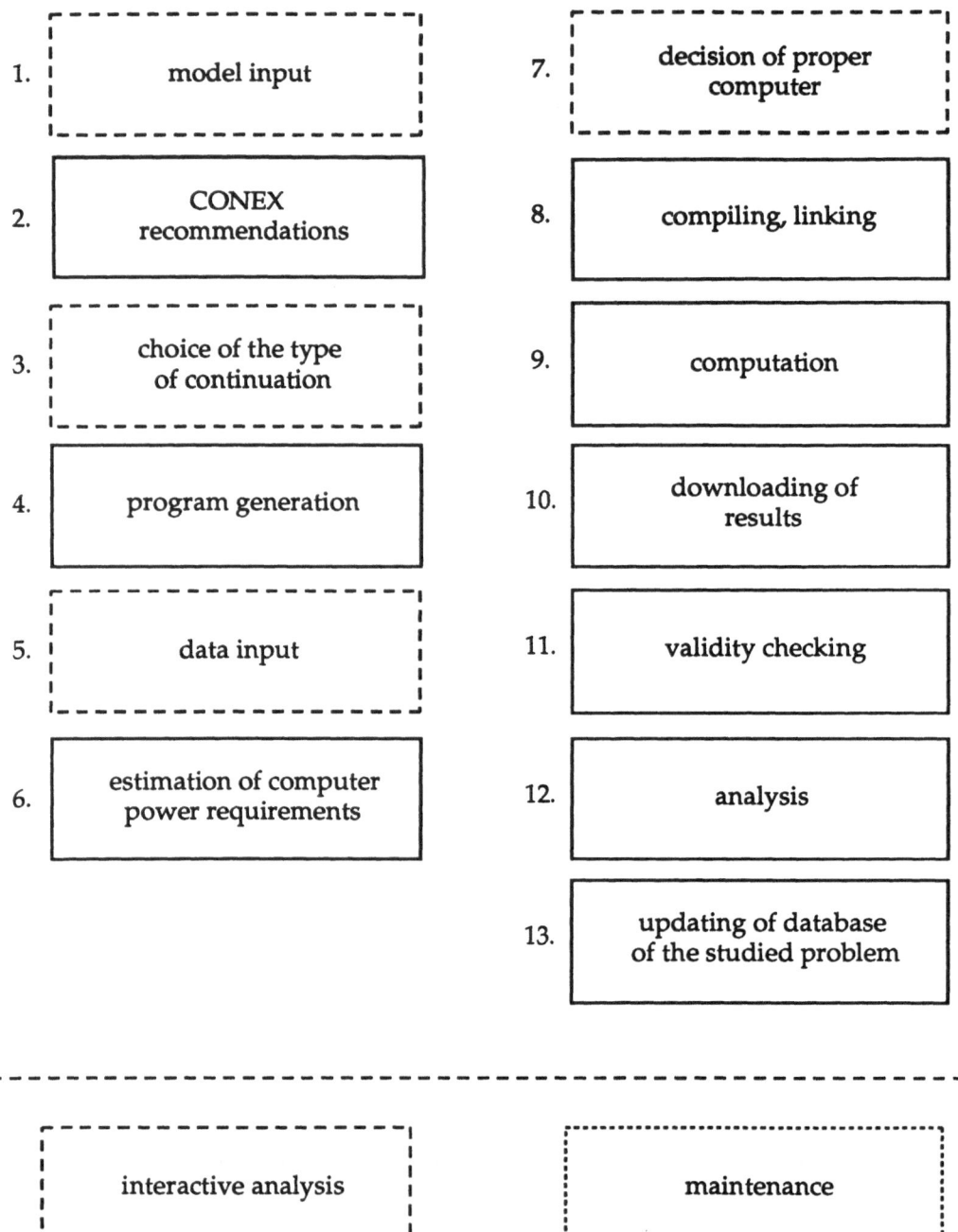

Figure 2. Users view of the C O N E X

9. Computations are executed either on the PC or on the host computer.
10. If the computations were performed on the host, results are downloaded.
11. Immediately after the computations, the run of the validity checking program ensures that the data in the output file are acceptable.
12. After this step the CONEX can proceed with an analysis of the output dataset and
13. Update the database of the studied problem.

The next step involves either an interactive analysis of the information stored in the database of the studied problem or taking recommendations from the CONEX on further course of computations.

Examples of results for two problems will be presented in the Section 3.

## 3. Examples

### 3.1. PROBLEM 1

The oxidation of carbon monooxide on the surface of the Pt/Al2O3 catalyst can be described by the following 4 ODE's:

$$x' = 1 - x + rm1 - r1$$
$$thco' = alfam * (r1 - rm1 - r3 - r5)$$
$$tho' = alfam * (2 * (1 - fco) * r2 / fco - r3 - r4)$$
$$thox' = alfam * (r4 - r5)$$

where

$$d = t0 / (q * fco * p)$$
$$s = 1 - tho - thco - thox$$
$$tq = t0 \wedge 0.5 / q$$
$$alfam = 0.14226554 / (d * q)$$
$$em1 = exp(-9000 / t0)$$
$$e3 = exp(-8000 / t0)$$
$$e4 = exp(-3468.607 / t0)$$
$$r1 = 0.00127782 * x * s * tq$$
$$rm1 = 6185616.0 * d * thco * em1$$
$$r2 = 3.58800000E-09 * s * s * tq$$
$$r3 = 358765.728 * d * thco * tho * e3$$
$$r4 = 0.0001113 * d * tho * (1 - thox) * e4$$
$$r5 = 0.0235 * d * thco * thox * e3$$

and x, thco, tho, thox are model variables (x is dimensionless concentration of CO; thco, tho, thox are dimensionless surface concentration of CO, oxygen and PtOx respectively). fco, t0, p and q are model parameters (fco is the input concentration of CO, t0 is the temperature, p pressure and q denotes the input feed rate).

Examples of continuation curves (dependences of variables x, thco, tho and thox on parameter fco) are shown on Fig.3. Solid lines denote stable solutions while dashed lines denote the unstable ones. Empty circles denote Hopf bifurcation points.

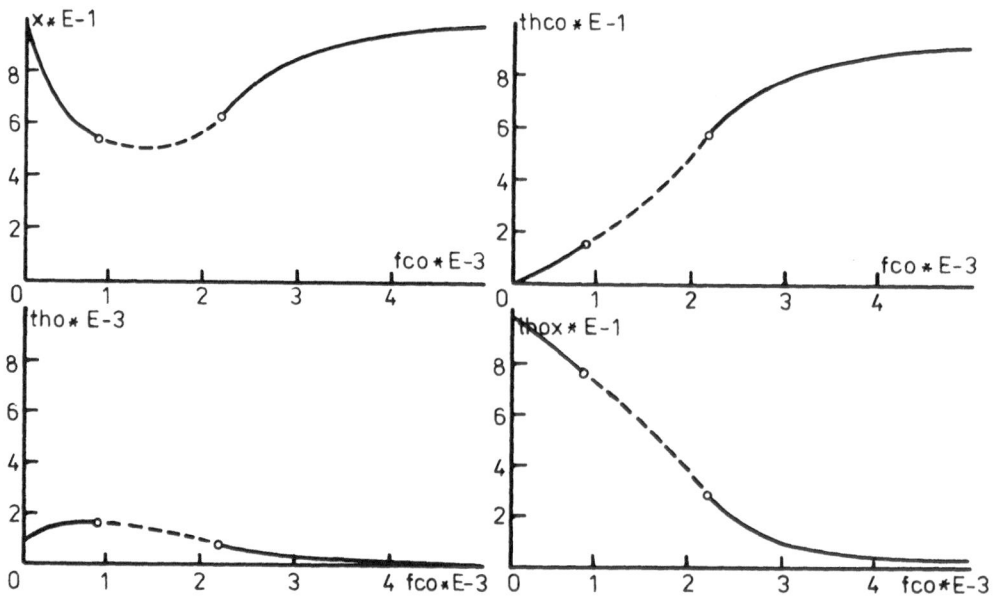

Figure 3. CONEX printout of the dependence of the stationary solution on parameter fco for Problem 1.
(p=100 kPa, q=3.0E-6m$^3$s$^{-1}$, T=411.4 K)
Full lines      - stable solutions
Dashed lines    - unstable solutions
Empty circles denote Hopf bifurcation points.

## 3.2. PROBLEM 2

A harmonic damped oscillator with impacts subjected to an external periodic force was studied in [10,11]. Equations of motion are in the form

$$x'' + \alpha x + x' \quad = A.\cos(\omega t) , \quad \omega = 2\pi/T , \quad x(t) > \delta$$

$$x'(t+) \qquad = - rx'(t-) , \qquad\qquad x(t) = \delta$$

(2)

where $\alpha, \delta, r, A$ were chosen to be constant parameters and qT - periodic solutions were studied in the dependence on the forcing period T . Every such solution can be characterized by an impact ratio p/q where p is the number of impacts occurring within the period qT . The model is piecewise linear and the main problem for the use of the continuation technique is the apriori unknown time interval between the impacts. This problem is resolved by considering the impact times as dependent on initial conditions for x and x'. Choosing a proper Poincare plane we can obtain the Poincare mapping and its derivatives with respect to x , x' and T by numerically integrating the model together with variational equations and then applying implicit differentiation rules. For this purpose the CONEX generates two subprograms, one for the free flight of the body and the other for the impacts. Examples of continuations are shown in Figs 4 and 5. Every curve is characterized by its impact ratio. There exist limit points and period doubling bifurcations in spite of the piecewise linear character of the equations. However, the branches of solutions with a given impact ratio terminate abruptly due to the piecewise linearity. Fig. 5 is a magnified portion of Fig. 4, where chaotic motions frequently occur [10]. The knowledge of periodic behaviour in this parameter region may elucidate some aspects of transitions to chaos.

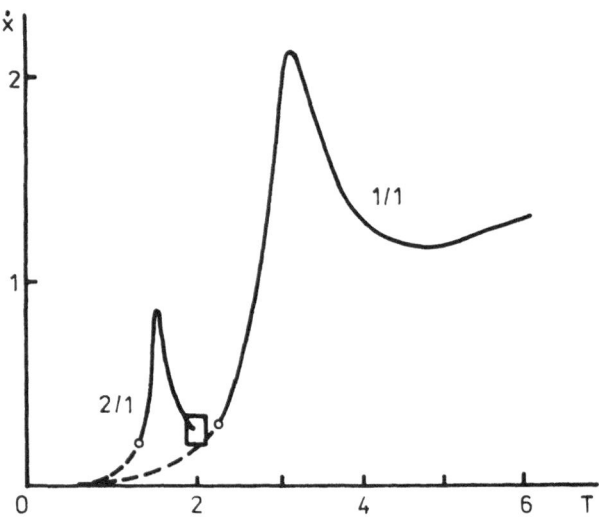

Figure 4. Dependence of the motion velocity x' on the forcing period T for Problem 2

Full lines     - stable solutions

Dashed lines   - unstable solutions

Empty circles denote period doubling bifurcation points

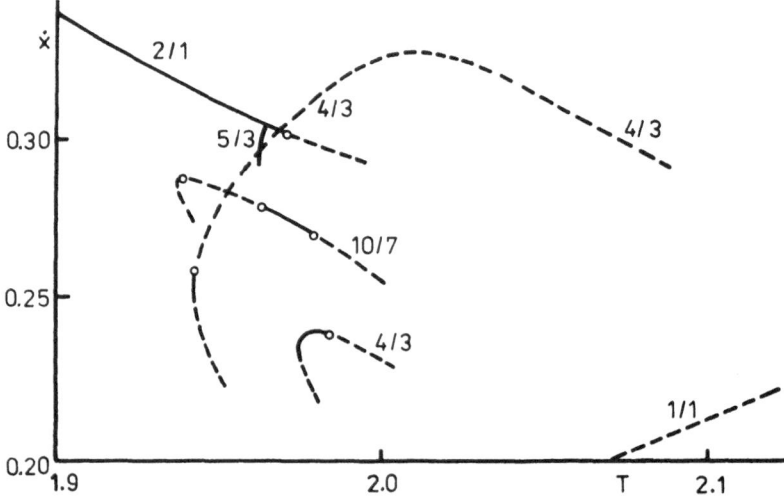

Figure 5. Enlarged part of Fig.4.

Full lines     - stable solutions

Dashed lines   - unstable solutions

Empty circles denote period doubling bifurcation points

# 4. Conclusions.

The CONEX can be used:

1. in the simplest form as a source of numerical subroutines for the construction of dependencies on parameters by continuation methods (the user can modify the generated program and use it either independently or in the CONEX environment),
2. as a tool for program generation, interactive computations and presentation of results in the form of graphs or tables,
3. in its full form where computed parametric dependencies are analyzed, inferences on obtained information and further course of computations suggested by the CONEX are utilized and the Problem Database is created.

The entire system is now in the stage of testing both by internal and external users.

# 5. Literature references

1. Kubicek, M. and Marek, M. (1983), Computational Methods in Bifurcation Theory and Dissipative Structures, Springer Verlag, Berlin
2. Holodniok, M., Klic, A., Kubicek,M. and Marek, M. (1986), Methods of Analysis of Nonlinear Dynamical Models, Academia, Prague, (in czech)
3. Rheinboldt, W.C (1986),Numerical Analysis of Parametrized Nonlinear Equations, J. Willey and Sons, New York
4. Doedel, E.J. and Kernevez, J.P. (1986), 'AUTO: Software for Continuation and Bifurcation Problems in Ordinary Differential Equations', Applied Mathematic Technical Report, California Institute of Technology
5. Küpper, T., Seydel, R. and Troger, H. (editors), (1987), Bifurcation : Analysis, Algorithms, Applications, Birkhauser Verlag, Basel
6. Seydel, R. (1988), From Equilibrium to Chaos, Elsevier Science Publ., New York
7. Marek, M. and Schreiber, I. (1989), Chaotic Behaviour of Deterministic Dissipative Systems, Cambridge Univ. Press, Cambridge, Academia Press, Prague
8. Information Processing Systems - Computer Graphics - Graphical Kernel System (GKS) - Functional Description, (1985), norm ISO 7942
9. Marek, M. and Schreiber, I. (1987), 'Formation of Periodic and Aperiodic Waves in Reaction-Diffusion Systems', in T. Kupper, R. Seydel and H.

Troger (eds.), Bifurcation: Analysis, Algorithms, Applications, Birkhauser, Basel, pp. 201-213

10. Shaw, S. and Holmes, P. (1983), 'A Periodically Forced Piesewise Linear Oscillator', J. Sound and Vibration 90, 129

11. Isomaki, A.M., Boehm, von J. and Raty, R. (1985), 'Devil's Attractors and Chaos of a Driven Impact Oscillator', Phys. Lett, 107A, 343

# LINLBF : A PROGRAM FOR CONTINUATION AND BIFURCATION ANALYSIS OF EQUILIBRIA UP TO CODIMENSION THREE

A.I. KHIBNIK
Research Computing Centre
USSR Academy of Sciences
Pushchino Moscow Region
142292 USSR

ABSTRACT. An approach which combines recent results on local bifurcation theory with the continuation method is discussed. This approach has been implemented in the LINLBF program. The program provides an advanced continuation strategy to analyze bifurcations of equilibrium points of ODEs up to codimension three singularities. We discuss the main principles, mathematics and numerics of the program, and illustrate this on a three-dimensional chemical kinetic model depending upon seven parameters. A new interactive version of the program for IBM PC compatibles is also described.

## 1. Introduction

This paper concerns the development of versatile continuation software to analyze bifurcations of nonlinear dynamical systems in a broad range of applications. During the last decade several programs emerged in this field [2,4,8,9,13-15,17].

Here we intend to discuss briefly the philosophy, mathematics and numerics of the LINLBF program. This is one of the program series LINLBF, LINBAS, LINBFP [13]. These programs provide continuation and bifurcation analysis of stationary (LINLBF) and periodic (LINBAS) solutions to ODEs, and of fixed points and periodic orbits of iterated maps (LINBFP).

It should be noted that each program of the series supports only one type of dynamical system and one type of solution (not many as AUTO [4] or PATH [8]). If one develops a separate code for each type of solution, more attention should be given to multi-parameter systems, with three or more control parameters, and to higher singularities. In particular, the LINLBF program provides an advanced continuation strategy to locate and continue some codimension three bifurcations in ODEs with four control parameters. (Note that for multi-parameter optimal control problems a similar continuation strategy is proposed by Doedel and Kernevez [4]).

All programs in the series above are based on the same principles and have similar design. Therefore, we will here discuss only the LINLBF program. A detailed description of this series and the theoretical approach to the continuation

*D. Roose et al. (eds.), Continuation and Bifurcations: Numerical Techniques and Applications, 283–296.*
© 1990 *Kluwer Academic Publishers.*

strategy in multi-parameter systems are given by Khibnik [10].

In the next section we formulate the problem and in Section 3 we discuss the basic principles of LINLBF. In Section 4 a list of bifurcations supported by LINLBF and their graph of adjacency are presented. Section 5 is devoted to so-called bifurcation functions used to locate and continue various singularities. Some specific features of the LINLBF continuation code are described in Section 6. In Section 7 we discuss the LOCBIF program, a PC interactive version of LINLBF. Finally, in Section 8 we illustrate LINLBF with a chemical kinetic model.

## 2. Formulation of the problem

Let us consider ordinary differential equations depending upon parameters

$$\dot{x} = F(x, \alpha), \tag{1}$$

where $x \in RR^n$ is a n-vector of phase variables, $\alpha \in RR^m$ is a m-vector of parameters, and assume that the function $F(x, \alpha)$ is smooth enough.

Let $M = \{(x, \alpha) \in RR^n \times RR^m : F(x, \alpha) = 0\}$ be an equilibrium manifold of (1) in the product of the phase and the parametric space. Suppose some equilibrium point in $M$ is given, and $M_0 \subset M$ is a maximal connected component of the manifold $M$ containing this point.

The question arises how the bifurcations of system (1) occurring with the elements of the manifold $M_0$ can be studied. To answer this, we should consider several closely related problems :

(i)   how to locate elements of $M_0$ having different singularities (in the sense of dynamical systems theory),

(ii)  how to analyze bifurcations of system (1) in a neighbourhood of singular elements of $M_0$,

(iii) how to describe a structure of the manifold $M_0$ called bifurcation structure, generated by the partition of $M_0$ by different singularities,

(iv)  how to describe a structure of the parametric space induced by the bifurcation structure of $M_0$.

## 3. The basic principles

The LINLBF program gives some numerical approach to the problem. It is applicable to systems with no more than four control parameters. Obviously, the system may have more than four parameters, but only the $k$-dimensional coordinate slices of the parametric space with $k \leqslant 4$ can be considered.

The program is based on a continuation technique combined with recent local bifurcation theory results (see [1,6]). The program is based on the following assumptions :

1.  equations considered are *generic* (i.e. without any symmetry conditions) ;

2. various bifurcations of codimension 1, 2, and 3 can be studied ; all of them are interconnected within the manifold of equilibrium points (see Section 4) ;

3. all the *degeneracy* points on the solution or bifurcation curves should be detected and located (the term *degeneracy* here means bifurcation of a higher codimension or transversality conditions broken) ;

4. the coefficients of an appropriate *normal form* for each degeneracy point should be calculated ;

5. each degeneracy point can be used as a *starting point* for continuation of a new branch ; the bifurcation continued may be of the same or higher by one codimension (the last case needs one more control parameter) ;

6. The determining system for continuation of codimension $\nu$ bifurcation contains a *minimal* number of equations : $N = n + \nu$ ;

7. the same *continuation code* should be used for all bifurcations studied ;

8. *low-dimensional* systems $(n \leqslant 10)$ are mainly regarded.

We should stress that the limitation of the dimension of the systems has a methodological, rather than a formal meaning. We think that "universal" bifurcation codes like LINLBF are valid only for low-dimensional systems, otherwise problems of efficiency will arise. In order to get an effective code in a high-dimensional case, one usually has to consider each bifurcation problem separately.

In order to provide a continuation strategy for different bifurcations, it seems quite natural to choose some space for the location and continuation of all bifurcation curves of interest. In the LINLBF program this is the phase-parametric space $RR^n \times RR^m$ containing the manifold $M_0$. Of course, this decision seems the simplest, and it actually is appropriate in almost all cases. In particular, it means that eigenvalues and eigenvectors are not used directly in the continuation procedure.

## 4. Bifurcation curves and the graph of adjacency

Here we describe the bifurcation curves which can be continued by the LINLBF. Let $\nu$ be the codimension of bifurcation, $k$ be the number of control parameters, and $\lambda$ be an eigenvalue of the equilibrium point. The curves are presented in Table 1.

TABLE 1. Curves continued by LINLBF.

*Generic cases :*

$\nu = 0 , k = 1$

A : Stationary solution

$\nu = 1 , k = 2$

| | |
|---|---|
| B : Turning point bifurcation | $(\lambda = 0)$ |
| C : Hopf bifurcation | $(\lambda = \pm i\omega)$ |
| D : Multiple eigenvalues | $(\lambda_1 = \lambda_2)$ |

$\nu = 2 , k = 3$

| | |
|---|---|
| E : Bogdanov-Takens bifurcation | $(\lambda_1 = \lambda_2 = 0$ with Jordan cell of order two) |
| F : Gavrilov-Guckenheimer bifurcation | $(\lambda = 0, \pm i\omega)$ |
| G : Cusp point bifurcation | $(\lambda = 0$ ; cubic case) |
| H : Degenerate Hopf bifurcation | $(\lambda = \pm i\omega$ ; first Lyapunov's value zero) |

$\nu = 3 , k = 4$

| | |
|---|---|
| I : Degenerate Bogdanov-Takens bifurcation, I | $(\lambda_1 = \lambda_2 = 0$, with Jordan cell of order two ; cubic case) |
| J : Degenerate Gavrilov-Guckenheimer bifurcation, I | $(\lambda = 0, \pm i\omega$ ; cubic case) |

*Nontransversal cases :*

$\nu = 1 , k = 3$

| | |
|---|---|
| K : Turning point bifurcation with transversality condition broken | $(\lambda = 0$ ; $det(.) = 0)$ |
| L : Hopf bifurcation with transversality condition broken | $(\lambda = \pm i\omega$ ; $det(.) = 0)$ |

$\nu = 2 , k = 4$

| | |
|---|---|
| M : Bogdanov-Takens bifurcation with transversality conditions broken | $(\lambda_1 = \lambda_2 = 0$, with Jordan cell of order two ; $det(.) = 0)$ |
| N : Gavrilov-Guckenheimer bifurcation with transversality conditions broken | $(\lambda = 0, \pm i\omega$ ; $det(.) = 0)$ |

The table needs some comments.

1. The curve D is included in the list to cover the critical case which is of importance in some applications. It specifies the boundary between equilibria of focus and node type, or in other terms, between oscillatory and aperiodic behaviour near an equilibrium point.

2. In fact, the curve C also represents a so-called neutral saddle case which means that the equilibrium point is of saddle type with a pair of positive and negative eigenvalues, equal in absolute value. Actually, this curve, called also neutrality curve, is defined by the condition that two eigenvalues with sum zero exist. Such extension of the Hopf bifurcation curve has two reasons. Firstly, it gives rise to a natural and smooth continuation of the original Hopf bifurcation curve through the end points, corresponding to Bogdanov-Takens bifurcation. In general, the neutrality curve has no end points at all. Secondly, the "saddle" part of the curve may contain critical points, corresponding to codimension two bifurcations of saddle separatrix. The same extension is made for the curves F, J, L and N.

3. The term "cubic case" means that the quadratic term in the branching equation for equilibrium points vanishes, and the cubic term becomes leading. In this case the equilibrium point has multiplicity three.

4. Basically, the Bogdanov-Takens and Gavrilov-Guckenheimer bifurcations may have different degeneracies. We here consider only one type of degeneracies, the "cubic case", which we have marked by number I (see curves I and J).

5. We denote the nontransversality condition by $det(.) = 0$. The transversality conditions do not hold if for the corresponding generic case bifurcation some minor of determining system vanishes.

6. Here are two additional bifurcations of codimensions two and three which are partially supported by the LINLBF :

$$\nu=2 , k=3$$

O : Double Hopf bifurcation $(\lambda_{1,2} = \pm i\omega_1 ,$
$\lambda_{3,4} = \pm i\omega_2)$

$$\nu=3 , k=4$$

P : Triple point bifurcation $(\lambda_1 = \lambda_2 = \lambda_3 = 0,$ with Jordan cell of order two)

The LINLBF can locate these bifurcations when tracing the branches of some other curves, such as selfcrossing points of these curves. But the program does not support continuation of the curves O and P. In the graph below we have marked this specific situation with a dotted line round the curve's name.

288

Now we can present the graph of adjacency which shows all the bifurcations described above and their interconnections (Fig. 1).

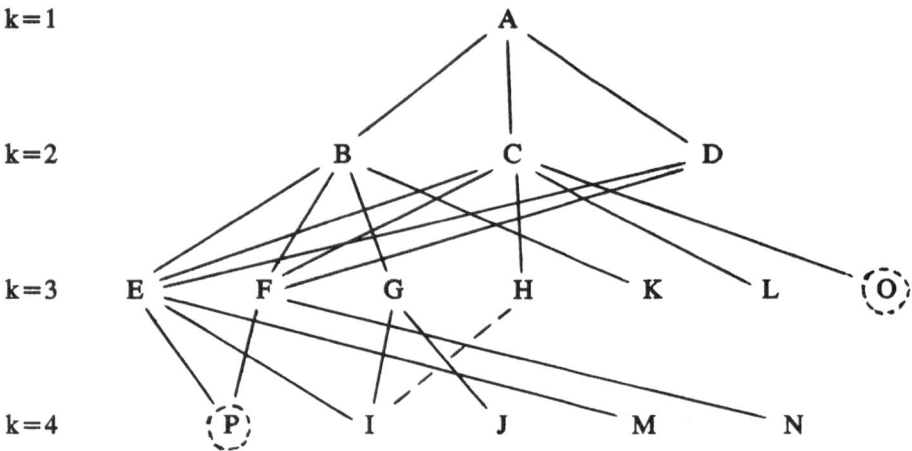

Figure 1. The graph of adjacency.

This graph demonstrates the LINLBF continuation strategy. For example, it shows three lines which connect A with B, C, and D. They mean that when tracing a branch of stationary solution curve A, the bifurcations of B, C and D types (turning points, Hopf bifurcations or neutral saddle points, and multiple eigenvalue points) will be automatically detected and located. Each bifurcation point found may be used to start tracing the relevant curve.

All bifurcations in the graph are ordered in accordance with the number of control parameters needed for their continuation. At each step of continuation strategy one should choose : 1) a bifurcation curve to study, 2) a starting point on it, and 3) control parameters (their number is determined by the curve). The starting point should be taken from some previously computed curves. The graph of adjacency shows the possible ways of passing over from one curve to another.

We should note some special case shown by the dotted line in the graph. The degeneracy point of I type does not lie on the bifurcation curve H, but appears as its limits. For such cases, the degeneracy points usually cannot be located accurately.

## 5. Bifurcation functions

By bifurcation functions we mean those functions which are used to define bifurcation curves and to find their degeneracy points. The majority of the curves mentioned may be described by the five following functions :

$$\phi_1 = \det A \tag{2}$$

$$\phi_2 = Res\ (P(\lambda), P(-\lambda)) \tag{3}$$

$$\phi_3 = Res\ (P(\lambda), P'(\lambda)) \tag{4}$$

$$\phi_4 = \frac{d^2}{d\xi^2} < v, F(x + \xi u, \alpha) > |_{\xi=0} \tag{5}$$

$$\phi_5 = L_1 \tag{6}$$

Here $A$ is the Jacobian matrix of system (1), $P(\lambda)$ is the characteristic polynomial of matrix $A$, $Res$ stands for the resultant of two polynomials, $u$ and $v$ are eigenvectors corresponding to $\lambda=0$, of matrices $A$ and $A^T$, with conditions $<u, u> = 1$ and $<u, v> = 1$, $<\cdot,\cdot>$ denotes scalar product, $L_1$ is the first Lyapunov's value for an equilibrium point with a pair of pure imaginary eigenvalues.

The condition $\phi_1 = 0$ implies that a zero eigenvalue exists. Function $\phi_2$ equals zero if and only if there are two eigenvalues with sum zero [5]. So, $\phi_2 = 0$ holds for an equilibrium point with two pure imaginary eigenvalues and for a neutral saddle. Obviously, the condition $\phi_3 = 0$ implies multiple roots of the characteristic polynomial.

The functions $\phi_4$ and $\phi_5$ relate to degeneracies in nonlinear terms of the Taylor expansion of $F(x, \alpha)$ in an equilibrium point for critical parameter values. The condition $\phi_4 = 0$ refers to the nonquadratic (generally, cubic) case for branching of equilibrium points with $\lambda = 0$. (Remark that only a zero eigenvalue with geometric multiplicity one is regarded here.)

The condition $\phi_5 = 0$ is the degeneracy condition for Hopf bifurcation. This condition separates supercritical $(L_1 < 0)$ and subcritical $(L_1 > 0)$ Hopf bifurcations, and specifies a border between soft and sharp losses of stability of a stationary solution when its eigenvalues pass through the imaginary axis. Note that for some unfoldings of the degenerate Hopf bifurcation two limit cycles appear. To compute the first Lyapunov's value $L_1$ on (or near) a Hopf bifurcation curve, we have used the ASIMPC program [13], which is similar to BIFOR2 [7].

In our notation, the curve B, which consists of turning points, is determined by the system

$$\begin{aligned} F(x, \alpha) &= 0, \\ \phi_1(x, \alpha) &= 0. \end{aligned} \tag{7}$$

It depends on the n-vector $x$ and on two control parameters chosen, $\alpha_i$ and $\alpha_j$ (the other parameters in vector $\alpha$ are assumed to have fixed values).

The determining systems for the curves C and D are derived in a similar way, replacing $\phi_1$ by $\phi_2$ and $\phi_3$ respectively. To determine the curves E or F, we use both functions $\phi_1$ and $\phi_2$, so the determining system has the form

$$\begin{aligned} F(x, \alpha) &= 0, \\ \phi_1(x, \alpha) &= 0, \\ \phi_2(x, \alpha) &= 0. \end{aligned} \tag{8}$$

Curves, functions which determine them, and functions which are used to find degeneracy points at the curves, are related as listed in Table 2.

TABLE 2. Bifurcation curves and bifurcation functions

| Codim | Curve | Determining functions | Testing functions and degeneracy points |
|-------|-------|-----------------------|------------------------------------------|
| 0 | A | – | $\phi_1$ – B<br>$\phi_2$ – C<br>$\phi_3$ – D |
| 1 | B | $\phi_1$ | $\phi_2$ – E,F<br>$\phi_3$ – E,F,...<br>$\phi_4$ – G |
| 1 | C | $\phi_2$ | $\phi_1$ – E,F<br>$\phi_3$ – E,F,...<br>$\phi_5$ – H |
| 1 | D | $\phi_3$ | $\phi_1$ – E,F,...<br>$\phi_2$ – E,F,... |
| 2 | E | $\phi_1,\phi_2$ | $\phi_4$ – I |
| 2 | F | $\phi_1,\phi_2$ | $\phi_4$ – J |
| 2 | G | $\phi_1,\phi_4$ | $\phi_4$ – I,J |
| 2 | H | $\phi_2,\phi_5$ | – |
| 3 | I | $\phi_1,\phi_2,\phi_4$ | – |
| 3 | J | $\phi_1,\phi_2,\phi_4$ | – |

The right column of this table represents testing functions whose zeros are detected and located when tracing a curve. For each testing function we indicate a singularity of higher by one codimension which occurs when this function vanishes. In particular, the functions $\phi_1$, $\phi_2$ and $\phi_3$ are tested on the solution curve A to find the points of B, C and D types (turning points, Hopf bifurcations or neutral saddle points, and multiple eigenvalue points). On the curve B the program tests functions $\phi_2$, $\phi_3$ and $\phi_4$ to locate points of E, F, G types (degeneracy points of some other types, which are not mentioned above, also can occur).

Basically, what do we require with respect to the bifurcation functions ? The testing function which we use to detect and locate degeneracy points on a curve, should be smooth or at least continuous on this curve, and change sign at such points. If for some curve we use the bifurcation function in the relevant determining system, this function must be defined in a neighbourhood of this curve in the space of phase variables and parameters, and should be smooth there.

One can see from Table 2 that we use the same functions for both locating the degeneracy points and building the determining system for curve continuation. Evidently, this implies constraints on the bifurcation functions used.

We want to emphasize that the LINLBF program does not use functions (2) – (5) directly. Some improvements were made in the implementation of the above ideas to expand the region of definition of these functions, and to provide that they are smooth and robust. This is of great importance, in particular since numerical derivatives are widely used in the program.

## 6. Continuation code

The present version of the LINLBF program makes use of an original continuation code. It is called BEETLE and has been made by Nikolaev and Khibnik. The routine is based on a Newtonian correction scheme similar to that in the PITCON program by Rheinboldt [16]. To find a new point on a curve, the program fixes a value of the variable which (locally) changes at the highest speed on the curve, and seeks, by a Newtonian iteration procedure, the point where the curve crosses the hyperplane. For prediction of the new point on the curve, a simple tangent predictor is used. The program provides step size control by local curvature.

An advantage of BEETLE is that it can automatically find points of the curve, where variables chosen by the user have local extrema, or the testing functions chosen pass through zero values. It can also detect and locate selfcrossing (bifurcation) points on the curve.

When we incorporate BEETLE into the LINLBF program, the actual role of user-defined testing functions is played by the bifurcation functions. LINLBF chooses them individually for each curve, in accordance with Table 2. Note that basically different testing functions are used for the different bifurcation curves.

## 7. Interactive LOCBIF program

In 1989, an interactive version of the LINLBF program for IBM PC compatibles has appeared. It is the LOCBIF program (Local Bifurcation analyzer), which has been developed by Kuznetsov, Levitin, Nikolaev and Khibnik at the Research Computing Centre in Pushchino.

In interactive mode it can be used to :

— choose a bifurcation curve for continuation ;

— pick out some control parameters from the parameters of the system studied ;

— set an initial point on the curve by numerical values, or by fitting a point from a previous curve ;

— choose an arbitrary two dimensional projection for graphical representation of the curve (phase variables, control parameters, eigenvalues and bifurcation functions may equally be used for this aim) ;

— continue the chosen curve in a step-by-step regime or automatically, displaying it graphically and numerically during continuation, indicating its degeneracy points (continuation can be interrupted at any point).

LOCBIF maintains archives of computed curves on a disk. A curve can be saved or loaded from archives. After it is loaded, it can be plotted on the screen, or can be used to take a point for starting some other curve.

The program allows the user to enter a new system during the session, by settings its right-hand-sides as a Fortran routine, or by modifying the right-hand-sides available.

## 8. Example

Now we want to illustrate the LINLBF (or LOCBIF) program with a chemical kinetic model [3]. The model describes CO oxidation reaction in the form

$$\dot{x} = k_1 z^2 - 2k_{-1}x^2 - k_3\,xy,$$
$$\dot{y} = k_2 z \quad - k_{-2}\,y \quad - k_3\,xy, \tag{9}$$
$$\dot{s} = k_4 z \quad - k_{-4}\,s,$$

where $x$, $y$, $z$, $s$ are the concentrations with $x + y + z + s = 1$ and $k_i$ stand for the reaction rate constants. We regard the coefficients $k_i$ as parameters, so the system (9) has seven parameters (note that only one of them may be eliminated by scaling).

For this model several local and nonlocal singularities of codimension two and three occur. By combining the related theory and the continuation method, 23 different phase portraits were found in the model [11,12]. When investigating (9) by the LINLBF (LOCBIF) program, we followed a continuation scheme which is represented by the graph of adjacency (see Fig. 1). Namely, we increased gradually the number of control parameters $k$, from one to four, and for each $k$ we computed the relevant bifurcation curves. It should be noted that only singularities with no more than two critical eigenvalues on the imaginary axis were found in the system. Actually, all bifurcations of Section 4, besides the F, J, N, O and P cases, and all connections between them were realized.

Some results of our study are shown in Fig. 2 and Fig. 3, as they were given by the LOCBIF program. The parameters $k_1, k_2, k_3, k_4, k_{-1}, k_{-2}$, and $K = k_{-4}/k_4$ are denoted here $Q_1$, $Q_2$, $Q_3$, $Q_4$, $Q_5$, $Q_6$, and $Q_7$ respectively.

Figure 2 shows three solution curves for control parameter $Q_2$ and different values of $Q_7$. The codimension one turning point and Hopf bifurcation curves B and C are plotted in this graph too, they correspond to control parameters $Q_2$ and $Q_7$. It can be seen in the figure that the bifurcation curves pass through the relevant bifurcation points on the solution curves. There are also some multiple eigenvalue points marked on the curves.

The Hopf bifurcation curve C is closed; it consists of two connected parts corresponding to Hopf and neutral saddle cases. These parts are separated by two Bogdanov-Takens bifurcation points. The first (Hopf) part of the curve also contains two degenerate Hopf bifurcation points.

The turning point curve B contains two Bogdanov-Takens bifurcation points mentioned and one cusp bifurcation point. In Fig. 2, the limit point on the curve B corresponds to the cusp bifurcation point.

In Fig. 3 the codimension one and two bifurcation curves are represented as projections onto the parameter plane $Q_2$ - $Q_7$. These are turning point and Hopf bifurcation curves B and C (codimension 1), as well as cusp point curve G, Bogdanov-Takens bifurcation curve E, and degenerate Hopf bifurcation curve H (all of codimension 2). To start computing the curves E, G and H, we took the relevant points on the curves B and C. For codimension two curves the parameters $Q_2$, $Q_7$ and $Q_1$ were assumed to be control parameters.

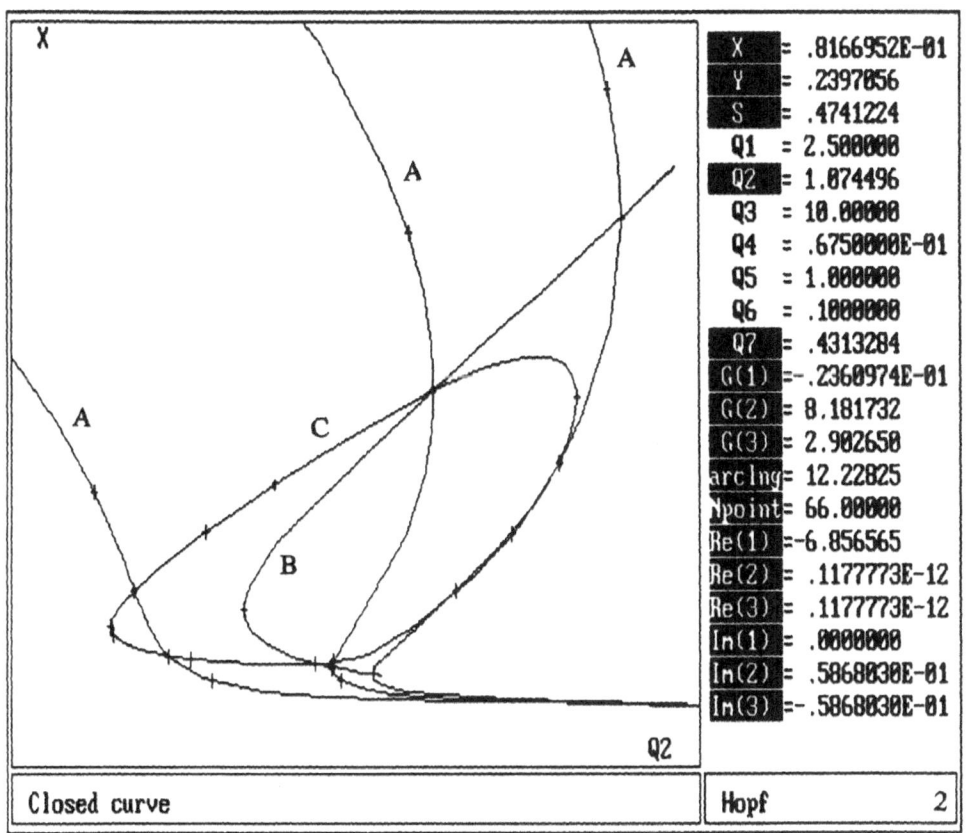

Figure 2. The solution and codimension one bifurcation curves in the equations (9) ( A - stationary solution curve, B - turning point curve, C - Hopf bifurcation curve).

294

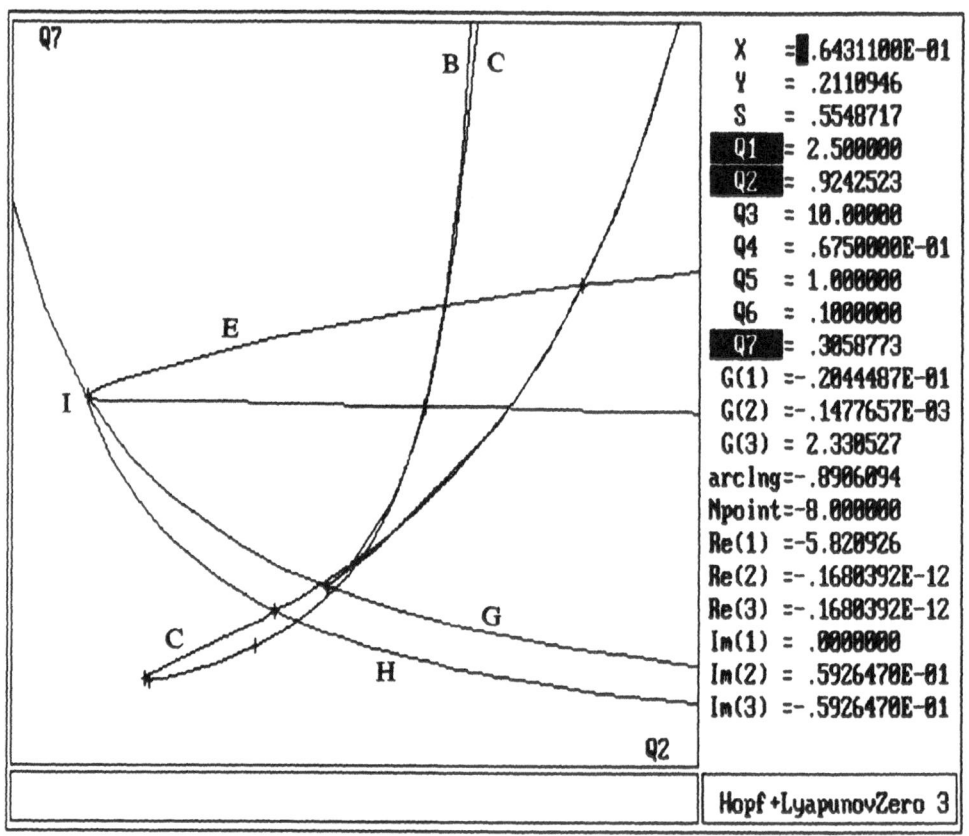

Figure 3. Codimension one and two bifurcation curves in the equations (9) ( B - turning point curve, C - Hopf bifurcation curve, E - Bogdanov-Takens bifurcation curve, G - cusp point curve, H - degenerate Hopf bifurcation curve, I - degenerate Bogdanov-Takens bifurcation point).

Note that the degenerate Bogdanov-Takens bifurcation point I has been located both on the curves E and G. It also has been found as the end (unreachable) point on the degenerate Hopf bifurcation curve H. The degenerate Bogdanov-Takens curve I, for control parameters $Q_2, Q_7, Q_1$ and $Q_5$ was computed too, but it is not shown in the graph.

## 9. Discussion

We have presented here the LINLBF program which is intended to continue stationary solution curves and several bifurcation curves in ODEs. The generic case equations only have been considered. The program can compute 13 different types of curves, they are listed in Table 1. They correspond to singularities of codimension one through three. Some singularities with nontransversality conditions are included in the list too.

The program provides an advanced continuation strategy to study systems with several control parameters, from one to four. This strategy is represented by the graph of adjacency (see Fig. 1) and illustrated with the three dimensional CO oxidation reaction model having seven parameters.

It should be noted that the list of bifurcations is not complete as far as higher singularities are considered. The problem is to append the list with some more known singularities of higher codimension. Recall that by our assumption the normal form coefficients should be computed for each singularity considered.

What we also want to stress is that interactive software is necessary for bifurcation analysis, especially when multi-parameter systems are considered. And our interactive version of the LINLBF, the LOCBIF program, is one of the ways of coping with this problem.

## 10. Acknowledgements

I thank Dr. E.V. Nikolaev for his contribution to the LINLBF continuation code. I am also grateful to Dr. V.V. Levitin and especially to Dr. Yu.A. Kuznetsov for their enthusiasm and great efforts in developing the interactive version of the LINLBF program.

## 11. References

[1] Afrajmovich, V.S., Arnold, V.I., Il'yashenko, Yu.S. and Shilnikov, L.P. (in press) 'Bifurcation theory', in V.I. Arnold (ed.), Dynamical Systems V. The Theory of Bifurcations and Catastrophes, Springer-Verlag, New York.

[2] Borisyuk, R.M. (1981) Stationary solutions of ordinary differential equations depending upon parameter, FORTRAN Software Series 6, Research Computing Centre, USSR Academy of Sciences, Pushchino (in Russian).

[3] Bykov, V.I., Yablonskii, G.S. and Kim, V.F. (1978) 'On one simple model of kinetic self-oscillations in catalytic reaction of CO oxidation', Dokl. Sov. Math., 637-639.

[4] Doedel, E.J. and Kernevez, J.P. (1986) 'AUTO : Software for continuation and bifurcation problems in ordinary differential equations', Report, Applied Mathematics, California Institute of Technology, Pasadena.

[5] Gantmacher, F.R. (1967) The Matrix Theory, Nauka, Moscow (in Russian).

[6] Guckenheimer, J. and Holmes, P. (1983) Nonlinear Oscillations, Dynamical Systems, and Bifurcations of Vector Fields, Springer-Verlag, New York.

[7] Hassard, B.D., Kazarinoff, N.D. and Wan, Y.-H. (1981) Theory and Applications of Hopf Bifurcation, London Mathematical Society Lecture Notes Series 41, Cambridge University Press, Cambridge.

[8] Kaas-Petersen, C. (1987) 'PATH - User's Guide', Centre for Nonlinear Studies University of Leeds, Leeds.

[9] Khibnik, A.I. (1979) Periodic solutions of n-dimensional ordinary differential equations, FORTRAN Software Series 5, Research Computing Centre, USSR Academy of Sciences, Pushchino, (in Russian).

[10] Khibnik, A.I. (in press) 'Numerical methods in bifurcation analysis of dynamical systems : a continuation approach', in A.D. Bazykin and Yu.G. Zarhin (eds.), Mathematics and Modelling, Research Computing Centre, USSR Academy of Sciences, Pushchino (in Russian).

[11] Khibnik, A.I., Bykov, V.I. and Yablonskii, G.S. (1986) 'Parametric portrait of the catalytic oscillator', Preprint, Institute of Catalysis, USSR Academy of Sciences, Novosibirsk (in Russian).

[12] Khibnik, A.I., Bykov, V.I. and Yablonskii, G.S. (1987) '23 phase portraits of a simplest catalytic oscillator', J. Fiz. Khim. 61, 1388-1390 (in Russian).

[13] Khibnik, A.I. and Shnol. E.E. (1982) Software for Qualitative Analysis of Differential Equations, Research Computing Centre, USSR Academy of Sciences, Pushchino (in Russian).

[14] Kubicek, M. and Marek, M. (1983) Computational Methods in Bifurcation Theory and Dissipative Structures, Springer-Verlag, New York.

[15] Kuznetsov, Yu.A. (1983) One-dimensional invariant manifolds of ordinary differential equations depending upon parameters, FORTRAN Software Series 8, Research Computing Centre, USSR Academy of Sciences, Pushchino (in Russian).

[16] Rheinboldt, W.C. (1986) Numerical Analysis of Parametrized Nonlinear Equations, Wiley-Interscience Publication, New York.

[17] Seydel, R. (1988) From Equilibrium to Chaos : Practical Bifurcation and Stability Analysis, Elsevier.

ON THE TOPOLOGY OF THREE-DIMENSIONAL SEPARATIONS,
A GUIDE FOR CLASSIFICATION

P.G. BAKKER
Dept. Aerospace Engineering
Delft University of Technology
P.O. Box 5058, 2600 GB  Delft
Fax: (015) 781822  The Netherlands

ABSTRACT

The topology of three-dimensional separated flow structures is examined for Navier-Stokes flows. A classification strategy is developed
so that viscous flow patterns with complicated topology can be analyzed in a systematic way. The method relies on the qualitative theory
of differential equations and on bifurcation theory. Some examples
illustrate the method.

1. INTRODUCTION, GENERAL APPROACH

The topology of three-dimensional (3D) separated flow patterns is a
subject of several contemporary studies in aerodynamics. Interesting
results are performed by Hornung (1983), Dallmann (1983) and Perry &
Chong (1986). Hornung introduces the vortex-skeleton model in order to
describe unambiguously the qualitative features of the structure in a
steady 3D flow. Dallmann analyses a special set of local solutions of
the Navier-Stokes (NS) equations describing the flow patterns with
some special symmetry conditions. Later on, Perry & Chong have developed an algorithm that enables them to generate local solutions of the
NS-equations numerically. Recently we have developed a classification
strategy along which viscous flow patterns with complicated topology
can be analyzed in a systematic way. The method relies on the qualitative theory of differential equations and the theory of bifurcations.
For 2D flows the method is outlined by Bakker (1988), the investigation of 3D flows is taken up by de Winkel (1988) and Kooij (1989).
Some results will be presented in this paper.

The flow is assumed to be steady, viscous, incompressible and it
satisfies no-slip boundary conditions on the wall. The topology of
such a flow is studied on the basis of local solutions of the NS-
equations. The streamline pattern is represented by the trajectories

297

D. Roose et al. (eds.), Continuation and Bifurcations: Numerical Techniques and Applications, 297–318.
© 1990 Kluwer Academic Publishers.

of a third-order dynamical system $\dot{x}=u(x,y,z)$, $\dot{y}=v(x,y,z)$, $\dot{z}=w(x,y,z)$ with u, v, w denoting the velocity components in an orthogonal cartesian reference system x,y,z; ($\dot{\,}$) denotes $\frac{d}{dt}$ and t is real time. The wall is represented by y=0. Since the NS-equations allow for analytical solutions, even in separated flow regions, see Dean (1950), the local solutions may be obtained by performing Taylor expansions of the velocity vector field.

Consider the flow pattern that results if the velocity vector field is expanded up to the Nth order near an arbitrary point P, then the trajectory pattern near P is governed by the system (S):

$$\dot{x} = u = \sum_{i=0}^{N} \sum_{j=0}^{N-i} \sum_{k=0}^{N-i-j} U_{i,j,k} x^i y^j z^k + O(N+1)$$

$$\dot{y} = v = \sum_{i=0}^{N} \sum_{j=0}^{N-i} \sum_{k=0}^{N-i-j} V_{i,j,k} x^i y^j z^k + O(N+1) \qquad\qquad (S)$$

$$\dot{z} = w = \sum_{i=0}^{N} \sum_{j=0}^{N-i} \sum_{k=0}^{N-i-j} W_{i,j,k} x^i y^j z^k + O(N+1)$$

where O(N+1) denote higher-order terms of at least order N+1, composed by powers of x, y and z. The coefficients $U_{i,j,k}$, $V_{i,j,k}$ and $W_{i,j,k}$ are treated as constants and are left unknown in a local analysis. Due to the flow equations and the boundary conditions some relationships between $U_{i,j,k}$, $V_{i,j,k}$ and $W_{i,j,k}$ exist, reducing the number of coefficients that can be chosen independently. Our main objective will be to give an unified description of all possible topologies near P if the remaining coefficients in (S) are varied independently.

At first sight this task seems very hard since system (S) contains an infinite number of terms.

To overcome this difficulty we content ourselves to study the phase portraits of a truncated system $(S_N)$ which is derived from (S) by omitting the O(N+1) terms. Since $(S_N)$ contains a finite number of terms the question arises whether it is true that systems (S) and $(S_N)$ have equal topological structure near P. Can N be chosen such that it suffices to study the trajectory pattern of system $(S_N)$ instead of that of the original system (S).

With regard to this question we know that phase portraits are mainly characterized by the number and position of singular points, the local trajectory pattern near these points and the position of separatrices.

Thus if singular points (and their local trajectory pattern) of system $(S_N)$ keep their qualitative character even if higher order terms are added, then $(S_N)$ suffices to obtain a qualitative description of the local flow topology of system (S). For isolated hyperbolic singularities (no eigenvalues have a vanishing real part) only linear terms are of importance so that the truncated system is already found by taking N=1.

For isolated non-hyperbolic singularities also a truncated system $(S_N)$ can be found provided that N is chosen sufficiently large so that O(N+1) terms do not disturb the topological character of the degenerate singularities. Normal form theory is used to examine which higher order terms play a significant role. This implies that for a given N a classification of degenerate singularities can be established, of which each of them represents a local flow topology which can occur in flows governed by the full NS-equations.

Apart from these relatively simple local flow patterns containing only one singularity, system $(S_N)$ enables us also to encounter flow patterns consisting of a cluster of singular points forming a coherent flow structure. In order to guarantee that such a cluster is a local solution of the NS-equations it will be sought as an unfolding of a non-hyperbolic singularity of system $(S_N)$.


2. TOPOLOGICAL ANALYSIS OF THE FLOW ON AND NEAR THE WALL

Consider the incompressible steady laminar flow in the neighbourhood of a plane wall at rest. The flow near the wall occurs in the upper halfplane y≥0. It satisfies:

the continuity equation : div $\underline{V}$ = 0

the NS-equations conserving
momentum : $(\underline{V}.\nabla)\underline{V} + \nabla p^* = \nu\Delta\underline{V}$

the no-slip boundary
conditions : $u(x,0,z) = v(x,0,z) = w(x,0,z) = 0 \ \forall x,z\epsilon R$

with $p^* = \dfrac{p}{\rho}$ the kinematic pressure, $\nu$ the kinematic viscosity, p the pressure and $\rho$ the mass density. Applying the continuity equation and the no-slip condition system (S) takes the form

$$\dot{x} = u = y \;\; \bar{u} \; (x,y,z)$$

$$\dot{y} = v = y^2 \; \bar{v} \; (x,y,z) \tag{1}$$

$$\dot{z} = w = y \;\; \bar{w} \; (x,y,z)$$

The plane y=0 is a plane with singular points (u=v=w=0) implying a quasi non-hyperbolic singular character at the wall surface. Since the trajectories of system (1) at y≠0 are identical with those of the equivalent system:

$$\dot{x} = y^{-1}u(x,y,z) = \bar{u} \; (x,y,z)$$

$$\dot{y} = y^{-1}v(x,y,z) = y\bar{v} \; (x,y,z) \tag{2}$$

$$\dot{z} = y^{-1}w(x,y,z) = \bar{w} \; (x,y,z)$$

the singular character of the plane y=0 can be removed by investi-gating (2) instead of (1). If $\bar{u}, \bar{v}$ and $\bar{w}$ are expanded in a Taylor series the equivalent system which governs the flow above the wall becomes:

$$\dot{x} = a_1 + a_2 x + a_3 y + a_4 z + a_5 x^2 + a_6 xy + a_7 xz + a_8 y^2 + a_9 yz + a_{10} z^2 + 0(3)$$

$$\dot{y} = y\{b_1 + b_2 x + b_3 y + b_4 z\} + 0(3) \tag{3}$$

$$\dot{z} = c_1 + c_2 x + c_3 y + c_4 z + c_5 x^2 + c_6 xy + c_7 xz + c_8 y^2 + c_9 yz + c_{10} z^2 + 0(3)$$

where $a_i$, $b_j$ & $c_k$ $\epsilon R$. To fulfil the continuity equation and the NS-equations relations between the coefficients $a_i$, $b_j$ & $c_k$ exist, for

details see de Winkel (1988). The one which will be often used reads
$a_2 + 2b_1 + c_4 = 0$.

The unknown coefficients occurring in (3) can be used for a proper embedding of a local flow structure into a surrounding main flow. The coefficients $a_i$, $b_j$ & $c_k$ can also be expressed in the physical quantities $p^*$ and $\tau, \sigma$. The latter are the components of the shear stress in x- and z-direction respectively, they are defined by

$$\tau = \mu \left(\frac{\partial u}{\partial y}\right)_{y=0} \quad \& \quad \sigma = \mu \left(\frac{\partial w}{\partial y}\right)_{y=0}$$

where $\mu$ is the dynamic viscosity.
A few relations between coefficients and stress quantities are:

$$\mu a_1 = \tau, \quad \mu a_2 = \tau_x, \quad 2\mu a_3 = p_x, \quad \mu a_4 = \tau_z$$

$$2\mu b_1 = p_y = -\tau_x - \sigma_z$$

$$\mu c_1 = \sigma, \quad \mu c_2 = \sigma_x, \quad 2\mu c_3 = p_z, \quad \mu c_4 = \sigma_z$$

Note that all the stress quantities have to be evaluated in the origin (0,0,0).

Considering system (3) we observe that the plane y=0 is now filled with solution curves described by

$$\dot{x} = \bar{u}(x,0,z,), \quad \dot{z} = \bar{w}(x,0,z)$$

or $\qquad \dfrac{dz}{dx} = \dfrac{\bar{w}(x,0,z)}{\bar{u}(x,0,z)}$

Since $\qquad \left(\dfrac{\partial u}{\partial y}\right)_{y=0} = \bar{u}(x,0,z)$ and $\left(\dfrac{\partial w}{\partial y}\right)_{y=0} = \bar{w}(x,0,z)$

these solution curves form a trajectory pattern which can be identified with the skin friction field on the wall defined by

$$\dot{x} = \tau(x,0,z) \quad , \quad \dot{z} = \sigma(x,0,z) \tag{4}$$

In conclusion, system (3) is the basic equation that enables us to study both the viscous flow pattern above the wall as well as the skin friction pattern on the wall.


## 3. SINGULAR POINTS ON THE WALL, JORDAN NORMAL FORMS

Since the phase portrait of a dynamical system is mainly characterized by the number, position and type of singular points, it is useful to start the investigation of system (3) by examining its singularities in some detail.
Singularities of system (3) can be interpreted in two ways, those lying above the wall represent stagnation points in the flow, those located at the wall have a vanishing shear stress vector indicating either flow separation from the wall or flow attachment to the wall. Since we are particularly interested in the topology of separation and attachment structures, we restrict the study to wall singularities and take the origin of the reference system in such a singularity. Then in system (3) the coefficients $a_1$ and $c_1$ disappear and we have

$$\dot{x} = a_2 x + a_3 y + a_4 z + a_5 x^2 + a_6 xy + a_7 xz + a_8 y^2 + a_9 yz + a_{10} z^2 + O(3)$$

$$\dot{y} = y(b_1 + b_2 x + b_3 y + b_4 z) + O(3) \tag{5}$$

$$\dot{z} = c_2 x + c_3 y + c_4 z + c_5 x^2 + c_6 xy + c_7 xz + c_8 y^2 + c_9 yz + c_{10} z^2 + O(3)$$

Or in shorthand notation with $\underline{x} = (x,y,z)^T$ we may write

$$\underline{x} = A\underline{x} + f(\underline{x}) + O(3) \quad \text{where } A = \frac{1}{\mu}\begin{pmatrix} \tau_x & \frac{1}{2}p_x & \tau_z \\ 0 & \frac{1}{2}p_y & 0 \\ \sigma_x & \frac{1}{2}p_z & \sigma_z \end{pmatrix} = \begin{pmatrix} a_2 & a_3 & a_4 \\ 0 & b_1 & 0 \\ c_2 & c_3 & c_4 \end{pmatrix} \tag{6}$$

and $f(x)$ is a vector polynomial containing the second and third order terms of system (3). To analyse singularities of the non-linear system (6) we may apply Hartman-Grobman's theorem which states that the character of a singularity is determined by the eigenvalues of the linear part of the system unless there are eigenvalues having real parts equal to zero, $Re(\lambda)=0$; for details see Guckenheimer & Holmes (1983). Points having all $Re(\lambda) \neq 0$ are called hyperbolic. Otherwise they are called <u>non-hyperbolic</u>, degenerate or higher-order points. The linear part $\dot{\underline{x}} = A\underline{x}$ yields the eigenvalues $\lambda_{1,3} = -b_1 \overset{+}{\mp} \sqrt{(a_2+b_1)^2 + a_4 c_2}, \lambda_2 = b_1$. The continuity equation gives the relation $\lambda_1 + 2\lambda_2 + \lambda_3 = 0$ reflecting the fundamental property that parts of the flow will enter and other parts of the flow will leave the neighbourhood of the singular point.

TABLE 1. Jordan normal forms of the linear part of system (6).

| Hyperbolic | Non-Hyperbolic |
|---|---|
| 1) all eigenvalues are real and different | 1) all eigenvalues are real and different |
| $\begin{pmatrix} \lambda_1 & 0 & 0 \\ 0 & \lambda_2 & 0 \\ 0 & 0 & -(\lambda_1 + 2\lambda_2) \end{pmatrix}$ (1a) | $\begin{pmatrix} 0 & 0 & 0 \\ 0 & \lambda_2 & 0 \\ 0 & 0 & -2\lambda_2 \end{pmatrix}$ (1b) $\quad\begin{pmatrix} \lambda_1 & 0 & 0 \\ 0 & 0 & 0 \\ 0 & 0 & -\lambda_1 \end{pmatrix}$ (1c) |
| 2) all of them are real and two are equal | 2) all of them are real and two are equal |
| $\begin{pmatrix} \lambda_1 & 0 & 0 \\ 0 & \lambda_1 & 0 \\ 0 & 0 & -3\lambda_1 \end{pmatrix}$ (2a) $\quad\begin{pmatrix} \lambda_1 & 0 & 0 \\ 0 & -\lambda_1 & 0 \\ 0 & 0 & \lambda_1 \end{pmatrix}$ (2b) | — — — — — — — — — — |
| $\begin{pmatrix} \lambda_1 & 1 & 0 \\ 0 & \lambda_1 & 0 \\ 0 & 0 & -3\lambda_1 \end{pmatrix}$ (2c) $\quad\begin{pmatrix} \lambda_1 & 0 & 1 \\ 0 & -\lambda_1 & 0 \\ 0 & 0 & \lambda_1 \end{pmatrix}$ (2d) | 3) all of them are real and equal |
| | $\begin{pmatrix} 0 & 0 & 1 \\ 0 & 0 & 0 \\ 0 & 0 & 0 \end{pmatrix}$ (3a) $\quad\begin{pmatrix} 0 & 1 & 0 \\ 0 & 0 & 0 \\ 0 & 0 & 0 \end{pmatrix}$ (3b) |
| 3) all of them are real and equal | $\begin{pmatrix} 0 & 0 & 1 \\ 0 & 0 & 0 \\ 0 & 1 & 0 \end{pmatrix}$ (3c) $\quad\begin{pmatrix} 0 & 0 & 0 \\ 0 & 0 & 0 \\ 0 & 0 & 0 \end{pmatrix}$ (3d) |
| — — — — — — — — — — | |
| 4) one eigenvalue is real and two are complex conjugated | 4) one eigenvalue is real and two are complex conjugated |
| $\begin{pmatrix} -\lambda_2 & 0 & \mathrm{Im}\,\lambda_1 \\ 0 & \lambda_2 & 0 \\ -\mathrm{Im}\,\lambda_1 & 0 & -\lambda_2 \end{pmatrix}$ (4a) | $\begin{pmatrix} 0 & 0 & \mathrm{Im}\,\lambda_1 \\ 0 & 0 & 0 \\ -\mathrm{Im}\,\lambda_1 & 0 & 0 \end{pmatrix}$ (4b) |

A suitable linear transformation $\underline{x}=T\underline{u}$ which does not affect the plane $y=0$ may be applied to bring the non-linear system (6) into the equivalent form

$$\dot{\underline{u}} = J\underline{u} + g(\underline{u}) \tag{7}$$

where $J = T^{-1}AT$ is one of the normal forms of A listed in Table 1 and $g(\underline{u}) = T^{-1}f(\underline{u})$.

Table 1 shows the equivalent Jordan forms of all possible linear flow structures. Obviously only these Jordan forms need to be analyzed to get a complete picture of all possible topologies that can appear in a viscous fluid flow.


## 4. ELEMENTARY SINGULARITIES, SKIN FRICTION PATTERNS.

The qualitative behaviour of trajectories around a hyperbolic point informs us about the possible types of 3D elementary singularities and the corresponding skin friction field. To classify them we use the Jordan normal forms as listed in Table 1. A singularity is called an attachment point $(\lambda_2 < 0)$ or a separation point $(\lambda_2 > 0)$ if a particular streamline above the wall can be found which enters or leaves the singular point on the wall. Since $\lambda_2 = b_1 = \frac{1}{2}p_y$ attachment points indicate a pressure decrease and separation points show a pressure increase when going in off wall direction. We find three basically different skin friction patterns near an attachment point; the singularity is either a saddle ($\tau_x$ and $\sigma_z$ have different sign), an unstable node (general, starlike or inflected, $\tau_x$ and $\sigma_z$ have the same sign) or an unstable focus $\left((\tau_x - \sigma_z)^2 + 4\tau_z\sigma_x < 0\right)$. Very similar skin friction patterns arise near a separation point, however the node and the focus must be stable in that case. The possible flow structures near elementary singularities in a three-dimensional viscous flow are shown in Fig. 1.

Some of these local flow structures can be recognized in Aerodynamics. The (general) node appears for instance on the surface of a body at the attachment point of the free stream. In case of a symmetrical body (like a sphere) one might expect the star node as an attachment point. The focus corresponds to a vortex standing on the surface of the body and saddles occur for example in flows with closed separation bubbles. The inflected node is not much familiar in aerodynamic applications because it may appear as a transition pattern between node and focus.

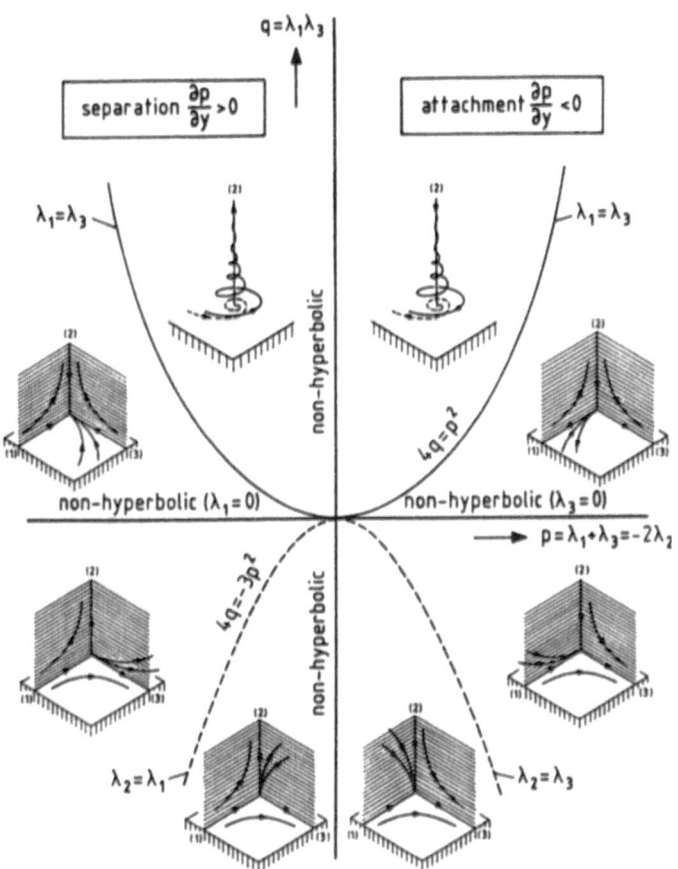

Figure 1. Elementary singularities in 3D viscous flows.


## 5. NON-HYPERBOLIC SINGULARITIES AND BIFURCATIONS

The proper determination of the local flow topology near a non-hyper-bolic singularity requires several dominant non-linear terms to be retained in the series expansion. These terms can be found by applying a near-identity transformation $\underline{u} = \underline{y} + P(\underline{y})$ yielding a normal form of system (7); $P(\underline{y})$ being a vector polynomial (at least of degree 3) mapping the wall onto $y_2=0$. Then a blow-up technique is used to analyze the structurally unstable flow patterns. They can fall apart when subjecting them to small disturbances. Bifurcation theory is applied to find the resulting structurally <u>stable</u> flow structures. To illustrate the procedure two cases of Table 1 namely (4b) and (3c) will be treated in the next paragraphs.

5.1. Flow structures associated with the Jordan normal form (4b)

Consider system (6) with Jordan form

$$
A = \begin{pmatrix} 0 & 0 & \omega \\ 0 & 0 & 0 \\ -\omega & 0 & 0 \end{pmatrix}
$$

implying $\tau=\tau_x=0$, $\sigma=\sigma_z=0$, $p_x=p_y=p_z=0$ and $\tau_z=-\sigma_x=\omega\neq0$ in the singularity at $(0,0,0)$.

Using normal form theory, applying the continuity equation and using cylindrical coordinates system (6) is equivalent to

$$
\dot{r} = r\{cy + er^2 + fy^2 + 0(3)\}
$$

$$
\dot{y} = y\{-\tfrac{2}{3}cy - 2er^2 - \tfrac{1}{2}fy^2 + 0(3)\} \tag{8}
$$

$$
\dot{\phi} = -\omega + by + gy^2 + jr^2 + 0(3)
$$

In (8) the $\dot{\phi}$-equation is decoupled from the $\dot{r}$-and $\dot{y}$-equation indicating that the local flow can be investigated in a meridional plane $\phi$=constant. Using Stokes' stream function concept it can be proven that the decoupled behaviour remains if higher order terms are added. The local flow in a meridional plane is found using standard methods such as the blow-up/down technique. See Andronov (1973) or Takens (1974). The method consists of a transformation that maps the singularity onto two or more singularities in the transformed vectorfield. These singularities are analyzed and if all of them are hyperbolic one knows which terms of the original system play an essential role and one can construct the phase portrait. If, after one blow-up, one or more of the singularities are non-hyperbolic, a second blow-up has to be performed.

Application of the blow-up method results in laborious calculations which will be deleted here. We confine ourselves to mention the main conclusions resulting from this blow-up proces.

Fig. 2 shows the meridional structures of degenerate 3D flow patterns that belong to a Jordan normal form (4b) and that have a rather low degree of degeneracy. All the observed patterns have saddle-type domains. At the lowest level of degeneracy, c.e≠0, the flow has a combined separation/attachment structure with a bowl-shaped separation surface tangent to the wall if c.e<0 (Fig. 2a) and a saddle type structure if c.e>0 (Fig. 2b). For c=0 the flow patterns have a higher degree of degeneracy; there appears a higher order saddle point if e.f>0 (Fig. 2b) or a higher order saddle point with two hyperbolic sectors if e.f<0 (Fig. 2c). In the latter case the separation surface is a conic having its vertex in the singularity.

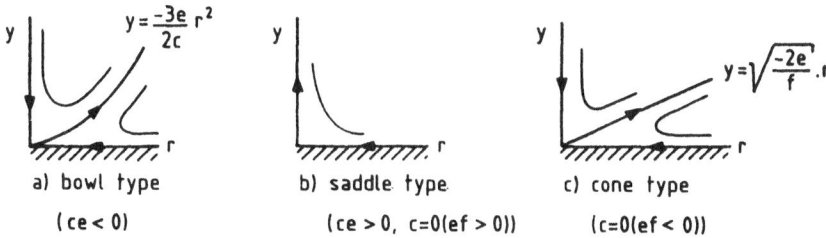

a) bowl type
(ce < 0)

b) saddle type
(ce > 0, c=0(ef > 0))

c) cone type
(c=0(ef < 0))

Figure 2. Meridional structures of degenerate 3D flow patterns having

a Jordan form $\begin{pmatrix} 0 & 0 & \omega \\ 0 & 0 & 0 \\ -\omega & 0 & 0 \end{pmatrix}$

For sake of convenience these structures are referred to as a bowl- (Fig. 2a), a saddle- (Fig. 2b) and a cone type (Fig. 2c) separation. Degenerate patterns with much higher degree of degeneracy occur at e.f=0. This case is not worked out here, a further expansion of the velocity components would then be necessary.

5.2. Burst of a 3D closed separation bubble, bifurcation of system (8)

As an example, the unfolding of the structurally unstable cone type separation (Fig. 2c) will now be analyzed. It can be obtained by using Mather's technique, as outlined by Shirer & Wells (1983). Then the leading terms of (8) are supplemented with lower order terms which contain the bifurcation parameters $\mu_1$ and $\mu_2$ and satisfy the governing flow equations. The unfolding in cylindrical coordinates reads supplementary terms          degenerate singularity

$$\dot{r} = \quad \mu_1 r + \mu_2 ry \qquad\qquad +r\{er^2 + fy^2$$

$$\dot{y} = \quad -\mu_1 y - \tfrac{2}{3}\mu_2 y^2 \qquad +y\{-2er^2 - \tfrac{1}{2}fy^2\} \qquad (9)$$

$$\dot{\phi} = \qquad\qquad\qquad\qquad -\omega + by$$

Since the $\dot{\phi}$-equation is decoupled and the bifurcation parameters do not appear in this equation the bifurcation can be analyzed in a plane $\phi$ = constant. The following bifurcation sets can be identified (Fig. 3):

higher order saddle (HS): $\mu_1 = 0$ & $\mu_2 \neq 0$

third order saddle (TS): $\mu_1 = \dfrac{2}{9f}\mu_2^2$ & $\mu_2 < 0$

cusp            (C) : $\mu_1 = \dfrac{8}{27f}\mu_2^2$ & $\mu_2 < 0$

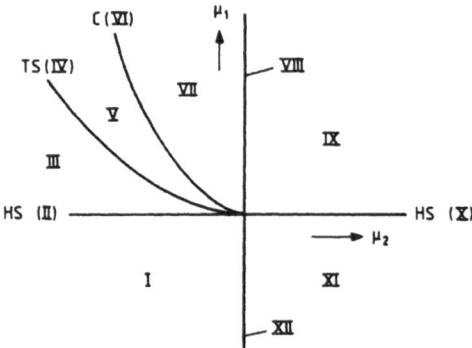

Figure 3. Bifurcation sets.

The various flow patterns that result from bifurcation are shown in Fig. 4. In region I we observe a flow pattern having two saddle points on the y-axis. The 3D flow reveals two separate flow domains which interact together in a rotational movement. There is a part A which moves on to the wall more or less in perpendicular direction. Next to it a part B exists which flows along the wall towards the axis of rotation. Both flows are separated by a 3D bowl-shaped separation surface fully embedded in the flow above the wall. No separation from the wall is observed. Going from region I towards III bifurcation of a higher order saddle at r=y=0 occurs. The local flow structure near this higher order singularity is already depicted in Fig. 2b, case c≠0. The higher order saddle bifurcates into a closed bubble fitted to the wall (region III). Due to the formation of the bubble flow separation in part B occurs along a closed separation line. This closed separation line appears as a stable limit cycle in the skin friction field on the wall. Inside the bubble particle paths lie on tori and they form a center in the meridional plane. Since the flow in the r-y-plane is governed by a hamiltonian system these centers remain also if higher order terms should be taken into account. This implies that mass is conserved inside the bubble.

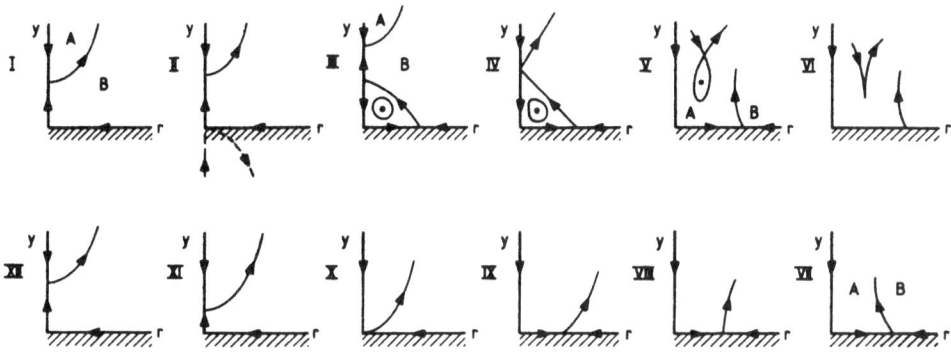

Figure 4. Unfolding of the degenerate cone type flow pattern.

Moving from region III to region V a third order saddle point appears on the y-axis above the wall. The closed bubble interferes with the bowl-shaped separation surface. Then the flow in part B is pushed away from the axis of rotation and an open separation surface standing on the wall results. The flow in part A, forces the bubble to leave the wall and generates an annulus of recirculating flow (with probably two commensurate frequencies) inside the flow A. The observed phenomenon may be referred to as bursting of the closed separation bubble. Moving to VII the annular region of circulating flow shrinks until we pass region IV where cusp bifurcation appears in the meridional plane. The cusp disappears and we end up with a flow having a 3D open separation surface originating from the wall (region VII-IX). Going from IX to XI the separation line on the wall moves to the origin (region X) where bifurcation causes lift off of the separation surface (region XI, XII, I).

5.3. Flow structures associated with the Jordan normal form (3c)

In this paragraph we consider system (6) with Jordan normal form

$$A = \begin{pmatrix} 0 & 0 & 1 \\ 0 & 0 & 0 \\ 0 & 1 & 0 \end{pmatrix}$$

implying $\tau=\tau_x=0$, $\sigma=\sigma_x=\sigma_z=0$, $p_x=p_y=0$ and $\tau_z \neq 0$, $p_z \neq 0$ in the singularity at $(0,0,0)$.

Using normal form theory and applying the continuity equation system (6) is equivalent to

$$\dot{x} = z + x^2 - \frac{2}{3} y^2 + 0(3)$$

$$\dot{y} = - xy + 0(3) \tag{10}$$

$$\dot{z} = y + cx^2 - \frac{1}{3} cy^2 + 0(3)$$

The local flow structures of this higher order singularity are for the first time discussed by Kooij (1989) for the case $c \neq 0$. The coefficient c is related to the local value of $\sigma_{xx}$ in the singularity; for $c>0$ and $c<0$ the 3D flow patterns are shown in Fig. 5.

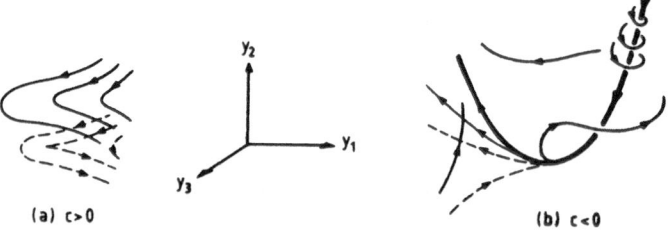

(a) c>0          (b) c<0

Figure 5. The local flow structures of system (10) near the origin.

The skin friction pattern is given by the dotted lines. The non hyperbolic case $c < 0$ implies $\sigma_{xx} < 0$; it represents a 3D degenerate flow
structure with a cusp singularity in the skin friction field. Above
the wall there appear two special spatial trajectories, both passing
through the origin. Near the singularity the trajectories form a
parabola in the plane $-3y + 2cz = 0$. One trajectory is an attachment
streamline whereas the other is a separation streamline; the wall
surface is touched tangentially.

The case $c > 0$ implies $\sigma_{xx} > 0$ in the singularity. Again a cusp is formed
in the skin friction pattern. The streamlines above the wall follow
more or less the skin friction lines. There is no separation nor
attachment in this particular structurally unstable flow situation.

A numerical calculation of the skin friction field in the vicinity of
the cusp shows a remarkable strong convergence of the skin friction
lines to the separatrices of the cusp singularity; see Fig. 6.

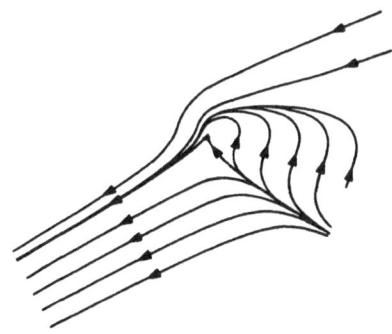

Figure 6. Strong convergence of skin friction lines near a cusp.

From this situation we observe that the line of convergence itself is
a skin friction line. In aerodynamic applications the strong convergence phenomenon is mostly interpreted as a reliable indication for
flow separation.

5.4. Development of an open separation surface (case $c < 0$), bifurcation
     of system (10)

The unfolding of the case $c < 0$ will now be analyzed. Applying Mather's
technique the unfolding of system (10) becomes

312

$$\dot{x} = \mu x + z + x^2 - \frac{2}{3} y^2$$

$$\dot{y} = -\frac{1}{2} \mu y - xy \qquad \qquad (11)$$

$$\dot{z} = \lambda \quad + y + cx^2 - \frac{1}{3} cy^2$$

where $\lambda$ and $\mu$ represent small perturbations of the shear stress quantities $\sigma$ and $\tau_x$ respectively. The bifurcation sets and corresponding flow patterns are shown in Fig. 7 and 8, respectively. The bifurcations that can occur are

Hopf ($p^+=0$)  $\quad : \lambda = \frac{1}{4} \mu^2 \ (\mu < 0)$

saddle-node ($D=0$ & $\bar{p}=0$)  $\quad : \lambda = 0 \ \& \ \lambda = \frac{1}{4} \mu^2 \ (\mu > 0)$

homoclinic cycle (S)  $\quad : \lambda = (\frac{7}{10})^2 \mu^2 + 0(\mu^3) \ (\mu < 0)$

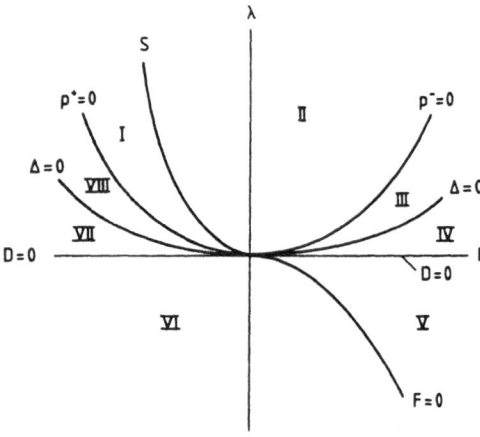

Figure 7. Bifurcation sets of system (11) for c<0.

Along $\lambda = \frac{1}{64} \mu^4 (\Delta=0)$ and $\lambda = -\frac{2}{9} \sqrt{3} \mu^{3/2}(F=0)$ foci and nodal points transfer into each other by passing an improper node.

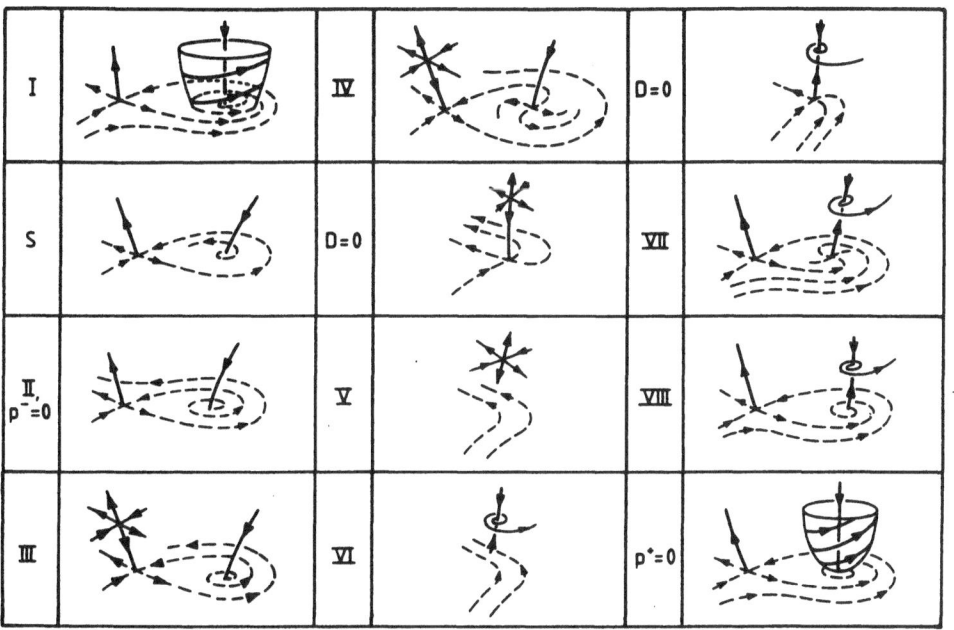

Figure 8. Flow patterns of system (11); c<0

Let us discuss is some detail the various flow patterns appearing from bifurcation.

Moving in the parameter plane onto various domains, a sequence of complicated flow structures in $R^3$ is encountered. The most salient features of them will be revealed briefly. If the degenerate state $\mu=0$, $\lambda=0$ is perturbed by taking $\lambda<0$ a regular skin friction field results (VI). Additionally there appears a stagnation point in the flow where two vortex tubes seem to terminate in a focal singularity which carries the fluid away from the tube axis. No separation from nor attachment to the wall surface is observed. The focal singularity is replaced by a singularity having a 1D stable eigenspace and 2D unstable eigenspace when moving to domain V. A perturbation to $\lambda>0$ can

314

result into flow patterns having separated flow regions. Two distinct singularities appear in the skin friction field. If separation is present then there is at least one singularity with an unstable manifold which carries the fluid away from the wall. The dimension of the unstable manifold is one or two. The first case corresponds with a stable focus (domain VIII) or with a saddle point (domains I, II, VII, VIII) in the skin friction field. If both singularities are present (domain VIII) then the well-known saddle-focus pattern as observed by Legendre (1965) results. A 2D unstable manifold occurs in correspondance with a higher-order stable fine focus ($p^+$=0) in the skin friction pattern. In this case the manifold is more or less bowl-shaped, it originates from the fine focus on the wall and it forms a separation surface in the flow. The flow pattern is structurally unstable. A perturbation via a Hopf bifurcation creates a structurally stable flow pattern containing a separation surface forming a closed separation line on the wall (domain I). This separation line appears as a limit cycle in the skin friction field and encloses an unstable focus. Such a limit cycle is a rather new phenomenon in skin friction patterns. Just as the open separation phenomenon as observed by K.C. Wang (1983) it is an example where a separation line has no terminating singular points.

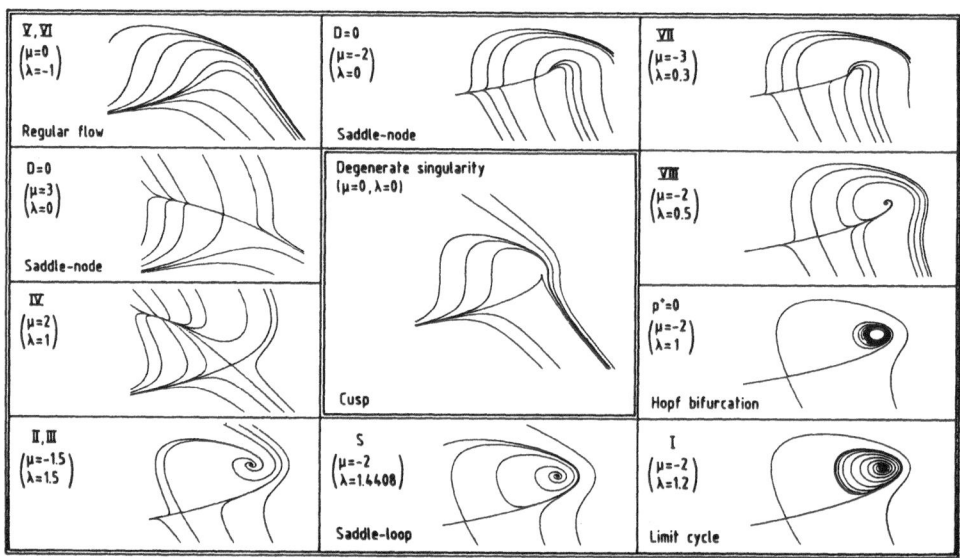

Figure 9. Skin friction patterns generated from bifurcation of a cusp singularity.

Results of a numerical calculation of the various skin friction patterns in the parameter plane, which are described by $\dot{x}=z+\mu x+x^2$ & $\dot{z}=\lambda-x^2$ so c=-1, are depicted in Fig. 9. The limit cycle (closed separation line) can grow onto a structurally unstable saddle-loop which is the union of a homoclinic cycle and a saddle point singularity. The saddle-loop can be perturbed by a global bifurcation (S) with the effect that the closed separation line vanishes.

## 5.5. Development of a closed three-dimensional separation bubble (case c>0), bifurcation of system (10)

The unfolding of system (10) as given by equation (11) will now be discussed for the case c>0. The bifurcation is governed by the skin friction quantities $\lambda=\sigma(0,0,0)$ and $\mu=\tau_x(0,0,0)$. The bifurcation sets together with the occurring flow patterns are depicted in Figs. 10 and 11 respectively

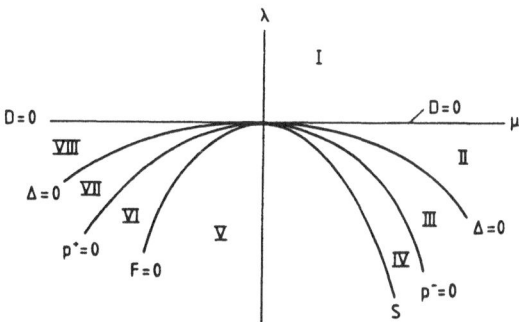

Figure 10. Bifurcation sets of system (11) for c>0.

The following bifurcation sets will appear

Hopf ($p^-$ = 0) : $\lambda = -\frac{1}{4}\mu^2$ ($\mu$>0)

saddle-node (D=0 & $p^+$=0): $\lambda = 0$ & $\lambda = -\frac{1}{4}\mu^2$ ($\mu$<0)

homoclinic cycle (S) : $\lambda = -\left(\frac{7}{10}\right)^2\mu^2$ ($\mu$>0)

316

This bifurcation has the very interesting property that in the parameter domain I the flow field has no singularities neither in the skin friction pattern on the wall nor in the flow region above the wall. It implies that this bifurcation can be of practical importance for the investigation of incipient separations in regular three-dimensional flow domains.

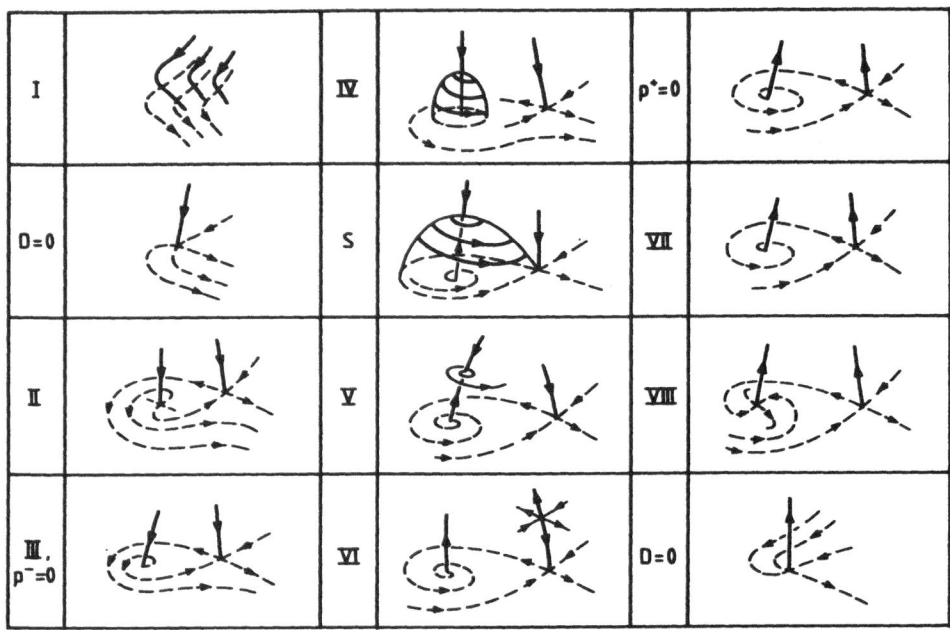

Figure 11. Flow patterns of system (11); c>0.

As we observe from this bifurcation, a regular flow field can be disturbed by a saddle-node bifurcation causing two attachment singularities. One of them is always a saddle point, the other can be either a nodal point (II, VIII) or a focus (III-VII). This focus can become structurally unstable ($\bar{p}=0$) so that via an unstable (in the Lyapunov sense) fine focus a Hopf bifurcation appears. The Hopf bifurcation creates a three-dimensional bubble-shape separation surface which forms a closed separation line on the wall (IV). Inside the bubble the fluid recirculates in a rotational movement; no singularities in the flow appear but a stable focus is formed in the skin friction pattern. A one-dimensional unstable manifold which originates from this focus transports the fluid on to the top of the bubble where it is spread out over the bubble surface. Then the fluid is conveyed by the closed separation surface again to the wall. The closed separation line on the wall appears as a limit cycle in the skin friction field. Via a global bifurcation (S) the bubble dissolves into the main flow and the closed separation line which has been grown into a saddle-loop (S) breaks up resulting into the disappearance of a closed separation line and leaving the ordinary skin friction pattern with the well-known saddle-focus pattern (V-VII) (Legendre-Werlé). This pattern remains until the focus is transformed into a nodal point (VIII). Via a saddle-node bifurcation both singularities vanish and a regular flow field is obtained.


6. CONCLUSIONS AND PROSPECTIVES

The method described in this paper yields local solutions of the NS-equations by considering bifurcations of degenerate singularities in the velocity vector field of a 3D flow. These solutions are of particular interest as they provide insight into complicated 3D separated flow structures. Several cases are still unresolved, especially the Jordan forms (3a) & (3d) have to be analyzed. At the moment case (3b) with additional symmetries is under consideration. Furthermore some basic experiments are planned in order to verify the theoretical results. Finally it seems possible to extend the method so that curved walls, moving walls and unsteady flow problems can be treated as well.


7. References

Andronov, A.A. et.al. (1973)    Qualitative theory of second order dynamical systems. Wiley, New York.

Bakker, P.G. (1988)    Bifurcations in flow patterns. PhD thesis, Delft University of Technology.

Dallmann, U. (1983)    Topological structures of three-dimensional vortex flow separation, AIAA-paper 1735.

Guckenheimer, J. (1983) & Holmes, P.J.    Nonlinear oscillations, dynamical systems and bifurcation of vector fields. Springer, New York, Berlin.

Hornung, H.G. (1983)    The vortex-skeleton model for three-dimensional steady flows. AGARD Conference Proc. 342.

Kooij, R.E. (1989)
& Bakker, P.G.

Three-dimensional viscous flow structures from bifurcation of a degenerate singularity with three zero eigenvalues, Report LR-572, Delft University of Technology.

Legendre, R. (1965)

Lignes de courant d'un ecoulement continu, La Recherche Aerospatiale, 105.

Perry, A.E. (1986)
& Chong, M.S.

A series-expansion study of the Navier-Stokes equations with three-dimensional separation points, J. Fluid Mech., vol. 173.

Shirer, H.N. (1983)
& Wells, R.

Mathematical structures of the singularities at the transitions between steady states in hydrodynamic systems, Lecture notes in physics, vol. 185, Springer, New York.

Takens, F. (1974)

Singularities of vector fields, Publ. Math. IHES, 43, 47-100.

Wang, K.C. (1983)

On the disputes about open separation, AIAA-paper 83-0296.

Winkel, M.E.M. de (1988)
& Bakker, P.G.

On the topology of three-dimensional viscous flow structures near a plane wall. A classification of hyperbolic and non-hyperbolic singularities on the wall, Report LR-541, Delft University of Technology.

# THE CONSTRUCTION OF CUBATURE FORMULAE
## USING CONTINUATION AND BIFURCATION SOFTWARE.

RONALD COOLS and ANN HAEGEMANS
*Department of Computer Science*
*Katholieke Universiteit Leuven*
*Celestijnenlaan 200A*
*B3030 Heverlee, Belgium*

ABSTRACT. An $n$-dimensional integral $I[f]$ can be approximated by a quadrature or cubature formula $Q[f]$. A quadrature or cubature formula of degree $d$ is by definition a solution of a system of polynomial equations. The region and the weight function only influences the constants in each equation. The purpose of this text is to illustrate that continuation and bifurcation software can be useful tools to construct cubature formulae. First, we consider integrals where the region or weight function contains a parameter. Standard regions can be embedded in a family of regions. Continuation methods are used to construct cubature formulae for different values of the parameter, starting from a solution for a specific value of the parameter. This approach can be used to construct new cubature formulae for standard regions and regions that are difficult to treat otherwise. We also study continua of cubature formula using continuation methods.

## 1. Introduction

Let $I$ be an integral over an $n$-dimensional region $\Omega$ with a positive weight function $w(\vec{x})$

$$I[f] = \int_{\Omega} w(\vec{x}) f(\vec{x}) \, d\vec{x} \quad , \quad \Omega \subset \mathbb{R}^n \ .$$

Let $Q$ be an approximation for this integral of the form

$$Q[f] = \sum_{i=1}^{N} w_i f(\vec{x}_i) \quad , \quad w_i \in \mathbb{R} \ , \ \vec{x}_i \in \mathbb{R}^n. \tag{1}$$

If $I$ is a one-dimensional integral, $Q$ is called a *quadrature formula*. If $I$ is a multi-dimensional integral, $Q$ is called a *cubature formula*. A quadrature or cubature formula is called *good* if

$$\vec{x}_i \in \Omega \text{ and } w_i > 0 \text{ for all } i \in \{0, 1, \ldots, N\}.$$

The vector space of all polynomials in $n$ variables of degree $\leqslant m$ is denoted by $\mathbb{P}_m^n$. A cubature formula which is exact for all $f \in \mathbb{P}_d^n$ but not for all

*D. Roose et al. (eds.), Continuation and Bifurcations: Numerical Techniques and Applications, 319–333.*
© 1990 *Kluwer Academic Publishers.*

$f \in \mathbb{P}^n_{d+1}$ is said to have degree $d$.

The theory of the construction of quadrature formulae is well known. A lower bound for the number of points $(= N)$ is known. The lower bound is attained by Gauss quadrature formulae with all points $x_i$ inside the region and all weights $w_i$ positive. The construction of cubature formulae is much more difficult. Lower bounds exist, but are seldom attained. It is clear nowadays that the minimum number of points depends not only on the region $\Omega$, but also on the weight function $w(\vec{x})$. All construction methods depend in one way or another on some lower bound.

Basically, there are two distinct approaches :

1) One can try to solve the system of polynomial equations that determines a cubature formula of degree $d$:

$$Q[f_i] = I[f_i] \quad , \quad i = 1,2,\ldots, \dim \mathbb{P}^n_d \ , \tag{2}$$

where the $f_i$ form a basis for $\mathbb{P}^n_d$.

This was usually done by using an iterative zero–finder, arbitrary starting values and some luck (e.g. [9], [12]).

2) One can translate the problem into an algebraic problem using the theory of polynomial ideals (e.g. [13], [14]). If the problem is not too large, it can be solved using computer algebra systems.

For larger problems, one has to solve large systems of polynomial equations (much smaller than (2)!) numerically (e.g. [4], [6]).

So, sooner or later the construction of cubature formulae requires that one can find a solution (preferably all) of a system of polynomial equations. Basically, there are two methods to find all solutions of a system of polynomial equations.

1) *Elimination methods* : The essential part in an elimination method is the elimination step. In one elimination step, the system of polynomial equations in $n$ variables is reduced to a system of equations in $n - 1$ variables. Well known is the method based on the determinant of Sylvester. (See e.g. [20].) The construction of a Gröbner basis for the ideal generated by a set of polynomials can also be considered as a way to do elimination steps [2].

2) *Continuation methods* : If all solutions of a system of polynomial equations $F(\vec{x}) = 0$ are desired the following idea is used. Suppose all solutions of another system of polynomial equations $G(\vec{x}) = 0$ with the same number of equations and unknowns as $F$ are known.

Using a parameter $t \in \mathbb{C}$, the family of systems

$$H(\vec{x},t) = tF(\vec{x}) + (1 - t)G(\vec{x}) = 0$$

is build. This is called an artificial parameter homotopy. One starts at $t = 0$ and changes $t$ continuously from 0 to 1. Intuitively one can expect that solutions to $H(\vec{x},0) = G(\vec{x}) = 0$ change continuously into solutions to $H(\vec{x},1) = F(\vec{x}) = 0$. The homotopy can be constructed so that the path tracking algorithm must not be very sophisticated. (See e.g. [16].)

In this paper we want to investigate continua of cubature formulae for some integrals and to construct cubature formulae for families of integrals. This leads to natural parameter homotopies : A system of nonlinear equations $F(\vec{x},\alpha) = 0$ is obtained where $\alpha$ is a parameter with a physical meaning. The solution branches must be dealt with as they are. Therefore, curve tracking becomes the main focus of the problem-solving effort. Continuation consists in numerically following a solution branch by a predictor-corrector method, starting from a regular solution on that branch. For a complete understanding of the structure of the solution set of the nonlinear system in function of the parameter $\alpha$, one should construct the complete bifurcation diagram (i.e. one should compute all solution branches ). In order to exploit the knowledge of a solution point on a branch as much as possible, the continuation procedure must detect bifurcation points lying on that branch and the bifurcating branches must be computed also.

Some software packages for continuation and bifurcation analysis exist. We have used AUTO [8], a package for the bifurcation analysis of differential equations

$$\frac{d\vec{y}}{dt} = F(\vec{y}(t),\alpha) \ .$$

In its search for steady-state solutions $\left| \dfrac{d\vec{y}}{dt} = 0 \right|$ it detects turning points and also bifurcation points and computes the bifurcating branches. In order to start the continuation procedure, one needs a regular solution of the system of equations.

## 2. Families of integrals

The left-hand sides of the nonlinear equations (2) that determine a cubature formula, are independent of the region and the weight function. The region and the weight function only influence the right-hand sides, i.e. the constants in each equation. In this section we consider cases where the region or the weight function depends on a parameter. The system of nonlinear equations that determines the cubature formulae can be written as

$$Q[f_i] = I_\alpha[f_i] \quad , \quad i = 1,2,\ldots,\dim\mathbb{P}_d^n$$

where the $f_i$ form a basis for $\mathbb{P}_d^n$. We assume that for one value of the parameter $\alpha$ a regular solution is known. Starting from this known solution, solutions for other values of $\alpha$ can be computed by a continuation procedure. If also some bifurcations occur, multiple solutions for a specific value of $\alpha$ can be found.

## 2.1 GAUSS–JACOBI QUADRATURE FORMULAE

Consider the family of integrals

$$I_\alpha[f] = \int_{-1}^{1}(1 - x^2)^\alpha f(x)dx \ , \ \alpha > -1. \tag{3}$$

The construction of a quadrature formula of degree 13 for the integral $I_\alpha[f]$ (3) for different values of $\alpha$ using continuation techniques will be discussed in detail. We have chosen this example because most readers of this text are more familiar with one–dimensional integration than with multi–dimensional integration. We know that there are much better ways to solve the problem than the way we will use. The purpose of this section is to show that continuation gives insight in the problem.

The following properties are well known [7]:

G1) If $I$ is a positive functional then the minimum number of points in a quadrature formula of degree $2k - 1$ is $k$. This lower bound is attained by the Gauss quadrature formulae.

G2) All points of a Gauss quadrature formula are inside the region of integration.

G3) All weights of a Gauss quadrature formula are positive.

The Gauss quadrature formula of degree 13 for the integral $I_\alpha[f]$ (3) uses 7 points. The interval is symmetric and the weight function is even. This symmetry is reflected in the quadrature formula. The 7 points can be chosen as

$$-x_3\ ,\ -x_2\ ,\ -x_1\ ,\ 0\ ,\ x_1\ ,\ x_2\ ,\ x_3$$

with the respective weights

$$w_3\ ,\ w_2\ ,\ w_1\ ,\ w_0\ ,\ w_1\ ,\ w_2\ ,\ w_3.$$

Demanding that the quadrature formula has degree 13 gives the following system of nonlinear equations.

$$
\begin{aligned}
w_0 + 2w_1 + 2w_2 + 2w_3 &= I_\alpha[1] \\
2w_1x_1^2 + 2w_2x_2^2 + 2w_3x_3^2 &= I_\alpha[x^2] \\
2w_1x_1^4 + 2w_2x_2^4 + 2w_3x_3^4 &= I_\alpha[x^4] \\
2w_1x_1^6 + 2w_2x_2^6 + 2w_3x_3^6 &= I_\alpha[x^6].
\end{aligned}
\tag{4}
$$

Only the constants in these equations depend on the parameter $\alpha$. As starting point we use the Gauss-Legendre formula (i.e. $\alpha = 0$) which satisfies properties (G1), (G2) and (G3). In Figures 1 and 2 is shown that the points and weights change smoothly if $\alpha$ changes. In Figure 1, lines can never intersect: an intersection would correspond to a quadrature formula with 5 points and that is impossible (G1). Notice in Figure 2 that for $\alpha = -0.5$ all weights are equal : this is the Gauss–Chebyshev formula. Notice also that if $\alpha$ approaches $-1$, a weight goes to infinity. This is expected because the integral $I_\alpha[f]$ (3) diverges for $\alpha \leqslant -1$.

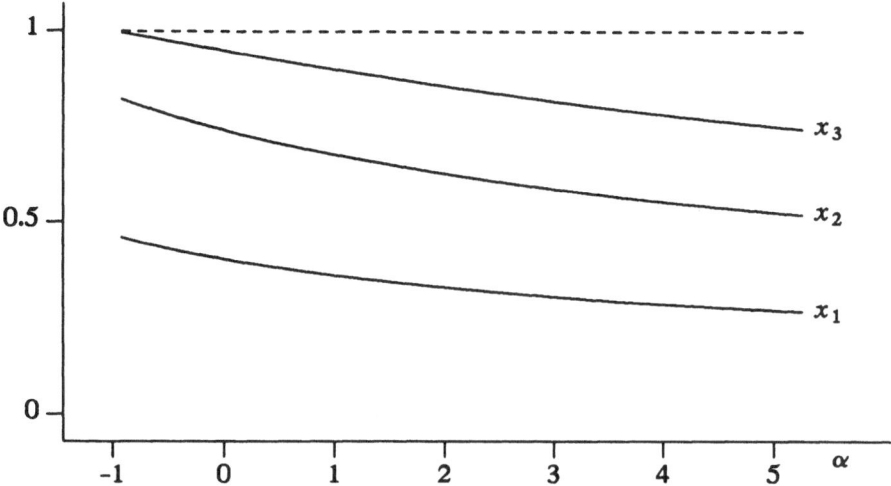

Figure 1. Points of the Gauss-Jacobi quadrature formula

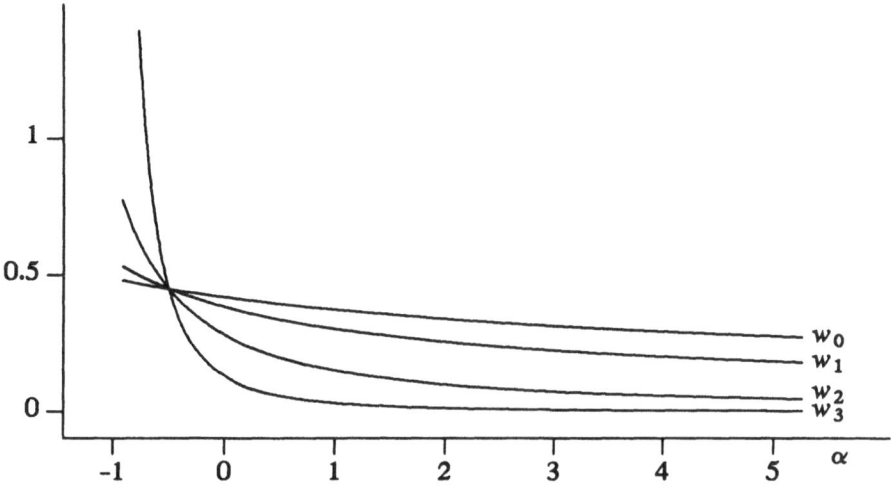

Figure 2. Weights of the Gauss-Jacobi quadrature formula

## 2.2  EXTENDED GAUSS-JACOBI QUADRATURE FORMULAE

In order to approximate an integral together with an error estimate, one frequently uses two quadrature formulae $Q_1$ and $Q_2$. Then $|Q_1[f] - Q_2[f]|$ can be used as an error estimate for the less precise formula, although in practice it is used as an error estimate for the more precise one.

324

In this section we will consider the following problem. Let

$$Q_1[f] = \sum_{i=1}^{n} w_i f(x_i) \tag{5}$$

be an $n$-point Gauss quadrature formula for the integral $I_\alpha[f]$ (3). We are interested in a quadrature formula

$$Q_2[f] = \sum_{i=1}^{n} \bar{w}_i f(x_i) + \sum_{i=n+1}^{m} w_i f(x_i) \tag{6}$$

that re-uses the points $x_1, x_2, \ldots, x_n$ of $Q_1$ and has a higher degree.

The following properties are known [15]:

K1) If an $n$-point Gauss quadrature formula is given, at least $n+1$ new points must be added to obtain a higher degree formula. This formula is called the extended Gauss quadrature formula.

K2) It has been proven that for a Gauss-Jacobi quadrature formula the extended formula exists, has all points inside the interval and all weights positive for

$$\alpha \in [-1/2 , 3/2]. \tag{7}$$

The additional points of (6) alternate with those of (5).

K3) It is also known that for a given degree, there exists a $\alpha^*$ such that the extended quadrature formula is not good for all $\alpha > \alpha^*$.

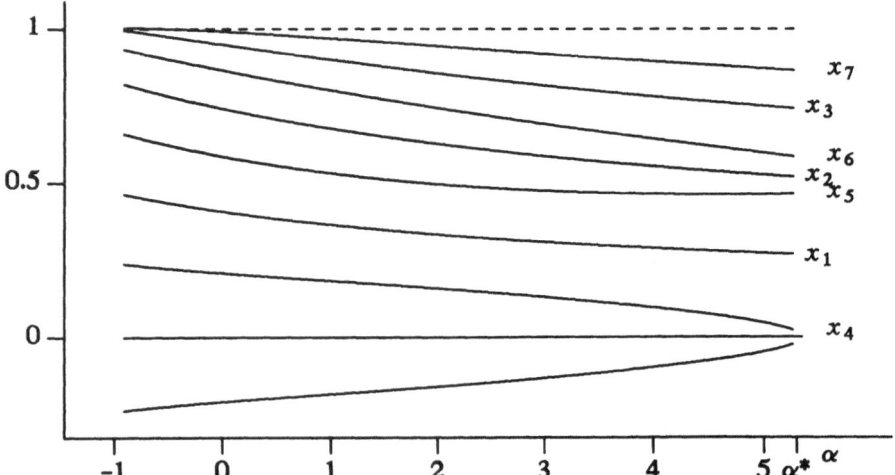

Figure 3. Points of the extended Gauss-Jacobi quadrature formulae.

We will construct quadrature formulae of degree 13 and there extensions for the integral $I_\alpha[f]$ (3) by solving a system of nonlinear equations similar to (4). As starting point for a continuation procedure we use the extended Gauss-Legendre formula with $7 + 8 = 15$ points, first constructed by Kronrod in 1964.

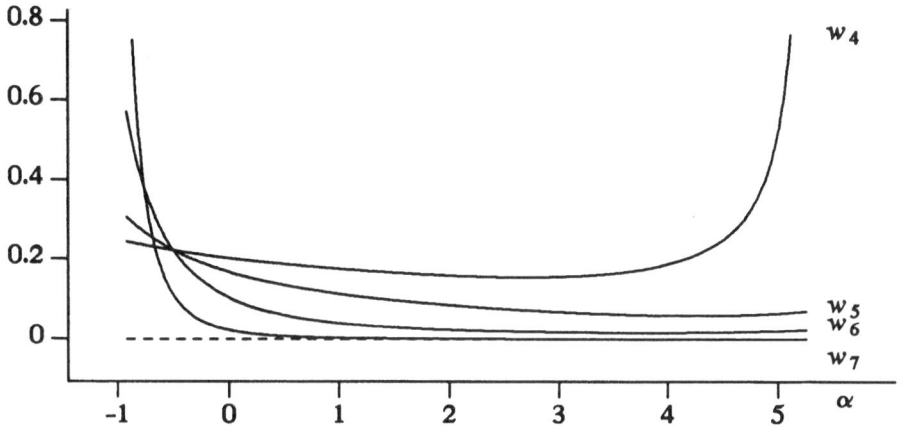

Figure 4. Weights of the new points in the extended quadrature formulae.

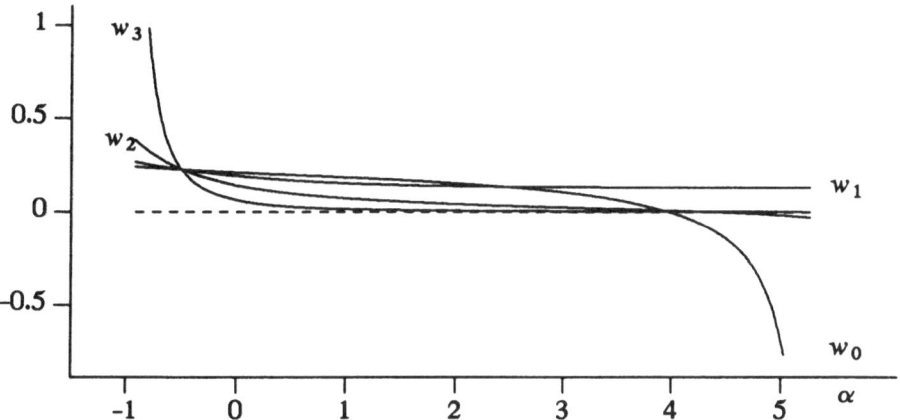

Figure 5. Weights of the fixed points in the extended quadrature formulae.

Figure 3 shonws that the new points lie between 2 given points. This is obvious, otherwise there would be a $\alpha'$ were a new point and a given point coincide which is impossible (K1). For $\alpha < -0.5$ the point $x_7$ is outside the interval. Thus the lower bound in (7) is sharp. For $\alpha = -0.5$ the Lobatto–Chebyshev formula of the first kind of degree 27 is obtained. For $\alpha = 0.5$ the extended Chebyshev formula of the second kind of degree 29 is obtained. The other extended quadrature formulae have degree 23. The upper bound in (7) is not sharp. For some $\alpha^* > 5.2$ the origin is a triple point. For $\alpha > \alpha^*$ the extended quadrature formula does not exist.

The behaviour of the weights is shown in Figures 4 and 5. It is expected that a weight tends to infinity if $\alpha$ approaches $-1$, because the integral $I_\alpha[f]$ (3) diverges for $\alpha \leqslant -1$. It is expected that for $\alpha = -0.5$ all weights are equal except the one that corresponds to the point on the boundary of the interval. For a $\alpha \in [3.968,3.980]$ a first weight becomes negative and for a $\alpha \in [4.393,4.402]$ a second weight becomes negative. Only the weights $\overline{w}_i$ of the fixed points in (6) can become zero or negative. If a weight $w_i$ of a new point could become negative, for some $\alpha'$ it would be zero and that is impossible (K1). If $\alpha$ approaches $\alpha^*$, a weight tends to $+\infty$ and another one to $-\infty$. This is the reason why with the approach we used, $\alpha^*$ cannot be calculated accurately. If the points are computed first (using orthogonal polynomials) and then the weights (by solving a system of linear equations), $\alpha^*$ can be computed as accurately as desired.

## 2.3 A FAMILY OF INTEGRALS OVER THE SQUARE

Consider the family of integrals

$$I_\alpha[f] = \int_{-1}^{1}\int_{-1}^{1} (1-x^2)^\alpha(1-y^2)^\alpha f(x,y)\,dxdy \ , \ \alpha > -1 \ . \tag{8}$$

The largest known lower bound for the number of points $N$ in a cubature formula (1) of degree $2k-1$ for the integral $I_\alpha[f]$ (8) is found by Möller [13]

$$N \geqslant N_{\min} = \frac{k(k+1)}{2} + \left[\frac{k}{2}\right] \quad ([x] \text{ is the integer part of } x) \ .$$

The existence of minimal cubature formulae (1) of degree $\leqslant 9$ for the integral $I_\alpha[f]$ (8) is investigated in detail in [18] for different values of $\alpha$. Until recent, minimal cubature formulae of degree 11 with 24 points, were only known for $\alpha = \pm 0.5$ [17] and $\alpha = 0$ [5]. These 3 formulae have a peculiar fact in common : they are all invariant with respect to the rotations of the square. In [21] we constructed cubature formulae of degree 11 with the same structure for $I_\alpha[f]$, $\alpha \in (-1,0.5]$ using AUTO. Here we only summarize some results.

The cubature formulae have the following form

$$Q[f] = \sum_{i=1}^{6} w_i Q_i[f] \tag{9}$$

where

$$Q_i[f] = f(x_i,y_i) + f(-y_i,x_i) + f(-x_i,-y_i) + f(y_i,-x_i)$$

is the sum of function values for all points of an orbit. The cubature formula is invariant under rotations for which the region is invariant. Demanding that the formula (9) has degree 11 leads to a system of 18 equations in 18 unknowns

$$F(\vec{u},\alpha) = 0 \ , \ \vec{u} = (w_1,x_1,y_1,\ldots,w_6,x_6,y_6). \tag{10}$$

The known cubature formula for $\alpha = 0$ was the only available regular solution, so this was used as starting point for AUTO.

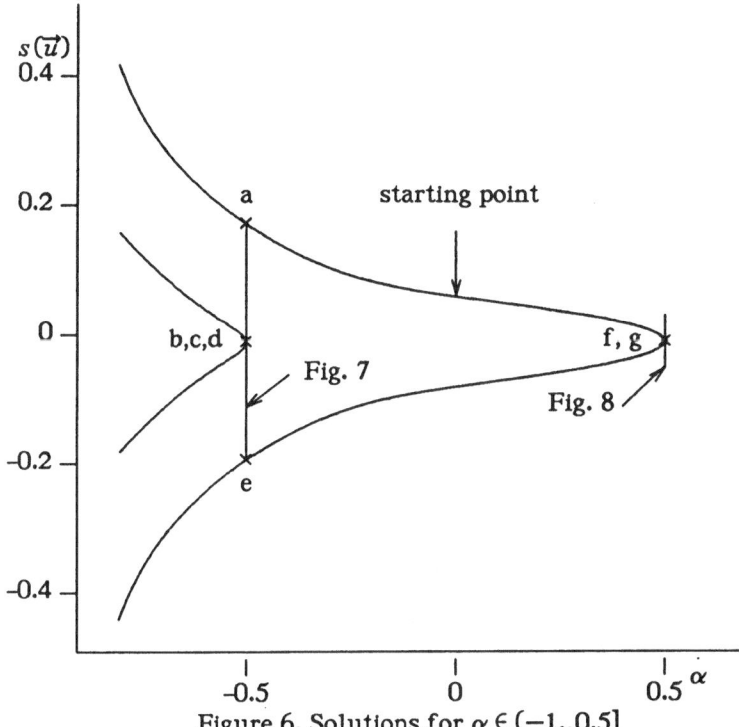

Figure 6. Solutions for $\alpha \in (-1.,0.5]$

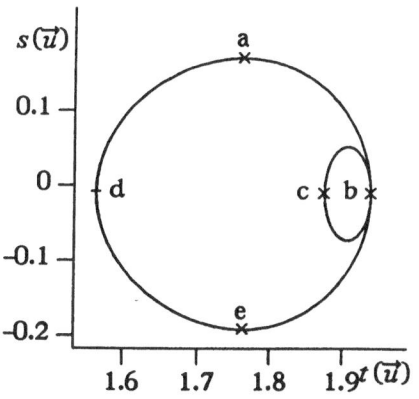

Figure 7. Solutions for $\alpha = -0.5$

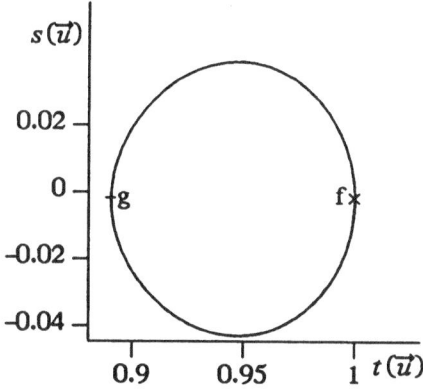

Figure 8. Solutions for $\alpha = 0.5$

Several solution branches are found by AUTO. To obtain a picture of these solution branches we draw for each value of $\alpha$ the value

$$s(\vec{u}) = \sum_{i=1}^{6} x_i y_i (x_i - y_i)(x_i + y_i)$$

for every computed solution $\vec{u}$ of $F(\vec{u}, \alpha) = 0$ (Figure 6). The motivation for this choice is the following :

— For each solution $\vec{u}$ of $F(\vec{u}, \alpha) = 0$, there exist in fact $6! \ 4^6 = 2949120$ equivalent solutions which can be obtained by changing the order of the 6 orbits and by taking another point for representing an orbit (e.g. $(-y_i, x_i)$ instead of $(x_i, y_i)$). All these different solutions yield the same cubature formula and the same value of $s(\vec{u})$.

— Let $g$ be the reflection $g(x, y) = (-x, y)$, $\vec{u}$ a solution yielding the cubature formula $Q[f]$ and $\vec{v}$ a solution yielding the cubature formula $Q[g(f)]$, then $s(\vec{u}) = -s(\vec{v})$. Thus, solutions $\vec{u}$ yielding cubature formulae which are invariant under $g$ (and thus invariant with respect to the symmetry group of the square) satisfy $s(\vec{u}) = 0$. Such formulae are known as "fully symmetric cubature formulae".

In Figure 6 the bifurcation points are marked with a "$\times$". We see that for $\alpha = \pm \ 0.5$ there are an infinite number of cubature formulae.
What happens in these cases, can be seen in Figures 7 and 8.
The abscissa value in Figures 7 and 8, $t(\vec{u}) = \sum_{i=1}^{6} x_i^2 y_i^2$, is also the same for the $6! 4^6$ equivalent solutions of (10).

The fully symmetric cubature formulae are unexpected. They are what Mantel and Rabinowitz [12] call "fortuitous" cubature formulae. Such formulae cannot be found using the first approach mentioned in the introduction!

## 2.4 A FAMILY OF INTEGRALS OVER THE ENTIRE PLANE

Consider the family of integrals

$$I_\alpha[f] = \int_{-\infty}^{\infty} \int_{-\infty}^{\infty} \exp(-(x^2 + y^2)^\alpha) f(x, y) dx dy \ , \quad \alpha > 0. \tag{11}$$

The family $I_\alpha[f]$ (11) contains 2 standard regions : for $\alpha = 0.5$ and $\alpha = 1$. In practice it turns out to be much more difficult to construct formulae for $\alpha = 0.5$ than for $\alpha = 1$:

— Cubature formulae for $\alpha = 0.5$ and $\alpha = 1$ of degree 5 (7 points) and degree 7 (12 points) are known. They are listed in [19].

— For $\alpha = 1$ a symmetric cubature formula of degree 9 with 18 points has been constructed by Haegemans and Piessens [11], but for $\alpha = 0.5$ such a formula has not been found.

— For $\alpha = 1$ a symmetric cubature formula of degree 11 with 25 points has been constructed by Haegemans and Piessens [10], but for $\alpha = 0.5$ such a formula has not been found.

— For $\alpha = 1$ we have constructed a symmetric cubature formula of degree 13 with 34 points [6]. We could not construct cubature formulae for $\alpha = 0.5$ with the same method due to practical problems.

If we change the parameter $\alpha$ continuously from 1 to 0.5, the cubature formula also changes continuously. So, starting from the known formulae for $I_1[f](11)$ we can hope to find cubature formulae for $I_{0.5}[f]$. The only things that can go wrong between $\alpha = 1$ and $\alpha = 0.5$ is that an orbit tends to infinity or becomes complex. To obtain a picture of these solutions, we draw for each value of $\alpha$ the maximum distance between a point and the origin. Figure 9 illustrates that Murphy also has influence in these things ! For degrees 9 and 11, we did not find a cubature formula for $I_{0.5}[f]$ but for degree 13 we obtained a new cubature formula with 34 points for $I_{0.5}[f]$. The formula is listed in [3].

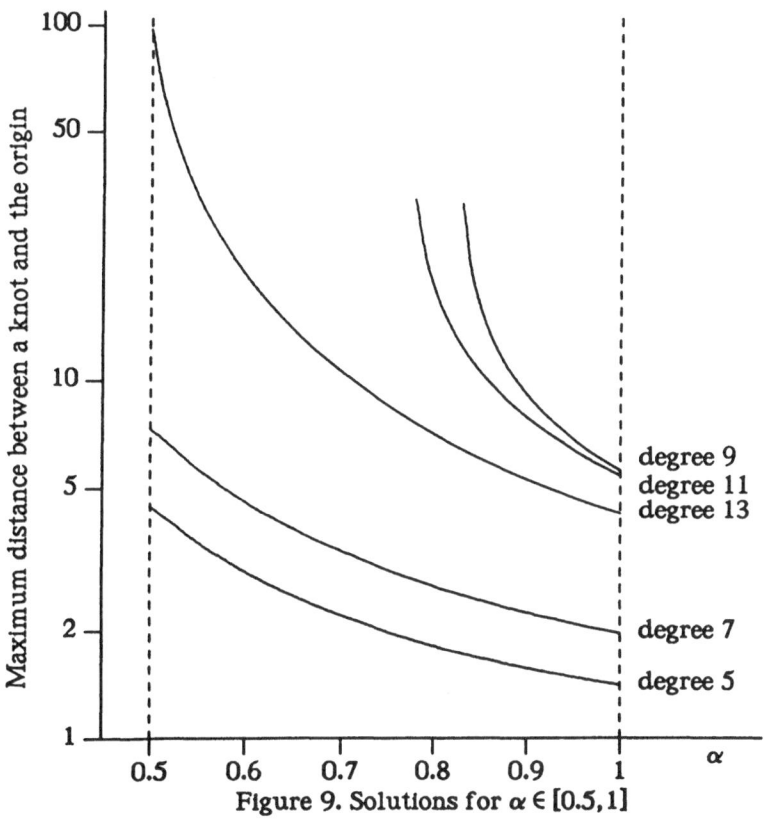

Figure 9. Solutions for $\alpha \in [0.5, 1]$

## 2.5 INTEGRALS OVER FAMILIES OF REGIONS

Consider the family of integrals

$$I_p[f] = \iint\limits_{\Omega} f(x,y)dxdy \tag{12}$$

with $\Omega = \{|x|^{2/p} + |y|^{2/p} \leqslant 1\}$. For $p = 0$, the region $\Omega$ is the square $[-1,+1]^2$. For $p = 1$, the region $\Omega$ is the circle with radius 1. For $p = 2$, the region $\Omega$ is again a square.

Using the family of integrals $I_p[f]$ (12) and a continuation program, cubature formulae for a circular region can be constructed if a cubature formula for a square region is known. Experiments has showed that classical formulae [19] for these 2 regions belong to the same solution branch. Of course this will not always work. For some $p \in (0,1)$ a turning point can occur or a point can tend to infinity.

## 3. Continua of cubature formulae

Consider the integral

$$I[f] = \int\limits_0^1 \int\limits_0^{1-x} f(x,y)dydx \tag{13}$$

Cubature formulae of degree 13 for this integral were first constructed by Dunavant [9]. He obtained a cubature formula with 37 points that has the same symmetry as the region : if $(x,y)$ is a point of the cubature formula then also $(x, 1 - x - y)$, $(y,x)$, $(y, 1 - x - y)$, $(1 - x - y,x)$ and $(1 - x - y,y)$ are points of the cubature formula.

The system of equations that determines such a cubature formula consists of 21 equations in 22 unknown coordinates and weights. So, one of the coordinates or weights can be seen as a parameter and an infinite number of solutions is expected. Dunavant only mentions one.

We chose the weight of the centroid $w_0$ as a free parameter. Starting from the known solution, we can investigate the continuum of cubature formulae. Figure 10 shows the result we obtained using AUTO. The starting point is marked with a "×".

It is interesting to look for a solution where $w_0 = 0$. Then a cubature formulae with 36 points is obtained. Initially $w_0 = 0.026260$. It can be decreased until $w_0 = 0.025808$ where a turning point occurs and $w_0$ increases. After a second turning point $w_0$ again decreases. One of the points is then already far out of the region of integration, which makes the result uninteresting.

Recently Berntsen and Espelid [1], at that time unaware of Dunavant's results, also constructed cubature formulae of degree 13 for the integral $I[f]$ (13). The 37-point formulae they obtained are marked with a "●" in Figure 10. They also computed the 36-point formula.

In [6] we obtained an infinite number of cubature formulae of degree 13 with 37 points for the integral over the square ($\alpha = 0$ in (8)), the circle ($p = 1$ in (11)) and the entire plane ($\alpha = 1$ in (10)). The weight of the origin $w_0$ was considered as a free parameter and the place in the continuum where $w_0 = 0$ was searched and found.

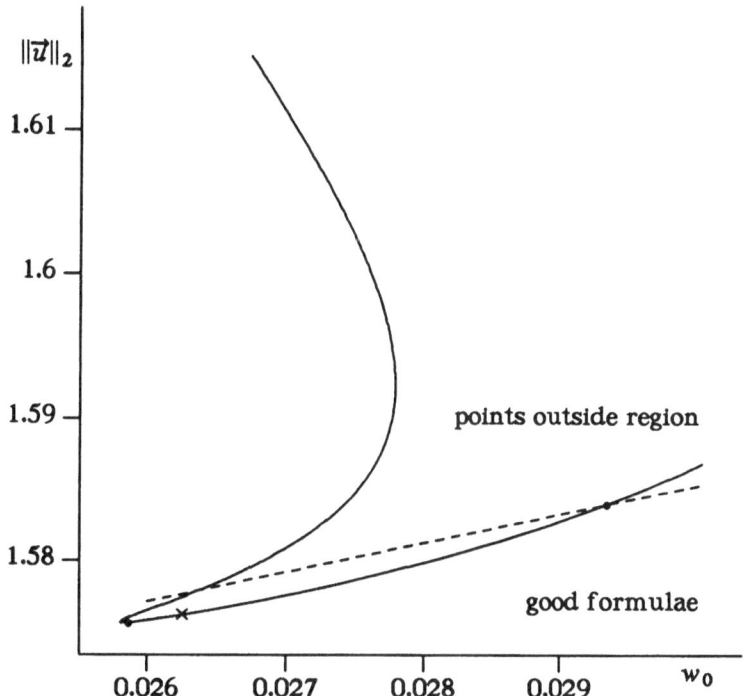

Figure 10. Continuum of cubature formulae for the triangle

## 4. Conclusion

In the previous sections is illustrated that continuation can be used to construct quadrature and cubature formulae. If a parameter dependent integral is given and for one value of the parameter a quadrature or cubature formula is constructed by hard labour, then formulae can be constructed for other values of the parameter with the aid of continuation software, i.e. without labour. Continuation is also the obvious technique to investigate continua of cubature formulae for a given integral.

We believe that continuation and bifurcation software like AUTO is a useful tool for anyone interested in quadrature or cubature formulae. Of course one should not forget that each tool is designed for a certain task and therefore has its limitations.

332

## References

[1] J. Berntsen and T. Espelid (1989) "Degree 13 symmetric quadrature rules for the triangle", *in preparation*.

[2] B. Buchberger (1985) "Gröbner bases: An algorithmic method in polynomial ideal theory", pp. 184-232 in *Multidimensional Systems Theory*, ed. N.K. Bose, D. Reidel Publ. Comp.

[3] R. Cools (1989) *The construction of cubature formulae using invariant theory and ideal theory*, Ph.D. thesis, Katholieke Universiteit Leuven .

[4] R. Cools and A. Haegemans (1987) "Construction of fully symmetric cubature formulae of degree $4k - 3$ for fully symmetric planar regions", *J. Comput. Appl. Math.*, Vol. 17, pp.173-180.

[5] R. Cools and A. Haegemans (1988) "Another step forward in searching for cubature formulae with a minimal number of knots for the square", *Computing*, Vol. 40, pp. 139-146.

[6] R. Cools and A. Haegemans (1988) "Construction of symmetric cubature formulae with the number of knots (almost) equal to Möller's lower bound", pp. 25-36 in *Numerical Integration III*, eds. H. Brass and G. Hämmerlin, Birkhäuser Verlag, Basel.

[7] P. Davis and P. Rabinowitz (1984) *Methods of numerical integration*, Academic Press, London.

[8] E. Doedel (1981) "Auto: A program for the automatic bifurcation analysis of autonomous systems", pp. 265-284 in Proc. 10th Manitoba Conf. on Num. Math. and Comp., Univ. of Manitoba, Winnipeg, Canada, 1980 Cong. Num. Vol. 30.

[9] D. A. Dunavant (1985) "High degree efficient symmetrical Gaussian quadrature rules for the triangle", *Internat. J. Numer. Methods Engrg.* , Vol. 21, pp 1129-1148.

[10] A. Haegemans and R. Piessens (1976) "Construction of cubature formulas of degree eleven for symmetric planar regions, using orthogonal polynomials", *Numer. Math.*, Vol. 25, pp. 139-148.

[11] A. Haegemans and R. Piessens (1977) "Construction of cubature formulas of degree seven and nine for symmetric planar regions, using orthogonal polynomials", *SIAM J. Numer. Anal.*, Vol. 14, pp. 492-508.

[12] F. Mantel and P. Rabinowitz (1977) "The application of integer programming to the computation of fully symmetric integration formulas in two and three dimensions", *SIAM J. Numer. Anal.*, Vol. 14, pp. 391-425.

[13] H.M. Möller (1976) "Kubaturformeln mit minimaler Knotenzahl", *Numer. Math.*, Vol. 25, pp. 185-200.

[14] H.M. Möller (1987) "On the construction of cubature formulae with few nodes using Gröbner bases", pp. 177-192 in *Numerical Integration*, eds. P. Keast and G. Fairweather, Reidel Publ. Comp.

[15] G. Monegato (1979) "An overview of results and questions related to Kronrod schemes", pp. 231-240 in *Numerische Integration*, ed. G. Hammerlin, Birkhäuser Verlag, Basel.

[16] A. Morgan (1987) *Solving polynomial systems using continuation for engineering and scientific problems*, Englewood Cliffs, N.J., Prentice Hall.

[17] C.R. Morrow and T.N.L. Patterson (1978) "Construction of algebraic cubature rules using polynomial ideal theory", *SIAM J. Numer. Anal.*, Vol. 15, pp. 953-976.

[18] H.J. Schmid (1983) "Interpolatorische Kubaturformeln" *Dissertationes Math.*, Vol. CCXX.

[19] A.H. Stroud (1971) *Approximate calculation of multiple integrals*, Englewood Cliffs, N.J., Prentice Hall.

[20] B.L. Van der Waerden (1950) *Modern algebra II*, Frederick Ungar Publishing Co., USA.

[21] P. Verlinden, R. Cools, D. Roose and A. Haegemans (1988) "The construction of cubature formulae for a family of integrals: a bifurcation problem", *Computing*, Vol. 40, pp. 337-346.

# DETERMINING AN ORGANIZING CENTER FOR PASSIVE OPTICAL SYSTEMS

G. DANGELMAYR and M. WEGELIN
*Institute for Information Sciences, University of Tübingen,*
*Köstlinstr. 6, D-7400 Tübingen, FR Germany*

ABSTRACT. The Maxwell-Bloch equations underlying passive optical systems are considered. Using MAPLE we determine a highly degenerate singularity which corresponds to the coalescence of a cusp and a degenerate Hopf bifurcation with broken transversality condition. The dynamics associated with the unfolding of this singularity gives rise to a rich variety of different behaviour including, for example, two types of infinite period bifurcations on tori.

## 1 Introduction

Passive optical systems have recently attracted increasing interest not only because of their potential applications as optical switching elements but also because the underlying model equations exhibit interesting dynamics [9,10]. Recent analytical and numerical work [8,10,11,12,13] reveals a variety of different kinds of behaviours occuring in different parameter regimes, including multistability, temporal oscillations and chaos. The most significant features are those of optical bistability and self-pulsing. They are, using the language of bifurcation theory, created through a cusp and a Hopf bifurcation, respectively. Attempting to gain some insight into the possible interactions between these two modes of behaviours, Armbruster [1] conjectured the presence of a highly degenerate singularity in the Maxwell-Bloch equations underlying passive optical systems. This singularity has codimension four and describes the coalescence of a cusp (or hysteresis) and a degenerate Hopf bifurcation. By using a classification of interacting Hopf and steady state bifurcations [2,3] Armbruster [1] analyzed the structurally stable bifurcation diagrams resulting from an unfolding of the singularity. His analysis was performed entirely in the context of imperfect bifurcation theory [6], that is, he confined the discussion to the branches of stationary and periodic solutions. This approach provides some insight into the interrelations between these branches when a distinguished bifurcation parameter is varied, but misses many of the interesting dynamical features organized by the codimension four singularity.

Although the conjecture of Armbruster was supported by numerical calculations of Lugiato et al. [9], neither a proof showing the existence of the singularity, nor a numerical calculation of the parameter values where it occurs have been presented yet. In this paper we show how the conditions for the singularity can be reduced analytically to a single polynomial equation of very high degree for one of the parameters. The calculations are performed by making extensive use of computer algebra (MAPLE). Solving this equation

*D. Roose et al. (eds.), Continuation and Bifurcations: Numerical Techniques and Applications,* 335–347.
© 1990 *Kluwer Academic Publishers.*

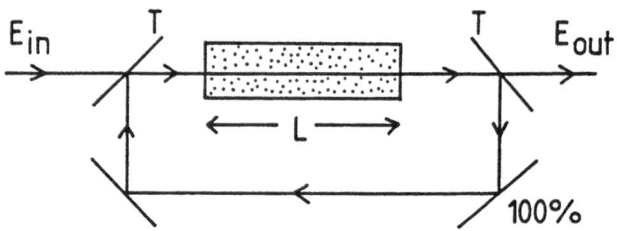

Figure 1: The ring cavity.

numerically then leads to the desired parameter values for which the singularity occurs. We begin, in Section 2, with a description of the physical configuration and introduce the Maxwell-Bloch equations. The necessary analytical and numerical calculations which determine the locus of the singularity described before are explained in Section 3. In the final Section 4 we consider the unfolded normal form corresponding to this singularity and present a two-dimensional section through the stability diagram, together with the resulting structurally stable phase portraits.

## 2  The Maxwell-Bloch equations

We consider the standard arrangement of a passive optical system shown in Figure 1. A coherent laser beam with frequency $\omega_{in}$ is injected into a ring cavity composed of two mirrors of transmittivity $T$, two perfect mirrors, and a homogeneously broadened two-level system with an usaturated absorption coefficient $\alpha$ which is distributed over a length $L$. In the mean field limit, defined by $\alpha L \to 0$, $T \to 0$ with $\alpha L/T \to$ const, the Maxwell-Bloch equations for the configuration of Figure 1 reduce to a system of ordinary differential equations [9,10] which, after a proper rescaling of the variables, can be written in the form

$$\dot{a} = -\rho_1\left[(1 + i\theta)a - A\right] - \sigma p \tag{1a}$$

$$\dot{p} = a(1 - d) - (1 + i\delta)p \tag{1b}$$

$$\dot{d} = \rho_2\left[\frac{1}{2}(a\bar{p} + \bar{a}p) - d\right], \tag{1c}$$

where the bar in (1c) denotes complex conjugation. Here, $a$, $p$ and $d$ are proportional to slowly varying envelopes of the outgoing electric field ($E_{out}$), the macroscopic polarization and the atomic inversion. The parameters $\rho_1$, $\rho_2$ are ratios of relaxation constants and $\sigma$ is a field-matter coupling constant which contains the atomic dipole moment as a factor. The quantities $\theta$ and $\delta$ are proportional to the atomic and cavity detuning, i. e., $\theta \sim \omega_c - \omega_{in}$ and $\delta \sim \omega_a - \omega_{in}$ where $\omega_c$ and $\omega_a$ are the cavity resonance and the atomic transition frequencies, respectively. In a typical experiment $\rho_1, \rho_2, \sigma, \theta, \delta$ are fixed whereas $A$, which is proportional to the amplitude of the incoming field ($E_{in}$), is varied. Thus one can consider $A$ as a distinguished bifurcation parameter. The equations (1) correspond to a single mode model in which only one cavity mode is taken into account. Since $a$ and $p$ are complex

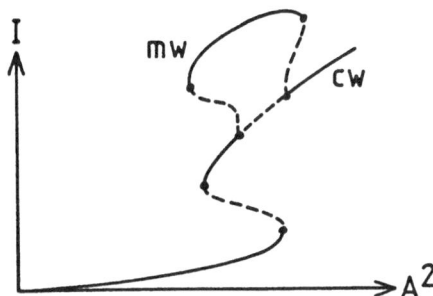

Figure 2: A typical bifurcation diagram showing optical bistability and self pulsing. Dashed branches are unstable.

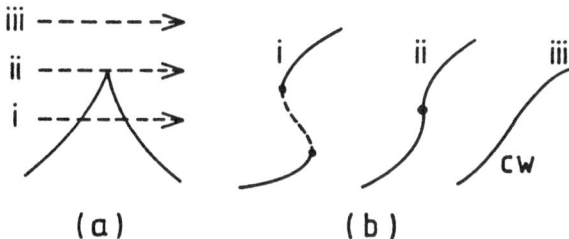

Figure 3: (a) A cusp in a two-dimensional section through the parameter space. The dashed paths i, ii, iii correspond to variations of $A^2$ for different fixed values of the remaining parameters. (b) Bifurcation diagrams associated with the paths of (a) in the $(I, A^2)$-plane.

and $d$ is real, (1) constitutes a five dimensional system of differential equations. We assume that $A \epsilon \mathbf{R}$, thus all parameters occuring in (1) are real.

A typical bifurcation diagram that can be extracted from [9] (see also [1]) is shown in Figure 2. In this diagram the intensity $I \equiv |a|^2$ of the outgoing field is plotted versus the intensity $A^2$ of the incoming field. The S-shaped branch cw is a branch of steady states (continuous wave transmisson) showing bistable behaviour. In the physical literature the upper and lower parts of the cw branch are denoted as high and low transmission branches, respectively. The bounded branch mw is a branch of periodic orbits which is created and annihilated in two Hopf bifurcations on the cw branch. It corresponds to a transmission in the form of an undamped sequence of pulses (modulated wave transmission), the so-called self pulsing state. Observe that a pair of periodic orbits exists, one stable and the other unstable.

The bifurcation diagram of Figure 1 can undergo two types of degeneracies. First there is the possibility of a cusp, that is, in a two-dimensional section through the parameter space we find the cusped curve shown in Figure 3(a). Inside this curve there exist three steady states, on the curve two of them coalesce and outside of it only one steady state

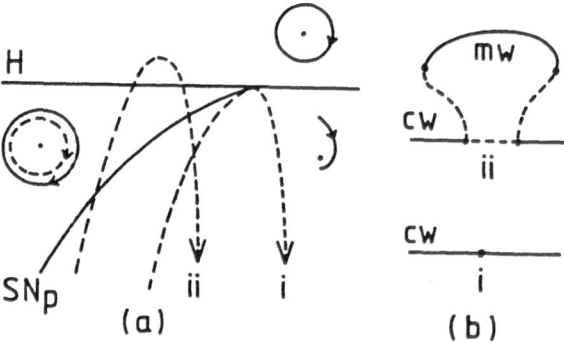

Figure 4: A two dimensional section through the parameter space containing a degenerate Hopf bifurcation point. Full (dashed) periodic orbits are stable (unstable). The paths i, ii correspond to variations of $A^2$. They induce the bifurcation diagrams as shown in (b).

survives. The bifurcation diagrams in Figure 3(b) demonstrate how the transition from bistability to montonic behaviour is organized by the cusp.

The second type of degeneracy is shown in Figure 4. In a two-dimensional section through the parameter space we encounter a line of saddle nodes for periodic orbits ($SN_p$) which terminates with a second order contact on a line of Hopf bifurcations ($H$). This leads to three open regions in the parameter space giving rise to two, one and no periodic orbit, respectively, as indicated in Figure 4(a). The point of contact of the lines $H$ and $SN_p$ is a degenerate (codimension two) Hopf bifurcation point. A path, obtained by variations of $A^2$, which is forced to hit the codimension two point will generically cross the $H$-line transversally. However, when a further parameter is varied, we may achieve that this path becomes tangent to the $H$-line like path i in Figure 4(a). This implies that the transversality condition for the degenerate Hopf bifurcation with respect to $A^2$ is broken, i. e., we encounter a further degeneracy. In the classification of degenerate Hopf bifurcations with a distinguished parameter [5] such a situation is referred to as $H(7)$. A slight perturbation of path i then produces a generic path like path ii which induces the local bifurcation diagram shown in Figure 4(b) (cf. also Figure 2). Further structurally stable bifurcation diagrams organized by $H(7)$ that correspond to other generic paths are presented in [5,6].

The occurence of $H(7)$, i. e., a degenerate Hopf bifurcation with a broken transversality condition was infered by Armbruster [1] from numerical results of Lugiato et al. [9]. The presence of a cusp can easily be shown analytically. In the next section we determine a point in the parameter space of the Maxwell-Bloch equations where the cusp and the $H(7)$-singularities coalesce.

# 3   Determining the codimension four bifurcation

Besides the conditions for a steady state the coalescence of $H(7)$ and a cusp requires five additional conditions which have to be imposed on the six parameters $(A, \rho_1, \rho_2, \sigma, \delta, \theta)$. One of these is for the Hopf bifurcation, two are for the cusp, and the remaining two produce the degeneracies associated with the Hopf bifurcation, namely, a broken transversality condition and a degenerate Hopf bifurcation that creates two periodic orbits nearby. Since we wish to determine isolated numerical values for the parameters, we have to impose one further condition. We do this by making the additional assumption that the detuning parameters are opposite, that is, we set

$$\theta = -\delta. \tag{2}$$

This restriction has the advantage that the calculations with (1) are greatly simplified. Analogeous calculations with the full set of parameters, i. e., determining a curve in the six-dimensional parameter space where the five conditions described before are satisfied, are still an open task.

Setting the time derivatives of (1) equal to zero and using the restriction (2) we obtain the following parametric representation for the steady states,

$$A = (1 - i\delta)(1 + 2C/K)a \tag{3a}$$

$$p = (1 - i\delta)a/K, \ d = I/K, \tag{3b}$$

where we have introduced the abbreviations

$$I \equiv |a|^2, \ K \equiv 1 + \delta^2 + I, \ C \equiv \sigma/2\rho_1.$$

Since $A$ is real, the phase $\psi$ of $a$, $a = \sqrt{I}e^{i\psi}$, satisfies the condition $\tan \psi = \delta$. Thus we regard $\psi$ as a function of $\delta$, but consider $|a|$, or the intensity $I$, as a free parameter that parametrizes the stationary states of (1). The parametric representation for the intensity $A^2$ of the incoming field is given by

$$A^2 = yI(1 + 2C/K)^2, \tag{3c}$$

where we have set

$$y \equiv 1 + \delta^2.$$

The conditions for a cusp are easily obtained from (3c) by setting the first and second derivatives of $A^2$ with respect to $I$ equal to zero. The resulting equations lead to simple expressions for the steady states, viz.,

$$p = \sqrt{3}/4, \ d = 3/4, \ I = 3y, \tag{4a}$$

and allow us to eliminate the parameters $A$ and $\sigma$ via

$$A^2 = 27y^2, \ \sigma = 8\rho_1 y. \tag{4b}$$

Observe that $p$ and $d$ attain numerical values whereas $I$, the phase $\psi$ of $a$ and $A$ vary with $\delta$. We note that such simple expressions like those occuring in (4) cannot be derived without the restriction (2). Although in the general case the conditions for a cusp also allow to express $A^2$ and $I$ in terms of $(\rho_1, \sigma, \theta, \delta)$, none of these parameters can be eliminated analytically.

We consider now the conditions for a Hopf bifurcation and for the broken transversality condition. Linearizing (1) around a steady state, parametrized by equations (3), and using the second relation of (4b) leads to the following characteristic polynomial for an eigenvalue $e$,

$$e^5 + \sum_{j=0}^{4} M_j(I; \rho_1, \rho_2, \delta) e^j = 0. \tag{5}$$

The coefficients $M_j$ are listed in the Appendix, equation (A.1). The condition for a Hopf bifurcation is that the real part $r(I)$ of an eigenvalue $e(I) = r(I) + i\omega(I)$ vanishes. For our purpose it is sufficient to evaluate this condition at the cusp point, thus we insert $I = 3y$ into the $M_j$ which leads to the expressions (A.2) and to $M_0 = 0$. Then, when $e = i\omega$ is substituted into (5), we find $\omega^2 = M_2/M_4$ and obtain a single equation for the parameters,

$$M_1 M_4^2 - M_2 M_3 M_4 + M_2^2 = 0. \tag{6}$$

If the $M_j$ are taken from (A.2), the equation (6) takes the form

$$q_0 + q_1 y + q_2 y^2 = 0, \tag{7}$$

where $q_0$, $q_1$, $q_2$ are polynomials in $(\rho_1, \rho_2)$ which are summarized in the Appendix, equation (A.4). The transversality condition for the Hopf bifurcation with respect to $A$ is broken if $\partial r(I)/\partial I = 0$. If we set $e = r + i\omega$ in (5) and then take the derivative of this equation with respect to $I$, we obtain four real equations determining $r$, $r'$ and $\omega$, $\omega'$ in terms of the $M_j$ and $M_j'$, where the prime denotes the $I$-derivative. The condition $r' = 0$ then leads to

$$(M_3 M_4 - 2M_2) M_2' + M_2 M_4 M_3' - M_4^2 M_1' = 0, \tag{8}$$

where we have made use of the fact that $M_0 = M_0' = M_4' = 0$. If now the $M_j$, $M_j'$ are taken from (A.2) and (A.3), (8) reduces to a linear equation for $y$,

$$l_0 + l_1 y = 0, \tag{9}$$

with certain polynomials $l_j(\rho_1, \rho_2)$ $(j = 0, 1)$ which are presented in equation (A.5). Eliminating $y$ from (7) and (9) leads to a single polynomial equation for $(\rho_1, \rho_2)$, i. e.,

$$P_1(\rho_1, \rho_2) = 0. \tag{10}$$

The polynomial $P_1$ is of degree 16 and will not be written down explicitly.

The final task in the determination of the codimension four bifurcation is to derive a further equation from the condition that the Hopf bifurcation be degenerate. This is a very tedious calculation, thus we will describe only the main steps. We collect the variables in (1) in a column vector $Z := (a, p, \bar{a}, \bar{p}, d)^T$ and denote by $Z_0$ the corresponding vector at a simultaneous Hopf and cusp bifurcation. Then we perform a linear transformation, $Z' = U(Z - Z_0) \equiv (z, \bar{z}, x, u, \bar{u})^T$ with an appropriate matrix $U$, such that the linear part of the $Z'$-system is in normal form, i. e.,

$$\dot{Z}' = L'Z' + \text{nonlinear terms}, \tag{11}$$

with

$$L' = \begin{pmatrix} i\omega & 0 & 0 & 0 & 0 \\ 0 & -i\omega & 0 & 0 & 0 \\ 0 & 0 & 0 & 0 & 0 \\ 0 & 0 & 0 & c_1 & c_2 \\ 0 & 0 & 0 & c_3 & c_4 \end{pmatrix}. \tag{12}$$

Here the $2 \times 2$-block with the entries $c_j$ corresponds to the stable manifold at the bifurcation point. The first component of (11) has the form

$$\dot{z} = i\omega z + Bxz + \text{other terms}. \tag{13}$$

The condition that the Hopf bifurcation be degenerate can now be formulated as

$$\text{Re } B = 0, \tag{14}$$

i. e., it involves the nonlinear terms. By using (4) and (9) to eliminate all parameters except $(\rho_1, \rho_2)$ we obtain from (14) a second polynomial equation of the form

$$P_2(\rho_1, \rho_2) = 0, \tag{15}$$

where $P_2$ is a polynomial of degree $\approx 50$.

We have derived two polynomial equations, (10) and (11), for two parameters $(\rho_1, \rho_2)$. To solve these we first compute the resultant of $P_1$, $P_2$ with respect to $\rho_1$. This leads to a single equation,

$$P(\rho_2) := \text{resultant}\,[P_1, P_2] = 0. \tag{16}$$

The degree of the polynomial $P$ is very high ($\approx 250$), but MAPLE was able to factorize it into polynomials of degree less than or equal to 27. We emphasize that the condition (16) for our codimension four bifurcation is exact. At this stage, however, we have to resort to a numerical calculation because the zeroes of $P$ cannot be determined analytically. It turns out that among all zeroes of $P$ there is only one with physically meaningful values. It is given by $\rho_2 = 1.02$ from which we can compute all other quantities of interest,

$$(\rho_1, \rho_2, \sigma, \delta) = (1.57, 1.02, 3.98, 212.20)$$
$$|a| = 7.12, \quad A = 87.69, \quad \omega = 10.15. \tag{17}$$

This completes the determination of the codimension four point.

## 4 The normal form and its dynamics

In the preceeding section we have determined an isolated point in the space of the parameters $(\rho_1, \rho_2, \sigma, \delta, A)$ where the Maxwell-Bloch equations (1) encounter a codimension four bifurcation in which a cusp and a degenerate Hopf bifurcation occur simultaneously. Because the linearized system has a three-dimensional center eigenspace at this point, the dynamics of (1) can be reduced, by means of a center manifold reduction, to a system of ordinary differential equations for three variables $(u_1, u_2, u_3)$. A near-identity transformation brings this system to the Poincare-Birkhoff normal form, which, using polar coordinates, $u_1 + iu_2 = re^{i\phi}$, and setting $x = u_3$, has the form [7],

$$\dot{r} = rg_1(r^2, x) \tag{18a}$$
$$\dot{x} = g_2(r^2, x) \tag{18b}$$
$$\dot{\phi} = \omega + g_3(r^2, x), \tag{18c}$$

where the $g_i$ are smooth functions that vanish at the origin and satisfy

$$g_{2,x}(0,0) = g_{2,xx}(0,0) = g_{1,x}(0,0) = 0. \tag{19}$$

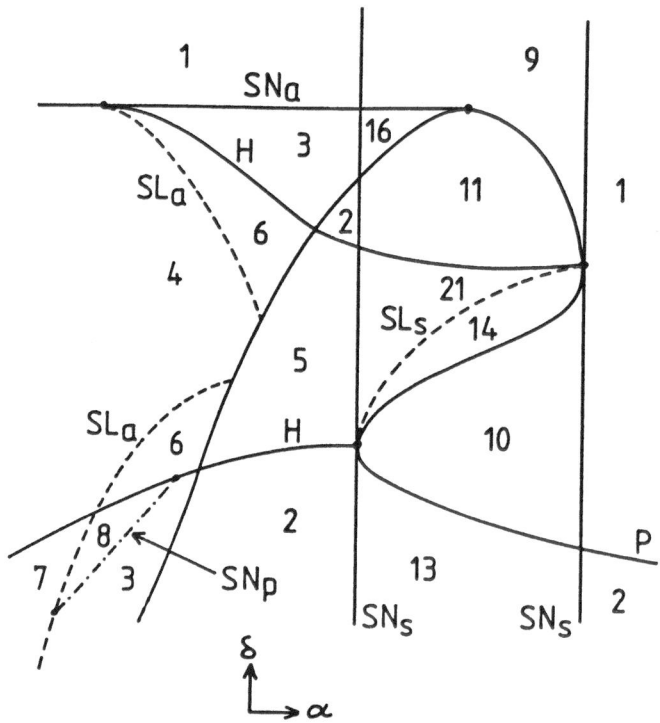

Figure 5: A $(\alpha, \delta)$-section throught the stability diagram of the unfolding (21). Solid lines are local and dashed or dashed-dotted lines are global bifurcations.

The normal form (18) has the $S_1$-symmetry $\phi \to \phi + \phi_0$ that is related to temporal translations along a periodic orbit. The first two and the third equation in (19) reflect the presence of a cusp and the degeneracy in the Hopf bifurcation, respectively. Further details of the center manifold reduction and the transformation to normal form will be discussed elsewhere.

Because the phase $\phi$ decouples from $(r, x)$ in virtue of the $S_1$-symmetry, it is sufficient to consider the closed two-dimensional system (18a,b) for $(r, x)$. In [4] we have shown that this system can be transformed further into the form

$$\dot{r} = -cr(x^2 + O(5)) \tag{20a}$$
$$\dot{x} = -(x^3 + r^2 - xr^2 + O(5)), \tag{20b}$$

where $O(5)$ denotes terms of order higher than or equal to 5 in $(r, x)$ and $c = 6.9$. The system (20) is valid exactly at the codimension four point, i. e., for the values of $(\rho_1, \rho_2, \sigma, \delta, A)$ summarized in (17). Varying the parameters in a neighborhood of this point induces a structurally stable unfolding of (20), given by

$$\dot{r} = -cr(x^2 + \gamma x + \delta) \tag{21a}$$

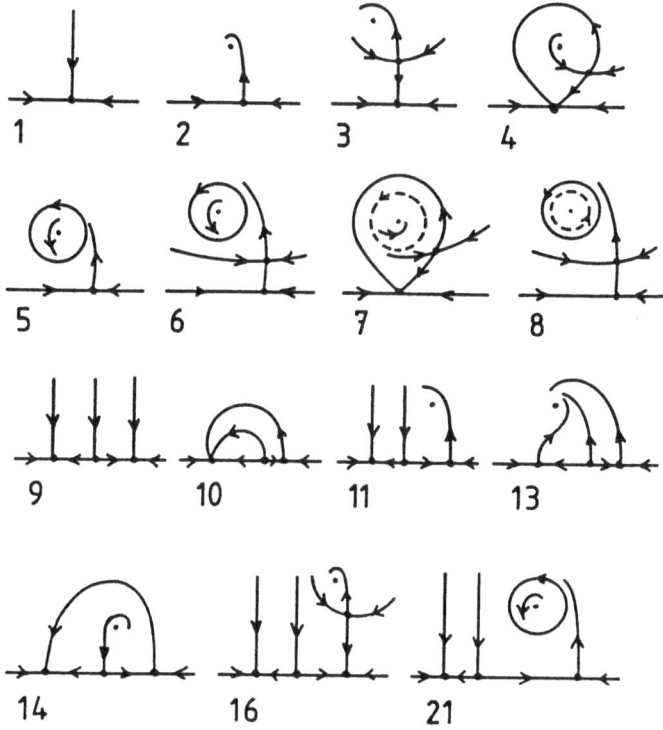

Figure 6: Phase portraits corresponding to the regions marked in Figure 5.

$$\dot{x} = -(x^3 + r^2 - xr^2 + \beta x + \alpha), \tag{21b}$$

where now the $O(5)$-terms have been neglected. Here, $\alpha, \beta, \gamma, \delta \epsilon \mathbf{R}$ are the unfolding pa-rameters. They can be related to the physical parameters through a certain mapping which still has to be determined, however, this computation is not yet completed. When supple-mented by the phase variation (18c), all phenomena exhibited by the unfolding (21) must also occur in the Maxwell-Bloch equations (1). It is therefore necessary to analyze the dynamics associated with the unfolded normal form (21).

In Figure 5 we have sketched a two-dimensional section through the stability diagram of the unfolding (21) that corresponds to fixed values of $(\beta, \gamma)$. The $(\alpha, \delta)$-plane is divided by a number of lines into various regions 1, 2, 3 etc. For fixed $(\alpha, \delta)$ in each of these regions a certain structurally stable phase portrait occurs in the $(r, x)$-plane. The phase portraits are summarized in Figure 6 where we have confined ourselves to the $(r \geq 0, x)$-half plane in virtue of the reflection symmetry $r \to -r$. The unfolding (21) organizes 22 different structurally stable phase portraits. Among these only the 15 portraits associated to the regions of Figure 5 are shown in Figure 6. Varying $(\beta, \gamma)$ induces changes in the $(\alpha, \delta)$-sections so that new regions are produced giving rise to the remaining phase portraits. A more comprehensive discussion of the stability diagram is in preparation. A description of

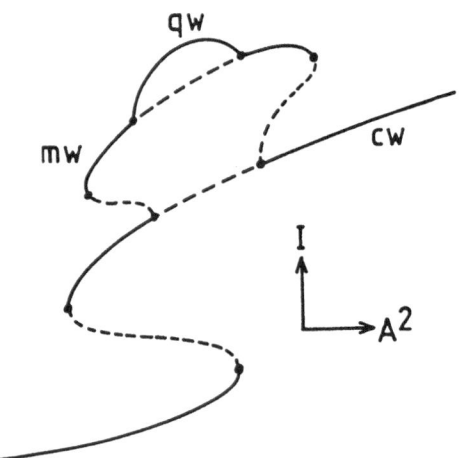

Figure 7: A bifurcation diagram derived from a path through the stability diagram of Figure 5 that traverses the regions 1, 9, 16, 11, 21, 5, 6, 3, 1 in succession.

all bifurcations of codimension smaller than four that occur as subsidiary bifurcations in the unfolding (21) can be found in [4].

On the lines in Figure 5 certain bifurcations of codimension one occur. The solid lines denote local bifurcations: on $P$ (pitchfork) an equilibrium of (21) with $r \neq 0$ bifurcates from an equilibrium with $r = 0$, on $SN_s$ ($SN_a$) two equilibria with $r = 0$ ($r \neq 0$) undergo a saddle node bifurcation, and $H$ is a line of Hopf bifurcations where a limit cycle is created from an equilibrium with $r \neq 0$. The dashed and dashed-dotted lines give rise to global bifurcations: on $SL_s$ a heteroclinic orbit occurs that connects two equilibria with $r = 0$, $SL_a$ is a line of homoclinic orbits for a saddle point with $r \neq 0$, and $SN_p$ denotes a saddle node bifurcation for limit cycles. So far we have considered the two-dimensional system (21). The dynamics in the full three-dimensional center manifold is obtained by supplementing (21) by the phase variation (18c). This means that the phase portraits of Figure 6 are to be rotated about the $x$-axis so that equilibria with $r \neq 0$ evolve into periodic orbits and limit cycles evolve into invariant tori. Thus, for the 3-d system, the pitchfork line $P$ is actually a line of Hopf bifurcations and the Hopf line $H$ is a line of torus bifurcations. Analogeous reinterpretations also hold for the other curves in Figure 5.

Of particular physical interest are the bifurcation diagrams which occur if only the distinguished parameter $A$ is varied for fixed values of the remaining parameters $(\rho_1, \rho_2, \sigma, \delta)$. To preserve the role of a distinguished bifurcation parameter [6] we write

$$(\alpha, \beta, \gamma, \delta) = (\alpha' - \lambda, \beta', \gamma', \delta' - \lambda^2),$$

where $(\alpha', \beta', \gamma', \delta')$ are new unfolding parameters that depend on $(\rho_1, \rho_2, \sigma, \delta)$, whereas $\lambda$ is the distinguished bifurcation parameter which depends on $(\rho_1, \rho_2, \sigma, \delta)$ and $A$. If $(\rho_1, \rho_2, \sigma, \delta)$ are fixed by (17), then $\lambda$ is proportional to $A - A_c$ where $A_c$ is the critical value of $A$ which is also defined in (17). For fixed $(\alpha', \beta', \gamma', \delta')$, i. e., fixed $(\rho_1, \rho_2, \sigma, \delta)$, we obtain a path in the $(\alpha, \delta)$-plane when $\lambda$ is varied. This path leads to a bifurcation diagram analogeously to that shown in Figure 2. For $(\alpha',\beta',\gamma',\delta') = (0,0,0,0)$ the path is given by $\delta = \alpha^2$, i. e., it is not transversal to the $\alpha$-axis. This non-transversality feature reflects the fifth condition imposed on the physical parameters, namely, that the transversality condition for the Hopf bifurcation with respect to $A$ is broken. The bifurcation diagram of Figure 2 can be derived from a path traversing the regions 1, 9, 16, 11, 2, 3, 1 in succession. This path does not hit a region with a limit cycle in the $(r, x)$-plane, i. e., a torus in the full 3-d center manifold. However, if the path is distorted so that the succession of 11, 2, 3 is replaced by 11, 21, 5, 6, 3, we obtain the bifurcation diagram of Figure 7 which possesses a torus branch qw (quasiperiodic wave transmission). Further structurally stable bifurcation diagrams are easily obtained by considering other generic paths through the unfolding space.

# Appendix

By using the parametrization (3) and the second relation of (4b), the coefficients $M_j$ of the characteristic equation (5) take the form

$$
\begin{aligned}
M_0 &= \rho_1^2\rho_2\left[yK + 48y^3/K + 128y^4/K^2\right]\\
M_1 &= \rho_1^2 y^2\left[1 + 16y/K + 64y^2/K^2\right]\\
&\quad + \rho_1\rho_2\left[(2+\rho_1 y)K + 7\rho_1 y(y-2) + 8(2+4\rho_1 - \rho_1 y)y^2/K\right]\\
M_2 &= \rho_2\left[(1+2\rho_1)K + 4\rho_1 - 10\rho_1 y + \rho_1^2 y + 24\rho_1 y^2/K\right] \qquad\text{(A.1)}\\
&\quad + 2\rho_1 y\left[1 + \rho_1 + 8(1+\rho_1)y/K\right]\\
M_3 &= \rho_2\left[I + 2(\rho_1+1)\right] + y(1+\rho_1^2) + 4\rho_1(1+4y^2/K)\\
M_4 &= \rho_2 + 2(1+\rho_1).
\end{aligned}
$$

Evaluating (A.1) at the locus of the cusp, i. e., for $I = 3y$ ($K = 4y$), yields $M_0 = 0$, and

$$
\begin{aligned}
M_1 &= 6\rho_1\rho_2(2-\rho_1)y + 9\rho_1^2(1+\rho_2)y^2\\
M_2 &= 4\rho_1\rho_2 + \left[\rho_2(2+\rho_1)^2 + 6\rho_1(1+\rho_1)\right]y \qquad\text{(A.2)}\\
M_3 &= 4\rho_1 + 2\rho_2(1+\rho_1) + (1+3\rho_2 + 4\rho_1 + \rho_1^2)y\\
M_4 &= \rho_2 + 2(1+\rho_1).
\end{aligned}
$$

The $M_j'$ are defined by the derivatives of $M_j$ with respect to $I$, evaluated at $I = 3y$. This leads to $M_0' = M_4' = 0$, and

$$
\begin{aligned}
M_1' &= \rho_1\rho_2(1-2\rho_1) + \frac{3}{2}\rho_1^2(\rho_2 - 2)y\\
M_2' &= \frac{1}{2}\rho_2(2+\rho_1) - \rho_1(1+\rho_1) \qquad\text{(A.3)}\\
M_3' &= \rho_2 - \rho_1.
\end{aligned}
$$

When the $M_j$ from (A.2) are substituted into the Hopf condition (6) we obtain (7), a quadratic equation for $y$, with the coefficients given by

$$
\begin{aligned}
q_0 &= -8\rho_1\rho_2(1+\rho_1)(2+\rho_2)(\rho_2+2\rho_1) \\
q_1 &= -48\rho_1^2(1+\rho_1)^2 - 16\rho_2\rho_1(1+\rho_1)(1+2\rho_1)^2 \\
&\quad -4\rho_2^2(4+6\rho_1+19\rho_1^2+15\rho_1^3+\rho_1^4) \\
&\quad -2\rho_2^3(4+8\rho_1+8\rho_1^2+\rho_1^3) \\
q_2 &= -12\rho_1(1-\rho_1^2)^2 - 2\rho_2(1-\rho_1^2)(4+21\rho_1+12\rho_1^2-\rho_1^3) \\
&\quad -6\rho_2^2(2+9\rho_1-5\rho_1^3) - 6\rho_2^3(2+2\rho_1-\rho_1^2).
\end{aligned} \tag{A.4}
$$

Finally, using (A.2) and (A.3) in (8) leads to (9), a linear equation for $y$, with coefficients

$$
\begin{aligned}
l_0 &= -8\rho_1^2(1+\rho_1)^2 - 4\rho_2\rho_1^2(1-\rho_1^2) + 2\rho_2^2(2+4\rho_1+5\rho_1^2+4\rho_1^3) \\
&\quad +\rho_2^3(2+6\rho_1+3\rho_1^2) \\
l_1 &= 2\rho_2(1-\rho_1^2)(1+2\rho_1+3\rho_1^2) - 2\rho_1(1-\rho_1^2)^2 \\
&\quad +\frac{1}{2}\rho_2^3(14+11\rho_1-\rho_1^2) + \frac{1}{2}\rho_2^2(14+33\rho_1+12\rho_1^2-11\rho_1^3).
\end{aligned} \tag{A.5}
$$

# References

[1] D. Armbruster, *An organizing center for optical bistability and self-pulsing*, Z. Phys. B **53** (1983), 157–166.

[2] D. Armbruster, G. Dangelmayr and W. Güttinger, *Imperfection sensitivity of interacting Hopf and steady state bifurcations and their classification*, Physica **16D** (1985), 99–123.

[3] G. Dangelmayr and D. Armbruster, *Classification of Z(2)-equivariant bifurcations with corank two*, Proc. London Math. Soc. **46** (1983), 517–546.

[4] G. Dangelmayr and M. Wegelin, *On a codimension four bifurcation ocurring in optical bistability*, in: *Proceedings of the Workshop on Dynamics, Bifurcations and Singularity Theory*, 10–14 July 1989, Warwick, to appear.

[5] M. Golubitsky and W. F. Langford, *Classification and unfoldings of degenerate Hopf bifurcations*, J. Diff. Eq. **41** (1981), 375–415.

[6] M. Golubitsky and D. Schaeffer: *Singularities and Groups in Bifurcation Theory*, Vol. I, Springer 1985.

[7] J. Guckenheimer and P. Holmes, *Nonlinear Oscillations, Dynamical Systems, and Bifurcations of Vector Fields*, Springer 1983.

[8] G. Hu and G. J. Yang, *Instability in injected lasers and optical bistable systems*, Phys. Rev. A **38** (1988), 1979–1989.

[9] L. A. Lugiato, V. Benza and L. M. Narducci, *Optical bistability, self-pulsing and higher-order bifurcations*, in H. Haken (ed.): *Evolution of Order and Chaos in Physics, Chemistry, and Biology*, Springer 1982.

[10] L. A. Lugiato, L. M. Narducci and R. Lefever, *Instabilities, spatial and temporal patterns in passive optical systems*, in R. Graham and A. Wunderlin (eds.): *Lasers and Synergetics. A Colloquium on Coherence and Self-Organization in Nature*, Springer 1987.

[11] L. A. Orozco, H. J. Kimble and A. T. Rosenberger, *Quantitative test of the single-mode theory of optical bistability*, Optics Comm. **62** (1987), 54–60.

[12] L. A. Orozco, H. J. Kimble, A. T. Rosenberger, L. A. Lugiato, M. L. Asquini, M. Brambilla and L. M. Narducci, *Single-mode instability in optical bistability*, Phys. Rev. A **39** (1989), 1235–1252.

[13] G. J. Yang and G. Hu, *Unstable regions of single-mode optical bistability*, Phys. Rev. A **39** (1989), 2514–2518.

# OPTIMIZATION BY CONTINUATION

J.P.KERNEVEZ and Y.LIU, *UTC, B.P.649, 60206 Compiègne Cedex, France,*
M.L.SEOANE, *Departemento de Matematica Aplicada, Facultad de Matematicas. Campus Universitario.15706 Santiago de Compostela,Spain* and
E.J.DOEDEL, *Department of Computer Science, Concordia University, Montréal, Québec H3G1M8, Canada*

ABSTRACT.   This paper presents two applications of continuation methods to the control of nonlinear systems.In the first one the aim is to control the position of isolas.This can be achieved by a continuation procedure to localize isola centers and  bifurcation points. The method also traces out curves of isola centers and perturbed bifurcation points. The technique has been applied to a reaction-diffusion biochemical system, for which numerical results are given, and to periodic solutions.The second application consists in viewing  optimal control problems as Cauchy problems which themselves can be solved by continuation techniques. As an example the method is applied in a general framework  to exact controllability of systems and in particular for a nonlinear distributed system.

Key words: isolas, center, continuation, Cauchy problems, optimal control, exact controllability, nonlinear elliptic system.

## 0.   Introduction

As is well-known, continuation methods are efficient tools for analyzing the behavior of a system as some parameter(s) vary. However, sometimes that is not enough and one would like to optimize the behavior of that system, to control it (in as an optimal way as possible), or check for controllability.We present in this paper two diffferent kinds of problems.The first one is about how to act upon isolas of solutions [14], and is solved by using two and more parameter continuation [1], [2], [5], [6], [8], [9], [12], [15], [16]. [5] and [16]  propose continuation methods for bifurcation points and cups.[5] prove that cusps and (nonsimple) non quadratic turning points are regular turning points of an augmented system

$$(1) \qquad f(y,\lambda,\tau) = 0, \qquad f_y v = 0, \ Lv-1 = 0,$$

where L is a linear operator. If n is the dimension of f, $(2n+1)$ is the dimension of (1) and the continuation of the turning points will need $2(2n+1)+1$ equations and a regularity at least $C^3$.Our method continues the same singularity  with a $2n+2$ system and $f \in C^2$ is enough.
The second problem is very general, since it is the problem to minimize a function $g(x)$ under constraints $f(x) = 0$.We show  in Section 3 how this problem can be formulated as a Cauchy problem, and in Section 2 how such a Cauchy problem can be solved by a continuation method.More details can be found in[11].In Section 4 we apply this methodology to controllability problems [4].

*D. Roose et al. (eds.), Continuation and Bifurcations: Numerical Techniques and Applications, 349–362.*
© 1990 *Kluwer Academic Publishers.*

## 1. Control of isolas

### 1.1. POSITION OF THE PROBLEM

We consider a system governed by the state-control equation

(1) $$f(y,\lambda,\tau) = 0, \; y \in Y, \;\; \lambda, \tau \in \mathbf{R}$$

where y is the state and $\lambda$ and $\tau$ are control parameters. It can occur that, for a given value of $\tau$, there exists an isola of solutions as $\lambda$ varies, and that these solutions are dangerous for the system. For example in Duffing equation, which models the ferroresonance phenomenon in electric lines, an isola of subharmonic solutions can exist, corresponding to high voltages dangerous for the lines. Another motivation for this work is the isola formation in reaction-diffusion biochemical systems due to perturbed bifurcation phenomena. There are also isolas of solutions which do not arise from perturbed bifurcation as described in [2],[8].

Consider a nonlinear operator $f : D \subset Y \times \mathbf{R}^2 \to Y'$ three times continuously differentiable in the Fréchet sense. Here Y denotes a reflexive real Banach space and Y' its dual. Our system is governed by a nonlinear equation

(1) $$f(y,\lambda,\tau) = 0, \; y \in Y, \;\; \lambda, \tau \in \mathbf{R}$$

which, for given $\lambda$ and $\tau$, can have multiple solutions. Conditions for the existence of isola centers in (1) are given in [1] and [8]., which leads us to take the following extended system for the continuation of isola centers:

(2) $$f(y,\lambda,\tau) = 0; \qquad {}^t f_y \, p = 0; \qquad {}^t f_\lambda \, p = \gamma; \qquad |p|_Y^2 = \zeta.$$

The extended system is also regular at perturbed bifurcation points but in this case we do not have isola formation. A justification of the following algorhitm can be found in [14]. Let us only remark that:

either      $f_y$ is non singular, in which case p = 0, and $\gamma = \zeta = 0$
or      $f_y$ is singular, in which case p $\neq$ 0, and $\zeta \neq 0$, and
        *either*    $\gamma \neq 0$, in which case we have a fold,
        or       $\gamma = 0$, in which case we have a center or a bifurcation point.

### 1.2. AN ALGORITHM TO LOCATE AND CONTINUE A CENTER OF ISOLA

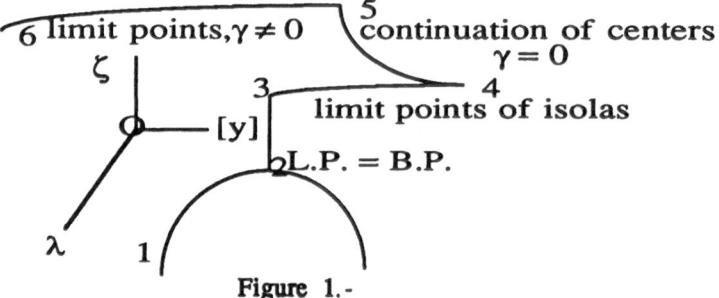

Figure 1.-

The algorithm described in this paragraph allows us to detect centers of isolas by continuation of the system (2) with respect to $\lambda$ and $\tau$. Once we have located this center we are able to follow a curve of these singularities as a function of a third free parameter. At any point on such a curve, it is sufficient to perturb the parameter $\tau$ in order to make an isola reappear. The path followed during steps 1-6 of the algorithm is indicated on Figure 1 by numbers 1-6.

*Step 1: Starting point.* We take a solution of (1). In the case of periodic solutions it is necessary to take as a starting point a nontrivial periodic solution; it can be obtained by means of a continuation procedure of the periodic solutions arising in a Hopf bifurcation[3]. In this step the variables p, $\zeta$ and $\gamma$ are zero.

*Step 2: Locate limit points of (1) with respect to $\lambda$ ($\tau$ fixed).* Starting with p, $\zeta$ and $\gamma$ equal zero, we follow a solution branch until we find a quadratic limit point of (1). It is detected as a bifurcation point of (2). At this point there exist two branches of solutions of system (2): one branch (1-2 in Fig.1) for which the variables p, $\zeta$, $\gamma$ are zero and another branch (2-3 in Fig.1) on which $\tau$ is kept fixed, y and $\lambda$ are fixed to their values at the limit point, and $p \neq 0$, $\zeta \neq 0$, $\gamma \neq 0$.

*Step 3: Computation of p.* Switch to the bifurcated branch until a value of $\zeta$ different from zero, for instance 1. We are in fact computing the null vector of ${}^t f_y$. At the same time $\gamma$ becomes generally nonzero.

*Step 4: Looking for $\gamma = 0$ in the equation (2iii).* Now free the perturbation parameter $\tau$ with $\zeta \neq 0$ fixed. The main continuation parameter in this step is $\gamma$. We trace out a branch of limit points since the variables p are no identically zero. The aim is to determine for which values of $\lambda$ and $\tau$ we have $\gamma = 0$. The point thus found is a candidate to be an isola center.

*Step 5: Continuation of centers.* If the initial equations (1) depend on more than two parameters, it is possible to continue the isola centers determined in step 4 with respect to a new free parameter. Now, p is nonzero, $\gamma$ is zero and each point of the branch is a center.

*Step 6: Recovery of isolas.* It is possible to obtain a family of isolas starting at a point of the curve of centers. Indeed, it is sufficient to continue, again, with respect to $\gamma$ until a value different from zero to recover a limit point on an isola.The continuation of (2) with p, $\zeta$ and $\gamma$ equal zero) will allow to trace it out completely.

*Remark.* The very nature of the continuation method implies that it is only possible to determine the centers belonging to the manifold of solutions in the space $(y,\lambda,\tau)$ on which the initial solution lies.

## 1.3. NUMERICAL RESULTS.

Isolas of solutions have been observed in several biochemical systems of reaction-diffusion type with inhibition due to excess of substrate ([10]). The concentrations of two chemical species in a membrane can be modeled by a system of two differential equations that describe the change of these concentrations in presence of an enzyme catalyzed reaction with diffusion inside the membrane and with a transport from an outside reservoir where the normalized concentrations are $s_0$ and $a_0$ :

$$s''(x) + \lambda[(s_0 + \varepsilon x - s(x)) - \rho R(s(x), a(x))] = 0$$

(3)

$$\beta a''(x) + \lambda[\alpha(a_0 - a(x)) - \rho R(s(x), a(x))] = 0$$

where $R(s, a) = as/(1 + s + ks^2)$ and $\alpha, \beta, \lambda, \rho$ are constants depending on the chemical species which play a role in the reaction, and on the geometry of the membrane. The parameter $\varepsilon$ is a perturbation term. An analytical and numerical study for $\varepsilon = 0$ has been carried out in [5] and [10].

We have used the package AUTO [3] for the continuation process. When the asymmetric perturbation term is different from zero then there exists an isola of solutions.

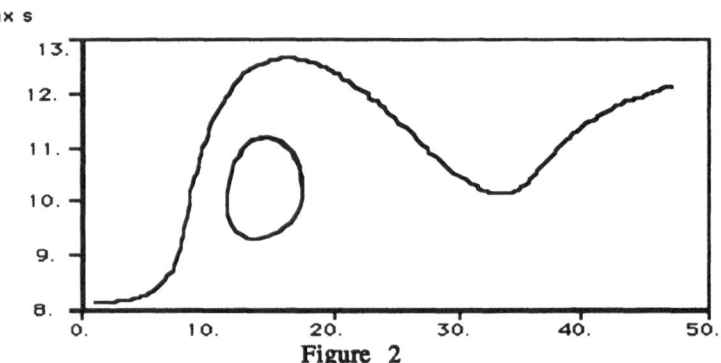

**Figure 2**

In Figure 2 we see the diagram of $\max\{s(x): x \in [0,1]\}$ as a function of the continuation parameter $\lambda$. Here and hereafter fixed parameter values are $\alpha = 1.45$ and $\beta = 5$. The other parameter values are $s_0 = 100$, $a_0 = 500$, $\rho = 13$, $\varepsilon = 0.15$, whereas $\lambda$ varies. Applying the method of continuation described in §1.2 we can locate two centers of isolas and two perturbed bifurcation points with $\varepsilon$ as a 2nd parameter. This is shown in Figure 3.

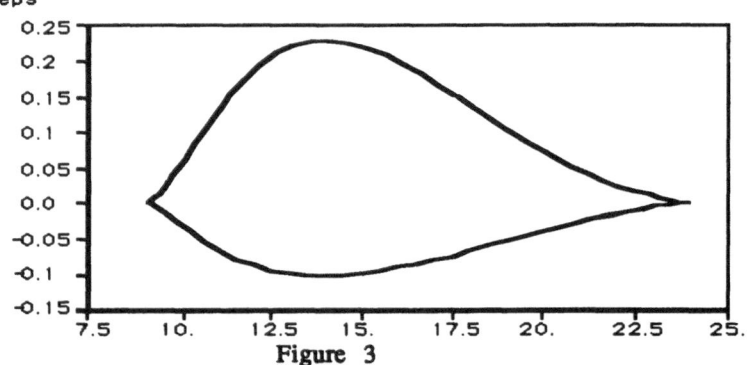

**Figure 3**

The continuation of one of these centers and one of these perturbed bifurcation points as $a_0$ varies together with $\lambda$ and $\varepsilon$ is shown in Figure 4, in 3-dimensional representation.

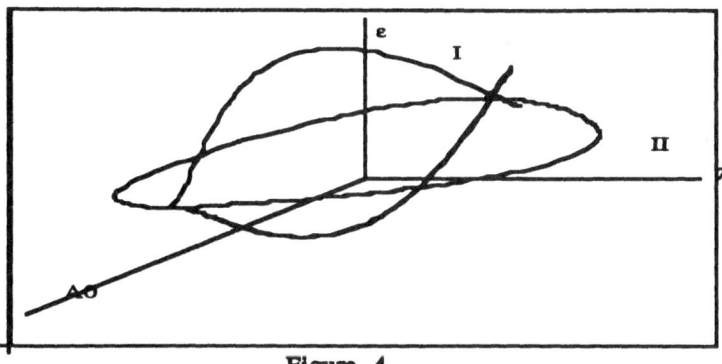

Figure 4

A zoom of the butterfly of Figure 4 is shown in Figure 5.

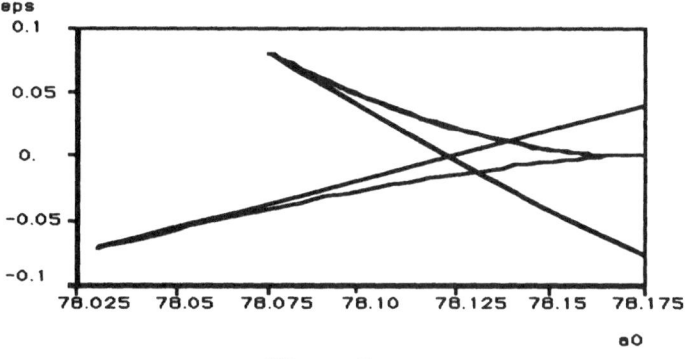

Figure 5

At last Figure 6 shows s as a function of x for an isola center (spatially non uniform curve) and for a perturbed bifurcation point (uniform profile, because it is a bifurcation from a "trivial" branch where s and a are independent of x).

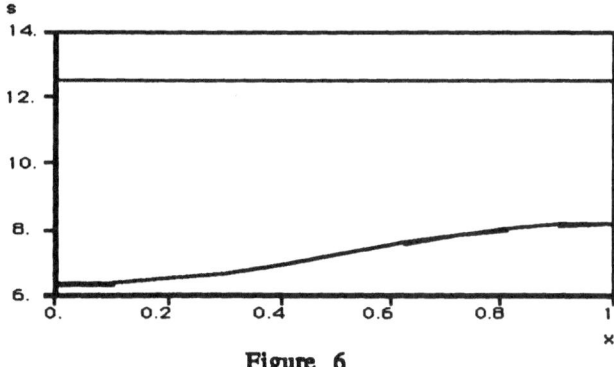

Figure 6

## 2. Solving Cauchy Problems by Continuation Methods

We only give here the principle of the method, refering to [11] for more details. We consider Cauchy problems of the form

(1) $\qquad F(y,y',t) = 0, \qquad y(0) = y_0$

where $F: \mathbb{R}^m \times \mathbb{R}^m \times \mathbb{R}$ is smooth enough. We consider t as a continuation parameter and rewrite (1) under the form

(2) $\qquad F(u,v,t) = 0,$

where $\dfrac{du}{dt} = v$, $u(0) = y_0$. Then $\dfrac{du}{dt} = v$ is approximated by a classical method, Crank-Nicholson for example, giving

(3) $\qquad g(u,v,t) = 0$, where $g(u,v,t) = \dfrac{u - u_k}{t - t_k} - \dfrac{1}{2}(v + v_k)$,

$(u_k, v_k, t_k)$ being an already found point. To equations (2) and (3) we adjoin the initial conditions

(4) $\qquad u_0 = y_0, t_0 = 0.$

(2), (3) and (4) are of the form of a continuation problem $\hat{F}(u,v,t) = \begin{pmatrix} F(u,v,t) \\ g(u,v,t) \end{pmatrix} = 0$, where t is the continuation parameter. The algorithm is the following (see [11] for more details, and in particular the calculation of singular solutions of (1)):

(i) INITIALIZATION.

Given $u_0$, $t_0$ calculate $v_0$ from (2) by Newton method and $w_0 = \dot{v}_0$ from (2) as indicated later in (15).

Calculate $\Delta \hat{t}_1$ (predicted time step) as indicated later in (14).

Then $t_1, u_1, v_1, w_1$ are calculated as follows:

*Prediction.* $\hat{u}_1 = u_0 + \Delta \hat{t}_1 v_0$, $\hat{v}_1 = v_0 + \Delta \hat{t}_1 w_0$

*Correction.*

$u_1, v_1$, and $t_1$ are calculated from $\hat{u}_1, \hat{v}_1$ and $\hat{t}_1$ by solving $\hat{F}(u_1, v_1, t_1) = 0$. $w_1$ is calculated by (15).

(ii) ITERATIONS.

After Initialization one knows two points $(t_{k-1}, u_{k-1}, v_{k-1}, w_{k-1})$ and $(t_k, u_k, v_k, w_k)$. To pass from iteration k to k+1 one does :

Calculation of $\Delta \hat{t}_k$ (predicted time step) as indicated later in (14).

*Prediction.*

We define the basis polynomials $p_{i,j}(t)$, $i = 1, 2, j = 0, 1$ by

(5) $\quad p_{1,0}(t_{k-1}) = 1,\ p_{1,0}(t_k) = 0,\ \dot{p}_{1,0}(t_{k-1}) = 0,\ \dot{p}_{1,0}(t_k) = 0;$

(6) $\quad p_{1,1}(t_{k-1}) = 0,\ p_{1,1}(t_k) = 0,\ \dot{p}_{1,1}(t_{k-1}) = 1,\ \dot{p}_{1,1}(t_k) = 0;$

(7) $\quad p_{2,0}(t_{k-1}) = 0,\ p_{2,0}(t_k) = 1,\ \dot{p}_{2,0}(t_{k-1}) = 0,\ \dot{p}_{2,0}(t_k) = 0;$

(8) $\quad p_{2,1}(t_{k-1}) = 0,\ p_{2,1}(t_k) = 0,\ \dot{p}_{2,1}(t_{k-1}) = 0,\ \dot{p}_{2,1}(t_k) = 1;$

Then the Hermite interpolation polynomial for $x = (u,v)$ can be written:

(9) $\quad \hat{x}(t) = x_{k-1} \cdot p_{1,0}(t) + \dot{x}_{k-1} \cdot p_{1,1}(t) + x_k \cdot p_{2,0}(t) + \dot{x}_k \cdot p_{2,1}(t),$

or

(10) $\quad \hat{x}(t) = x_{k-1} \cdot p_{1,0}(\Delta t_{k-1}, \Delta t) + \dot{x}_{k-1} p_{1,1}(\Delta t_{k-1}, \Delta t) +$

$\quad\quad\quad + x_k \cdot p_{2,0}(\Delta t_{k-1}, \Delta t) + \dot{x}_k \cdot p_{2,1}(\Delta t_{k-1}, \Delta t),$

where $\Delta t = t - t_k$, $\Delta t_{k-1} = t_k - t_{k-1}$, because $p_{i,j}$ only depends upon the relative lengths of $\Delta t$ and $\Delta t_{k-1}$. So knowing $\Delta t_{k-1}$ and a predicted arc-length $\Delta \hat{t}_k$ we have a prediction $\hat{x}_{k+1}$ of $x_{k+1}$. Accordingly we have

(11) $\quad \hat{u}_{k+1} = u_{k-1} \cdot p_{1,0}(\Delta t_{k-1}, \Delta \hat{t}_k) + v_{k-1} \cdot p_{1,1}(\Delta t_{k-1}, \Delta \hat{t}_k) +$

$\quad\quad\quad + u_k \cdot p_{2,0}(\Delta t_{k-1}, \Delta \hat{t}k) + v_k \cdot p_{2,1}(\Delta t_{k-1}, \Delta \hat{t}_k),$

(12) $\quad \hat{v}_{k+1} = v_{k-1} \cdot p_{1,0}(\Delta t_{k-1}, \Delta \hat{t}_k) + w_{k-1} \cdot p_{1,1}(\Delta t_{k-1}, \Delta \hat{t}_k) +$

$\quad\quad\quad + v_k \cdot p_{2,0}(\Delta t_{k-1}, \Delta \hat{t}_k) + w_k \cdot p_{2,1}(\Delta_{k-1}, \Delta \hat{t}_k).$

The method to calculate $\Delta \hat{t}_k$ is given in (14).

*Correction.* One calculates $t_{k+1}$, $u_{k+1}$ et $v_{k+1}$ from $\hat{t}_{k+1}$, $\hat{u}_{k+1}$ and $\hat{v}_{k+1}$ by solving
$\hat{F}(u_{k+1}, v_{k+1}, t_{k+1}) = 0$
One calculates $w_{k+1}$ by (15) and $\Delta t_{k-1}$ by $\Delta t_{k-1} = t_{k+1} - t_k$

*Determination of* $\hat{t}_{k+1}$. Let us write the Taylor development of u(t) at point $u_k$:

(13)     $u(t) = u_k + (t-t_k).\dot{u}k + \frac{1}{2}.(t-t_k)^2 \ddot{u}k + O((t-t_k)^3).$

Then $\Delta f_k$ is determined by

(14)     $\Delta \hat{t}_k \|\ddot{u}_k\|^{1/2} = $ some constant.

*Calculation of* $\ddot{u}_k$ . Derivating in (3) one obtains $\frac{\partial F}{\partial u}\dot{u} + \frac{\partial F}{\partial v}\ddot{u} + \frac{\partial F}{\partial t} = 0$ or , with $v = \dot{u}$, and $w = \ddot{u}$

(15)     $\frac{\partial F}{\partial v}w = -\frac{\partial F}{\partial t} - \frac{\partial F}{\partial u}v$

If $\frac{\partial F}{\partial v}$ is nonsingular Gauss elimination gives w. If $\frac{\partial F}{\partial v}$ is singular, twice differencing the Hermite

interpolation polynomial gives an estimation of $\ddot{u}$.

## 3.    Solving optimal control problems by continuation methods.

From the preceding Section it is enough to show that an optimal control problem can be reduced to a Cauchy problem for differential equations. To simplify the presentation we limit ourselves to the finite dimensional case, the extension to infinite dimensional spaces being straightforward, as shown in the example in next Section.

### 3.1. REDUCTION TO A CAUCHY PROBLEM.

We consider the state-control equation

(1)                 $f(y,v) = 0$           where $y \in R^m$ and $v \in R^n$

and $f : Y \times V \to Y$ is $C^1$, together with the cost function

(2)                 $\omega = g(y,v)$           where $g : Y \times V \to R$ is $C^1$.

(3)                 The problem is to solve inf g(y,v) under the constraint f(y,v) = 0.

The principle of the method is to define a curve $\Gamma$ contained in the manifold $M : f(x) = 0$ where $x = (y,v) \in Y \times V = X$, starting from a given point $x_0 = (y_0, v_0) \in M$ and ending at a local optimum of g. To achieve this goal we parametrize this curve by a parameter s such that

$$|\dot{x}(s)| = 1 \qquad (|.| \text{ norm in } R^{m+n}).$$

and $\dot{x}(s)$ is such that the function

(4)                 $j(s) = g(x(s))$

undergoes steepest descent. We assume that

(5)                 f is full rank (i.e. m)

Therefore $\dot{x} \in N(f_x(x))$. By hypothesis dim $N(f_x(x)) = n$. Let $\varphi_i$, $i = 1,..,n$, be an orthonormal basis of $N(f_x(x))$. Then $\dot{x}$ can be written in a unique way under the form:

(6)
$$\dot{x} = \sum_{i=1}^{n} \alpha_i \cdot \varphi_i.$$

We define

(7)
$$\alpha = {}^t(\alpha_1, \alpha_2, ....., \alpha_n).$$

To determine $\alpha$, we consider the problem:

(8)
$$\begin{cases} \min \left[ (\nabla J, \dot{x}) = (c, \alpha) \right] \\ (\alpha, \alpha) = 1, \end{cases}$$

where $(.,.)$ denotes the scalar product and $c \in \mathbf{R}^n$ is such that $c_i = (\nabla J, \varphi_i)$.

Writing the necessary optimality conditions of problem (8) we find:

(9)
$$\alpha = \pm \frac{c}{\|c\|}.$$

The sign "-" means that $\dot{x}$ defines a descent direction and the sign "+", an ascent direction. The solution of problem (8) then is

(10)
$$\alpha = - \frac{c}{\|c\|}.$$

From the preceding we find

(11)
$$\dot{x} = F(x,s), \text{where } F(x,s) = \sum_{i=1}^{n} \alpha_i \varphi_i$$

To system (11) we adjoin the initial condition

(12)
$$x(0) = x_0 \in X$$

3.2. CONVERGENCE TOWARDS A LOCAL MINIMUM

Theorem [11]. - We suppose that:

(13)         J and f are twice continuously differentiable, and dim $N(f_x(x)) = n$,

(14)         $J(x) \to \infty$ when $\|x\| \to \infty$,

(15)     $\nabla J(x_0) \notin R(f_x^T(x_0))$.

Then the associated Cauchy problem admits a unique (maximal) solution x(s), defined and continuously differentiable in some interval $[0, s^*[$. Moreover we have

(16)     $\nabla J(x(s^*)) \in R(f_x^T(x(s^*)))$.

3.3 CONTINUATION OF THE INTEGRAL CURVE

Let x(s*) be a local minimum of problem (3). The integral curve of the Cauchy problem (11) and (12) can be continued beyond s* (if s* < +∞) and it satisfies to

(17)     $\dot{x} = -F(x,s)$,     $s \in ]s^*, s^{**}]$,

where F(x,s) is defined in (11).

We remark that $\dot{x}$ now is an ascent direction. This process can be repeated if $s^{**} < +\infty$.

Several methods for the detection and localization of a local minimum are described in [11].

## 4.   A Continuation Algorithm for Exact Controllability

4.1 THE GENERAL CASE

The state-control equation relating the state y and control v is

(1)     $f(y,v) = 0$

where $f : Y \times V \to Y'$ is a $C^1$ map , Y and V being Hilbert spaces, Y' the dual of Y and V' is identified with V. The cost function is

(2)     $J(y,v) = \frac{1}{2} | Cy - z_d |_Z^2$

where $C: Y \to Z$ is a linear continuous map from the space of states Y to the Hilbert space of "observations" Z, $|.|_Z$ denotes the norm in Z and $z_d$ is a given observation. It is assumed that $z_d$ exactly corresponds to an unknown control $v_d$ which we are looking for (inverse problem). A possible algorithm consists in following a curve $\Gamma: R_+ \to Y \times V$ parameterized by arclength s : $s \mapsto ( y(s), v(s) )$, $(s \geqslant 0)$ with $\dot{v} = \frac{dv}{ds}$ such that

(3)     $| \dot{v}(s) |_V = 1$          and

(4)     $\frac{dj}{ds}$ is minimum (i.e; negative with largest absolute value )

where

(5) $\qquad$ $j(s) = J(y(s), v(s))$ , $f(y(s), v(s)) = 0.$

Thus we start from some given pair of Cauchy data

(6) $\qquad$ $y(0) = y_0, v(0) = v_0,$ such that $f(y_0, v_0) = 0$

and define a curve

(7) $\qquad$ $x(s) = (y(s), v(s))$

on the manifold $f(x) = 0$ going by steepest descent from the starting point towards an eventual minimum of $j(s)$.

*Proposition.* - The curve $\Gamma$ is defined by

(8) $\qquad$ $f(y, v) = 0$

(9) $\qquad$ $u = M(Cy - z_d)$

(10) $\qquad$ $\dot{v} = -\dfrac{Ku}{\lceil Ku \rceil_V}$

(11) $\qquad$ $f_y \dot{y} + f_v \dot{v} = 0$

(12) $\qquad$ $\dot{z} = C\dot{y}.$

where M is the canonical isometry from Z onto Z',

(13) $\qquad$ $K : Z' \rightarrow V$ is the transposed operator of L,

(14) $\qquad$ $L : V \rightarrow Z$ is defined by $\forall h \in V, Lh = Cw \Leftrightarrow f_y w + f_v h = 0. (w \in Y)$

*Proof.* Derivating in (5) with respect to s, we have (11) and

$$\frac{dj}{ds} = (Cy - z_d, C\dot{y})_Z = <M(Cy - z_d), C\dot{y}>_{Z',Z.}$$

Let u denote

(15) $\qquad$ $u = M(Cy - z_d).$

Then $\dfrac{dj}{ds} = <u, C\dot{y}>_{Z',Z} = <u, L\dot{v}>_{Z',Z} = ({}^t Lu, \dot{v})_V = (Ku, \dot{v})_V$

and it is minimum (i.e. (3) and (4) are satisfied) for

(16) $\qquad$ $\dot{v} = -\dfrac{Ku}{\lceil Ku \rceil_V}.$

Finally at each point $(y,v)$ of the curve $\Gamma$ the tangent $(\dot{y},\dot{v})$ is defined by (8),(9),(10),(11) (and we have (12)).

## 4.2 EXACT CONTROLLABILITY OF A NONLINEAR DISTRIBUTED SYSTEM

The state-control equation is a diffusion-reaction equation governing enzyme kinetics in a membrane $\Omega$. Here $\Omega$ is an open bounded set in $\mathbf{R}^2$ with smooth boundary $\Gamma = S \cup S'$ like in Figure 7:

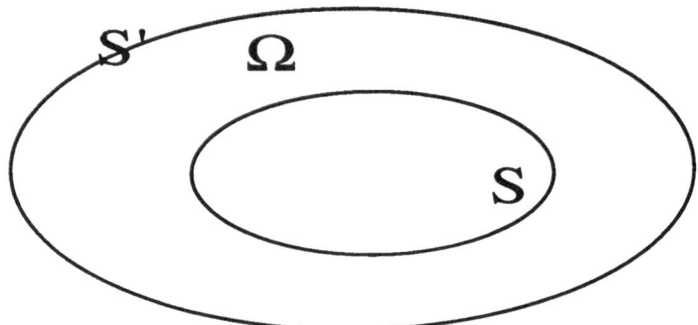

Figure 7

(17) $\qquad -\Delta y + F(y) = 0$ in $\Omega$, $y = 0$ on S, $y = v$ on S' where

(18) $\qquad F(y) = \sigma\, \dfrac{y}{1+|y|+ky^2} \qquad (k \geqslant 0)$.

The observation z is

(19) $\qquad z = Cy = \dfrac{\partial y}{\partial v}$ (outward normal derivative) on S $\qquad (Cy \in Z)$

In this example $Y = H^{1/2}(\Omega)$, $V = L^2(S')$, $Z = H^{-1}(S)$.

Then $M : H^{-1}(S) \to H^1(S)$ satisfies $(z, \hat{z})_{H^{-1}(S)} = \; <Mz, \hat{z}>_{H^1(S), H^{-1}(S)}$ where $M = (-\Delta_S + I)^{-1}$, $u = Mz \Leftrightarrow -\Delta_S u + u = z$, $-\Delta_S$ Laplace-Beltrami operator, and $L : L^2(S') \to H^{-1}(S)$ is defined by

$-\Delta\Psi + F'(y)\Psi \quad = 0$ in $\Omega$, $\Psi = 0$ on S and $\Psi = h$ on S', $Lh = \dfrac{\partial\Psi}{\partial v}$ on S , and

$K = {}^{t}L : H^1(S) \to L^2(S')$ by

$-\Delta\Phi + F'(y)\Phi \quad = 0$ in $\Omega$, $\Phi = g$ on S and $\Phi = 0$ on S', $h = Kg = \dfrac{\partial\Phi}{\partial v}$ on S'.

One easily checks that

$$(Kg, \hat{h})_{L^2_{(S')}} = \ <g, L\,\hat{h}>_{H^1(S), H^{-1}(S)}.$$

Indeed

$$\int_\Omega [-\Delta\Phi + F'(y)\Phi]\,\hat{\Psi}\,d\Omega - \int_\Omega [-\Delta\hat{\Psi} + F'(y)\hat{\Psi}]\,\Phi\,d\Omega = 0 =$$

$$-\int_\Gamma \frac{\partial\Phi}{\partial\nu}\,\hat{\Psi}\,d\Gamma + \int_\Gamma \frac{\partial\hat{\Psi}}{\partial\nu}\,\Phi\,d\Gamma \text{ (Green Formula)} = -\int_S \frac{\partial\Phi}{\partial\nu}\,\hat{\Psi}\,dS' + \int_S \frac{\partial\hat{\Psi}}{\partial\nu}\,\Phi\,dS =$$

$(-Kg,\hat{h}) + <L\hat{h}g>$ (duality between $H^{-1}(S)$ and $H^1(S)$ ).

Therefore the conditions (8)-(12) can be rewritten

(8')         $-\Delta y + F(y) = 0$ in $\Omega$, $y = 0$ on S, $y = v$ on S',

(9')         $-\Delta_S u + u = Cy - z_d$ on S, $Cy = \dfrac{\partial y}{\partial\nu}$ on S

h = Ku defined by

$-\Delta\Phi + F'(y)\Phi = 0$ in $\Omega$, $\Phi = u$ on S and $\Phi = 0$ on S', $h = Ku = \dfrac{\partial\Phi}{\partial\nu}$ on S'

(10')         $\dot{v} = -\dfrac{h}{\lVert h\rVert_{L^2_{(S')}}}$

(11')         $-\Delta\dot{y} + F'(y)\dot{y} = 0$ in $\Omega$, $\dot{y} = 0$ on S and $\dot{y} = \dot{v}$ on S'

(12')         $\dot{z} = C\,\dot{y} = L\,\dot{v}$

## 5.    Conclusion

We have seen in this paper that continuation methods can be applied to a large class of problems, from the control of isolas to controllability, including Cauchy problems and optimization problems.Admittedly these methods are expensive in computer time. However they could be useful or even necessary for some problems.

362

REFERENCES:

[1] Dellwo,D.-Keller,H.B.- Matkowsky, B.J.-Reiss,E.L.: "On the birth of isolas". SIAM J.App. Math., 2, 1982, pp. 956-963.

[2].Dellwo, D.: "A Constructive Theory of isolas supported by parabolic cusp, centers and bifurcation points". SIAM J.Appl.Math., 46, 1986, pp.740-764.

[3] Doedel, E.J.and Kernévez, J.P.: "AUTO: Software for continuation problems in ordinary differential equations with applications", Technical Report, Appl. Math., California Institute of Technology, 1989.

[4] Lions,J.L. (1989) "Exact controllability, stabilization and perturbations for distributed systems" SIAM Review

[5] Jepson, A. and Spence, A. (1985) "Folds in solutions of two parameter systems and their numerical calculation.Part 1. SIAM J.Numer.Anal.., 22, 347-368.

[6] Keener, J.P.and Keller, H.B.: "Perturbed Bifurcation Theory", Arch. Rat. Mech. Anal., 50, 1973, pp. 159-175.

[7] Keller, H.B.: "Numerical Solution of Bifurcation And Non Linear Eigenvalue Problems". Application in Bifurcation Theory, P.H. Rabinowitz, (ed.), Academic Press, 1977, pp.359-384.

[8] Keller, H.B.: "Two New Bifurcation Phenomena". Rapport de Recherche, 369. INRIA, 1979.

[9] Keller, H.B.: "Isolas and perturbed Bifurcation Theory", in "Non Linear Partial Differential Equations in Engineering and App. Sci.", Stenberg, Kalinowski, Papadakis (eds.), Marcel Dekker, 1980, pp. 45-52.

[10] Kernévez, J.P.: "Enzyme Mathematics". North-Holland, Amsterdam, 1980.

[11] Liu, Y. (1989) Thesis, Compiègne.

[12] Moore, G. and Spence, A. (1980) "The calculation of turning points of nonlinear equations" SIAM J.Numer.Anal.,. 17, 567-576.

[13] Rabinowitz, P.H. (ed.) : "Applications in bifurcation theory". Academic Press, New York, 1977.

[14] Seoane, M.L. (1990) Thesis, Santiago de Compostela.

[15]Spence, A. and Jepson (1984) "The numerical calculation of cusps, bifurcation points and isola formation points in two parameter problems", in Kupper, Mittelmann, Weber (eds.), Numerical methods for bifurcation problemss, Birkhauser, pp.502-514.

[16] Spence, A. and Werner, B. ()" Nonsimple turning points and cusps" IMA J. Numer.Anal., 2, 413-427.

# STABILITY OF MARANGONI CONVECTION IN A MICROGRAVITY ENVIRONMENT

HANS D. MITTELMANN
*Department of Mathematics*
*Arizona State University*
*Tempe, AZ 85287-1804*

ABSTRACT. In a recent paper of SHEN *et al.* energy-stability bounds were computed for the thermocapillary convection in a model of the float-zone crystal-growth process. The main application is expected to be the production of high-quality semiconductor material in low-gravity environments. Here we outline the physical and mathematical background and then describe in detail the numerical method used to solve the resulting nonlinear eigenvalue problem. Some information on the performance of the method is given and numerical results are presented for the zero gravity case.

## 1. Introduction

Thermocapillary convection is a fluid motion driven by surface-tension gradients on a liquid-gas interface, where these gradients arise from surface-temperature gradients and the temperature dependence of surface tension. This type of convection plays an important role in many technological and scientific applications; interesting examples may be found in the field of materials processing, particularly crystal-growth processes in which bulk melts are found. One such crystal-growth process is the so-called "float-zone" technique by which high-purity electronic materials (notably silicon) can be produced. In this method a rod of poly-crystalline material is moved slowly through a heating device which melts a portion of it. Ideally, as the melt re-solidifies it does so as a single crystal which is then used as substrate for building micro-electronic devices. Since the method is containerless, the possibility of contamination by contact with other materials is reduced. However, because surface-tension forces must support the weight of the material contained in the zone, the size of the resulting crystal is limited in *earth-based* production; in fact, some materials have properties which prevent this process from manufacturing crystals of reasonable size. Consequently, a microgravity environment such as that provided by the Space Shuttle has been suggested as a possible site for growing bigger, and hopefully better, crystals.

In addition to allowing larger crystals to be grown, a microgravity environment would also significantly reduce the magnitude of convection induced by buoyancy forces. This convection was once thought to be at least partly responsible for the

363

*D. Roose et al. (eds.), Continuation and Bifurcations: Numerical Techniques and Applications, 363–377.*
© 1990 *Kluwer Academic Publishers.*

presence of undesirable non-uniformities in material properties called *striations* observed in float-zone material. Recent speculation, however, is that the onset of *time-dependent thermocapillary convection* (PREISSER, SCHWABE & SCHARMANN 1983) is actually responsible for the appearance of these striations. Since this mode of convection will exist in any gravitational environment, the stability properties of thermocapillary convection are of possible technological importance; the identification of a region in an appropriate parameter space which is free of oscillatory thermocapillary convection is particularly relevant. In this work we exclusively consider the case of zero gravity.

The flow domain of an actual float zone is very complicated. It is bounded by non-planar melting and freezing solid-liquid interfaces and a deformable free surface, all of which are influenced by the translation of the material through the heater. Forced convection, due to independent rotation of the feed/seed material which is utilized to reduce asymmetry in the external heating, may be present. Model experiments, however, have been designed to eliminate several of the features in an actual float zone (i.e., translation, rotation, and melting/freezing interfaces) and to minimize the influence of buoyancy relative to thermocapillarity. These experiments show that, under certain critical conditions, steady thermocapillary convection undergoes an abrupt transition to an unsteady, oscillatory-flow state. Hence, understanding of such a transition in an actual float-zone may be furthered by first studying its occurrence in this simpler situation.

One of the problems associated with model experiments is the inability to model all of the parameters relevant to an actual float-zone. Perhaps the most significant example is the Prandtl number $Pr \equiv \nu/\kappa$, where $\nu$ and $\kappa$ are the kinematic viscosity and thermal diffusivity, respectively. Prandtl numbers for materials grown by the float-zone process are $O(10^{-2})$ while those for the fluids typically used in model experiments are $O(10)$. The principal reason for this is that crystal-growth melts are optically opaque, while transparent fluids are desired for model experiments so that flow visualization can be employed. Another problem associated with laboratory experiments is the degree of uncertainty in the material properties themselves. One of these in particular, the rate of change of surface tension with respect to temperature (used in the construction of the dimensionless Marangoni number), is not known to a high degree of accuracy for most materials.

Unlike linear stability theory, energy-stability theory provides a sufficient condition for stability of a given basic state to disturbances of arbitrary amplitude. This technique is equivalent to the stability analysis utilizing a Lyapunov function, see SINHA & CARMI (1976) and the literature cited there.

In order to gain further understanding into the stability properties of thermocapillary convection in a cylindrical geometry, we have chosen to model the half-zone examined experimentally by others. We also choose to employ energy-stability theory rather than linear-stability theory for our first analysis of this basic state under the rationale that the identification of regions of *stability* can be of possible technological importance. If the crystal grower is able to adjust the process to stay below this limit, then it may be possible to grow striation-free material. An unconventional approach is used in the application of energy theory to our numerically determined basic state; rather than solving an eigenvalue problem defined by the Euler-Lagrange equations (corresponding to the standard variational problem), we shall attack the variational problem more directly by finite-difference discretization of the functional which appears.

The basic state computation and the stability analysis are outlined in §2 and §3. The numerical method and some results are presented in §4 and §5. Comparison with available experimental results for model half-zones shows that the energy-theory results for axisymmetric disturbances do not appear to be overly conservative (small).

## 2. Basic state

The basic state of interest is one of swirl-free thermocapillary convection in a model half-zone of $O(1)$ aspect ratio, see Fig. 1. Two coaxial cylindrical rods of radius $R$ with planar ends are oriented a distance $H$ apart with their axes in the direction of gravity. A liquid zone is created between the ends and the rods are heated differentially, with the upper rod being at a higher temperature $(T_H)$ than the lower $(T_C)$. Since the imposed axial temperature gradient within the liquid is vertically upward (and the radial temperature gradients are small) the half-zone is stably stratified from the standpoint of buoyancy. However, the temperature gradient which exists along the free surface causes motion along the free surface from the hot cylinder to the cold one, thereby driving bulk thermocapillary convection with the sense of circulation shown in the figure. The flow and temperature properties of the half-zone are meant to approximate the situation in the lower half of an actual float-zone melt. The flow and temperature fields are two-dimensional and must therefore be obtained numerically for the nonlinear cases of interest.

We model the liquid zone as a Newtonian, Boussinesq fluid and choose scales for length, velocity and pressure to be $R$, $\gamma(T_H - T_C)/\mu$, and $\gamma(T_H - T_C)/R$, respectively. The quantity $\mu$ is the dynamic viscosity coefficient and $\gamma > 0$ is the rate of decrease of surface tension $\sigma$ with temperature as defined by

$$\sigma = \sigma_m - \gamma(T - T_m), \tag{2.1}$$

where $T_m = \frac{1}{2}(T_H + T_C)$ is the mean temperature of the two solid cylinders and $\sigma_m$ is the surface tension at temperature $T_m$. The velocity scale is the Marangoni velocity (SEN & DAVIS 1982) obtained by balancing the surface-tension gradient along the interface with the jump in shear stress. A dimensionless temperature is defined by

$$\Theta \equiv \frac{T - T_m}{T_H - T_C}. \tag{2.2}$$

The resulting dimensionless governing equations for the basic-state velocity $\mathbf{U} \equiv (U, 0, W)$, pressure $P$ and temperature fields are

$$\frac{1}{r}(rU)_r + W_z = 0, \tag{2.3}$$

$$Re\,(UU_r + WU_z) = -P_r + \nabla^2 U - \frac{U}{r^2}, \tag{2.4}$$

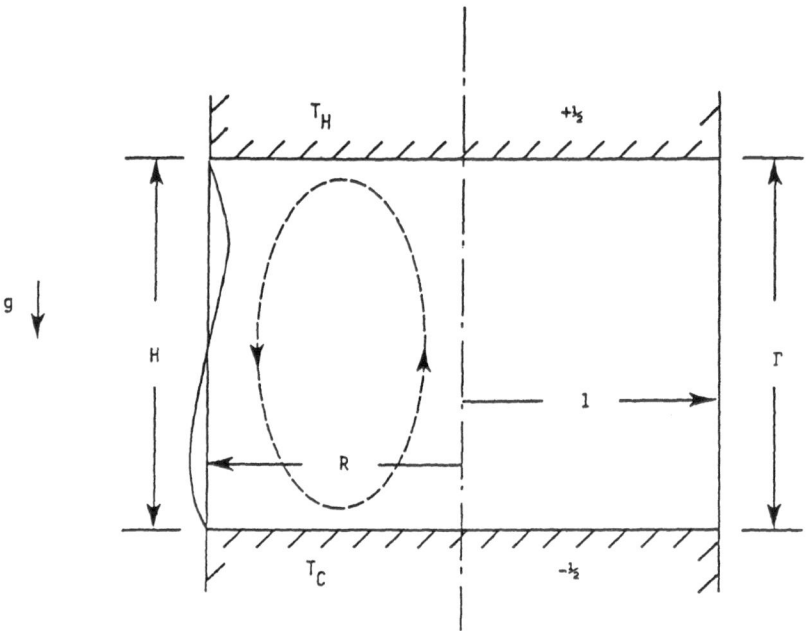

Fig. 1. The half-zone, showing geometric and thermal conditions, both in dimensional and dimensionless forms.

$$Re\,(UW_r + WW_z) = -P_z + \nabla^2 W, \qquad (2.5)$$

$$Ma(U\Theta_r + W\Theta_z) = \nabla^2\Theta, \qquad (2.6)$$

where

$$\nabla^2 \equiv \frac{1}{r}\frac{\partial}{\partial r}(r\frac{\partial}{\partial r}) + \frac{\partial^2}{\partial z^2}.$$

The two dimensionless parameters which appear are:

Reynolds number $\quad Re = \dfrac{\gamma(T_H - T_C)R}{\mu\nu},$

Marangoni number $\quad Ma = \dfrac{\gamma(T_H - T_C)R}{\mu\kappa},$

where $\alpha$ is the coefficient of volumetric expansion and the conventional subscript notation has been used to denote partial differentiation. The Prandtl number is obtained from the quotient $Ma/Re$.

We assume, as a first approximation, that the free surface is not permitted to deform, and so is fixed at $r = 1$. This corresponds to requiring that the volume of liquid in the half zone is $\pi\Gamma$ and that the mean surface tension, $\sigma_m$, is asymptotically large. The boundary conditions applied to complete the problem specification are:

$$U = W = 0 \quad , \quad \Theta = -1/2; \qquad z = 0 \qquad\qquad (2.7a - c)$$
$$U = W = 0 \quad , \quad \Theta = 1/2; \qquad z = \Gamma \qquad\qquad (2.8a - c)$$

$$U = 0, \quad U_z + W_r = -\Theta_z \quad , \quad -P + 2U_r = \frac{\sigma}{\sigma_m Ca},$$
$$\Theta_r = Nu[\Theta - \Theta_a(z)]; \qquad r = 1 \qquad\qquad (2.9a - d)$$

$$U = W_r = \Theta_r = 0 \quad ; \quad r = 0. \qquad\qquad (2.10a - c)$$

Equations (2.7) and (2.8) express the kinematic and no-slip conditions and the requirement of isothermal surfaces, while (2.9a) is the kinematic condition on the free surface. Equations (2.9b,c) represent the shear and normal-stress balances. Equation (2.9b) results from the relationship between the shear stresses and the gradient of the surface tension which due to (2.1) and (2.9) transforms into the given right-hand side. Symmetry conditions at the axis of symmetry are given by (2.10). The additional parameter appearing in (2.9d), which models the heat-transfer mechanism at the free surface, is the Nusselt number,

$$Nu = hR/k$$

where $h$ is a heat-transfer coefficient and $k$ is the thermal conductivity of the liquid. Since $h$ may vary with $z$, in general, $Nu = Nu(z)$. For the majority of the calculations $\Theta_a(z) = -1/2$, i.e., the environment was assumed to be at a constant temperature equal to that of the cold cylinder at $z = 0$. Condition (2.9c) contains the capillary number

$$Ca = \frac{\gamma(T_H - T_C)}{\sigma_m}$$

which vanishes in the limit of a non-deformable free surface. Hence, this condition is not required in the present analysis.

The numerical solution of this problem is accomplished by first transforming to a stream-function/vorticity form, thereby eliminating the pressure. The resulting equations are solved using a modification of the predictor-corrector multiple iteration (PCMI) technique employed successfully by NEITZEL & DAVIS (1981) and NEITZEL (1984) to study centrifugally unstable flows in cylindrical geometries; the reader is referred to these papers for details.

All computed half-zone flows consist of a single toroidal cell with flow at the surface in the direction opposite to that of the surface temperature gradient, as expected. As previously mentioned, one of the difficulties associated with extrapolation of the results of model experiments to cases of interest to crystal growers is the fact that the Prandtl number is different by a couple of orders of magnitude in the two situations. A case with moderate Marangoni number and exterior environment temperature which is linear in $z$ is shown in figure 2; in figure

2a, $Pr = 0.01$, while in figure 2b, $Pr = 10$. These Prandtl numbers are roughly representative of molten silicon and sodium nitrate (the material used in the model experiments of PREISSER *et al.* (1983)), respectively. The distortion, at high $Pr$, of the isotherms from the nearly conductive low-$Pr$ state and shift in the center of the eddy are clearly evident in the figure.

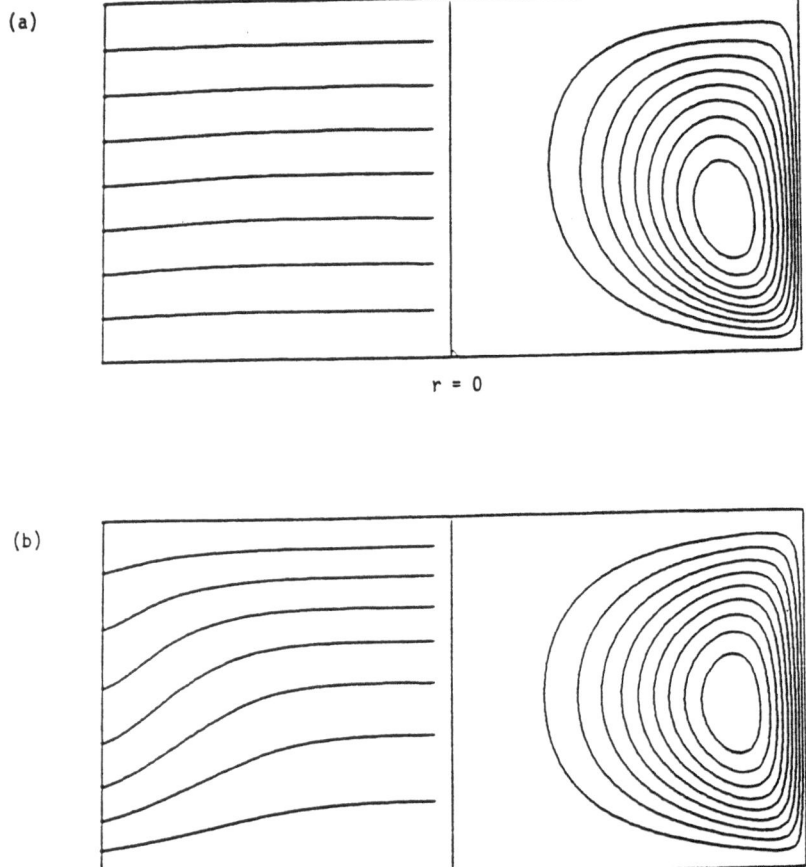

(a)

$r = 0$

(b)

$r = 0$

Fig. 2. Basic-state isotherms and streamlines for $\Gamma = 1$, $Ma = 100$, $Nu = -0.3$ and $\Theta_a(z) = z - 1/2$. (a) $Pr = 0.01$ (b) $Pr = 10$.

## 3. Energy-stability analysis

We begin the energy-theory analysis of the basic state in the usual fashion by deriving the energy identity. We assume there exists a solution $[u, p, T]$ to the governing equations ((2.1), plus the unsteady analogs of (2.2)–(2.4)) which is an *axisymmetric* perturbation to the axisymmetric basic state, i.e.,

$$
\begin{aligned}
[u, p, T] &= [U(r, z), 0, W(r, z), P(r, z), \Theta(r, z)] \\
&+ [u'(r, z, t), 0, w'(r, z, t), p'(r, z, t), T'(r, z, t)].
\end{aligned}
\tag{3.1}
$$

While the assumption of axisymmetric disturbances may limit the applicability of the results, available experimental evidence indicates that such an assumption is relevant for certain values of some of the parameters. Hopefully, the results of this analysis, when coupled with subsequent results for three-dimensional disturbances, will identify these values.

Substitution of (3.1) into the governing equations and boundary conditions leads to a system of equations for the disturbance quantities. We then take the inner product of the disturbance momentum equation with $u'$, add to this the disturbance energy equation multiplied by $\lambda Pr T'$, and integrate over the volume

$$
V = \{(r, \theta, z) \mid 0 \le r \le 1, \ 0 \le \theta \le 2\pi, \ 0 \le z \le \Gamma\}
$$

occupied by the liquid, using the disturbance boundary conditions. The result is the exact disturbance-energy evolution equation

$$
\frac{dE}{dt} = -Pr D - Ma I + Pr J,
\tag{3.2}
$$

where

$$
E = \frac{1}{2} \int_V \left(u' \cdot u' + \lambda Pr T'^2\right) dV, \quad D = \int_V \left(\nabla u' : \nabla u' + \lambda \nabla T' \cdot \nabla T'\right) dV,
$$

$$
I = \int_V \left(u' \cdot \mathbf{D} \cdot u' + \lambda Pr T' \nabla \Theta \cdot u'\right) dV, \quad J = \int_S \left(-w' T_z' + \lambda Nu T'^2\right) dS
$$

and $S$ is the free surface $r = 1$. The velocity and temperature disturbances have been joined by a positive *coupling parameter* $\lambda$ (JOSEPH 1976) to form a generalized disturbance energy, $E$, and the quantity $\mathbf{D}$ in the production integral $I$ is the *symmetric* basic-state deformation-rate tensor,

$$
\mathbf{D} = \begin{bmatrix} U_r & 0 & \frac{1}{2}(U_z + W_r) \\ & U/r & 0 \\ & & W_z \end{bmatrix}.
$$

Employing the re-formulated energy theory of DAVIS & VON KERCZEK (1973), (3.2) is divided by the positive-definite functional $E$ and an upper bound is constructed for the resulting right-hand side, viz.,

$$\frac{1}{E}\frac{dE}{dt} = \frac{1}{E}(-PrD - MaI + PrJ)$$

$$\leq \nu = \max_{H}\left(\frac{-PrD - MaI + PrJ}{E}\right),$$

(3.3)

where the maximum is taken over the space of kinematically admissible functions,

$$H = \{\mathbf{u}', T' \mid \mathbf{u}' = T' = 0 \text{ at } z = 0, \Gamma; \ u' = 0 \text{ at } r = 0, 1; \ \nabla \cdot \mathbf{u}' = 0\}.$$

We choose to formulate the problem so that the Marangoni number is the stability parameter. For fixed values of the other parameters associated with the problem, the smallest value of $Ma$ which corresponds to the condition $\nu = 0$ will be called $Ma^*(\lambda)$.

Since $\lambda$ is a free parameter, the maximum value of $Ma^*$ for positive values of $\lambda$ is sought (JOSPEH 1976). In the general case of three-dimensional disturbances, this value would be the energy-stability limit, $Ma_E$, i.e., the flow would be asymptotically stable in the mean for $Ma < Ma_E$. Since we have restricted attention to two-dimensional disturbances, this value is, at worst, an upper bound to $Ma_E$. With this restriction in mind, we will use the notation $Ma_E$ to refer to the value obtained from the present analysis,

$$Ma_E = \max_{\lambda > 0} Ma^*.$$

(3.4)

In many analyses, the search for this maximum is not performed. Rather, the variable $\lambda$ is arbitrarily set to some value, say $\lambda = 1$, and the result is accepted as a lower bound to the actual energy-stability limit. It will be seen that, for the problem of interest here, the effort necessary to determine $Ma_E$ is extremely worthwhile.

It is convenient to consider a slightly different functional than $\nu$ in (3.3) which incorporates the divergence constraint by means of a Lagrange multiplier. Hence, the maximum problem to be solved is expressed as

$$\nu = \max_{h}[-PrD - MaI + PrJ + 2\int_{V}\pi\nabla\cdot\mathbf{u}'\,dV],$$

(3.5)

where $\pi(r, z)$ is a Lagrange multiplier and $h$ is the extension of $H$ obtained by removing the divergence constraint. We are therefore interested in the variation of a discretized version $F_D$ of the quadratic functional

$$F = -PrD - MaI + PrJ + 2\int_{V}\pi\nabla\cdot\mathbf{u}'\,dV.$$

For details we refer to SHEN et. al.

A stationary value of $F_D$ is located by differentiating it with respect to each unknown and setting each of these derivatives to zero, i.e.,

$$\frac{\partial F_D}{\partial q_n} = 0 \quad q_n = u_{i,j}, w_{i,j}, \phi_{i,j} \text{ or } \pi_{k,l}.$$

(3.6)

This process yields a *generalized algebraic eigenvalue* problem. We seek the minimum positive eigenvalue of this system as the approximate (subject to discretization error) value of $Ma^*$. Calling the vector consisting of the unknowns on all grid points $X$, we rewrite (3.6) in the matrix form

$$AX = \rho B(Ma)X, \qquad (3.7)$$

where $A$ and $B$ are *indefinite, symmetric* matrices with $A$ having a banded structure and $B$ depending on the basic-state deformation-rate tensor $D$. The symmetry of the discrete problem is consistent with that of the variational problem (3.5). The dependence on the basic state, which depends in turn on the Marangoni number $Ma$, complicates the calculation of $Ma^*$. For a given value of $Ma$, denote the smallest positive eigenvalue of the generalized eigenvalue problem (3.7) by $\rho^*$. If $\rho^* \neq Ma$, then a new $Ma$ is chosen, the basic state re-computed, and the eigenvalues re-calculated. This process is repeated until $\rho^* = Ma$, in which case, $Ma^* = \rho^*$. This, of course, assumes all other parameters, including the coupling parameter $\lambda$, are fixed, necessitating further computation to find $Ma_E$ according to (3.4).

## 4. Numerical procedure for finding $Ma_E$

Equation (3.7) represents a nonlinear generalized eigenvalue problem. The matrices $A$ and $B$ are symmetric and sparse, but, in general, indefinite. In addition to the basic-state dependence of $B$ mentioned above, $A$ and $B$ depend on the other parameters of the problem, namely, $Pr$ and the coupling parameter, $\lambda$. We first address the case that all these parameters are fixed, reducing (3.7) to the generalized eigenvalue problem

$$AX = \rho BX, \qquad \|X\| = 1, \qquad (4.1)$$

where $\| \cdot \|$ denotes the Euclidean norm.

The eigenvalues $\rho$ of (4.1) may be real of either sign or be complex-conjugate pairs. To each null vector of $B$ corresponds an "infinite" eigenvalue (at least $N$ of these are known to occur), while the null vectors of $A$ yield zero eigenvalues. Standard linear-algebra software packages provide implementations of the $QZ$-algorithm for eigenvalue problems of the form (4.1). These packages compute *all* of the eigenvalues of the system (our stability result requires only a *single* eigenvalue), do not exploit the sparseness of $A$ and $B$ and are therefore suitable only for relatively coarse discretizations of the underlying continuous problem. The method adopted for the present computations makes use of both the symmetry and sparseness of $A$ and $B$ and computes only the eigenvalue of interest—the smallest, positive one.

The algorithm of choice for finding a single eigenvalue of (4.1) appears to be some form of *inverse iteration*. The technique used here is a generalization of that employed by BANK & MITTELMANN (1986) in the program PLTMG for the simpler problem of finding the smallest eigenvalue of a positive-definite matrix. For this a starting vector $X_0$, $\|X_0\| = 1$, is needed. Initially, this inverse iteration process is started with a random vector. Subsequent iterations use previously computed eigenvectors corresponding to nearby parameter values. A first approximation for $\rho$ is obtained through the *Rayleigh quotient*

$$\rho_0 = X_0^T A X_0 / X_0^T B X_0. \qquad (4.2)$$

In the unlikely event that the denominator is zero, a different $\mathbf{X}_0$ has to be chosen.

Given this initial pair $\mathbf{X}_0$ and $\rho_0$, the inverse iteration procedure is performed as follows:

1. Solve $(\mathbf{A} - s\mathbf{B})\overline{\mathbf{Y}} = (\rho_k\mathbf{B} - \mathbf{A})\mathbf{X}_k$ and define $\mathbf{Y} = \dfrac{\overline{\mathbf{Y}} - \overline{\mathbf{Y}}^T\mathbf{X}_k}{\|\overline{\mathbf{Y}} - \overline{\mathbf{Y}}^T\mathbf{X}_k\|}$.

2. Form $\mathbf{Q} = [\mathbf{X}_k|\mathbf{Y}]$ and solve the $2 \times 2$ problem

$$\mathbf{Q}^T\mathbf{AQZ} = \tau\mathbf{Q}^T\mathbf{BQZ}$$

for the eigenvalues $\tau_1, \tau_2$ and associated normalized eigenvectors $\mathbf{Z}_1, \mathbf{Z}_2$. Without loss of generality let $\tau_1$ be the *smallest positive* eigenvalue.

3. Set $\rho_{k+1} = \tau_1$, $\mathbf{X}_{k+1} = \mathbf{QZ}_1$ and check for convergence. If not converged, increment iteration index $k$ and repeat.

Several remarks are in order on the above algorithm. The quantity $s$ is a positive real number which has to be closer to the desired eigenvalue than to any of the other eigenvalues. While, in some applications this "shift parameter" may have to be adjusted during the computation in order to satisfy this requirement, this was not necessary in the present case. Earlier computations with the $QZ$-algorithm for moderate-size problems had shown that, for the cases considered, there were no complex-conjugate pairs that were smaller in modulus than $\rho^*$. Also, the negative eigenvalue of smallest modulus was similar in modulus to $\rho^*$. It was thus relatively easy, with some rough knowledge of $\rho^*$, to find a value for $s$.

The eigenvalue problem in step 2 is basically an orthogonal projection of the original problem into the subspace spanned by the columns of $Q_k$. Simpler inverse iteration algorithms are indeed available; however, their application to the present problem did not yield satisfactory results. In general, of course, this $2 \times 2$ eigenvalue problem may have complex eigenvalues, as well as real ones. While several precautions for this and other cases were put into the program, they will not be described here, being a rather technical detail. Eventually $\tau_1$ will be positive and approximate $\rho^*$ while $\mathbf{QZ}_1$ approximates the associated eigenvector. $\mathbf{X}_{k+1}$ and $\rho_{k+1}$ are related through the Rayleigh quotient (4.2).

While steps 2 and 3 need no further explanation, the solution of the linear system in step 1 represents a nontrivial problem. The matrix on the left is symmetric but indefinite. The newest version of the FORTRAN subroutine SYMMLQ, part of the NAG library, was used. It applies a conjugate-gradient method and permits preconditioning by a positive-definite matrix. No attempt was made to find a near optimal choice for the preconditioner. In all computations it was taken as the diagonal matrix with the $i$-th element equal to the Euclidean norm of the $i$-th column of the matrix $\mathbf{A} - s\mathbf{B}$.

The convergence of the above inverse iteration procedure is linear with a factor asymptotically equal to

$$\left|\frac{s - \rho^*}{s - \rho_n}\right| < 1$$

where $\rho_n$ is the next nearest eigenvalue of (4.1) to $s$. Choosing $s$ close to $\rho^*$ will thus speed up convergence of the inverse iteration while, in general, requiring more conjugate-gradient iterations for the nearly singular system matrix. The essential computational requirement per conjugate-gradient iteration is one matrix-vector multiplication with the system matrix.

In addition to this method for solving the eigenvalue problem, two outer iterations are needed to determine the energy-stability limit $Ma_E$. The requirement that $\rho^*(Ma) = Ma$ suggests a fixed-point iteration. The second requirement, that $Ma_E$ is found as the maximum of all these $\rho^*$ with respect to $\lambda$ suggests an optimization procedure. No attempt was made to simultaneously attack both these problems. Both possibilities of a successive solution were used, the fixed-point iteration as either an inner or outer iteration. In the first case, say, when $\lambda$ is temporarily fixed, the following would be the well-known Picard iteration

$$\mathbf{A}\mathbf{X}_{k+1} = \rho_{k+1}\mathbf{B}(\rho_k)\mathbf{X}_{k+1}, \quad k = 0, 1, 2, \cdots \tag{4.3}$$

where $\rho_{k+1}, \mathbf{X}_{k+1}$ is the solution found through the inverse iteration procedure defined above. This iteration will only converge if $|\varphi'(Ma^*)| < 1$, where $\varphi$ denotes the relationship between $\rho_{k+1}$ and $\rho_k$ given through (4.3) and if started close enough to $Ma^*$. A simple acceleration procedure due to Aitken (1926) was implemented to guarantee convergence:

$$\rho_k^{(0)} = \rho_k,$$

$$\rho_k^{(i)} = \varphi(\rho_k^{(i-1)}), \quad i = 1, 2, \tag{4.4}$$

$$\rho_{k+1} = \rho_k + \frac{(\rho_k^{(1)} - \rho_k^{(0)})^2}{(\rho_k^{(1)} - \rho_k^{(0)}) - (\rho_k^{(2)} - \rho_k^{(1)})}.$$

The sequence $\{\rho_k\}$ converges quadratically if $\varphi$ is twice continuously differentiable and $\varphi'(Ma^*) \neq 1$.

For the sake of completeness we also outline how the maximization of $Ma^*$ with respect to $\lambda$ was accomplished. Starting from an initial value $\lambda_1$ and corresponding value $\rho_1$ (or $Ma_1$) two additional pairs of values are computed with their $\lambda$-values in the vicinity of $\lambda_1$. Through these 3 points a quadratic parabola is fit and the point corresponding to its maximum replaces one of the points. The parabola need not have a maximum; in the event that a minimum occurs, some modification is required. The details will again not be given here since they are straightforward. As is well known, maximization through successive quadratic interpolation has a convergence order of about 1.3.

The iterative procedure described above provides a relatively efficient method to calculate $Ma_E$. After computation of the first basic state, subsequent basic-state computations need fewer relaxations if they are done for a convergent sequence of Marangoni numbers. Analogously, the inverse iteration only requires several (more than 1-2) iterations when initiated, i.e., with the random vector $\mathbf{X}_0$. It is thus not surprising that the entire computation of $Ma_E$ took only a few times the amount of work needed for the first basic state and $\rho^*$-computation. It should be noted that, to minimize inaccuracies introduced by differentiation of basic-state quantities, the basic state was computed on a grid with *twice* the resolution used for the stability calculations.

## 5. Results

The calculations of $Ma_E$ described in §4 have been performed for a variety of parameters on both the Arizona State University IBM 3090-500E/VF supercomputer and on an Ardent Titan mini-supercomputer in the Advanced Research Computing Facility of the Department of Mathematics. A number of tables and figures were presented in SHEN *et al*. We summarize here only some information not given there on the performance of the numerical method. The basic state was computed on grids of size $139 \times 139$, $197 \times 99$ etc. using the relaxation type methods of NEITZEL (1984). Through differencing the basic state quantities entering the functional $I$ in (3.3) were obtained. On coarser grids of size $70 \times 68$, $99 \times 48$ etc. the functional $F_D$ in (3.6) was then obtained resulting in a dimension of about 20,000 for this system.

For fixed $\lambda$ and $Ma$ the solution of eigenvalue problem (3.7) required 3–10 inverse iterations. The solution of the linear system in the first step of this method using the cg-algorithm took about 5–15 percent of the number of variables. This was accomplished by preconditioning the matrix by a diagonal matrix whose $i$-th entry was the Euclidean norm of the $i$-th column. Through the integration of the five nested iterations: relaxation for basic state, fixed-point iteration on $Ma$, optimization with respect to $\lambda$, inverse iteration for $\rho$, and cg within the latter, the overall computation became rather effective requiring between 10 and 50 inverse iterations and a few recomputations of the basic state. Here, again the subsequent computations together typically were less costly than the initial iteration.

In Table 1 we list the critical Marangoni number as a function of $\lambda$ exhibiting a maximum for $\lambda \approx 10^{-7}$. For this value of $\lambda$ and different aspect ratios and Prandtl numbers stability bounds are given in Table 2. For a comparison to the results obtained with model experiments we refer to SHEN et al. What is clear from these results is that the computed energy-stability limits are not conservative in the sense of yielding an absurdly low boundary for guaranteeing stability. Calculations of energy stability limits allowing for non-axisymmetric disturbances and a deformable free surface are clearly warranted. Ultimately, it is hoped that energy-stability calculations for a realistic model of an actual float-zone allowing for melting/freezing and radiant heating will yield results which allow the crystal grower to better control the process and grow striation-free material.

*Acknowledgement.* This work was partially supported by the US Air Force Office of Scientific Research under the grant AFOSR-840315.

Table 1. Variation of $Ma^*$ with $\lambda$. For all cases, $\Gamma = 1$, $Pr = 1$, $Nu = 0.3$, $\Theta_a(z) = -0.5$ and a $70 \times 68$ grid was used.

| $\lambda$ | $Ma^*$ $(Gr = 0)$ |
|---|---|
| 2 | 62.9 |
| 1 | 90.6 |
| 0.5 | 131.9 |
| 0.25 | 194 |
| 0.1 | 333 |
| 0.05 | 525 |
| 0.025 | 900 |
| 0.01 | 1565 |
| 0.006 | 1654 |
| 0.003 | 3031 |
| 0.001 | 4492 |
| $10^{-4}$ | 6015 |
| $10^{-5}$ | 6247 |
| $10^{-6}$ | 6253 |
| $10^{-7}$ | 6302 |
| $10^{-8}$ | 6275 |
| $10^{-9}$ | 6274 |

Table 2. Energy-stability results for axisymmetric disturbances. For all cases, $Nu = 0.3$ and $\Theta_a(z) = -0.5$.

| $\Gamma$ | $Pr$ | $Re^*$ | $Ma^*$ | grid |
|---|---|---|---|---|
| 1 | 0.01 | 2595 | 26 | $70 \times 68$ |
| | 0.05 | 2996 | 150 | |
| | 0.10 | 3689 | 369 | |
| | 0.50 | 5488 | 2744 | |
| | 1.0 | 6302 | 6302 | |
| | 1.1 | 6430 | 7073 | |
| | 1.2 | 6578 | 7894 | |
| | 1.3 | 6701 | 8711 | |
| | 1.4 | 6914 | 9680 | |
| | 1.5 | 7017 | 10526 | |
| | 1.6 | 7170 | 11472 | |
| | 0.1 | 3708 | 371 | |
| | 0.1 | 3734 | 373 | |
| | 0.5 | 5397 | 2699 | |
| | 0.5 | 5462 | 2731 | |
| | 1.0 | 6163 | 6163 | |
| | 1.0 | 6082 | 6082 | |
| 0.5 | 0.1 | 6719 | 672 | $99 \times 48$ |
| 0.67 | 0.1 | 4985 | 499 | $76 \times 49$ |
| 1.33 | 0.1 | 2110 | 211 | $55 \times 71$ |
| 2.0 | 0.1 | 1363 | 136 | $50 \times 97$ |
| 0.5 | 1.0 | 10799 | 10799 | $99 \times 48$ |
| 0.67 | 1.0 | 8249 | 8249 | $76 \times 49$ |
| 1.33 | 1.0 | 4509 | 4509 | $55 \times 71$ |
| 2.0 | 1.0 | 3601 | 3601 | $50 \times 97$ |

# REFERENCES

AITKEN, A. C. (1926) On Bernoulli's numerical solution of algebraic equations, *Proc. Roy. Soc. Edinburgh* **46**, 289.

BANK, R. & MITTELMANN, H. D. (1986) Continuation and multigrid for nonlinear elliptic systems. In *Multigrid Methods II* (ed. W. Hackbusch & U. Trottenberg) *Lect. Notes Math.* **1228**, Springer.

DAVIS, S. H. & von KERCZEK, C. (1973) A reformulation of energy stability theory. *Arch. Rat. Mech. Anal.* **52**, 112.

JOSEPH, D. D. (1976) *Stability of Fluid Motions I, II*, Springer-Verlag, Berlin.

NEITZEL, G. P. (1984) Numerical computation of time-dependent Taylor-vortex flows in finite-length geometries. *J. Fluid Mech.* **141**, 51.

NEITZEL, G. P. & DAVIS, S. H. (1981) Centrifugal instabilities during spin-down to rest in finite cylinders. Numerical experiments. *J. Fluid Mech.* **102**, 329.

PREISSER, F., SCHWABE, P. & SHARMANN, A. (1983) Steady and oscillatory thermocapillary convection in liquid columns with free cylindrical surface. *J. Fluid Mech.* **126**, 545.

SEN, A. K. & DAVIS, S. H. (1982) Steady thermocapillary flow in two dimensional slots. *J. Fluid Mech.* **121**, 163.

SHEN, Y., NEITZEL, G. P., JANKOWSKI, D. F. & MITTELMANN, H. D. (1989) Energy stability of thermocapillary convection in a model of the float-zone, crystal-growth process. Submitted to J. Fluid Mech.

SINHA, S. C. & CARMI, S. (1976) On the Liapunov-Movchan and the energy theories of stability. *J. Appl. Math. Phys.* **27**, 607–612.

# COMPUTING WITH REACTION–DIFFUSION SYSTEMS : APPLICATIONS IN IMAGE PROCESSING

C.B.Price  P.Wambacq  A.Oosterlinck
*ESAT-MI2*
*Katholieke Universiteit Leuven*
*Kardinaal Mercierlaan 94*
*B-3030 Heverlee, Belgium*

We recently proposed the use of Reaction-Diffusion (RD) Systems as a Computational Paradigm, with specific applications in Image Processing [3]. Here we briefly outline how we engineer and use RD–Systems to perform *image enhancement* and *analysis* tasks, and describe our bifurcation analysis of one equation, comparing this with experimental results.

## 1  Introduction

Important applications of image processing are found in image *enhancement* where the relative amount of desired image information is raised while lowering noise levels, and in image *analysis* where specific features are sought for within the image data. Within these application areas, an important class of images are those containing quasi–periodic information, like textures (eg textiles) and fingerprints. It is well–known that certain RD–systems exhibit a stationary–time spatially–periodic solution class, the *Turing Structures* [7], where the spatial period is defined by the system parameters and not by the boundary conditions. We engage this pattern–forming capability to selectively and adaptively enhance quasi–periodic image detail, this being the first engineering application of RD–systems. The design of a particular system involves matching two characteristics of the image to the RD–equations, namely the generated spatial frequencies and the generated pattern symmetries. The matching of spatial frequency proceeds from results of linear analysis alone. Non–linear analysis enables the *structure* of the RD-equations to be chosen to generate the desired symmetries, zero or one–dimensional patterns (blobs or lines). For example texture images (e.g., see Fig.1) have a quasi–periodic 0–dimensional structure while fingerprints (Fig.4) have a quasi–periodic 1–Dimensional structure. Our non–linear analysis suggests that a pure cubic nonlinearity will produce the latter structure, while a mixed quadratic–cubic nonlinearity will produce the former.

## 2  Equation and Parameter Selection

The general n–component RD–system is written as a nonlinear vector PDE;

$$\dot{\underline{C}}(\underline{x}, t) = \underline{F}(\underline{C}, \lambda) + \underline{\underline{D}}\nabla^2\underline{C}(\underline{x}, t) \tag{1}$$

where $\underline{C}$ is the concentration vector, $\underline{\underline{D}}$ the diagonal diffusion matrix and $\lambda$ a suitable bifurcation parameter. Unlike classical Reaction–Diffusion work, where equations are derived by

379

*D. Roose et al. (eds.), Continuation and Bifurcations: Numerical Techniques and Applications, 379–387.*
© 1990 *Kluwer Academic Publishers.*

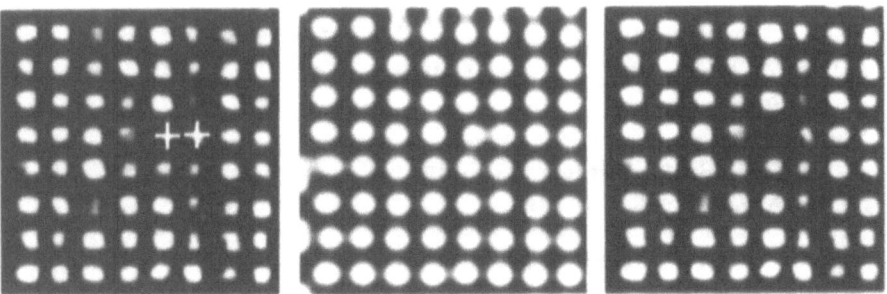

Figure 1: Processing of a Texture Image by a RD–System. Left is a typical texture image with defect, input as a perturbation to a RD–system. The evolution produces the centre pattern where the defect has been regenerated. The lower amplitude pattern near the defect enables it to be easily labelled, shown in the right image

modelling the underlying chemical, biological or ecological situation, we are free to choose any non–linear function (the *reaction 'kinetics'* consistent with the results above. The exact procedures used to build suitable equation sets is detailed in [4], where one criterion we implicitly invoked was to remain close to plausible chemical/biochemical systems, with the view to actually constructing a chemical computer. The resulting non–dimensionalized equations resulting from our design procedure are, for zero–dimensional quasi–periodic structures;

$$\frac{\partial E}{\partial t} = \frac{E^2}{I} - bE + \nabla^2 E$$
$$\frac{\partial I}{\partial t} = c + E^2 - I + D\nabla^2 I \tag{2}$$

which is a Meinhardt–Gierer type equation [1], and for one–dimensional quasi–periodic structures;

$$\frac{\partial E}{\partial t} = f(E) - I + D\nabla^2 E$$
$$\frac{\partial I}{\partial t} = \frac{1}{\epsilon}\left(E - I + \nabla^2 I\right)$$
$$f(E) = \alpha E(E + a)(a - E) \tag{3}$$

which resembles the Fitzhugh–Nagumo equation for pulse propagation along the nerve axon, (see [6]). Here the two system components are labelled $E$, the *excitory* component and $I$, the *inhibitory* component. The equation parameters $b,c,D$, and $\alpha,\epsilon,a,D$, are fixed using measurements of the spatial frequency composition of the image ensembles, so that the RD–system generates patterns of these frequencies. Details of this procedure, which starts from the linearized equations, are given in [4]. The unique feature of our systems is the operation far from the primary bifurcation point. Remember that Turing instability appears when the determinant of the Jacobian of the kinetic terms goes negative. For equation(3) this defines the *marginal stability curve*;

$$(f' - Dk^2)(1 + k^2) = 1 \tag{4}$$

shown in Fig.2, calculated for $D = 0.2101$ . Points above the curve correspond to Turing instability. The horizontal line at $f' = 0.7275$ corresponds to the parameter used in the enhancement processing shown in Fig.4. It is clear that the system has a range of frequencies where structure may grow. One reason for this is to allow growth of patterns in images where a range of spatial frequencies is present, our definition of *quasi*-periodicity. This large characteristic *activation band* suggests that a bifurcation analysis beyond the first bifurcation is required to fully catalogue the system behaviour, or alternatively, numerical computation of the full bifurcation diagram for a large 2–dimensional system. Our systems are large in the sense that the pattern wavelength is usually around 1/10'th of the linear system size.

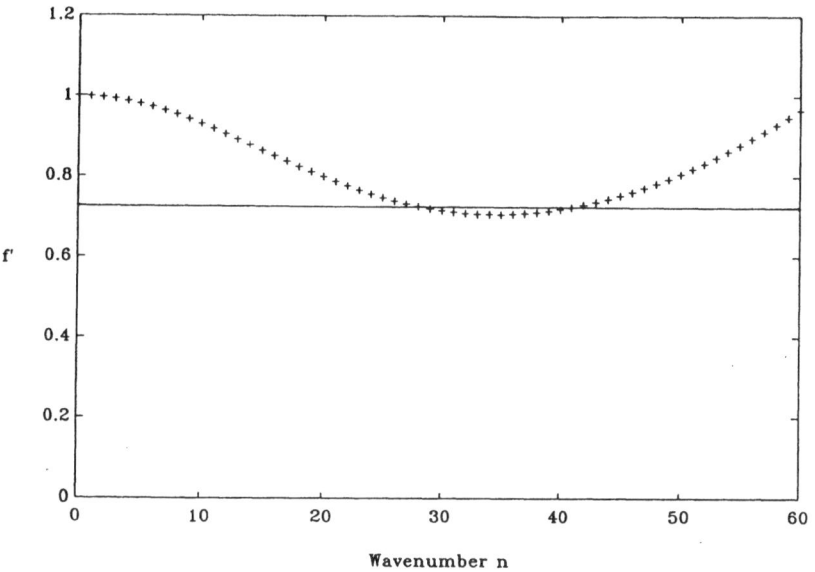

Wavenumber n

Figure 2: Marginal Stability Curve, equation (4), for a 100x100 image. Points above the curve correspond to unstable modes with wavevector $k = n\pi/100$. The indicated value of the bifurcation parameter $f'$ defines a broad activation band of unstable modes, used to enhance the fingerprint image (Fig.4)

# 3 Running a RD–Image Processing System

The equation parameters are set from measurements of the spatial frequency content of a particular image ensemble. Then the simulated RD–system is initially configured in a state of homogeneous equilibrium, and the input image is applied as a perturbation to one component (usually $E$). The system is allowed to evolve to a stable asymptotic state, which is then read out as the processed image. At the moment, simulations are run on a discrete 100x100 grid using an adaptive–timestep Runge–Kutta integrator. Run times on a VAX 8530 vary between 1 and 3 hours depending on the required parameter choice and ensuing stiffness of the system. From an algorithmic point of view, this processing paradigm may be classed as parallel and distributed. Its elegance stems from the adaptive self–organizing behaviour of this class of RD–systems where no complicated sequential programming or homunculi are required.

# 4 General Bifurcation Analysis

We have made two approaches to a non–linear analysis of the system. Firstly, an 'amplitude equation' approach has been taken where the PDE's are modelled as a small set of interacting plane–wave modes, described by coupled non–linear ODE's . This approach enables us to obtain a good qualitative feel for the system, and explains many observed pattern evolution scenarios. In particular, it shows how our RD–systems give differential growth of image detail near defects in the original periodicity, e.g., as shown in the texture image in Fig.1. Here, the lower growth rate of the 0–dimensional structure near the defect enables us to directly discover the location of the defect. In image–processing terms, this is a noteworthy result. Details of this work will appear elsewhere [4].

Our second approach to non–linear analysis is an analytical bifurcation analysis of the general n–component RD–system, which gives some important directives on how to build and use RD–systems in image processing. This is outlined in this section for the pure–cubic system (3) where the bifurcating solution is constructed, and its stability investigated. The analysis is carried out for a general RD–system on the infinite plane which leads to the choice of plane–waves as basis functions (see below). The reason for this was twofold. Firstly, experimental results indicated that with a suitable parameter choice leading to a wide activation band of unstable linear modes, the influence of the boundaries remains restricted to within one characteristic wavelength of the boundary. Evolving patterns in the body of the medium, and most important, their symmetries, were found to be independent of the boundary geometry. Secondly, we did not wish to impose any *a priori* restrictions on the nature of the bifurcating solutions.

Looking for solutions around the trivial homogeneous solution $\underline{C}_0$ satisfying

$$\underline{F}[\underline{C}_0(\lambda), \lambda] = 0 \tag{5}$$

we define the linear operator $\underline{\underline{L}}(\underline{C}_0)$

$$\underline{\underline{L}}(\underline{C}_0) = \underline{\underline{F}}_c(\lambda) + \underline{\underline{D}}\nabla^2 \tag{6}$$

or in $k$–space;

$$\underline{L}(k^2, \lambda) = \underline{F}_{\underline{c}}(\lambda) - k^2 \underline{D} \qquad (7)$$

where $\underline{F}_{\underline{c}}$ is the Jacobian of $\underline{F}$, evaluated at $\underline{C} = \underline{C}_0$. Assuming plane–wave solutions which are appropriate for our infinite medium, introducing the corresponding eigenfunctions of $\underline{L}(k^2, \lambda)$,

$$\underline{\Phi}_{n,k} = e^{i\underline{k}\cdot\underline{r}}\,|n\lambda k^2 > \qquad (8)$$

the linearized problem reduces to the eigenvalue problem,

$$[\underline{F}_c + \underline{D}\nabla^2]\Phi(\underline{x}) = \sigma\Phi(\underline{x}) \qquad (9)$$

where the eigenvectors satisfy,

$$\underline{L}(k^2, \lambda)\,|n\lambda k^2 > \,=\, \sigma_n(k^2, \lambda)\,|n\lambda k^2 > \qquad (10)$$

We also introduce the left–hand eigenvectors of the operator;

$$\sigma_n(k^2, \lambda) < n\lambda k^2| \,=\, < n\lambda k^2|\,\underline{L}(k^2, \lambda) \qquad (11)$$

The notation is taken from Nitzan and Ortoleva [2], the contents of the ket indicate the dependence of the eigenvectors on $k^2$ and $\lambda$, $n$ indicating the eigenvalue number. For a two–component system, for example $n = 0, 1$.

We now carry out the standard procedure of bifurcation analysis according to Sattinger [5], writing

$$\begin{aligned}
\underline{c} &= \epsilon\underline{\phi}_0 + \epsilon^2\underline{\psi}_0 + ..., \\
\lambda &= \epsilon\lambda_1 + \epsilon^2\lambda_2 + ...,
\end{aligned} \qquad (12)$$

and expanding the rate function $\underline{F}$ as,

$$\underline{F}(\underline{C}_0 + \underline{c}; \lambda) = \underline{F}_{\underline{c}}\underline{c} + \lambda\underline{F}_{\underline{c}\lambda}\underline{c} + \frac{1}{2}\underline{F}_{\underline{c}\underline{c}}\underline{c}\underline{c} + \frac{\lambda}{2}\underline{F}_{\underline{c}\underline{c}\lambda}\underline{c}\underline{c} + \frac{\lambda^2}{2}\underline{F}_{\underline{c}\lambda\lambda}\underline{c} + \frac{1}{6}\underline{F}_{\underline{c}\underline{c}\underline{c}}\underline{c}\underline{c}\underline{c} \qquad (13)$$

Substituting equations (12) and collecting terms of equal powers in $\epsilon$ generates the usual sequence of linear inhomogeneous equations;

$$\underline{L}\underline{\phi}_0 = 0$$

$$\underline{L}\underline{\psi}_0 + \lambda_1\underline{F}_{\underline{c}\lambda}\underline{\phi}_0 + \frac{1}{2}\underline{F}_{\underline{c}\underline{c}}\underline{\phi}_0\underline{\phi}_0 = 0$$

$$\underline{L}\underline{\psi}_1 + \underline{F}_{\underline{c}\underline{c}}\underline{\phi}_0\underline{\psi}_0 + \underline{F}_{\underline{c}\lambda}\underline{\psi}_0\lambda_1 + \lambda_2\underline{F}_{\underline{c}\lambda}\underline{\phi}_0 + \frac{\lambda_1}{2}\underline{F}_{\underline{c}\underline{c}\lambda}\underline{\phi}_0\underline{\phi}_0 + \frac{\lambda_1^2}{2}\underline{F}_{\underline{c}\lambda\lambda}\underline{\phi}_0 + \frac{1}{6}\underline{F}_{\underline{c}\underline{c}\underline{c}}\underline{\phi}_0\underline{\phi}_0\underline{\phi}_0 = 0$$

For the case of a pure cubic nonlinearity, all terms involving $\underline{F}_{\underline{c}\underline{c}}$ disappear, leaving a simpler sequence of equations,

$$\underline{L}\underline{\phi}_0 = 0$$

$$\underline{L}\underline{\psi}_0 + \lambda_1\underline{F}_{\underline{c}\lambda}\underline{\phi}_0 = 0$$

$$\underline{L}\underline{\psi}_1 + \lambda_1\underline{F}_{\underline{c}\lambda}\underline{\psi}_0 + \frac{\lambda_1^2}{2}\underline{F}_{\underline{c}\lambda\lambda}\underline{\phi}_0 + \frac{1}{6}\underline{F}_{\underline{c}\underline{c}\underline{c}}\underline{\phi}_0\underline{\phi}_0\underline{\phi}_0 = 0 \qquad (14)$$

From the first–order equation, and given the degeneracy of the null eigenvalue, we look for solutions as sums of plane–waves;

$$\phi_0 = \sum_{s=1}^{N/2}(A_s e^{i\underline{k}_s \cdot \underline{r}} + A_s^* e^{-i\underline{k}_s \cdot \underline{r}})|00k_0^2> \tag{15}$$

The eigenvector $|00k_0^2>$ corresponds to the principal eigenvalue of $\underline{L}$ which rises to zero at the bifurcation point $\lambda = 0$. Here $s$ gives the direction of the wavevector $\underline{k}_s$, and all $|\underline{k}_s|$ are equal to the critical wavevector magnitude $k_0$. The above sum is over $N/2$ directions. Application of the usual solvability condition to the second–order equation yields $\lambda_1 = 0$. The function $\phi_0$ remains undetermined in this order, but is obtained by application of the solvability condition to the third–order equation as;

$$\underline{c} = \frac{2}{\sqrt{N-1}}\sqrt{\frac{-2\lambda P}{C}}\sum_s^{N/2} cos(\underline{k}_s \cdot \underline{r})|00k_0^2> \tag{16}$$

where numbers $P$ and $C$ are given by;

$$\begin{aligned} P &= <00k_0^2|\underline{F}_{c\lambda}|00k_0^2> \\ C &= <00k_0^2|\underline{F}_{ccc}|00k_0^2>|00k_0^2>|00k_0^2> \end{aligned} \tag{17}$$

If $C < 0$ then the bifurcation is supercritical and the bifurcating solution is stable (Fig.3). This turns out to be the case when this general analysis is specified to the pure–cubic equation (3), as described below.

Concerning the stability of these plane–wave solutions, the system was linearized around the bifurcating branch $[\underline{c}(\epsilon), \lambda(\epsilon)]$, and a solution obtained at the second order. Testing stability of modes within the above set of $N/2$ plane waves, the secular determinant yields one eigenvalue $CN/2$, and $N/2 - 1$ zero eigenvalues. Thus a single plane–wave is stable per se, but a set of $N/2$ plane waves has $N/2 - 1$ neutral waves, at least at this order. Since our experimental results showed exclusively single plane–wave solutions, this analysis was not pushed any further. Testing stability of the modes against other plane–waves external to the above set of $N/2$ waves, showed the set to be stable to perturbations by these, provided that $C < 0$. This too provided sufficient analysis for our system.

## 5 The Pure Cubic Equation

As mentioned above, the pure cubic equation gives $C < 0$, yielding the first bifurcating solution;

$$\underline{c} = \frac{2}{\sqrt{N-1}}\sqrt{\frac{f'-f_0'}{3\alpha}}\sum_s^{N/2} cos(\underline{k}_s \cdot \underline{r})|00k_0^2> \tag{18}$$

where

$$\begin{aligned} k_0^2 &= \frac{1}{\sqrt{D}} - 1 \\ f_0' &= 2\sqrt{D} - D \end{aligned} \tag{19}$$

Results from several trial simulations in 2–dimensions provided us with a set of amplitudes for various values of the bifurcation parameter $f'$. The stable solution in all cases was a single plane–wave, as shown in the fingerprint image in Fig.4. These experimental results are also plotted on Fig.3 and show good agreement with the function (18), with expected divergence at higher amplitudes where higher–order terms will be required. This is pleasantly surprising

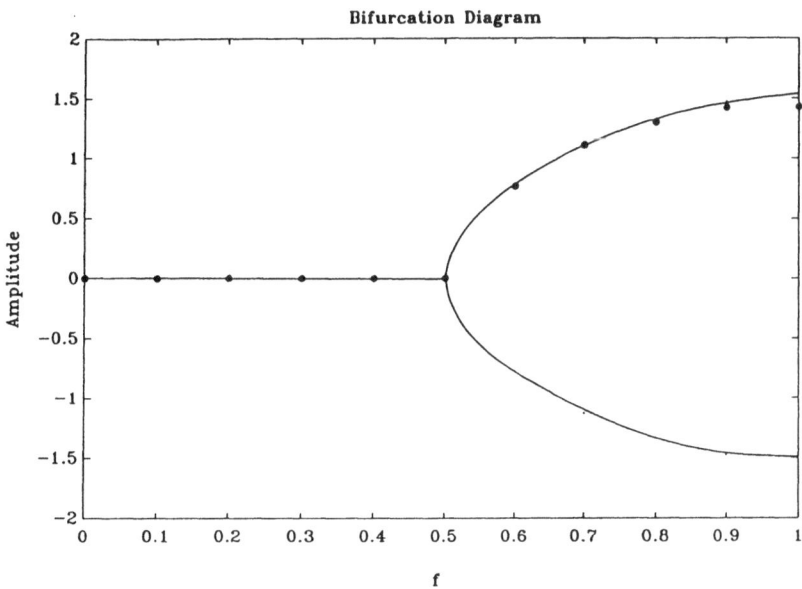

Figure 3: Bifurcation diagram for the pure cubic system (3). The solid curve is produced from equation (18) and the dots are experimental results

since our systems were run with $f'$ way past the primary bifurcation point, in a region where we would expect many successive and secondary bifurcations, as suggested by the marginal stability curve (Fig.2). We emphasize that no continuation (or other) techniques were used to produce these amplitudes, which were available from various unconnected simulations. The parameters used to produce these points are; $D = 0.1$, $a = 2.0$, $f'_0 = 0.5325$, $\epsilon = 0.25$, $\alpha = 0.1331$.

# 6 Conclusions

Reaction–Diffusion systems are able to provide robust, adaptive processing of image data. Linear and non–linear analysis provide complementary design tools enabling to engineer such systems. While the amplitude equation approach has shown excellent modelling of the equations we have studied, we feel that a fuller (numerical) bifurcation approach is needed, especially in the case of equation (2). At the moment, for our large 2–dimensional systems,

386

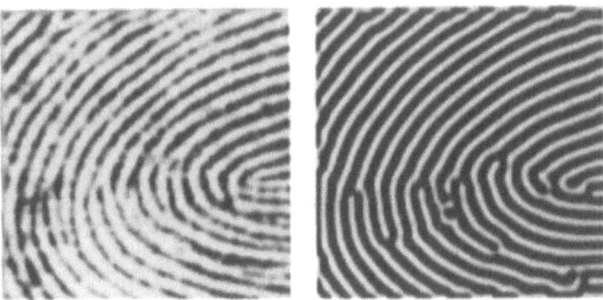

Figure 4: Enhancement of Fingerprint image using equation (3). Left is the noisy input image and right the final processed image, which is almost binary in nature. The locations of dislocations in the original image are well–preserved.

numerical computation of a full bifurcation diagram seems out of the question.

# References

[1] Meinhardt H.
*Models of Biological Pattern Formation*
Academic Press, London, 1982.

[2] Nitzan A., Ortoleva P.
*Scaling and Ginzburg criteria for critical bifurcations in nonequilibrium reacting systems*
Phys.Rev.A. Vol.21 Nr.5 (1980) 1735-1755

[3] Price C.B., Wambacq P., Oosterlinck A.
*Applications of Reaction–Diffusion Equations to Image Processing* IEE 3rd Int. Conf.
Image Processing and its Applications (1989) 49-53

[4] Price C.B., Wambacq P., Oosterlinck A.
*Image Enhancement and Analysis with the Reaction–Diffusion Paradigm*
IEE Proceedings Communications, Speech and Vision. (submitted).

[5] Sattinger D.H.
*Topics in Stability and Bifurcation Theory*
Lecture Notes in Mathematics Vol 309 (1973)

[6] Segel L.
*Mathematical Models in Molecular and Cellular Biology*
Cambridge (1980)

[7] Turing A.M.
*The Chemical basis of Morphogenesis*
Philos.Trans.Roy.Soc.London Ser B 237 (1952) 37-72

# Bifurcation of Codimension 2 for a discrete map

A. Steindl
Inst. of Mechanics
Techn. Univ. Vienna
Wiedner Hauptstraße 8-10
A-1040 Wien

ABSTRACT: We investigate the nonlinear stability of the periodic motion of a simple robot. In a 2-dimensional parameter space the boundaries of linear stability are computed. At the intersection of two boundary lines an interaction between a transcritical and a Flip-bifurcation takes place. Calculating the discrete Time-T-map and reducing the non-linear equations by Center Manifold projection and Normal Form theory a $Z_2$-symmetric bifurcation equation is obtained which is simplified further by symmetry preserving equivalences. The discussion of the final system of 2 quadratic equations yields a partition of the parameter space into open sets of qualitatively similar bifurcation diagrams.

## 1 Introduction

Bifurcation theory has developed a powerful toolset to analyze complicated systems by condensing them to rather simple equations that reproduce the essential behavior of the complete system. The Poincaré-map or Time-T-map can be used very efficiently to reduce a nonlinear system of periodic differential equations to a small set of simple algebraic equations. The necessary computations can be performed by standard numerical software. We apply this method to the control problem of a simple robot.

## 2 Mechanical model

We consider a planar robot consisting of 2 rigid arms connected by hinges (Fig. 1). The robot is controlled by drive moments $M_1^d$ and $M_2^d$ acting at its hinges. Its endpoint $x_G$ is intended to move on a circle with constant angular velocity $\omega$. We investigate the stability of this motion under variation of $\omega$ and a characteristic value of the controller.

The 2 degrees of freedom of the robot are given by the angles $\varphi_1$ and $\varphi_2$; the equations of motion can easily be derived by Lagrange's formula [1]:

$$
\begin{pmatrix} \mu/2 + 5/4 + \cos\varphi_2 & -(1 + 2\cos\varphi_2)/4 \\ -(1 + 2\cos\varphi_2)/4 & 1/3 \end{pmatrix} \begin{pmatrix} \ddot{\varphi}_1 \\ \ddot{\varphi}_2 \end{pmatrix} +
$$

$$
+ \begin{pmatrix} k_1\dot{\varphi}_1 + 1/2\,\dot{\varphi}_2(\dot{\varphi}_2 - 2\dot{\varphi}_1)\sin\varphi_2 \\ k_2\dot{\varphi}_2 + 1/2\,\dot{\varphi}_1^2\sin\varphi_2 \end{pmatrix} +
$$

$$
+ k_g \begin{pmatrix} \mu/2 + 1 + 1/2\,\cos(\varphi_2 - \varphi_1) \\ -1/2\,\cos(\varphi_2 - \varphi_1) \end{pmatrix} = \begin{pmatrix} M_1 \\ M_2 \end{pmatrix} \tag{1}
$$

389

D. Roose et al. (eds.), Continuation and Bifurcations: Numerical Techniques and Applications, 389–396.
© 1990 Kluwer Academic Publishers.

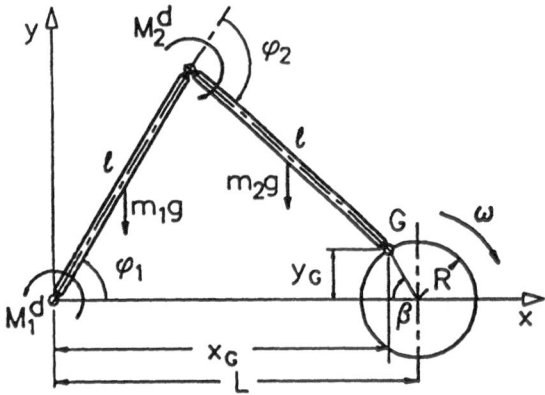

Figure 1: Mechanical model of the robot

where the following nondimensional parameters

$$\mu = m_1/m_2, \qquad M_i = M_i^d/m_2\ell^2\omega^2, \qquad k_i = K_i/m_2\ell^2\omega,$$

$$k_g = g/\omega^2\ell, \qquad t = \tau\omega, \qquad\qquad (\dot{.}) = \partial(.)/\partial t.$$

are used.

The drive moments $M_i$ consist of two parts:

$$M_i = M_i^0 + \Delta M_i. \tag{2}$$

The portion $M_i^0$ produces the nominal circular motion $\varphi^0$ of the robot. It is calculated by inserting the appropriate functions $\varphi_i^0(t)$ and their derivatives into (1). The control portion $\Delta M_i$ is introduced to correct deviations of the actual motion from the prescribed one. We stipulate a linear position dependent control law (*PD-controller*)

$$\boldsymbol{\Delta M} = R_k \begin{pmatrix} 1 & 0 \\ 0 & 1/3 \end{pmatrix} (\varphi - \varphi^0). \tag{3}$$

The constant $R_k$ will serve as distinguished bifurcation parameter.

After a shift of coordinates $\psi_i = \varphi_i - \varphi_i^0$ the differential equations are transformed to a periodic first order system:

$$\dot{y} = f(t, y) \tag{4}$$

with $f(t + 2\pi, y) = f(t, y)$.

**Remarks**

1. Due to the construction of the drive moments the trivial function $y \equiv o$ is a solution of (4) for all parameter values.

2. Since some quantities have been neglected in our investigation, (eg. dry friction, backlashes, restrictions from the control mechanism) we have to introduce imperfection parameters and unfold our equations to account for the differences

$$\dot{y} = F(t, y, \alpha), \tag{5}$$

where $\alpha \in \mathbf{R}^p$ denotes the imperfection parameters and $F(t, y, \mathbf{o}) \equiv f(t, y)$. Remark 1 is valid only for $\alpha = \mathbf{o}$. The further investigations are carried out for the perfect system (4).

## 3  Calculation of Time-T-map

Since we want to study a dynamical system near a periodic orbit we simplify our computations by using the discrete Time-T-map, which assigns to each initial state $\eta$ at $t = 0$ the state after one period:

$$P : \eta \mapsto P(\eta) = y(T; 0, \eta), \tag{6}$$

where $y(t; t_0, \eta)$ denotes the solution of (4) with initial conditions

$$y(t_0) = \eta.$$

The coefficients of the Taylor-expansion of P are calculated by solving a moderately large system of linear inhomogeneous differential equations: The constant part $P(\mathbf{o})$ vanishes since $y \equiv \mathbf{o}$ is a periodic solution of (4). To calculate the linear part we differentiate (4) w.r.t. $\eta_i$ and reverse the order of differentiation:

$$\frac{d}{dt}\frac{\partial y}{\partial \eta_i} = \frac{\partial f}{\partial y}(t, \mathbf{o})\frac{\partial y}{\partial \eta_i}, \tag{7}$$

$$\frac{\partial y_j}{\partial \eta_i}(0) = \delta_{ij}, \tag{8}$$

or shortly with $y_i = \partial y / \partial \eta_i$

$$\dot{y}_i = f'(t, \mathbf{o})y_i, \tag{9}$$
$$y_i(0) = e_i. \tag{10}$$

The solution vectors $y_i(T)$ yield the columns of the linearization matrix of $P$, which is also well known as *Fundamental Solution Matrix*.

The partial derivatives of second order are determined by the solution vectors of the system

$$\dot{y}_{ij} = f'(t, \mathbf{o})y_{ij} + f''(t, \mathbf{o})(y_i, y_j), \tag{11}$$
$$y_{ij}(0) = \mathbf{o}, \tag{12}$$

with $y_{ij} = \partial^2 y / \partial \eta_i \partial \eta_j$.

Applying this procedure again we obtain the third order derivatives. The amount of computation seems to grow very rapidly with the order of expansion, but it turns out that only a small part of coefficients is actually needed in the bifurcation equations. By choosing the eigenvectors of the linearized map as basis the resulting system of equations can be reduced considerably.

## 4    Stability of the fundamental solution

The stability of the prescribed motion $y = o$ is determined by the eigenvalues $\mu$ of the linear map $P_1 = P'(o)$: If all eigenvalues of $P_1$ are located in the interior of the unit circle the motion is stable.

There exist 3 simple cases of loss of stability:

- $\mu = 1$: Transcritical bifurcation. A solution of period $T$ branches off the trivial solution. The robot performs a periodic motion different from the prescribed one.

- $\mu = -1$: Flip bifurcation. The motion of the robot closes after two periods.

- $\mu = e^{i\vartheta}$: Hopf bifurcation. It leads to an quasiperiodic motion on a Torus.

By varying the driving frequency $\omega$ and the control parameter $R_k$ we obtain the map in Fig. 2. All simple types of bifurcations are found for our model. The boundaries of the

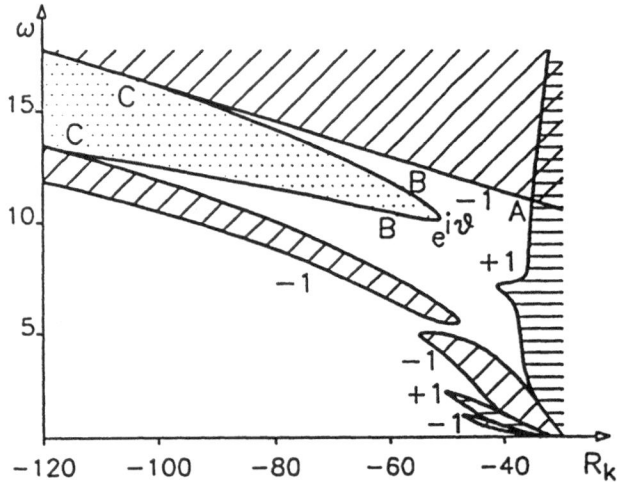

Figure 2: Stability diagram for different values of the driving frequency $\omega$ and controller constant $R_k$.

instability regions were computed by path following methods [4].

The instability regions in the lower right corner are due to a nearby unstable static solution, there seem to exist infinitely many tongues accumulating at $\omega = 0$.

Besides these bifurcations of codimension 1 several bifurcations of codimension 2 are found:

- Along the Hopf bifurcation border higher order resonances appear, for example resonances of order 3 ($\vartheta = 2\pi/3$) at points B.

- The Hopf bifurcation curve meets the Flip curves resulting in a semi-simple eigenvalue $-1$ at points C.

- At point A the linear map becomes unstable due to eigenvalues 1 and $-1$. The further investigations are carried out for that particular case.

## 5  Reduction to Normal Form

All methods of bifurcation theory can be applied to simplify the investigation of the non-linear behavior of the discrete Time-T-map to obtain a manageable system of algebraic equations which describe any small solution of the original problem.

First we perform a linear change of coordinates to bring the linear map into Jordan Normal Form. Second we use *Center Manifold theory* to eliminate the stable solution components. The 2-dimensional map on the Center Manifold is denoted by

$$w_{n+1} = Jw_n + \tilde{f}(w_n), \tag{13}$$

with

$$J = \begin{pmatrix} 1 & 0 \\ 0 & -1 \end{pmatrix}.$$

Now we apply classical *Normal Form Theory* to simplify the nonlinear terms in $\tilde{f}(w)$: By a near identity transformation

$$w = z + h(z) \tag{14}$$

we introduce new coordinates $z$ and insert in (13)

$$z_{n+1} + h(z_{n+1}) = J(z_n + h(z_n)) + \tilde{f}(z_n + h(z_n)). \tag{15}$$

Collecting the lowest order nonlinear terms we obtain the linear equation for the coefficients of $h$

$$h(Jz_n) - Jh(z_n) = \tilde{f}(z_n). \tag{16}$$

The kernel of the homological Operator $H(h)(z) = h(Jz) - Jh(z)$ consists of functions that commute with $J$. The Normal Form consists of those functions $\tilde{f}$ in (16), which are not in the image space of the operator $H$. Since $H$ acts diagonally on vectors of monomials, the complement of the image of $H$ is spanned by $J$-equivariant functions. Therefore the Normal Form generates a mathematical symmetry in the equations. The simplified map is denoted by

$$z_{n+1} = Jz_n + \tilde{g}(z_n)$$

Now the matrix $J$ is a standard representation for $Z_2$-symmetry. So we can use Golubitsky's results on mode interactions with $Z_2$-symmetry [3] to continue our studies.

Before we can do that we have to create a bifurcation equation for our system. Since we expect orbits of period T and 2T to branch off the trivial solution we set up an equation for 2T-periodic solutions:

$$
\begin{aligned}
z_n &= z_{n+2} = J(Jz_n + \tilde{g}(z_n)) + \tilde{g}(Jz_n + \tilde{g}(z_n)) \\
&= z_n + g(z_n)
\end{aligned} \tag{17}
$$

Dropping the index and denoting the dependence of this equation on the parameters explicitly we arrive at the $Z_2$-equivariant bifurcation problem

$$g(z, \lambda) = 0. \tag{18}$$

## 5.1 $Z_2$-EQUIVALENT NORMAL FORM

By using $Z_2$-equivalences [3] equation (18) can be simplified further to give

$$\begin{aligned}
\varepsilon_1 x^2 + \varepsilon_2 y^2 + \varepsilon_3 \lambda x - \alpha &= 0, \\
(\rho x + \varepsilon_4 \lambda - 2\beta)y &= 0
\end{aligned} \tag{19}$$

with

$$\varepsilon_1 = \mathrm{sgn}(g_{1,xx}), \quad \varepsilon_2 = \mathrm{sgn}(g_{1,yy}), \quad \varepsilon_3 = \mathrm{sgn}(g_{1,x\lambda}),$$

$$\varepsilon_4 = \mathrm{sgn}(g_{2,y\lambda}), \quad \rho = 2\frac{g_{1,x\lambda}g_{2,xy}}{g_{1,xx}g_{2,y\lambda}}$$

where $\lambda$ denotes the distinguished parameter, $\alpha$ and $\beta$ are imperfection parameters and $\rho$ is a modular parameter.

While $\beta$ can be related to physical parameters in our model, the imperfection $\alpha$ has to be calculated from perturbations to the model, for example a small friction coefficient or an additional mass applied at the endpoint.

Generically the system (19) has 2 different types of solution:

1. The symmetric branch of $2\pi$-periodic solutions ($Jz = z$)

$$\begin{aligned}
\varepsilon_1 x^2 + \varepsilon_3 x\lambda - \alpha &= 0, \\
y &= 0
\end{aligned} \tag{20}$$

yields the zeros of the simple transcritical bifurcation. The $Z_2$-symmetry forces the Jacobian at the zeros to be diagonal:

$$dg = \begin{pmatrix} 2\varepsilon_1 x + \varepsilon_3\lambda & 0 \\ 0 & \rho x + \varepsilon_4\lambda - 2\beta \end{pmatrix}. \tag{21}$$

2. The mixed-mode solution branch

$$\begin{aligned}
(\varepsilon_1 - \varepsilon_3\varepsilon_4\rho)x^2 + \varepsilon_2 y^2 + 2\varepsilon_3\varepsilon_4\beta x - \alpha &= 0, \\
\lambda &= \varepsilon_4(2\beta - \rho x)
\end{aligned} \tag{22}$$

consists of solutions with period $4\pi$. The Jacobian turns out to be

$$dg = \begin{pmatrix} \varepsilon_1 x + \varepsilon_3\lambda & 2\varepsilon_2 y \\ \rho y & 0 \end{pmatrix}. \tag{23}$$

The zeros of (19) are stable, if the eigenvalues of the matrix $I + dg$ are inside the unit circle, where $I$ denotes the unit matrix. For small zeros of (19) this requires the eigenvalues of $dg$ to lie in the left half plane.

## 5.2 NUMERICAL RESULTS

If we fix the angular velocity $\omega$ and vary the controller constant $R_k$ the quantities in (19) are as follows:

$$\varepsilon_1 = 1, \quad \varepsilon_2 = 1, \quad \varepsilon_3 = 1$$
$$\varepsilon_4 = 1, \quad \rho \approx 1.5$$

We provide a short sketch of the discussion of the bifurcation equations, a complete classification can be found in [3].

The symmetric solution set consists of two hyperbolas in $(x, \lambda)$-space. If $\alpha > 0$ one hyperbola is throughout stable, the other one unstable wrt. the eigenvalue in the $Z_2$-invariant subspace $y = 0$. A secondary bifurcation occurs, if one of the hyperbolas intersects the straight line $\lambda = 2\beta - \rho x$. The parabola

$$\alpha = \frac{\beta^2}{\rho - 1}$$

separates the 3 possible cases.

If $\alpha < 0$ both hyperbolas contain a stable and an unstable part, connected by *Fold points* at

$$\lambda = \pm 2\sqrt{-\alpha},$$
$$x = -\lambda/2.$$

A secondary bifurcation occurs at the intersections of the straight line $\lambda = 2\beta - \rho x$ with the hyperbolas. Along the parabola $\alpha = -4\beta^2/(2 - \rho)^2$ one of these intersection points coincides with a fold point.

The $4\pi$-periodic solutions are always unstable, because $\det(dg) = -\rho y^2 < 0$ along this branch.

The $(\alpha, \beta)$-parameter plane consists of 6 regions with qualitatively different bifurcation diagrams. One of the most interesting cases occurs in region 3. It corresponds to a horizontal path in the $(R_k, \omega)$-plane in Fig. 2 slightly above point A; the external imperfection parameter $\alpha$ is considered to be negative and much smaller than $\beta$, which is proportional to $\omega - \omega_0$. Starting in the stable region the robot reaches the flip-bifurcation border, where an unstable $4\pi$-periodic solution branches off subcritically from the slightly perturbed prescribed solution. The $2\pi$-periodic unstable branch continues until it undergoes

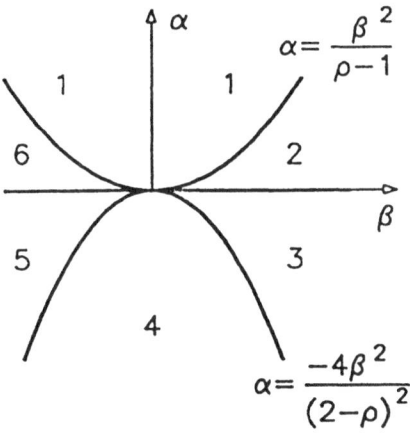

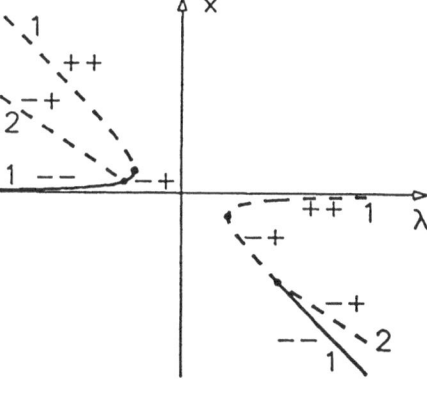

Figure 3: Transition varieties giving a partition of the $(\alpha, \beta)$-plane into regions of qualitatively different bifurcation diagrams.

Figure 4: Bifurcation diagram in region 3 of Fig. 3. Stable eigenvalues are indicated by a '$-$'-sign. Bifurcation points are indicated by dots.

the saddle-node bifurcation. After a small gap, where no small periodic solutions exist, 2 new solution branches are created by a second saddle-node and one of these branches gains stability by a second Flip-bifurcation.

## 6  Conclusions

Imperfections play an important role when the robot operates near a stability limit. Also in parameter regions of stable operating conditions they could lead to intolerable deviations from the required behavior. Therefore the influence of all possible perturbations on the motions should be estimated and the control mechanism has to be improved to avoid instability phenomena in the domain of possible working conditions.

Numerical computations of the T-map are very useful to investigate the dynamics near the instability regions. They could also be used to calculate the influence of different perturbations and control mechanisms on the performance of the equipment.

## References

[1] Lindtner E.: Anwendung der Verzweigungstheorie zur Bestimmung des überkritischen Stabilitätsverhaltens von Roboterbewegungen. VDI 18. VDI-Verlag

[2] Lindtner E., Steindl A., Troger H.: Generic one-parameter bifurcations in the motion of a simple robot. Journal of Computational and Applied Mathematics 26 (1989) 199–218 North Holland

[3] Golubitsky M., Schaeffer D.: Singularities and Groups in Bifurcation Theory, vol. 2, Applied Math. Sciences, Springer-Verlag, New York, Heidelberg 1985.

[4] Seydel R.: 'A Continuation Algorithm with step control.' International Series of Numerical Mathematics, Vol. 70, 1984, Birkhäuser Verlag Basel

# Bifurcation of periodic solutions in PDE's: Numerical techniques and applications

Holodniok M., Kubíček M. and Marek M.
Prague Institute of Chemical Technology, 166 28 Praha 6, Czechoslovakia

The method of lines with different types of space discretization has been used for the study of the behaviour of solutions of two parabolic PDE's with Brusselator reaction scheme in the form

$$\frac{\partial x}{\partial t} = \frac{D_x}{L^2} \frac{\partial^2 x}{\partial z^2} + x^2 y - (B + 1)x + A \qquad (1a)$$

$$\frac{\partial y}{\partial t} = \frac{D_y}{L^2} \frac{\partial^2 y}{\partial z^2} + Bx - x^2 y \quad . \qquad (1b)$$

with boundary conditions of the Dirichlet type

$$x(0,t)=x(1,t)=A \qquad y(0,t)=y(1,t)=B/A \qquad (2)$$

More details about the model can be found e.g. in [1]. The spatial derivatives in the system (1) have been approximated by a three-point discretization formula and analysis of resulting system of ODE's has been made in the form of the construction of the dependence of periodic solutions on the characteristic parameter (length of the system L) by means

D. Roose et al. (eds.), Continuation and Bifurcations: Numerical Techniques and Applications, 397–399.
© 1990 Kluwer Academic Publishers.

of continuation techniques. The point of the bifurcation of an invariant torus (L $\doteq$ 1.3676) on the branch of periodic solutions has been localized for the following parameter values: A = 2, B = 5.45, $D_x$ = 0.008, $D_y$ = 0.004. Discretizations based on a five-point formula and orthogonal collocation formula have been also used for the analysis of periodic solutions. Comparison of results for all three methods of discretization and varying number of mesh points was also discussed in [2] and [3]. The comparison shows that the simplest three-point discretization formula with the number of mesh points N = 21 gives qualitatively good results (the number of ODE's is in that case 38).

The invariant torus found by a dynamic simulation in the neighbourhood of the bifurcation point is depicted in Fig. 1 in the form of the trajectory of the Poincaré map constructed as an intersection of the trajectory of the system with the 37-dimensional hyperplane defined as $x(\tilde{t},0.3) \sim x_6(\tilde{t})=2$. The development of this quasiperiodic solution and its desintegration into a chaotic attractor via

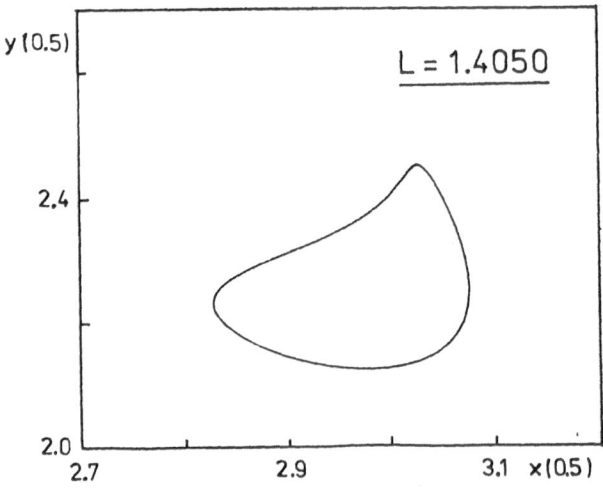

Fig. 1: Poincaré map of a simple torus, L = 1.4050

a cascade of the torus doublings were presented e.g. in [3].
An example of the Poincaré map of the double-torus which is
in the form of a connected two-folded invariant circle in
the hyperplane is depicted in Fig. 2.

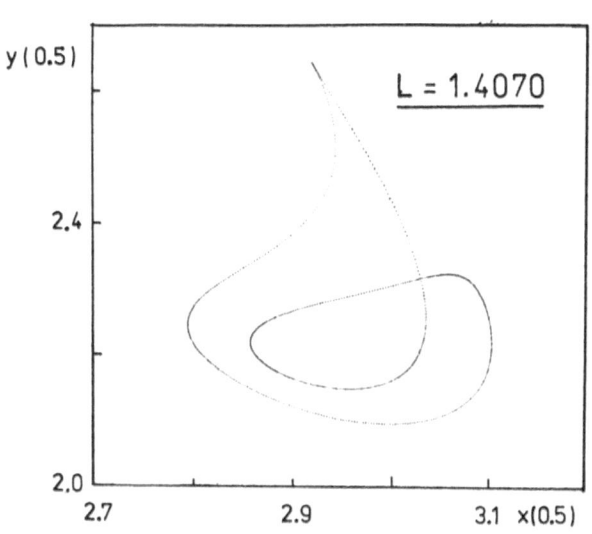

Fig. 2: Poincaré map for a double-torus, L = 1.4070

Literature
----------

[1]  Kubíček M., Marek M.: Computational Methods in
     Bifurcation Theory and Dissipative Structures.
     Springer Verlag, New York 1983.

[2]  Holodniok M., Kubíček M.: Numerical aspects of
     computation of periodic and quasiperiodic solutions in
     systems of partial differential equations. Proceedings
     of the International Conference EQUADIFF-7, Prague
     1989, in press.

[3]  Holodniok, M., Kubíček, M., Marek, M.: Desintegration
     of an invariant torus in a reaction - diffusion system.
     Teubner Texte zur Mathematik, Teubner Leipzig 1990, in
     press.

# BIFURCATION AND CHAOS IN CHUA'S CIRCUIT FAMILY

L. O. CHUA
University of California, Berkeley
Department of Electrical Engineering
and Computer Sciences
Berkeley, CA 94720
U.S.A.

ABSTRACT: This paper surveys a wide variety of bifurcation and chaotic phenomena that can be observed from the normal form equation associated with a simple family of electronic circuits as described in Wu [1]. The equation consists of 3 ordinary differential equations whose only nonlinearity is a piecewise-linear function of one variable. A great deal of experimental, numerical, and analytical results concerning this equation has been published [2-23] the highlights of which will be surveyed. In the context of the theme of this workshop, it is interesting to point out that in addition to some new numerical techniques applicable only to piecewise-linear equations, the continuation method developed in [24-25] has played an important role in deriving the many 2-parameter bifurcation diagrams.

*D. Roose et al. (eds.), Continuation and Bifurcations: Numerical Techniques and Applications, 401–403.*
© 1990 *Kluwer Academic Publishers.*

## REFERENCES

[1] Wu, S. (8/1987) "Chua's Circuit Family", *Proceeds of the IEEE*, 1022-32

[2] Matsumoto, T. (12/1984) "A Chaotic Attractor From Chua's Circuit", *IEEE Trans. Circuits Sys.*, Vol. CAS-31, 1055-1058

[3] Zhong, G.-Q, and Ayrom, F. (1/1985), "Experimental Confirmation of Chaos from Chua's Circuit", *Int. J. Circuit Theory Appl.*, Vol 13, 93-98

[4] Zhong, G.-Q. and Ayrom, F. (5/1985) "Periodicity and Chaos in Chua's Circuits", *IEEE Trans. Circuits Syst.*, Vol. CAS-32, 797-818

[5] Matsumoto, T., Chua, L.O, and Komuro, N., (8/1985) "The Double Scroll", *IEEE Trans. Circuits Syst.*", Vol CAS-32 797-818

[6] Matsumoto, T., Chua, L.O., and Komuro, M., (8/1986) "Double Scroll via a Two-Transistor Circuit", *IEEE Trans. Circuits Syst.*, Vol. CAS-33, 828-835

[7] Matsumoto, T., Chua, L.O., and Komuro, M. (4/1986) "The Double Scroll Bifurcations", *Int. J. Circuit Theory Appl.*", Vol 14, 117-146

[8] Chua, L.O., Komuro, M., and Matsumoto, T. (11/1986) "The Double Scroll Family," *IEEE Trans. Circuits Syst.*, Vol CAS-33, 1072-1118

[9] Mees, I., and Chapman, P.B., (9/1987) "Homoclinic and Heteroclinic Orbits in the Double Scroll Attractor", *IEEE Trans. Circuits Syst.*, Vol CAS-34, no. 9

[10] Broucke, M.E. (2/1987) "One-Parameter bifurcation Diagram for Chua's Circuit", *IEEE Trans. Circuits Syst.*, Vol. CAS-34 208-209

[11] Yang, L., and Liao, L. (4/1987) "Self-similar Bifurcation Structures from Chua's Circuit", *Int. J. Circuit Theory Appl.*, 189-192

[12] Parker, T.S., and Chua, L.O., (9/1987) "The dual double scroll equation" *IEEE Trans. Circuits Syst.* Vol. CAS-34 no.9

[13] Kahlert, C., (9/1987) "The chaos producing mechanism in Chua's circuit and related piecewise-linear dynamical systems." *Proceedings of the 1987 European Conference on Circuit Theory and Design*

[14] Kahlert, C., and Chua, L.O., (1/1987) "Transfer maps and return maps for piecewise-linear three-region dynamical systems" *Int. J. Circuit Theory Appl.*, Vol.15(1) 23-49

[15] Matsumoto, T., Chua, L.O., and Komuro, M. (1987) "Birth and Death of the Double Scroll" *Physica 24D*, 97-124

[16] Matsumoto, T., Chua, L.O., and Tokunaga, R., (3/1987) "Chaos via Torus Breakdown" *IEEE Trans. Circuits Syst.*, Vol. 34, 240-253

[17] Kahlert, C. (1988) "The chaos producing mechanism in Chua's circuit" *Int. J. Circuits Theory Appl.*, Vol. 16 227-232

[18] Kahlert, C. (1988) "The ranges of transfer and return maps in three-region piecewise-linear dynamical systems" *Int. J. Circuit Theory Appl.* Vol 16 11-23

[19] Bartissol, P., and Chua, L.O., (12/1988) "The Double Hook" *IEEE Trans. Circuits Syst.*, 1512-1522 1512-1522

[20] Kahlert, C., (1989) "Dynamics of the inclusions appearing in the return maps of Chua's Circuit — I. The creation mechanism" *Int. J. Circuit Theory Appl.*, Vol. 17, 20-46

[21] Kahlert, C. (1989) "Dynamics of the inclusions appearing in the return maps of Chua's Circuit — II. The Annihilation Mechanism," *Int. J. Circuit Theory Appl.*, Vol 17, in press

[22] Kahlert, C. (1990) "Heterodinic Orbits and Scaled Similar Structures in the Parameter Space of the Chua Oscillator" to be published in G. Baier and M. Klein (eds.), A Chaotic Hierachy *World Scientific*, Singapore

[23] Parker, T.S. , and Chua, L.O., (1989) "Practical Numerical Algorithms for Chaotic Systems, Springer Verlag, New York

[24] Chua, L.O. and Ushida, A., (7/1976) "A Switching-parameter algorithm for finding multiple solutions of nonlinear resistive circuits" *Int. J. Circuits Syst.*, Vol.4, 215-239

[25] Ushida, A. and Chua, L.O. (1/1984) "Tracing Solution Curves of Nonlinear Equations with Sharp Turning Points" *Int. J. Circuits Syst.*, Vol. 12, 1-21

# CONTINUATION AND COLLOCATION FOR PARAMETER DEPENDENT BOUNDARY VALUE PROBLEMS *

G. BADER and P. KUNKEL
Universität Heidelberg
Institut für Angewandte Mathematik
Im Neuenheimer Feld 294
D–6900 Heidelberg
West Germany

ABSTRACT. The numerical approximation of boundary value problems in the context of one–parameter continuation is considered. A collocation discretization of parameter dependent boundary value problems is suggested. The convergence properties are analysed and a stable algorithm is presented. Tangent continuation and steplength control are combined with a Gauss–Newton type corrector, which is shown to be quadratically convergent. Based on this analysis, a general purpose continuation code COLCON has been implemented. Numerical examples demonstrating the efficiency of the approach are presented.

The present work has been supported by the Deutsche Forschungsgemeinschaft.

* This paper is published in : SIAM J. Sci. Stat. Comput., 10, 72–88, (1989)

D. Roose et al. (eds.), Continuation and Bifurcations: Numerical Techniques and Applications, 405.
© 1990 Kluwer Academic Publishers.

# POROUS MEDIUM COMBUSTION

H.M. BYRNE
Oxford University Computing Laboratory
Numerical Analysis Group
8-11 Keble Road
Oxford OX1 3QD
England

ABSTRACT. I shall study a model of porous medium combustion, derived by Norbury and Stuart [1], restricting attention to the class of travelling wave solutions. Under this simplifying assumption the model reduces to a fifth-order system of coupled, nonlinear ordinary differential equations with a discontinuous forcing term (this corresponds to the reaction rate and is discontinuous because of the high energy activation asymptotics involved) and three free parameters $(\lambda,\mu,a)$. The form of the reaction rate enables us to deduce the existence of only four types of solution to the model, each solution characterised by the manner in which the reaction is activated and terminated.

The aim of my talk will, then, be to describe how the path-following package AUTO can be used to determine the degeneracy-, or bifurcation-, surfaces which separate the different classes of solutions in the three-dimensional parameter space. In this way it should then be possible to predict what type of solution will occur for given values of the parameters $(\lambda,\mu,a)$.

Reference

[1] Norbury, J. and Stuart, A.M., *A Model for Porous Medium Combustion*, Q. Jl. Mech. appl. Math., Vol. 42, Pt.1, 1989, pp. 159-178.

D. Roose et al. (eds.), Continuation and Bifurcations: Numerical Techniques and Applications, 407.
© 1990 Kluwer Academic Publishers.

# BLOCK ELIMINATION AND THE COMPUTATION OF SIMPLE TURNING POINTS

W. GOVAERTS
Department of Mathematics
University of Ghent
Krijgslaan 281
9000 Gent
Belgium

ABSTRACT. We present the algorithm GMBE to solve a block linear system

$$
\begin{array}{c} n \\ m \\ l \end{array}
\begin{bmatrix} A_{11} & A_{12} & A_{13} \\ 0 & A_{22} & A_{23} \\ A_{31} & A_{32} & A_{33} \end{bmatrix}
\begin{bmatrix} x_1 \\ x_2 \\ x_3 \end{bmatrix}
=
\begin{bmatrix} f_1 \\ f_2 \\ f_3 \end{bmatrix}
\begin{array}{c} n \\ m \\ l \end{array}
$$

or $Mz = h$ which appears in numerical computation of turning points and symmetry-breaking bifurcation. GMBE principally uses 1 solve with $A_{11}^T$ and $A_{22}^T$ each and 2 solves with $A_{11}$ and $A_{22}$ each. M must be well-conditioned but $A_{11}$ and $A_{22}$ may be arbitrarily ill-conditioned, in fact singular to machine precision. The error analysis requires that the solvers for $A_{11}, A_{22}, A_{11}^T$ and $A_{22}^T$ are stable (in the sense of backward projection of the errors) and that the solvers for $A_{11}, A_{22}$ are bounded (in a sense to be made precise). Both properties are typically possessed in practice by solvers based on either direct or iterative methods. GMBE is the first algorithm that solves $Mz = h$ by using the solvers for $A_{11}$ etc. as 'black boxes'.

*D. Roose et al. (eds.), Continuation and Bifurcations: Numerical Techniques and Applications*, 409.

# BIFURCATION INTO GAPS IN THE ESSENTIAL SPECTRUM

T. KUPPER
Institut für Angewandte Mathematik
Universität Hannover
Welfengarten 1
D-3000 Hannover 1
West Germany

ABSTRACT. In this lecture bifurcation from any boundary point of the essential spectrum is studied for a class of nonlinear operators generalizing results concerning bifurcation from the lowest point of the continuous spectrum. The study is motivated by a nonlinear Hill's equation as a typical example for a problem with gaps in the purely continuous spectrum.

By a generalized Lyapunov-Schmidt reduction we derive a variational problem for an auxiliary functional on a manifold in a space related to a spectral block of the linearized operator. Typically, this space is infinite dimensional estimates with suitable testfunctions are required to prove that supremum resp. infimum are attained by the functional.

Details are given in the following papers which will appear in the Journal für die Reine und Angewandte Mathematik :

T. Küpper - C.A. Stuart : Bifurcation into gaps in the essential spectrum
T. Küpper - C.A. Stuart : Gap-Bifurcation for nonlinear perturbations of Hill's
equation

*D. Roose et al. (eds.), Continuation and Bifurcations: Numerical Techniques and Applications,* 411.
© 1990 *Kluwer Academic Publishers.*

# APPLICATION OF NUMERICAL CONTINUATION IN AEROSPACE PROBLEMS

E. RIKS
N.L.R.
Vestiging Noordoosterpolder
Voorsterweg 31
P.B. 153
NL-8316 PR Marknesse
The Netherlands

ABSTRACT. A basic element in the structural design of an aircraft, rocket booster or space craft is the thin walled shell. Shells are applied for their high stiffness to weight ratio, for their high strength to weight ratio and last but not least for aerodynamic reasons.

When compressive loading conditions prevail the load carrying capacity of shell structures is often determined by loss of stability of equilibrium. For these and related problems of nonlinear structural behavior several methods of analysis are available. The most sophisticated tools of analysis are the large scale finite element codes and it is in these codes that the methods of continuation come into play.

Indeed, without path following techniques, the calculation of the loss of stiffness or loss of structural integrity for specific loading conditions can in general not be carried out. We will focus on the scope of a well known finite element code STAGS ; a code specifically designed for nonlinear shell analysis. We will give an overview of the path following techniques that are part of this code, including a bifurcation processor. Finally, we will show some practical problems that were solved by it.

D. Roose et al. (eds.), Continuation and Bifurcations: Numerical Techniques and Applications, 413.
© 1990 Kluwer Academic Publishers.

# APPLICATION OF A REDUCED BASIS METHOD IN STRUCTURAL ANALYSIS

H. SCHIPPERS and M. KOOLSTRA
N.L.R.
Informatics Division
Voorsterweg 31
P.B. 153
NL-8316 PR Marknesse
The Netherlands

ABSTRACT. In this lecture the application of the reduced basis method in the finite element system B2000 will be discussed. This system is presently under development at the Polytechnical University of Lausanne and the National Aerospace Laboratory NLR. It is designed for the analysis of nonlinear elastic structures.

The B2000 system comprises two continuation procedures for this type of analysis. In order to improve the performance of these procedures the reduced basis method was implemented. The performance of this method was investigated for two testcases : 1. A cylindrical panel with a central point load and 2. A cylindrical panel in axial compression. The basis vectors are defined by the path derivatives that are determined by numerical differentiation. The order of differentiation formulas and the magnitude of the stepsize is discussed as well as their effects on the load displacement diagrams. For the second testcase it appears to be difficult to find a robust computational strategy for switching between the reduced basis and the full basis.

D. Roose et al. (eds.), Continuation and Bifurcations: Numerical Techniques and Applications, 415.
© 1990 Kluwer Academic

# SOME APPLICATIONS OF BIFURCATION THEORY IN ENGINEERING

H. TROGER
Institut für Mechanik
Technische Universität Wien
Wiedner Hauptstrasse 8-10 / 325
A-1040 Wien
Austria

ABSTRACT. Classical bifurcation theory (center manifold and Ljapunov-Schmidt reduction, normal form theory, universal unfolding, calculation of bifurcation diagrams) has become an important and very useful means in the solution of nonlinear stability problems in many branches of engineering. In this talk we explain by means of several examples from mechanical engineering why *nonlinear* stability investigations are necessary. This is because quite frequently the stability behavior of practical engineering systems is not adequately described by the linear limit but only by a description of the behavior in the post-bifurcation regime. We start with some problems of rail and road vehicle dynamics. Then the periodic motion of a simple robot and the motion of a fluid carrying tube with an elastic support which can be considered to be a model of a fire man operating a fire hose will be presented. This latter problem also addresses bifurcation phenomena for symmetric systems. Furthermore, static buckling phenomena are considered, where the limitations of the classical bifurcation theory are pointed out performing a comparison of buckling problems of plates versus shells. Whereas for plates classical bifurcation theory gives good results, this is not the case for instabilities of equilibria of shells where in the post buckling regime localization phenomena occur which must be treated by different methods, for example singular perturbation theory.

*D. Roose et al. (eds.), Continuation and Bifurcations: Numerical Techniques and Applications,* 417.
© 1990 *Kluwer Academic Publishers.*

# HIGHER ORDER PREDICTORS IN NUMERICAL PATH FOLLOWING SCHEMES

K. ULRICH
Institut für Angewandte Mathematik
Universität Hannover
Welfengarten 1
D-3000 Hannover 1
Germany

ABSTRACT. A new class of polynomial higher order predictors for numerical path following schemes is presented. Explicit computation of first order derivatives only is required, no process of numerical differentiation is involved to obtain the additional higher order terms. The new predictor method presented here offers a variety of advantages and possible applications. The method is hierarchical and adaptive : predictors of varying order may be used on following a solution curve ; the method allows to detect the presence as well as to efficiently compute the exact locations of limit points ; it allows to monitor the approximation quality by means of a forward-backward strategy.

To improve quality in the subsequent corrector process a new ellipse normalization condition is suggested which provides an automatic step length and direction adjustment in the iterative process ; it can be used in combination with the common predictors based on tangent directions as well.

419

*D. Roose et al. (eds.), Continuation and Bifurcations: Numerical Techniques and Applications, 419.*
© 1990 *Kluwer Academic Publishers.*

# Index

---

Only full papers are indexed. For most keywords, only the first appearance in a particular paper is listed.